W9-BMY-593

Technology and Society
A Bridge to the 21st Century

Second Edition

Edited by

Linda S. Hjorth

Barbara A. Eichler

Ahmed S. Khan

John A. Morello
DeVry University—DuPage Campus

Prentice Hall

Upper Saddle River, New Jersey
Columbus, Ohio

Library of Congress Cataloging in Publication Data

Technology and society: a bridge to the 21st century / edited by Linda S. Hjorth.—2nd ed.
    p. cm.
  Includes bibliographical references and index.
  ISBN 0-13-092475-X
  1. Technology—Social aspects. I. Hjorth, Linda S.
  T14.5 .T44168 2003
  303.48′3—dc21

                                                                    2002066309

Editor in Chief: Stephen Helba
Assistant Vice President and Publisher: Charles E. Stewart, Jr.
Production Editor: Alexandrina Benedicto Wolf
Production Coordination: Clarinda Publication Services
Design Coordinator: Diane Ernsberger
Cover Designer: Jeff Varick
Cover art: Corbis Stockmarket
Production Manager: Matthew Ottenweller
Marketing Manager: Ben Leonard

This book was set in Times Roman by The Clarinda Company. It was printed and bound by R.R.
Donnelley & Sons Company. The cover was printed by The Lehigh Press, Inc.

Pearson Education Ltd.
Pearson Education Australia Pty. Limited
Pearson Education Singapore Pte. Ltd.
Pearson Education North Asia Ltd.
Pearson Education Canada, Ltd.
Pearson Educación de Mexico, S.A. de C.V.
Pearson Education—Japan
Pearson Education Malaysia Pte. Ltd.
Pearson Education, *Upper Saddle River, New Jersey*

**Copyright © 2003, 2000 by Pearson Education, Inc., Upper Saddle River, New Jersey
07458.** All rights reserved. Printed in the United States of America. This publication is protected by
Copyright and permission should be obtained from the publisher prior to any prohibited reproduction, storage in a retrieval system, or transmission in any form or by any means, electronic, mechanical, photocopying, recording, or likewise. For information regarding permission(s), write to: Rights
and Permissions Department.

10 9 8 7 6 5 4 3 2

ISBN 0-13-092475-X

Preface

ISSUES EXPLORED IN THE TEXT

One of the underlying issues explored in *Technology and Society: A Bridge to the 21st Century* is whether we are in charge of technology or whether technology controls us. At what point does technological dependency cause social problems? And to what extent are we, as caring social beings, concerned about technological impact? This text encourages readers to analyze and reflect on technology's impact on the global village economically, politically, and environmentally. The new century will usher in an urgent challenge to resolve the conflicts among our technological, environmental, and social worlds. The ability to understand the impact of technology on our lives and on succeeding generations will be essential to reaching the goals of survival, peaceful coexistence, ethical living, safety, and prosperity.

The chapters in this text are designed to stimulate, inspire, and provoke awareness of technology's impact on society. They are supported by a variety of features intended to supplement and complement learning, critical analysis, and social awareness.

FEATURES

- *Flowcharts:* One of the unique features of the text is the use of flowcharts as logical, interactive maps that emphasize the problems, possible solutions, and points of direction and significance of the chapters and case studies. Many of the flowcharts have been class-tested and we have found that students like them because they appeal to their kinesthetic and problem-solving learning style. The visual process of flowcharting the information presented in the readings seems to increase insight and critical thinking skills.
- *Internet Exercises:* The Internet exercises enable students to (a) become familiar with the knowledge dissemination mode of the ever-expanding Internet; and (b) incorporate the Internet's multimedia resources to enhance learning.
- *Useful Web Sites:* Considering the vast scope and rate of change of technological growth, it is difficult to cover all facets of technology and related issues of its impact on society. Therefore, each part concludes with a list of Web sites for the reader to use to supplement and enhance the content.
- *Questions:* Each article concludes with questions to integrate knowledge and synthesize understanding of social issues impacted by technology. The questions also create

excitement and wonderment about the tenacity of technology and its impact on the quality of life.

- *Statistics:* Boxes containing statistics, percentages, and bar graphs enhance many readings. They allow the reader to comprehend the magnitude of the social issue presented from a numerical format.

ORGANIZATION OF THE TEXT

Part I encourages students to clarify and sort out the many ways we look at and define technology. It confronts the need for ethical behavior in the use of technology. By reading this section, students will gain a greater understanding of the personal and social responsibility that accompanies technological development and implementation. Readers are asked to consider the mismatch between people and this newly created world of technology. This part presents ethical theory and frameworks for ethical decision-making, which are understood more easily through the cases of Roger Boisjoly and Whistle-Blowing.

Part II offers a brief look at the history of technology and its involvement in human development. It provides an extensive timeline, highlighting technology from the most primitive to the most sophisticated forms. It also features a glimpse of humanity at a time when technology and the skills to use it were few and far between. There's also a whimsical look at what new devices would have done to our ancestors if they had what we have and knew what we know.

Part III investigates the impact of technology on energy development and conservation. This section provides a survey of renewable and nonrenewable energy technologies and discusses the issues and challenges of economic growth within the domains of a sustainable environment.

Part IV presents a variety of ecological issues and challenges to help the reader understand how technological developments correlate to a healthy environment. The part begins with an overview of the major ecological risks for the 21st century, followed by three articles that provide a global view of the consequences of increased and intense resource use. Tables and graphs promote empirical understanding. Specific case studies address more specialized and local issues of the environment. We hope that the reader will understand the urgency and sensitivity of environmental decisions in the new century.

Part V offers a series of articles and case studies on the topic of population growth. This section gives the reader an understanding of the global exponential population growth that is facing the world and its associated problems, with some suggested strategies for improving this dilemma. Special problems of Third-World countries are included in this discussion.

Part VI, "Health and Technology," is a far-reaching survey of some of the greatest accomplishments of modern medicine and its greatest ethical concerns. A medical timeline illustrates the rapid growth of medical technology. It is followed by articles that persuade the reader to reconsider the use of some medical technologies; question the overuse of antibiotics; evaluate the use of animals for organ transplants (xenotransplants); and ponder physician-assisted suicide.

Part VII explores the state and impact of technology in developing countries and provides an insight into the intrinsic and extrinsic problems and their potential solutions.

Part VIII presents a time-and-space approach, enabling the reader to reflect on the issues presented in previous parts and to project their development and applications to the future. The purpose of this approach is to refocus and converge the thoughts across the "bridge of now" so that they guide us to the "bridge to the 21st century."

Part VIII begins by crossing over the bridge into the 21st century and envisioning a preferred future. Specific case studies of new technological applications that will affect our lives in health, medicine, and the military—as well as future predictions—guide this chapter. "Bridges to Your 21st Century Understanding" highlight special progress on topics mentioned above.

The book concludes with authors' commentaries presenting our own interpretation as we walk across the bridge to the 21st century.

The eight parts are organized in a pattern that encourages a developmental understanding of technology and its impact on energy use, population, ecology, social responsibility, medicine, Third-World countries, and the future. A psychologist, an anthropologist, an engineer, and a historian worked long and hard to make the book interesting. This diversity has added depth, scope, and encompassing views that we hope will play a positive and decisive role in the decisions you make regarding the use of technology in the future.

We hope that readers will better understand the complex relationship between technology and society, as well as the reasons that technological development sometimes causes society to worry and wonder about its impact—and that they will marvel, at the same time, at its potential for doing good.

ACKNOWLEDGMENTS

We gratefully acknowledge the valuable input of the following reviewers: Julian Thomas Euell, Ithaca College; Raymond A. Eve, University of Texas at Arlington; Samuel A. Guccione, Eastern Illinois University; and Gerald Harris, DeVry, Chicago. In addition, special thanks are extended to Charles Stewart for his support.

If you are thinking a year ahead, sow seed. If you are thinking 10 years ahead, plant a tree. If you are thinking 100 years ahead, make people aware. By sowing seed once, you harvest once. By planting a tree, you will harvest tenfold. By opening the minds of people, you will harvest a hundredfold.

CHINESE PROVERB

To the students who, in their quest for knowledge, consistently reinforce our desire to present varied ideas within the field of technology and society.

LINDA S. HJORTH

To my family, my students, and colleagues whose very capable adoption of technology, along with their arguments for the value of humankind, always inspire me.

BARBARA A. EICHLER

To Tasneem, my parents, my students, and all seekers of truth and wisdom.

AHMED S. KHAN

To our students and our families for their patience, support, and prayers.

JOHN MORELLO

Contents

Technology and Society

PART

I

Ethics and Technology

A Paul Klee Look-alike.
Painting by Ahmed S. Khan.

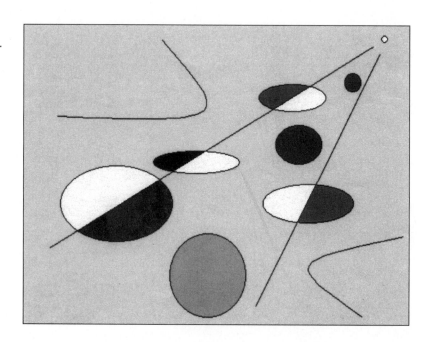

2

OBJECTIVES

Part 1 will help you to

- Understand consequentialist, utilitarianism and deontological ethical theories.
- Become aware of science- or technology-based ethical issues and conflicts.
- Understand science- or technology-based rights.
- Evaluate the challenge of contemporary science and technology to traditional ethical theories
- Appreciate the relationship between ethics and technology
- Develop a precise definition of technology
- To evaluate the way in which technology is created and functions in a culture
- To understand Kohlberg's Model of Cognitive Moral Development

INTRODUCTION

Every technology is both a burden and a blessing; not either-or, but this-and-that.
NEIL POSTMAN, AUTHOR, *Technopoly: The Surrender of Culture to Society*

As technologies are embraced as "blessings," this part resonates concerns that techno-users need to partake with immense responsibility so that newly developed technologies will not mature into societal burdens. Technology is more than a mechanism or tool; it is also a catalyst for societal changes. Once technology is introduced, life, thinking, behavior, and social norms change. When cars, microwaves, computers, birth-control pills, penicillin, and other technologies were invented, life as we knew it drastically changed. Our point of reference changed, and our behaviors were altered. With the advent of technology, we arrive places faster, communicate globally, cook quickly, and even look to medical technologies to get us to sleep swiftly. Technologies have been invented, for the most part, to make life "better," but do they also carry a societal or psychological cost?

For hundreds of years, the Kaiapo Indians of the lower Amazon basin were considered to be "skilled farmers and hunters and the fiercest warriors of central Brazil" (Henslin; Simons, 1995, p. 463). They eventually sold gold and mahogany and became a wealthy village. With the newfound wealth, Chief Kanhonk bought a small satellite, which the Indians called the "big ghost," so that they could watch television. Prior to the satellite purchase, the Kaiapos would meet at night to tell stories and share ancestral customs. After the purchase of the satellite, however, children were found straying from their ancestral storytelling traditions to watch Western cartoons instead (Henslin 1995, 464). When studying various cultures, we often find that each has its own story to tell about the impact of technology on the lives, culture, and values of its people.

As technology permeates through all cultures and timespans, sociological and behavioral adaptations develop. For example, in the 1930s, many American children surrounded the radio in their living room to let their imaginations go while listening to Orphan Annie, Terry and the Pirates, or Jack Armstrong and the All American Boy. The words emitted from the radio allowed children to create the characters and backgrounds within the world of their imaginations. No visual cues limited what their minds created. This all changed for their children when television dominated the living rooms of the 1950s. Children no longer relied on their imaginations, as the television presented every visual and auditory detail. Additionally, parents worried that their kids would end up with damaged eyes or receive large amounts of radiation from sitting too close to the television.

This part of the book, ethics and technology, discusses the link between technology and its implications for society. Many of the issues presented are more serious than those faced by the Kaiapos or American children in the 1930s and 1950s. But the issues remain the same: With each technology introduced at any time in any culture, adaptations seem to follow. Many of the cases presented in this part encourage users of technology to evaluate the technologies with respect and with their own analyses of personal and social responsibility.

Most technologies work to provide progress to those who use them. However, there are times when technologies are misused and harm or death befalls their users. Therefore, it is befitting to start this part of the book with ethics and its place in the development of technology.

As we bridge our way into a new century and prepare to hand our world over to a younger generation, it will be important to consistently evaluate the technologies that we use and their impact on society and the world in which we live. This part of the book supports that endeavor.

REFERENCES

Henslin, J. (1995) *Sociology: A down-to-earth approach.* Needham Heights, MA. Allyn and Bacon.
Simons, Marlise. (1995) "The Amazon's savvy Indians." *Down to earth sociology: Introductory readings.* 8th ed. New York, Free Press.

Ethics
ROBERT McGINN

INTRODUCTION

For at least the last two decades, many of the most divisive ethical issues debated in Western societies have been precipitated by developments in science and technology, including advances in reproductive, genetic, weapons, and life-prolonging technologies. The adoption and alteration of public policy for regulating science- or technology-intensive practices, such as the provision of access to exotic medical treatment, the disposal of toxic waste, and the invasion of individual privacy, have also raised perplexing ethical issues. This chapter is devoted to the analysis of such conflicts.

There is no universally shared criterion for deciding when a conflict of values falls within the province of ethics rather than, say, law. However, the issues and conflicts discussed in this chapter involve values widely regarded as integral to the enterprise of ethics in contemporary Western societies; such values include freedom, justice, and human rights such as privacy. Disputes over whether an agent's freedom should be limited prospectively or its prior exercise punished, whether justice has been denied or done to some party, or whether someone's human rights have been protected or violated, are widely regarded in Western societies as specifically *ethical* disagreements, thus marking them as human value issues or conflicts of special importance in these societies.

Science, Technology and Society by McGinn, Robert © 1991. Reprinted by permission of Pearson Education, Inc., Upper Saddle River, N.J.

As a prelude to analysis of science- or technology-based ethical issues, we will describe a quartet of basic considerations centrally involved in judgments about and the playing out of such conflicts. We will then characterize and analyze a number of important *kinds* of ethical issues and conflicts associated with contemporary science and technology. Where appropriate, we will indicate noteworthy sociocultural factors that, in concert with the technical developments in question, help generate the issue or conflict. All this will pave the way for a key conclusion reached in this chapter: that developments in contemporary science and technology are calling into question the adequacy of traditional ethical thinking. A more comprehensive and sensitive kind of ethical analysis is now required, one more adequate to the complexity and consequences of contemporary scientific and technological processes and products.

CLARIFICATION OF ETHICAL ISSUES AND CONFLICTS

Ethical issues and conflicts, whether or not they are associated with developments in science and technology, can often be usefully clarified if four kinds of considerations pertinent to ethical decision- and judgment-making about controversial actions, practices, and policies are kept in mind.[1]

The Facts of the Matter

One consideration is that of determining, as scrupulously as possible, the facts of the situation underlying or surrounding the conflict in question. Doing so may require

5

ferreting out and scrutinizing purportedly factual assumptions and allegedly empirical claims made by disputants about the conflict situation in question. It may also require ascertaining whether any persuasive accounts of the facts of the matter have been developed by neutral parties. In such efforts, important concerns include unmasking pseudo-facts and factoids posing as bona fide facts, ensuring that the credibility attributed to an account of the facts reflects the reputation and interests of its source, and setting the strength of the evidence required to warrant acceptance of an account of the facts at a level proportional to the gravity of the issue or conflict in question.

Affected Patients and Their Interests

A second kind of clarificatory consideration in thinking about an ethical issue or conflict is that of identifying all pertinent "patients"—that is, all affected parties with a legitimate stake in the outcome of the dispute. Further, all protectable interests of each stakeholder should be delineated and their relative weights carefully and impartially assigned.

Key Concepts, Criteria, and Principles

A third kind of consideration is that of identifying the key concepts, criteria, and principles in terms of which the ethical issue or conflict in question is formulated or debated. For example, the abortion issue hinges critically on the protagonists' respective concepts of what it is to be a "human being" and a "person" as well as what is meant by a "viable" fetus. The ethical (and legal) debate over the withdrawal of technological life-support systems turns sharply on what is meant by "killing" someone as well as on which criteria implicitly or explicitly govern protagonists' use of the key terms "voluntary consent" (to withdrawal or withholding of treatment) and "death."

Ethical Theories and Arguments

A fourth kind of basic consideration to be kept clearly in mind is that ethical disputes often involve two quite distinct kinds of decision-making theories and arguments. *Consequentialist* ethical theories and arguments make determination of the rightness or wrongness of actions and policies hinge exclusively on their estimated *consequences*. The most familiar consequentialist ethical theory is utilitarianism—the view that an action or policy is right if and only if it is likely to produce at least as great a surplus of good over bad, or evil, consequences as any available alternative. There are different versions of utilitarianism, depending on, among other things, what a given thinker understands by "good" and "bad," or "evil," consequences. For example, so-called hedonic utilitarians, of whom nineteenth century British reformer Jeremy Bentham is perhaps the best known, took pleasure to be the only good, and pain the only bad, or evil. "Ideal utilitarians," such as the early twentieth century British philosopher G. E. Moore, construed "good" and "bad" quite differently, including things like friendship and beauty among goods and their absence or opposites, such as alienation and ugliness, among bads, or evils.[2]

The second kind of ethical theory and argument that often enters into ethical disputes is called deontological. *Deontological* ethical theories and judgments hold that certain actions or practices are inherently or intrinsically right or wrong—that is, right or wrong in themselves, independent of any consideration of their consequences. Different deontological theorists and thinkers point to different things about actions and policies, in light of which they are judged to be right or wrong. Some point to supposedly intrinsic moral properties of actions and policies falling into one or another category. For example, actions such as telling a lie or breaking a promise may be regarded as intrinsically wrong. Others emphasize that a certain course of action is obligatory or impermissible because it is approved or disapproved, or unconditionally mandated or prohibited by some authority, perhaps a deity. As we will see in this chapter, several of the most important kinds of ethical issues and conflicts engendered by developments in science and technology arise from or are exacerbated by the fact that partisans of one position on an issue are consequentialists while their opponents are deontological thinkers (for convenience, deontologists).*

KINDS OF SCIENCE- OR TECHNOLOGY-BASED ETHICAL ISSUES AND CONFLICTS

We now turn to examination of science- and technology-based ethical issues and conflicts. Given the purpose of this book, our objective here will not be to provide detailed discussions of—much less solutions to—even a select number of the long list of such problems. Rather, we will

---

*In reality, the ethical thinking of denizens of contemporary industrial societies is rarely so black and white. It often incorporates consequentialist, deontological, and perhaps other considerations in uneasy or unstable combinations.

identify and critically analyze the limited number of qualitatively distinct kinds of such disputes. (Eight are considered here.) The aforementioned quartet of basic concerns—facts; patients and interests; concepts, criteria, principles; and ethical theories and arguments—will be used to shed light on the sociotechnical roots and intractability of many of these problems.

Violations of Established World Orders

Some ethical conflicts arise from the fact that scientific or technological breakthroughs make possible actions or practices that, in spite of what some see as their benefits, others believe violate some established order of things whose preservation matters greatly to them. The order of things in question may be regarded as "natural" or as "sacred."

For example, much opposition to recent achievements in biomedicine and genetic engineering flows from beliefs that employing such techniques is *unnatural.* Thus some oppose the technique of *in vitro* fertilization as involving the unnatural separation of human reproduction from sexual intercourse. In a similar vein, the production of farm animals with genes from at least two different animal species ("transgenic animals") is viewed by some critics as a transgression of natural animal-species boundaries, while the use of genetically mass-produced bovine growth hormones to substantially increase the volume of milk produced by cows is opposed by some as "chang[ing] the natural behavior of animals" or as "interrupt[ing] the naturalness of [farmers'] environment."[3]

Opposition to technological violations of natural orders is, however, sometimes based on concern about the long-term consequences of intervention for human or other animal well-being or for ecosystem integrity. For example, some oppose the production of transgenic animals because they believe that the resultant animals will suffer physically as a result of being maladapted. Similarly, some critics of the use of bovine growth hormone to raise milk production levels are primarily concerned with the safety of such milk for young children. The plausibility of such consequentialist ethical thinking hinges on the details of the particular case under consideration, including the estimated magnitudes, likelihoods, and reversibility of the projected consequences of intervention.

Deontological ethical arguments against such intervention as intrinsically wrong take two forms. First, the intervention-free order of nature is regarded as natural and intrinsically "good," while technology is not viewed as part of the natural order but rather as artificial. Therefore, it is concluded, attempts to use technology to intervene in the natural order are improper. A second argument notes that something has existed or has been done in the so-called natural way from time immemorial and concludes that therefore it should be done or continue to be done in that same way—without technological intervention. Is either the "unnaturalness" or the "longevity" argument persuasive?

It is unclear why the development and transformative use of technology on nature should be seen as "unnatural." The claim that because God created the natural order it should not be "tampered with" by humans is suspect for two reasons. Those holding this idea presumably also believe that the human being was created by the deity, in which case the human is no less "natural" a creation than the "natural order" and indeed is properly regarded as part of that order. Moreover, they also presumably believe that God endowed humans with creative powers, including the ability to devise technics, thereby enabling them to intervene in the natural order. If so, why is it unnatural for natural creatures to use their God-given powers to intervene in the natural order? It seems implausible that the deity would endow its natural creatures with an unnatural power or with a power whose use was unnatural. If it is not the very use of technology to transform nature that is unnatural but only certain uses of it, then these opponents of technological violations of natural orders must clarify what it is that makes some technological interventions "violations" of those orders and others simply harmonious interventions in them.

As for the argument that the existing way of doing something is the proper way because of its longevity, it too fails to convince. That a practice is long-standing may make it familiar or seem natural. But long-standingness cannot by itself justify the view that the practice is proper. That would be drawing an *evaluative* conclusion from a purely *factual* premise. Conversely, a particular technological intervention in a long-established natural order might initially be resisted because it is unfamiliar or deemed unnatural. However, the fact that something runs counter to a long-standing practice does not suffice to show that it is improper. If that were so, then the abolition of slavery would have been improper. In fact, opposition to a practice initially regarded as unnatural because of its novelty or strangeness often diminishes over time as the new way becomes increasingly familiar. Some innovative technical practices eventually come to seem natural and quite proper, as has been the case with the use of antibiotics.

Is there, then, nothing to the concerns about technological violations of established orders as unnatural? Even if the deontological arguments examined here fail to hold water, the concerns they express still warrant serious consideration, for deontological thinking and argument are sometimes disguised or compressed versions of what are at bottom consequentialist thinking and argument. Reference to an innovative practice as unnatural and therefore as intrinsically wrong may be a powerful if deeply misleading way of expressing concern over its possible elusive long-term consequences.

Other scientific and technological developments have made possible practices that are seen by some groups as violations of world orders viewed not so much as natural but as *sacred*. Thus, in the Hasidic community centered in Brooklyn, New York, birth control is forbidden, supposedly on the basis of the Torah.[4] For the Wahabi, a fundamentalist Arabian Muslim sect, television, with its image-reproducing capacity, violates the sacred order related in the Koran. Opposition to certain technologies or technological ways of doing things as violations of sacred orders is less likely to ebb in the minds of such opponents, for the sacred way is apt to be regarded as God's way and, as such, as immutable and thus as something that ought not adapt itself to human technological change.

Violations of Supposedly Exceptionless Moral Principles

Other ethical issues arise from the fact that the use, failure to use, or withdrawal of particular scientific procedures or items of technology is seen by some as violating one or another important moral principle that its adherents believe to be exceptionless. For example, some people are categorically opposed to nuclear weapons on the grounds that their use will inevitably violate the supposedly exceptionless principle that any course of action sure to result in the destruction of innocent civilian lives in time of war is ethically impermissible.

Similarly, the supposedly exceptionless principle that "life must never, under any circumstances, be taken"—put differently, that "life must always be preserved"—is clearly an important ground of the categorical judgment that withdrawal of life-sustaining technologies, whether mechanical respirators or feeding and hydration tubes, is ethically wrong. A third example is the opposition by some to the "harvesting" of fetal tissue for use in treating Parkinson's or Alzheimer's disease, even in a relative. This opposition is often rooted in the supposedly exceptionless principle that "a human being must never be treated merely as a means to some other end, however worthwhile in itself."

Sociologically, opposition to certain scientific and technological developments on the grounds that they involve or may involve violations of some special order of things or some supposedly exceptionless moral principle, reflects a fundamental fact about modern Western industrial societies. While much has been written about their secularization, there remain in such societies significant numbers of people whose ethical thinking is deontological in character, whether or not religiously grounded.

The world views of such individuals contain categories of actions that are strictly forbidden or commanded. For them, the last word on a particular science- or technology-based ethical issue or conflict sometimes hinges solely on whether the action or practice in question falls into one or another prescribed or proscribed category. While deontological thinkers may resort strategically to consequentialist arguments in attempting to change the views of consequentialist opponents, the latter's arguments against their adversaries' deontologically grounded positions usually fall on deaf ears, however well documented the empirical claims brought forward as evidence. Deontological fundamentalists and consequentialist seculars are, in their ethical judgment and decision making, mutual cultural strangers.

An interesting consequence of deontological appeals to supposedly exceptionless moral principles in the context of potent contemporary technologies is the appearance of moral paradoxes. For example, Gregory Kavka has shown that the situation of nuclear deterrence undermines the venerable, supposedly exceptionless "wrongful intention principle"—namely, the principle that "to form the intention to do what one knows to be wrong is itself wrong."[5] Launching a nuclear strike might well be ethically wrong (because of the foreseeable loss of innocent civilian lives). But what about forming the intention to do so if attacked? According to the wrongful intention principle, forming the intention to carry out that wrong action would itself be *wrong*. However, since forming that intention might well be necessary to deter an attack and thus to avoid launching an impermissible retaliatory strike, it might well be ethically *right*. One and the same action—that of forming the intention to retaliate—is therefore both right and wrong, a moral paradox. Thus can technological developments compel reassessment of supposedly exceptionless ethical principles.

Distributions of Science- or Technology-Related Benefits

Some contemporary ethical issues and conflicts arise from the fact that the benefits of developments in science and technology are allocated in ways that do not seem equitable to one or another social group. This is particularly so with respect to medical benefits, whether they be diagnostic tests, surgical procedures, or therapeutic drugs, devices, or services.

Concerns over whether an allocation of such benefits is "distributively just" often emerge when demand for the benefit exceeds its supply and decisions must be made about who will receive the benefit and who will not—sometimes tantamount to deciding "who shall live and who shall die." For example, in the early 1960s, the supply of dialysis units available to the Artificial Kidney Center in Seattle, Washington, was insufficient to meet the needs of those with failed kidneys.[6] Criteria were selected to use in deciding who would be granted access to this beneficial scarce technical resource. Today, the demand for various kinds of human organs often exceeds available supplies. The criterion of "need" is thus by itself insufficient to make allocation decisions. Criteria such as "likelihood of realizing a physiologically successful outcome" seem promising, but are quite problematic, for, as Ronald Munson has argued,

> [T]he characteristics required to make someone a "successful" dialysis patient are to some extent "middle-class virtues." A patient must not only be motivated to save his life, but he must also understand the need for the dialysis, be capable of adhering to a strict diet, show up for scheduled dialysis sessions, and so on. As a consequence, where decisions about whether to admit a patient to dialysis are based on the estimates of the likelihood of a patient's doing what is required, members of the white middle class have a definite edge over others. Selection criteria that are apparently objective may actually involve hidden class or racial bias.[7]

Other criteria, such as "probable post-treatment quality of life" and "past or likely future contribution of the treatment candidate to the community" are no less problematic. Hence, some believe that for such allocations to be distributively just, once need and physiological compatability have been established, a lottery should determine access to the scarce benefit.

On other occasions, it is not the shortage but the high cost of a medical treatment and the inability of all needy patients who want the treatment to afford it that engenders ethical conflict. Science and technology are often central factors in these high costs, for such costs may reflect the high purchase price of a machine or drug paid by a care unit, something which may in turn reflect the device's or substance's high research and development costs. Such situations help pose the contentious ethical issue of whether access to some needed expensive drug or procedure should be permitted to hinge on whether a prospective patient can afford to pay the going market price.

Deontological ethical thinkers who have come to think of medical care as a basic human right may find it morally unthinkable that a person be denied access to such treatment simply because of not being able to afford it (or because, for example, of being "too old"). In contrast, consequentialists, some of whom find the concept of an "absolute right" potentially dangerous, may believe that a particular technically exotic treatment is so expensive that granting everyone in life-or-death need unlimited access to it will effectively preclude many more individuals from getting less expensive, more beneficial, non-life-or-death treatments. Diverging accounts of "the facts of the matter" and different criteria for what makes a treatment "exotic" often bulk large in such judgments.

Consequentialists are apt to believe that individuals do *not* have a moral right to draw without limit on public or insurance-company funds to have their or their family members' lives extended regardless of the quality of the sustained life and the prognosis for its improvement. They may even hold that the financial and social consequences of doing so create a moral obligation to *terminate* such treatment, or at least to cease drawing on public funds to pay for continued treatment. In such ways have advances in science and technology as well as people's varying concepts (e.g., of a life worth living) and divergent ethical theories become intertwined in complex ethical disputes over distributive justice, rights, and obligations.

Infliction of Harm or Exposure to Significant Risk of Harm Without Prior Consent

A fourth category of ethical issues and conflicts engendered by developments in science and technology arises from activities that, while undertaken to benefit one group, inflict harm or impose significant risk of harm on another without the latter's prior consent. Examples of this sort of phenomenon abound and include some research on animals; production of cross-border and multi-generational pollution; the maintenance of carcinogen-containing workplaces; and the operation of "hair-trigger" military defense systems. As with earlier categories, the ethical issues and conflicts here have both technical and social roots.

Most parties to such disputes would agree that, other things being equal, it is always unethical to subject a morally pertinent party to undeserved harm or serious risk of same without the party's freely given prior consent. Let us examine how the consent issue plays out in the four above-mentioned kinds of cases.

In laboratory experimentation on sentient animals, the issue of consent bulks large in the persistent ethical conflict. Consequentialist proponents hold, on cost-benefit grounds, that activities that promise future benefits (including avoidance of suffering) for humans but that (supposedly unavoidably) inflict suffering on existing animals are ethically permissible, perhaps even obligatory. Those carrying out such activities may proceed because since animals cannot consent to anything, they are different in a morally relevant respect from human beings.[8] Hence the consent condition, precluding similar treatment of humans, is, in the case of animals, legitimately waived. Some opponents of such research, often deontologists, also subscribe to the prior-consent principle, but they see animals such as rabbits and monkeys as no less morally relevant patients than are humans. They draw a diametrically different conclusion: Since the consent of laboratory animals cannot be obtained, research activity that produces suffering for animals is ethically wrong or impermissible, even if the suffering is "unavoidable"—computer models that make animal tests unnecessary may not be available— and the benefits of the research could plausibly be shown to exceed its costs. It is not difficult to see why resolution of this disagreement is unlikely to be forthcoming.

Explanation of the rise of ethical conflict over cases of cross-border and multigenerational pollution (e.g., acid rain, dumping toxic chemical or metal waste into multinational bodies of water, and the possibly insecure disposal of nuclear waste), must heed technical factors as well as the problematic issue of consent. But for the capacity of contemporary scientific and technological activities to produce potent geographically and temporally remote effects, these ethical disagreements would simply not arise. Moreover, this "action-at-a-distance" capacity contributes to the tendency either to neglect or to assign modest weights to the legitimate interests of affected patients at considerable geographical or temporal remove. This in turn facilitates proceeding with the activities in the absence of consent of such endangered parties, or, in the case of not-yet-born members of future generations, impartial reflection on whether they would consent if they were informed and in a position to give or withhold it. The facts that the human capacity for empathy tends to diminish rapidly the more remote the injured or endangered party and that the world is organized into a weak international system of sovereign states both contribute to the genesis of such ethical conflicts.

In ethical disputes over workplaces made dangerous because of scientific or technological activities or products, the issue of consent rears its head in a different form. Historically, employers or their representatives argued that maintenance of a dangerous workplace was not unethical because a worker who accepted a job in one thereby voluntarily consented to exposure to all its attendant risks. To the extent that workplace hazards in the early industrial era were primarily threats to worker safety and that a worker had other less dangerous employment alternatives, such a viewpoint might seem at least minimally plausible. However, as twentieth century industrial workplaces became pervaded with thousands of industrial chemicals of uncertain bearing on worker health, the notion that workers could meaningfully consent to the imposition of any and all risks that their workplaces posed to their health began to ring hollow. Workers had to make decisions to take or keep jobs in ignorance of what, if any, risks they would be or were being exposed to. Put differently, management could proceed with its risk-imposing activities without their workers' informed consent.

This situation came to be viewed by some as a violation of the prior-consent principle, and hence as unethical. Others saw it as ethically permissible because the benefits (to both company and workers) of proceeding in this way supposedly outweighed the (typically undervalued) costs of doing so. The main attempt to mitigate this situation has taken the form of right-to-know legislation: Workers have a right to a safe workplace but not to a risk-free (in particular, carcinogen-free) one. They are, however, entitled to that which is deemed necessary for their giving informed consent to imposition of workplace risks; specifically, they are entitled to "be informed about" all carcinogenic and other toxic substances used in their workplaces. Whether the extensive technical information provided and the way in which it is communicated to workers suffice to ensure their "informed consent" remains an open factual and criteriological question at the core of a persistent ethical issue.

Ethical conflict over the operation of "hair-trigger" military defense systems is also driven by both technical and social factors. Such systems are called "hair-trigger" because they can be set to "fire" on being subjected to slight pressures. Their risk arises from the enormity of destruction that can be unleashed by slight pressure on the sensitive firing mechanism; such pressure can be brought by mistaken "sightings" or misinterpretations of data. The 1988 downing of an Iranian commercial aircraft by the high-tech-equipped U.S.S. *Vincennes* because of misinterpretation of radar and electronic data is a tragic case in point, albeit on a relatively small scale. There have been numerous occasions on which American retaliatory nuclear forces have been activated and on the verge of being unleashed because of what turned out to be mistaken technological indications that a Soviet attack had been launched.

While technological "progress" is partly behind such ethical conflicts, consent is also a factor. In the case of hair-trigger military defense systems (e.g., ones operating on a "launch-on-warning" basis), the public has not been afforded an opportunity to explicitly give or withhold its consent, informed or otherwise, to the imposition of such grave risks. For opponents of such systems, this alone makes them ethically objectionable. For proponents, the astonishing speed of current or emerging offensive military technologies makes the risk of *not* employing hair-trigger defense systems exceed the risk of relying upon them. Moreover, such systems are morally justified, proponents argue, since the people have indirectly consented to such risks by voting democratically for the government that imposes them. The fact that civilian aircraft are permitted to fly over populous areas without their residents having first voted on whether to accept the associated risks does not suffice to show that the people do not consent to the risks imposed on them by this practice. Hence, it would be argued, the consent condition has not been violated and ethical impropriety has not been demonstrated. The same would be true in the case of hair-trigger defense systems. However, the greater the magnitude of the danger involved—enormous in the case of the nuclear war—the lower the risk of its accidental occurrence must be if the explicit securing of consent is to be reasonably set aside as having been implicitly given. The upshot is that the lack of shared criteria for deciding whether citizen consent has been effectively obtained in such cases is central to this acute ethical dispute.

Two morals of this kind of ethical conflict deserve attention. First, the problematics of consent are, to a significant degree, science- and technology-driven. Second, the potency of much contemporary scientific and technological activity is pressuring sensitive ethical analysts to enlarge the domain of morally pertinent patients whose interests are to be taken into account in assessing the ethical propriety of current or proposed actions or policies. This situation is reflected in ongoing struggles over whether to include various kinds of previously excluded stakeholders, such as those far afield who are nevertheless deleteriously affected by potent "spill-over" effects and future citizens whose legitimate interests may be jeopardized by activities designed first and foremost to benefit the presently living. Here, too, the contours of the evaluative enterprise of ethics are being subjected to severe stress by developments in contemporary science and technology.

Science- or Technology-Precipitated Value Conflicts

A fifth kind of science- or technology-based issue or conflict arises when a scientific or technological advance allows something new to be done that precipitates a value conflict, not necessarily between the values of opposed parties, but *between two or more cherished values of one and the same party*. For example, life-extending technologies have engendered situations in which family members are compelled to choose between two values; to both of which they owe allegiance: human life preservation and death with dignity. The crucial point about this increasingly frequent kind of value conflict is that the parties plagued by such conflicts would not be so but for scientific or technological advances.

Most recently, genetic tests allowing those with access to their results to know something of a sensitive nature about the health-related state or genetic predispositions of the person tested have proliferated. This has given rise to value conflicts between testers' or policymakers' concern for the protection of human *privacy* regarding disclosure of test results and their concern for *fairness* to one or another interested party other than the testee.

For example, in 1986 an adoption agency was trying to place a 2-month-old girl whose mother had Huntington's disease, a progressive, irreversible neurological disorder. The prospective adoptive parents indicated that they did not want the girl if she was going to develop the disease. The agency asked a geneticist to determine whether the child had the gene that would sooner or later manifest itself in the disease. The geneticist, while presumably sympathetic to the would-be adoptive parents' seemingly reasonable request, declined to do the testing. He reasoned that since many victims of the disease have claimed that they would have preferred to have lived their lives without knowing they had the "time-bomb" gene for the disease, it would be unethical to test someone so young, that is, at a point before she could decide whether to exercise her right to privacy in the form of *entitlement not to know* that she had the fatal gene.[9]

Tests for various genetic disorders, such as Down's syndrome, sickle cell anemia, and Tay Sachs disease, have been available for some time. In the foreseeable future, however, it is expected that tests will become available for identifying genetic traits that predispose people to more common health problems, such as diabetes, heart disease, and major forms of mental illness: Thus, according to Dr. Kenneth Paigen, "We are going to be able to say that somebody has a much greater or much less than average chance of having a heart attack before age 50 or after age 50."[10]

The potential implications of such advances for matters such as employment eligibility, life insurance qualification, and mate selection are formidable. Will employers with openings for positions with public safety responsibilities (e.g., commercial airline pilots) be permitted to require that applicants take genetic tests that will disclose whether they are predisposed to heart disease or to a genetically based mental disorder such as manic depression? Will insurance companies be permitted to require applicants for life or health insurance to take genetic tests predictive of life expectancy or diabetes? Will prospective spouses come to expect each other to be tested to determine their respective genetic predispositions and whether they are carriers of certain traits of genetic diseases and to disclose the test results?

In the case of companies recruiting for jobs with public safety responsibilities, prohibition of such tests to protect applicant privacy could impose significant safety costs on society. In the insurance case, preventing mandatory disclosure of test results in the name of individual privacy would spread the cost of defending this cherished value over society at large in the form of increased insurance premiums for those *without* life- or health-threatening genetic traits or predispositions. In the case of mate selection, declining to pursue and disclose the results of reliable genetic tests could set up partners for severe strains on their relationship should presently identifiable genetic disorders manifest themselves in the partners or their offspring at a later date.

It remains to be seen how society will resolve the public policy questions raised in such cases by the ethical value conflict between privacy and fairness. However, it is already clear that advances in genetic science are going to pose powerful challenges to society's commitment to the right of individual privacy. The knowledge about the individual afforded by such tests is likely to be of such pertinence to legitimate interests of other parties that the protection afforded individual privacy may be weakened out of concern for fairness to those parties, perhaps to the point of recognizing that under certain conditions they have a right to that knowledge. Put differently, technology is here bringing micro, or personal, justice and macro, or societal, justice into conflict.

Science- or Technology-Engendered "Positive" Rights

Besides conflicts over the values of freedom and justice, issues of "rights," especially "human rights," bulk large in contemporary Western ethics. In recent decades, advances in science and technology have engendered a new ethical issue: that of how best to respond to newly recognized so-called *positive rights.*

In the modern era, some claims have come to be widely recognized as "human rights"—that is, as irrevocable entitlements that people supposedly have simply because they are human beings. Human rights are thus contrasted with civil or institutional rights—rights that people have because they are delineated in specific revocable legal or institutional documents. Among the most widely recognized human rights are "life" and "liberty."

These rights, and some derived from them—privacy, for example, is widely thought to be a kind of special case of liberty—have traditionally been viewed as what philosophers call "negative" or "noninterference" rights—that is, as entitlements *not to be done to* in certain ways. Thus, the right to life is construed as entitlement not to be deprived of one's life or physical integrity. The right to liberty is construed as entitlement not to have one's freedom of action physically constrained or interfered with, unless its exercise has unjustifiably harmed another's protectable interests (e.g., those in life, limb, property, reputation) or poses an unreasonable risk of doing so.

As scientific and technological progress has gathered momentum in recent decades, several rights traditionally viewed as negative have given birth to a number of correlative "positive" rights—in other words, entitlements *to be done to* in certain ways. Consider three examples. Many believe that the right to life, traditionally construed as a negative human right, must, *in the context of new life-*

preserving scientific and technological resources, be regarded as having taken on a positive component: entitlement to be provided with whatever medical treatment may be necessary to sustain one's life (independently of ability to pay for it). According to this way of thinking, affirmation of the right to life in the contemporary scientific and technological era requires affirmation of a positive right of access to whatever means are necessary to sustain life. Thus, for example, denial of costly life-sustaining drugs to a patient in need of them on any grounds save scarcity, including concern over the aggregate high cost to society of providing them, is seen by many as a violation of the patient's right to life. Hence, depending on whether the ethical analyst is a deontological or consequentialist thinker, failure to provide these drugs would be deemed categorically or prima facie ethically impermissible.

The right to privacy has traditionally been viewed as a noninterference right, entailing, among other things, entitlement not to have one's home broken into by government authorities without a search warrant issuable by a court only upon proof of "probable cause." However, the computer revolution has put individual privacy interests at risk. To compensate for this, legislation in the United States and other countries entitles citizens to be provided with certain categories of information being kept on them in computerized files. For example, the U.S. Fair Credit Reporting Act of 1970 entitles each citizen to review and correct credit reports that have been done on them and to be notified of credit investigations undertaken for purposes of insurance, mortgage loans, and employment. Given the exceptional mobility of this information and the potential for severely damaging individual privacy that this creates, protection of the right of individual privacy in the computer era is held to require acknowledgement of countervailing positive rights: entitlement to know what exactly about oneself is contained in computerized government files and to have one's record rectified if it is shown to be erroneous.

A third, somewhat more speculative example involves a special case of the general right to liberty—namely, the traditional negative right of reproductive freedom: entitlement not to be interfered with in one's procreative undertakings, be they attempts to have or to avoid having offspring, including via "artificial" contraceptive means. It remains to be seen whether, in the context of the recent and continuing revolution in reproductive science and technology, the negative right of individual reproductive freedom will engender a positive counterpart: entitlement of those with infertility problems to be provided with (at least some) technical reproductive services enabling them to attempt to have offspring, where access does not hinge on a client's ability to pay or even perhaps on marital status.

As such scientifically and technologically generated positive rights expand, ethical tension will mount. Some deontologists, believing that human rights are "absolute," may conclude that their corresponding positive rights are likewise, hence inviolable. Others, including most consequentialists, while treating rights as claims that always deserve society's sympathetic consideration, may conclude that they cannot always be honored without regard to the social consequences of doing so. The most important consequence of this ethical tension may be that the day is drawing closer when society will have to come decisively to grips with the consequences of philosophical commitment to a concept of rights as "absolute" and "immutable."

Public Harms of Aggregation

Suppose that each of a large number of people carry out essentially the same action. Suppose further that, considered individually, each of these actions has at worst a negligible negative effect on a social or natural environment. Finally, suppose that the aggregated effect of the large number of people doing the same thing is that substantial harm is done to the environment in question. Let us call such outcomes *public harms of aggregation.* Many, if not most, such harms are possible only because, to an unprecedented degree, modern production, communication, and transportation methods have made many scientific and technological processes and products available on a mass basis. A curious moral aspect of such situations is that as the individual acts were assumed to be of negligible negative impact, they are typically regarded as being ethically unobjectionable. Hence, the aggregate effect of a mass of ethically permissible actions may nevertheless turn out to be quite ethically problematic. In ethics, sometimes numbers *do* count.[11]

Consider, for example, the pollutants emitted by each of the approximately 400 million automobiles in the world. The aggregate negative environmental effect of individually innocuous, hence seemingly ethically unproblematic, effects is known by now to be substantial. To the extent that this aggregate effect can be shown to harm people's health, particularly groups at special risk of being affected (e.g., the elderly, young children, and those with respiratory problems), the aggregate effect would begin to be judged as ethically unacceptable and unjust, and the individual pollution-emitting activities might begin to appear as something other than ethically neutral.

An analogy may help clarify this novel ethical situation. Suppose that a populous nation experiences a devastating depression in which many of its people suffer. Suppose that after the fact it is plausibly shown that an important contributing cause of the depression's occurrence was the fact that each family in the country had accumulated a substantial but individually manageable level of consumer debt. If the country was fortunate enough to recover its economic health, would not the new accumulation of a substantial but still manageable amount of consumer debt by an individual family be likely to be regarded as an ethically irresponsible thing to do? If so, the same could be said of an individual car owner's emission of pollution or an individual consumer's failure to recycle.

A somewhat futuristic example of the same pattern from the biomedical realm is that of predetermination of the sex of one's offspring. Given the fact that reproductive freedom is widely viewed as a human right, it is safe to assume that attempts of individual couples—at least married ones—to avail themselves of the latest scientific or technological means to determine the sex of their offspring will be regarded as ethically permissible. However, suppose that in a populous society with a culture biased in favor of male offspring a significant number of couples opt for predetermination and that a significant sexual imbalance of male over female offspring results. Suppose further that at least some of the ethically problematic consequences envisioned as resulting from this state of affairs come to pass (e.g., increased crime committed by men or heightened male aggressiveness in competition for scarcer female mates).

The upshot of this situation is that twentieth century science and technology may be pushing society toward reevaluation of "permissive" ethical judgments traditionally made about individual actions that are at worst "negligibly harmful." Consequentially speaking, the threshold of harm necessary to activate negative ethical judgments may be in the process of being reduced by the aggregative potential of modern science and technology in populous societies.

Practitioner Problems

The kinds of science- and technology-based ethical problems considered thus far have something in common: While spawned by developments within the spheres of science and technology, the resultant issues and conflicts unfold, not primarily within those spheres, but in society at large. In contrast, problems in the final category considered here, while related to concerns of society at large, arise primarily *inside* the communities of practitioners of science and technology. They are ethical problems associated with the concrete processes and practices of scientific and technological activity, both those in which these activities unfold and those in which their results are communicated. Such problems are sometimes viewed as falling within the province of "professional ethics," meaning that they are ethical problems that arise in the course of professional practice.

Problems of Execution   Edward Wenk has identified three kinds of ethical issues faced by practicing engineers in their work.[12]

(1) *Distributive Justice.*   The first is essentially an issue of distributive justice, involving as it does an allocation of costs, benefits, and risks. This kind of problem arises when an engineer must decide whether to embark upon or proceed with a feasible project that he or she recognizes is likely to expose people to a non-trivial degree of risk to their safety, health, or property without their consent. Beyond answering the question "Can it be done?" about the contemplated project, the would-be ethical engineer must confront the quite different question "Ought it be done?" For example, from an ethical point of view, should an engineering company accept a lucrative contract to erect a potentially hazardous structure, such as a hydroelectric dam, in a geologically unstable area near a rural village in the absence of the informed consent of its inhabitants?

Other things being equal, if the degree of risk—understood as a function of the estimated magnitude of the harm that could occur and the estimated likelihood of its occurrence—is substantial, then it would be ethically wrong to proceed. If it is negligible, then it would be ethically permissible, perhaps even obligatory, to do so. One problem with this kind of situation is that determination of what constitutes an "acceptable risk" is not a strictly technical question but a social and psychological one. Among other things, the answer to it depends on *what* members of the population at risk believe to be at stake, on *how highly* they value it at the time in question, and on *how seriously* they would regard its loss.

Meridith Thring has extended this analysis in the case of engineers who are independent operatives doing work in research and development. For years, Thring, a university professor of mechanical engineering, had been doing research on industrial robots. However, he eventually came to believe that "the primary aim [of such work] is to displace human labour." For this reason he abandoned work on industrial robots and decided to work only on

> applications where the aim is to help someone to do the job he does now without actually exposing his body to danger or discomfort; or where we need to amplify or diminish his skill and strength. A good example is "telechirics," . . . artifacts that allow people to work artificial hands and arms and operate machines in hazardous or unpleasant environments as if they were there, while they are in fact in comfortable and safe conditions.[13]

Thring implies that it is also ethically incumbent on engineers to consider whether the work they contemplate—here, a research and development project—poses an unacceptable risk to any important nonsafety interest of patients likely to be affected by it—for example, that of not being rendered redundant. At bottom, this too is an issue in distributive justice. For Thring, the ethical engineer must first carefully estimate the costs, benefits, and risks likely to be associated with a possible technological endeavor and then ask "Are those benefits, costs, and risks likely to be allocated to the affected parties in a way that is distributively just?" The engineer may then proceed with the work only if he or she can answer that question in the affirmative.

Similar ethical constraints apply to the initiation or continuance of scientific experiments that pose significant undisclosed risks to the safety, health, or property interests of people participating in or likely to be affected by them. Three of the most ethically repugnant scientific experiments carried out in or on behalf of the United States during this century are of this character and warrant brief description.

Beginning in 1932, U.S. Public Health Service researchers administered placebos to 431 black men in Tuskegee, Alabama. Each experimental subject, induced to come in for blood tests supposedly as part of an area-wide campaign to fight syphilis, was tested for and found to have syphilis. However, *the subjects were neither told that they had the disease nor treated for it.* The purpose of the experiment was to obtain scientific knowledge about the long-term effects of syphilis on mental and physical health. Nontreatment continued for 40 years, long after it became known that penicillin was a cure for syphilis and was widely available. Following press exposure in 1972, the experiment was terminated.[14] A Public Health Service investigation disclosed that of 92 syphilitic patients examined at autopsy, 28 men (30.4 percent) died from untreated

syphilis—specifically, from syphilitic damage to the cardiovascular or central nervous systems. Hence, the total number of men who died as a result of nontreatment may have exceeded 100.[15]

In the 1950s, the U.S. Central Intelligence Agency solicited and funded a series of "mind-control experiments." Among the techniques used on experimental subjects were sensory deprivation, electroshock treatment, prolonged "psychological driving," and the administering of LSD and other potent drugs. By one estimate, at least 100 patients went through one series of brain-washing procedures.[16] Many participants in the experiments suffered long-term physical and mental health problems. In 1953, one subject was given a glass of liquor laced with LSD. He developed a psychotic reaction and committed suicide a week later.[17]

As part of its Biological Warfare program, the U.S. Army secretly sprayed bacteria and chemicals over populated areas of the United States (and Panama) during a 20-year period beginning in 1949. At least 239 such tests were carried out. The objective was to determine the country's vulnerability to germ warfare by simulating what would happen if an adversary dropped certain toxic substances on the United States. One frequently used microorganism was *Serratia marcescens.* Its safety was questioned prior to 1950, and strong evidence that it could cause infection or death existed in the late 1950s. Nevertheless, it continued to be used in tests over populous areas for the next decade. Four days after a 1950 spraying over the San Francisco Bay Area, a patient was treated at the Stanford University Medical Center for infection caused by *Serratia,* the first case ever recorded at the hospital. Within the next five months, ten more patients were treated at Stanford for the same infection. One of them died.[18]

Their ethically reprehensible character aside, such cases serve the useful purpose of showing that "freedom of scientific inquiry" is not an absolute, unconditional, or inviolable right. While clearly an important human value, "freedom of scientific inquiry" may, under certain conditions, have to take a backseat to other important values, such as protection of the dignity and welfare of each and every individual human being.

(2) *Whistle Blowing.* Wenk's second kind of ethical issue in engineering is that of "whistle blowing." Engineers—or scientists—may become aware of deliberate actions or negligence on the part of their colleagues or employers that seem to them to pose a threat to some component of the public interest (e.g., public safety, the effective expenditure of public monies, and so on). If the worried practitioner's "in-house" attempts to have such concerns addressed are rebuffed, then he or she must decide whether to "go public" ("blow the whistle") and disclose the facts underlying the concerns.

Problematic phenomena of the sort that impel some practitioners to consider such a course of action are often driven by the huge profits and professional reputations at stake in modern research and development activity. These phenomena can be associated with any of a number of phases of engineering or science projects. Consider, for example, misleading promotional efforts to secure public funding; cheap, unreliable designs; testing shortcuts; misrepresented results of tests or experiments; shoddy manufacturing procedures; intermittently defective products; botched installation; careless or inadequate operational procedures; or negligent waste disposal. A significant number of such cases have come to light in recent years, of which three follow.

At Morton Thiokol, Inc. (MTI), several engineers working on the *Challenger* space shuttle booster project tried to convince management that the fateful January 1986 launch should not be authorized since the already suspect O-ring seal on the booster rocket had not been tested at the unusually cold temperatures prevailing on the day of the tragedy. MTI senior engineer Roger Boisjoly testified before the presidential committee investigating the disaster about what led up to the decision by the company and the National Aeronautics and Space Administration (NASA) to authorize the launch, a process at whose turning point MTI's general manager told his vice president of engineering to "take off his engineering hat and put on his management hat."[19] For his candid testimony, Boisjoly was allegedly subjected to various forms of mistreatment within the company and was placed on extended sick leave.[20]

In 1972, three engineers employed by the San Francisco Bay Area Rapid Transit (BART) system, after receiving no response to their in-house memos of concern, went public about subsequently confirmed safety-related deficiencies that they had detected in the design of BART's Automated Train Control System. They were summarily dismissed for their trouble.[21]

A senior engineer at the Bechtel Corporation, part of a task force assigned to plan the removal of the head of the failed nuclear reactor vessel at Three Mile Island after the famous 1979 accident, became concerned about short-cuts allegedly being taken by his company in testing the reliability of the crane to be used to remove the vessel's 170-ton lid. When he protested the alleged shortcuts, he was relieved of many of his responsibilities. He then went public, was suspended, and later fired.[22]

Sociologically speaking, several things are noteworthy about such cases. Technical employees who find themselves in situations in which they are asked or required to do things that violate their sense of right and wrong are not an endangered species. Results of a survey of 800 randomly selected members of the National Society of Professional Engineers published in 1972 disclosed that over 10 percent felt so constrained. A "large fraction" were sufficiently fearful of employer retaliation that they acknowledged they would rather "swallow the whistle" than become whistle blowers.[23] A major 1983 study of technical employees found that 12 percent of respondents "reported that, in the past two years, they have been in situations in which they voiced objection to, or refused to participate in, some work or practice because it went against their legal/ethical obligations as engineers or their personal senses of right and wrong."[24]

For a number of reasons, engineers have traditionally been loath to criticize their employers publicly. Most obviously, those who feel compelled to "go public" enjoy precious little legal or professional-association protection against employer retaliation, often in the form of firing. However, as Wenk argues, some reasons that discourage whistle blowing are sociocultural in nature:

> For engineers, a problem arises because most work in large organizations and many find excitement of participation in mammoth undertakings. They learn to value formal social structures, an attitude akin to "law and order." They adopt the internal culture and values of their employer and are likely to be allied with and adopt the perspectives of the people who wield power rather than with the general population over whom power is held. As a consequence, there is less tradition and inclination to challenge authority even when it is known to be wrong in its decisions which imperil the public.[25]

Not without reason, engineers—and, increasingly, scientists in large industrial organizations—tend to see themselves as employees with primary obligations to their employers, not the public. Moreover, the notion that employees retain certain citizen rights—for example, freedom of expression—when they enter the workplace is a relatively new notion in American industrial history. In a classic 1892 opinion, Oliver Wendell Holmes, then Massachusetts Supreme Court justice, wrote:

> There are few employments for hire in which the servant does not agree to suspend his constitutional rights of free speech as well as idleness by the implied terms of the contract. The servant cannot complain, as he takes the employment on the terms which are offered him.[26]

Only in the late twentieth century has this traditional attitude begun to be reversed, partly because the costs of such enforced silence are now viewed as unacceptable to society.

Ethically speaking, the obligation of technical professionals to blow the whistle when it is appropriate to do so arises from several factors. First, much contemporary research and development is supported by public money, as is the graduate education of many scientists and engineers (through government fellowships and loans). Second, the scale of the possible harm to the public interest at stake in many contemporary technical activities is large. The third factor is the following ethical principle of harm prevention: "[W]hen one is in a position to contribute to preventing unwarranted harm to others, then, other things being equal, one is morally obliged to attempt to do so."[27] An engineer or scientist sometimes possesses personal, specialized, "insider" knowledge about a troubling facet of a technical activity or project. Coupled with the credibility attached to testimony provided by authoritative technical professionals (as opposed to claims made by nonspecialist activists), that knowledge puts the scientist or engineer in a special position to make a possibly decisive contribution to preventing unwarranted harm to others or at least to preventing its repetition. This gives rise to a moral obligation to blow the whistle—once all other reasonable steps to rectify the situation "in house" have been taken and failed.

Some have urged that the obligation to responsibly blow the whistle be emphasized during the formal education of scientists and engineers—by the use of actual case studies, for example.[28] Others have stressed the importance of effecting structural and policy changes in the organizations in which technical professionals work and in their professional associations so that whistle blowers are not required to choose between remaining silent and becoming self-sacrificing "moral heroes."[29] A third approach is that of legislation. A measure of protection for whistle-blowers has been built into some federal environmental and nuclear laws, and roughly half the states prohibit the firing of employees who have blown the whistle on their employers for practices violating existing public policies. However, some advocates for whistle-blowers see the need for comprehensive federal legislation allowing whistle-blowers who suffer reprisals to initiate legal action against their employers up to two years after such occurrences.[30]

(3) *Consideration of Long-Term Effects.*   Wenk's third and final category of ethical issues confronting engineers in daily practice involves "managing the future." He argues that engineers have a tendency to focus on designing, producing, or installing "hardware" without adequately "anticipat[ing] . . . longer term effects." This is an abdication of the engineer's "professional responsibility." In terms of the "quartet of basic concerns" that we have utilized in this chapter, given the scale and scope of the effects of many contemporary engineering products and projects, engineers who fail to scrutinize their projects with comprehensive critical vision, both with respect to its likely consequences (including possible longer-term ones) and its likely patients or "stakeholders" (including, where appropriate, members of future generations) are guilty of unprofessional and ethically irresponsible conduct. Uncritical allegiance to the deontological dictum "if it can be done, it should be done" no longer confers immunity from charges of ethical impropriety on technical practitioners.

Problems of Communication   Other ethical issues faced by technical practitioners have to do not with possible effects of scientific and technological projects on the safety, health, or property of those who may be affected by them, but with problematic aspects of *practitioners' communication* of and about their work.[31] Issues in this category, most notably ones involving *fraud* and *misrepresentation,* pertain to publication or presentation of claimed findings and to work-related interactions with nontechnical funding or policy-making officials. Cases of fraud have come to light in recent years in which experiments reported on in published papers were in fact never carried out, crucial data were fabricated, and conclusions were drawn from data allegedly known not to support them.[32]

(1) *Fraud.*   Falsification of scientific data may not be as infrequent as normally supposed. June Price Tangney surveyed researchers in the physical, biological, behavioral, and social sciences at a large American university. Of 1,100 questionaires distributed, 245 were completed and returned. Half of the respondents were senior researchers with the rank of full professor. Not surprisingly, the survey revealed that 88 percent of the respondents believe that scientific fraud is uncommon. However, 32 percent reported that they had a colleague in their field whom they had at some time suspected of falsifying data.[33] This figure, while suggestive, must be interpreted cautiously. It is not proof that one third of all scientists engage in such misconduct, for not only may suspicions be mistaken, but multiple respondents could have had the same individual in mind.

Whatever the extent of fraud in science, scientists see a number of factors contributing to the phenomenon. Tangney's respondents identified the following as major motivations for fraud: desire for fame and recognition (56 percent), job security and promotion (31 percent), firm belief in or wish to promote a theory (31 percent), and "laziness" (15 percent).[34] Underlying many such factors, she contends, is *the highly competitive nature of contemporary science:* the pressure to publish, the shortage of desirable jobs, and the fierce competition for funds. Beyond contributing to fraud, Tangney argues that such pressures can negate "what might otherwise be a fairly adequate self-policing mechanism in the scientific community." Indeed, the results of her survey call into question the common wisdom about the self-correcting nature of science, via processes like refereeing and publication, for of the aforementioned 32 percent who had suspected a colleague of falsifying data, over half (54 percent) reported that they had taken no action to confirm or disconfirm their suspicions.[35] The competitive nature of contemporary science may have biased the reward system in the profession *against* undertakings aimed at uncovering fraud.

In a highly competitive academic environment, many researchers may feel that, if they raise questions about serious misconduct, their own reputations will be tarnished and their own chances for resources and advancement will be diminished. A scientist may be rewarded for uncovering "legitimate" flaws or shortcomings in a rival's work. However, there generally is little to be gained and much to be lost by attempting to expose a fraud.[36]

(2) *Misrepresentation.* Misrepresentation takes a number of forms in the communication of research findings, including both failure to credit or fully credit deserving contributors and crediting or overcrediting undeserving contributors. It might seem that such acknowledged species of misconduct, however regrettable, do not deserve to be called unethical, except perhaps by deontologists, for whom they fall into forbidden categories of actions regardless of the gravity of their consequences for science or society. However, consequentialists may also selectively regard such practices as unethical, since they can in fact result in serious public harm. In May 1987, a scientist was accused by an investigative panel of scientists appointed by the National Institutes of Health of "knowingly, willfully, and repeatedly engag[ing] in misleading and deceptive practices in reporting results of research."[37] The pertinence of these practices to consequentialist ethical judgment making becomes clear from the panel's finding that the scientist's publications had influenced drug treatment practices for severely retarded patients in facilities around the United States.

Presentations of research findings to groups of peers also offer opportunities for ethically problematic behavior. Such presentations are sometimes used to establish claims of priority in the conduct of certain kinds of research. However, given the intensely competitive nature of the contemporary scientific research enterprise, if the research work is still in progress, it is understandable that scientists may opt to disclose just enough of their findings to serve their priority interests but not enough to reveal their overall strategies or the next steps in their "battle plans." However, quests for priority and resultant recognition may go beyond being unprofessional and become unethical if deliberately misleading or outright false information is disseminated in hopes of sending rivals "off on wild goose chases," diverting them from paths believed potentially fruitful. While making such an ethical judgment may appear open only to a deontological thinker, doing so can also be defended on consequentialist grounds, by, for example, appealing to the harm that such deception can inflict on knowledge-sharing institutions like the peer seminar which have usefully served scientific progress and thus, indirectly, human welfare.

The interaction of scientists and engineers with public funding agencies or policy-making officials can also be ethically problematic. Institutional or organizational pressures to obtain funding for research and development ventures or units with significant prestige or employment stakes can induce applicants to resort to various forms of misconduct in hopes of increasing the chances of favorable action by a funding agency. Among these are use of false data, misrepresentation of what has been accomplished to date on a project in progress, and gross exaggeration of what can be expected in the grant period or of the scientific or social significance of the proposed work.

Dealings with makers of public policy often lend themselves to such hyperbole, for policymakers are typically individuals with nontechnical backgrounds who are unable to assess critically the plausibility of the claims made about current or proposed research or development projects. If a prestigious researcher deliberately misrepresents the potential or state of development of a pet project to an influential policymaker in order to enhance the project's funding prospects, then insofar as approval is secured through this deception and at the cost of funding for other worthwhile projects, consequentialist thinkers may join deontologists in judging the practitioner guilty of unethical conduct. This they may do not least on grounds of the long-term consequences for the welfare of society of undermining the integrity of the research funding process.

THE CHALLENGE OF CONTEMPORARY SCIENCE AND TECHNOLOGY TO TRADITIONAL ETHICAL THEORIES

The foregoing discussion of categories of ethical issues and conflicts engendered by developments in science and technology strongly suggests that these forces are putting growing pressure on traditional absolutistic ethical thinking. There are several reasons that the validity and utility of such thinking are being called into question in an era of rapid scientific and technological development.

As we saw, many such theories condemn or praise particular kinds of actions if they but fall into one or another category of supposedly intrinsically good or bad deeds. However, as noted, an action or practice condemned as "unnatural" can come to seem less so over time, especially if the original ethical judgment was predicated on the act's or practice's being unusual or unfamiliar when it first came to attention. Similarly, traditional absolutistic ethical outlooks are sometimes based on static world views born of their subscribers' limited experience. As a culture or subculture dominated by such a world view overcomes its geographical or intellectual isolation and interacts more with the rest of the world, supposedly immutable categories or rules pertaining to "sacred" things or ways tend to loosen up. Adherents of such world views may come to recognize that respectable members of different cultures or subcultures think and act differently than they do about the same matters. Further, new products, processes, and practices can, as noted, have long-term hidden effects. Their eventual eruption and empirical confirmation sometimes call for revision of ethical judgments made before recognition that such subterranean effects were at work. However, such reevaluation is not congenial to absolutistic thinking, which purports to base its unwavering ethical judgments and decisions on something other than consequences. Considerations such as

these make it increasingly difficult to sustain absolutistic ethical theories and outlooks in contemporary scientific and technological society.

Besides challenges to its intellectual tenability, contemporary science and technology are calling into question the utility of traditional absolutistic ethical theories—that is, their ability to serve as intelligent guides to action in a world of rapid and profound technical and social change. Such categorical theories and outlooks are helpless when confronted with ethical issues and conflicts of the sorts discussed in the third section of this chapter. For example, such theories shed no light on cases involving the distribution of benefits, costs, and risks associated with scientific and technological developments; intrapersonal conflicts between two venerable ethical values; public harms of aggregation; or the situations of technical professionals torn between loyalty to employers, concern for their families' well-being, and devotion to promoting the public interest. Finally, traditional deontological approaches to ethical thinking offer no incentives to agents to consider whether, in the face of possible unforeseen effects of a technical innovation, expansion of the domain of pertinent patients or the list of their protectable interests might be in order.

This is not to imply that consequentialist theories and thinking are immune from difficulties when confronted by contemporary scientific and technological developments. For example, uncertainty about possible elusive or projected long-term consequences of scientific and technological innovations and developments makes ethical judgments based on such assessments provisional and open to doubt. However, that is a price that must be paid if ethical judgments and decisions are to be made on an empirical rather than an a priori basis and are to be focused on the bearing of scientific and technological developments on human harm and well-being.

One conclusion of this chapter, then, is that developments in contemporary science and technology call for revisions in traditional ethical thinking and decision-making. One kind of ethical theory deserving serious consideration we will call *qualified neo-consequentialism*. Under this theory, ethical judgments about actions, practices, and policies hinge first and foremost on assessments of their likely consequences. In particular, these assessments must have the following *neo-consequentialist* qualities. They should be

1. *Focused on harm and well-being*—directed to identifying and weighing the importance of consequences

likely to influence the harm or well-being of affected patients;*

2. *Refined*—designed to detect or at least be on the look-out for subtle effects that, although perhaps hidden or manifested only indirectly, may nonetheless significantly influence stakeholder harm or well-being;

3. *Comprehensive*—designed to attend to *all* harm- and well-being-related effects—social and cultural as well as economic and physical in nature—of the candidate action, policy, or practice on *all* pertinent patients, remote as well as present, "invisible" as well as influential;

4. *Discriminating*—designed to enable scientific and technological options to be examined critically on a case-by-case basis, in a manner neither facilely optimistic nor resolutely pessimistic, and such that any single proposal can emerge as consequentially praiseworthy and be adopted or as consequentially ill-advised and be rejected in its present form if not outright; and

5. *Prudent*—embodying an attitude toward safety that, as long as a credible jury is still out or if it has returned hopelessly deadlocked, is as conservative as the magnitude of the possible disaster is large.

Further, the assessments sanctioned by our proposed ethical theory must also meet certain conditions. If an action, policy, or practice is to earn our theory's seal of approval, its projected outcome must not only be likely to yield at least as large a surplus of beneficial over harmful consequences as that of any available alternative, but it must also meet certain additional *qualifications,* two of which will now be briefly discussed.

It is scarcely news that contemporary scientific and technological activities unfold in societies in which those who stand to benefit greatly from their fruits are rarely the same as those likely to bear their often weighty costs and risks. We therefore stipulate that to be ethically permissible or obligatory, the allocation of the scrupulously projected benefits, costs, and risks of a technical undertaking among the various affected stakeholders must also be *distributively just.*

---

*The reader will note that no account has been presented here of exactly what is meant by human "harm" and "well-being." That substantial task must be left for another occasion. Suffice it to say here that for this writer, "harm" is not reducible to considerations of physiological deprivation, physical injury, and property damage or loss; nor is "well-being" reducible to considerations of material abundance, financial success, and high social status.

Various criteria have been put forth for evaluating the justice of such distributions.[38] One that deserves serious consideration is John Rawls's famous "difference principle."[39] Imagine, says Rawls, a group of people in "the original position"—that is, convened to formulate from scratch the rules that will govern the first human society, one shortly to come into being. Suppose that these deliberations take place behind "a veil of ignorance"—that individual group members have no knowledge whatsoever of whom or what they will turn out to be (e.g., male or female, black or white, Asian or North American, physically handicapped or not) or of their eventual economic well-being or social status. Then, Rawls contends, the group would eventually reach agreement that it was in each member's best interest that the following rule of justice be adopted: an unequal distribution of any social or economic "good" will be permitted in the society-to-be only if there is good reason to believe that it will *make everyone, including the less fortunate, better off.* Indeed, Rawls eventually offered a stronger version of his principle according to which an unequal distribution of such a good is just only if there is good reason to believe that it will make everyone better off *and* that it will *yield the greatest benefit to those currently worst off.*[40] If either version of this rule is found compelling, it would have to be applied to each predominantly beneficial but unequal distribution of projected science- or technology-based benefits, costs, and risks before the conclusion could be reached that it was ethically permissible or obligatory to proceed with the action, project, or practice in question.

Our neo-consequentialist ethical theory has a second qualification. A science- or technology-related course of action may sometimes be denied ethical approval even if all of the foregoing conditions are met. Even then, it may be proper to withhold ethical approval if the projected harmful consequences (1) *exceed some substantial quantitative threshold*—either in a single case or when aggregated over multiple cases of a similar sort—and (2) are not *greatly* outweighed by their positive counterparts. In such situations, the decision-making party may decide that it would be prudent to decline the admittedly greater projected benefits offered by the option under consideration.

Ethical decision making that takes no account of the absolute magnitude of an option's projected negative consequences even if they are outweighed by their positive counterparts, or of how the outweighed negative consequences of an individual course of action may aggregate over multiple instances, is deeply flawed. Indeed, allowing "yielding a positive balance of benefit over harm" by itself to compel ethical approval of individual courses of action may even be unjustified on consequentialist grounds, for following that criterion consistently in multiple instances may over time lead to unacceptable public harms of aggregation. For example, assessing the impact on traffic of individual proposed downtown high-rise office buildings solely in terms of the modest number of additional cars each structure may attract into the city may allow the aggregate effect on traffic of approval of a large number of such projects to go unreflected in individual decision-making processes.

CONCLUSION

Hopefully, the reader will find the qualified neo-consequentialist approach to thinking about ethical issues and conflicts just sketched more adequate to the realities of contemporary scientific and technological practice. In any event, in this chapter we have characterized a number of different kinds of science- and technology-based ethical issues and conflicts, indicated some noteworthy technical and social roots of such problems, and argued that important traditional ethical concepts and modes of thinking are being subjected to increasing pressure by scientific and technological changes in contemporary society. While this stress is being strenuously resisted in some quarters, it is likely in the longer run to lead to major transformations of ethical ideas and thinking.

ENDNOTES

1. I owe my initial awareness of this framework to a 1972 lecture at Stanford University by ethicist Dr. Karen Lebacqz.
2. See, e.g., William Frankena, *Ethics,* Foundations of Philosophy series, 2nd ed. (Englewood Cliffs, N.J.: Prentice-Hall, 1973), chaps. 2 and 3. See also Mary Warnock, *Ethics Since 1900* (London: Oxford University Press, 1960), chap. 2, pp. 48–52.
3. *Wall Street Journal,* May 4, 1989, p. B4.
4. Stephen Isaacs, "Hasidim of Brooklyn: Fundamentalist Jews Amid a Slumscape," *Washington Post,* "Outlook" section, February 17, 1974, Section B, p. 1.
5. Gregory Kavka, *Moral Paradoxes of Nuclear Deterrence* (Cambridge: Cambridge University Press, 1987), pp. 19–21.
6. Ronald Munson, *Intervention and Reflection* (Belmont, Calif.: Wadsworth, 1979), p. 398.
7. *Ibid.,* pp. 399–400.
8. For example, Carl Cohen argues that animals are not members of any "community of moral agents." Incapable of,

among other things, giving or withholding consent, animals, unlike humans cannot have rights thereby precluding involuntary experimentation on them. See Cohen's "The Case for the Use of Animals in Biomedical Research," *New England Journal of Medicine,* 315, no. 14, 1986, 867.

9. Gina Kolata, "Genetic Screening Raises Questions for Employers and Insurers," *Science,* 232, no. 4748, April 18, 1986, 317.
10. *New York Times,* August 19, 1986, p. 21.
11. John M. Taurek, "Should the Numbers Count?" *Philosophy and Public Affairs,* 6, 1977, 293–316.
12. Edward Wenk, Jr., "Roots of Ethics: New Principles for Engineering Practice," American Society of Mechanical Engineers Winter Annual Meeting, Boston, Massachusetts, December 1987, 87-WA/TS-1, pp. 1–7.
13. Meredith Thring, "The Engineer's Dilemma," *The New Scientist,* 92 no. 1280, November 19, 1981, 501.
14. *New York Times,* July 26, 1972, p. 1.
15. *New York Times,* September 12, 1972, p. 23. For a detailed account of this episode, see James H. Jones, *Bad Blood: The Tuskegee Syphilis Experiment* (New York: Free Press, 1981).
16. Harvey Weinstein, *A Father, a Son, and the CIA* (Toronto: James Lorimer and Co. Ltd., 1988).
17. Leonard A. Cole, *Politics and the Restraint of Science* (Totowa, N.J.: Rowman and Allanheld, 1983), p. 111.
18. *Ibid.,* pp. 112–114.
19. Roger Boisjoly, "Ethical Decisions: Morton Thiokol and the Space Shuttle *Challenger* Disaster," American Society of Mechanical Engineers Winter Annual Meeting, Boston, Massachusetts, December 1987, 87-WA/TS-4, p. 7.
20. *Ibid.,* p. 11.
21. Stephen H. Ungar, *Controlling Technology: Ethics and the Responsible Engineer* (New York: Holt, Rinehart and Winston, 1982), pp. 12–17.
22. Rosemary Chalk, "Making the World Safe for Whistle-Blowers," *Technology Review,* January 1988, p. 52.
23. Rosemary Chalk and Frank von Hippel, "Due Process for Dissenting 'Whistle-Blowers.'" *Technology Review,* June/July 1979, p. 53.
24. Chalk, "Making the World Safe," pp. 56–57.
25. Wenk, "Roots of Ethics," p. 3.
26. Chalk and von Hippel, "Due Process," p. 53.
27. Compare this principle with Kenneth Alpern's "principle of due care" and "corollary of proportionate care" in his "Moral Responsibility For Engineers," *Business and Professional Ethics Journal,* 2, Winter 1983, 40–41.
28. Boisjoly, "Ethical Decisions," p. 12.
29. Richard DeGeorge, "Ethical Responsibilities of Engineers in Large Organizations: The Pinto Case," *Business and Professional Ethics Journal,* 1, no. 1, 1981, 12.
30. Chalk, "Making the World Safe," pp. 55–56.
31. For a useful bibliography on this aspect of the problem, see Marcel Chotkowski LaFollette, "Ethical Misconduct in

Research Communication: An Annotated Bibliography," published under NSF Grant No. RII-8409904 ("The Ethical Problems Raised by Fraud in Science and Engineering Publishing"), August 1988.

32. See, e.g., William Broad and Nicholas Wade, *Betrayers of the Truth* (New York: Simon & Schuster, 1982), pp. 13–15; and Nicholas Wade, "The Unhealthy Infallibility of Science," *New York Times,* June 13, 1988, p. A18.

33. June Price Tangney, "Fraud Will Out—Or Will It?" *New Scientist,* 115, no. 1572, August 6, 1987, 62.

34. *Ibid.*

35. *Ibid.*

36. *Ibid.,* p. 63.

37. *New York Times,* April 16, 1988, p. 6.

38. For cogent discussion of various criteria of distributive justice, see Joel Feinberg, *Social Philosophy,* Foundations of Philosophy Series (Englewood Cliffs, N.J.: Prentice Hall, 1973), chap. 7.

39. John Rawls, *A Theory of Justice* (Cambridge, Mass.: Harvard University Press, 1971), chaps. 1–3, especially pp. 75–78.

40. For discussion of alternate versions of Rawls's difference principle, see Robert Paul Wolff, *Understanding Rawls* (Princeton, N.J.: Princeton University Press, 1977), pp. 40–41.

QUESTIONS

1. How would you define ethics?

2. List three ways that ethics correlates to social responsibility.

3. There are four kinds of considerations that are pertinent to ethical decisions and judgment making. List and describe them.

4. What does the phrase "violations of supposedly exceptionless moral principles" mean?

5. Name a new technology and then provide an example of value-based conflicts that it has created.

The Relationship Between Ethics and Technology

PAUL ALCORN

- -

I did not move a muscle when I first heard that the atom bomb had wiped out Hiroshima. On the contrary, I said to myself, Unless now the world adopts nonviolence, it will spell certain suicide for mankind.

MOHANDAS "MAHATMA" GANDHI

In an evolving universe, who stands still moves backward.

R. ANTON WILSON

INTRODUCTION

In this chapter, we will develop a precise definition of technology to avoid misunderstanding when the term is used in this text. We will discuss the way in which technology is created and functions in a culture and then develop an understanding of how this concept of technology relates to ethical behavior. Finally, we will discuss how to use technology ethically, that is, do technology in a way that works.

DEFINITION OF TECHNOLOGY

Essentially, technology is that whole collection of methodology and artificial constructs created by human beings to increase their probability of survival by increasing their control over the environment in which they operate. Technology includes and is essentially a means of manipulating natural laws to our benefit by constructing objects and methodology that increase our efficiency and reduce waste in our lives. The objects we create are artifacts, literally artificial constructs, that have been manufactured for specific uses and purposes. Everything that we use that is not as it comes to us in nature falls under the heading of technology. This is a very broad definition. All of the phys-

ical objects of our lives that were in any way altered from the way they appeared in nature represent technology. A sharpened stick is technology, as is a dollar bill or a caterpillar tractor; they merely have different functions and have been produced through a different series of steps, usually through the use of other technology.

It may be noted that human beings are not the only animals that create artifacts, and for that reason, the mere creation of artifacts does not in and of itself constitute technology. Birds build nests, chimpanzees use sticks as tools to gather food, and bees build elaborate hives. What is missing in these artifacts that separates them from what we mean by technology is the matter of choice. A bee contributes to the development of a hive because of genetic encoding. It is a process that is "hard wired," as an electrical engineer would say. It has no choice about what it is doing. The same is true of a bird building a nest or an otter using a rock to open a clam by resting the clam on its stomach as it floats and hammering it with a stone. Such behavior is instinctual. But not all methodology used by living creatures other than humans is instinctual. Some higher primates, chimpanzees, for example, are capable of reasoning through problems and using objects to create methodology for solving those problems. They have been observed experimentally under controlled conditions learning to attach telescoping rods together to gather food that is otherwise out of reach. Yet they have very limited capacity in this regard and do not pass this information on

PRACTICAL ETHICS FOR A TECHNOLOGICAL WORLD by Alcorn, Paul A., © Reprinted by permission of Pearson Education, Inc., Upper Saddle River, NJ.

to others in a cultural way. What truly separates humans from the other members of the animal kingdom in this regard is our incredible power of choice.

TECHNOLOGY AND CHOICE

With humans, the technology we choose to build and the manner in which we use it is totally a matter of choice. We have an infinite capacity to produce technological goodies, within the boundaries of natural law, and we can accept or reject an idea as we choose. Thus, at one point in time, we may choose to develop the use of fire for cooking and at another decide to develop the art or science of architecture for the purpose of providing ourselves with shelter. Additionally, at one point we may decide to use dome-shaped hovels as shelter and at another time and place opt for alabaster palaces or multistory office buildings. The choice is all ours. It is in that choice of what artifacts to produce and the range of artifacts that we are capable of producing that we find the true nature of technology. And, as nearly as we can tell, that choice seems to be the sole province of human activity.

TECHNOLOGY AND EVOLUTION

In *Social Issues in Technology: A Format for Investigation,* I offered a detailed explanation of technology and the technological process. In this book I offer a general understanding of technology and why it exists in our lives. Technology is a vital part of what it is to be human; in order to understand our world, it is necessary to understand the purpose, the source, and the processes of our technological world.

For a human being, doing technology is a natural process. It represents one of the chief capacities with which nature has provided us for our survival. As with any other creature, Homo sapiens has certain characteristics that allow the species to perpetuate itself and successfully compete with other species for a niche in the natural world. Ecologically, we are an integral part of a much larger system that is designed to grow, develop, and maintain itself as an extensive living structure.

Every element in that system has the capacity to survive based on certain characteristics. For human beings, those *survival traits,* as these characteristics are called, include our capacity to create and use technology. There are specific and overwhelming advantages to this ability. Because we use artificial structures for our survival rather than develop the necessary characteristics through genetic alteration to

our being, we are able to develop and adapt at a much higher rate than other animals or plants. We have effectively externalized the process of evolutionary development.

As an example, consider the characteristics of other animals versus those of a human being. Other animals have the advantage of speed, or claws, or special poisons that they can inject into their prey. Herbivores have specially designed digestive systems that allow them to consume large amounts of cellulose, a very difficult substance to break down, and turn it into useful energy. Some animals fly, others are very fleet of foot, others have incredible capacities to blend into the environment, and still others design complex living environments (e.g., hanging basket nests or colonized networks of tunnels). Each species has specific characteristics that offer it an advantage.

Now compare this with a human being. We do not have armored bodies covered with scales or shells. We cannot run particularly fast (though genetically we do have incredible stamina compared to most animals, a characteristic that allowed our hunter ancestors to follow game for days until the game was exhausted). Nor can we take to the air, with wings on our backs, or glide on membranes built into our bodies as bats or flying squirrels do. Yet we are capable of moving at a rate of speed far beyond that of a cheetah or other fleet-footed animal. We are able to fly across the face of the planet and into the outer reaches of our world and beyond. We can live underwater in craft that outperform the largest fish and exist in environments in which the extremes of temperature or altitude would kill most other creatures. We do it all in spite of the fact that we have at our disposal not a single physical trait that allows us to do so.

That is because the nature of our evolution has been external to our bodies. Instead of developing the eyes of a hawk, we develop binoculars and telescopes. Instead of becoming fleet of foot, we build automobiles and locomotives and airplanes. Instead of wings on our back, we have the wings of air transports and helicopters and the lifting power of balloons and dirigibles. Our characteristics are external to our physical being. It is in this ability to artificially create what we need for survival that we find our chief advantage. Like other animals, we use the laws of nature to aid us in our survival, but whereas other species do this through genetic alteration, a process that takes thousands if not millions of years, we manufacture the alterations quickly and efficiently. We find ourselves at last at a point at which we do not adapt to nature, we adapt nature to us! Such capacity is unparalleled in nature.

But with this capacity comes a problem. Nature is an experimenter. Nature will try numerous variations on a theme to find the combination of characteristics that allow a given organism to survive in a competitive world. If one alteration does not work, such as growing extra wings or limiting the number of eyes of a species to one, then that version fails and does not survive long enough to create progeny, or pass on the undesirable trait. If a variation offers superior opportunities for survival, many more of that version survive to pass on the characteristics to offspring, and eventually, that version predominates. Thus, through evolutionary mutation and survival of the fittest, we arrive at a creature that is perfectly adapted to its environment.

This is also true of humans, but with one exception. Since we are producing change through the creation of technology rather than trial-and-error mutation, we can very quickly generalize a new "trait" over the entire population in a relatively short period of time. In a matter of generations rather than millennia, a new technological device such as the bow and arrow or the chariot can come into general use by everyone who sees it. If it offers a very great advantage to those who have it, everyone either perishes or soon learns to use the new technology. There is little time for experimentation and testing here.

This has been seen often in the past with sometimes devastating results. The practice of agriculture is an excellent example if we look at the relationship between climatic change and the extensive use of agriculture in a region. Some of the most arid regions of the globe were once great forests or grasslands that were cleared for agriculture. Unfortunately, with the deforestation came a host of environmental changes that led to everything from soil erosion to changes in weather patterns. This is just a single example of the problems that can arise from moving too quickly to embrace a technology. Other examples include the virtual lack of forests in Lebanon today, where once stood vast woodlands of cedar, a prized wood traded all over the Mediterranean, from North Africa to Egypt to ancient Israel, and the cliff dwellers of the southwestern United States, who flourished toward the end of the first millennium and then abandoned their cities when they could not adjust to climactic changes in growing cycles.

What if the governments of the world in the last half of the twentieth century had decided that since nuclear weapons were the ultimate in destructive power, they would embrace that technology as is and abandon other means of war? We would have been left with no alternative but to create a nuclear holocaust in case of threat or attack. We are perhaps now in a similar predicament with biological and chemical weapons of mass destruction; they are cheap, effective, and easily produced and delivered. A single strain of a deadly bacterium or virus could cause a reduction of population around the world that would bring civilization as we know it to an end. And the tragic event would be the result of industrial and technological processes at work.

TECHNOLOGY AND RESISTANCE TO CHANGE

Because of this danger to our well-being, these seeds of destruction within our success, nature has also equipped us with another trait. That other trait is a resistance to changes in our culture. *Homeostasis*, as it is known, represents a fear of the unknown that extends to any technological device that may come along. Any new idea or new technology is initially suspect to most of the population because it is untested, unfamiliar, and therefore considered a potential threat. This is as much a survival mechanism as the capacity to create that technology in the first place. Because of homeostasis, time is a necessary ingredient for a given advance in technology to be generalized over the whole society. It is first embraced by a small section of the population eager to try new things and ideas, but the rest of society either initially ignores it or cautiously watches to see where it will lead. Should the new idea not be a particularly good one, that is, should it not increase the probability of individual and group survival, it tends to go by the wayside without much further ado. On the other hand, if it is actually a valuable idea, the new technology will continue to exist long enough for people to get used to it or to lose their initial fear of it, and then they are more willing to try this new gizmo. This is particularly true if those who first accept it have illustrated its value. Eventually, the acceptance and use of the new technology spreads throughout the culture.

This process can be easily seen in the case of the computer. Less than a century old, this device, once a curiosity used for certain esoteric operations by scientists and government, has become one of the primary tools of a modern technological society. It has been viewed as an oddity, feared, mystically couched in arcane terminology and given unrealistic assumptions of power by the uninitiated, seen as the subject of hobbyists and gadgeteers, embraced by big business, then small business, and finally accepted as an unavoidable way of life. The process took time while the population figured out how to use the new technology and how to configure it so that it was useful for their needs. It took time to gain acceptance and overcome the natural tendency of human beings to do things in the "same old way." It grew in popularity and use as a solution to a range of problems over the life of its development. All of that time was a gestation period for society to absorb and gain benefit from the new technology. Every invention goes through the same process, affected by a number of factors such as complexity, range of application, expense, and the degree of societal resistance.

The point to remember is that that resistance is necessary and natural, a safety net built into us by nature that allows us to take time to differentiate between new ideas that are truly beneficial and those that are potentially or truly dangerous to our survival. It is all part of the same natural process of creation and use of technology.

Human beings cannot help being creative. It is an element of our makeup that cannot be changed. Creativity and technological expertise require nurturance, but the tendency to learn the laws of nature and apply them to creating artificial constructs to enhance our lives comes as natural to us as breathing.

TECHNOLOGY AND ETHICS

Given that creating technology is natural and that within the limits of our understanding as to the nature of the universe, we can choose what technology to use and how to use it, where do the ethics of the process arise? If you remember back to our working definition, ethics is the process of doing what works. Apparently, from the history of the human race, using technology tends to work. This is evidenced, if in no other way, by the predominance and domination of our species over the face of the earth. We are incredibly successful as a species, reflecting incredibly successful natural traits, and that includes technology and its use. Apparently, technology works for us, or we would not include the capacity to create it in our repertoire of survival traits in the first place. By definition, then, in and of itself, it must be ethical.

That's a nice idea, and it would certainly be a blessing for all of us if that were true. Unfortunately, it is not as simple as that. Technology, as it turns out, is neither ethical nor unethical; it is merely a tool to be used or misused as we choose. Thus, we are back to the choice of action again, the one control we have in our lives.

Each technology and each application of technology raises ethical issues with which we must deal. Each new device or application of what we know requires some consideration of whether the use of that device will work for us or not. To further muddy the issue, we often cannot even say with certainty whether a technology will benefit us or not. In fact, in most cases, technology turns out to be a double-edged sword, with both costs and benefits in its use, and this in turn requires us to determine whether or not the benefits

are worth the costs. And that's assuming we can even actually determine the costs accurately in the first place.

Also, we need to consider the idea that the use of technology may benefit some while costing others. This is not an uncommon occurrence, particularly where one technology replaces another, as in the case of the automobile replacing the horse-drawn buggy or the word processor replacing the "steno pool."

As you can see, this cost-benefit situation creates quite a dilemma. Just knowing that the ethical thing to do is to do what works is not very useful as a guide to behavior if we do not know what works in the first place. This is not a new idea. It is a problem that we as a species have been wrestling with off and on for ten thousand years or more, particularly when new technologies and new ways of manipulating the world present themselves. A few examples will clarify this point nicely.

When the automobile was first introduced, it was hailed not only as a solution to transportation problems within cities but also as a defense against growing pollution. That may seem quite confusing from our perspective as citizens of the world at the beginning of the twenty-first century, but a century ago, the pollution problems faced by industrial urban dwellers was decidedly different. At that time, at the birth of the automobile age, the chief means of transportation was the horse. Anyone who wasn't walking or traveling by train within an urban environment was traveling on foot or by horse. Carriages, drays (freight wagons), and specialized coaches were all horse drawn. With the horses came horse dung, and it was everywhere. The streets were pocked with piles of dung to be cleaned up, dung that ran into the sewers and that produced a prodigious number of flies. And with the flies came disease. We do not think of horses and horse dung as being a major health hazard in our lives today, but a hundred years ago, it was a major problem. Thus, the "horseless carriage" was hailed as the eliminator of the "hay burner" technology of equestrian transportation.

Yet today, we view the automobile as a chief air pollution source; it dumps tons of carbon monoxide and other pollutants into the atmosphere, promoting global warming and creating smog in any city of size. Hence the solution becomes the issue. At the present time, there is a movement toward nonpolluting electric cars. California has gone so far as to mandate a 10 percent noncombustion engine vehicle quota for the state. Electric cars are the obvious noncombustion engine choice, and as the number of electric vehicles rises, replacing gasoline engine automobiles, it is believed substantial improvement in the environment will result. And so another solution has been found.

This being the case, should we not expect these electric vehicles to create other dilemmas? At the present time, nearly all electric automobiles are powered by heavy lead-acid batteries, deep charged and able to deliver power at sufficient rates for a reasonable amount of time. And much research is being done to develop better and more powerful batteries that will charge more quickly and deliver more power for even longer periods of time. Thus it appears, at least for the foreseeable future, that a dependence on lead-acid batteries will be dominant. But a new problem arises: What will we do with the spent batteries? Batteries are already seen as a pollution problem, with only one per car. What will happen when the number of batteries per vehicle rises to twelve or twenty? Could we be exchanging one form of pollution for another? It is not just lead-acid batteries that present this type of dilemma as we progress and change technology.

With any technological change and any acceptance of a new technology as standard, there is always a cost. There is never a free lunch, though payment can be deferred for some length of time. Yet in the end, someone has to pay, and I'm sure it comes as no surprise that delaying payment until our children or grandchildren are making the rules is not a very efficient way to operate. Intuitively it is unethical to use this approach, though economically or politically it may be expedient.

To what extent should we consider the future payment for our exploitation of technology and technological possibility? Though we do not always know (indeed, seldom do we know) the true cost of a technological development, there are certainly some issues that we do know will need to be handled. History offers numerous examples of what to expect from technological change. How far does our responsibility go? One school of thought says not to worry about the future consequences because we have always been able to deal with what comes along. Still newer technology will solve the problem. New ideas and alternative ways of handling the issues will arise naturally out of necessity. We need only utilize what is available to us now, and let the future generations worry about how to handle the problems that arise. These are the attitudes that led to the destruction of environments in the ancient world. As agriculture and population exploded beginning some ten thousand years ago, whole civilizations were destroyed by resultant drought and crop failure. Whole ecosystems were altered, turning fertile plains into deserts and lush forests into arid wasteland. Solutions were found, but what was the cost? The people of these transitional periods endured starvation and being uprooted as their productivity collapsed.

On the other hand, consider the approach of the Five Nations of the Iroquois Confederation. These Native Americans of the north-eastern United States banded together in a peaceful structure that allied independent nations, building a greater confederation. The Cuyahoga, Seneca, Onondaga, Mohican, and Oneida nations agreed to work together for the betterment of all and for their mutual defense against their unfriendly neighbors, chiefly the Algonquin. This amazing group of people elected fifty men from among their number to collectively make the decisions for the whole group. (Interestingly, it was the women of the tribe who actually chose the fifty men to head the joint council.) They always considered the future consequences of those decisions, *for seven generations hence!* No decision that was merely expedient was acceptable. Compare this approach with the political process present in most industrialized countries today. How many decisions are made on the basis of how the people will be affected a century and a half in the future? It appears we could learn a great deal from these Native American tribes. (Incidentally, it would not be a wise idea to embrace the wisdom of the Native Americans without exception. The Iroquois, for example, are noted for their horrific treatment of prisoners of war, whom they first honored and then tortured for as long as possible without killing them, then ritually ate them, not for the food value, but to absorb some of their bravery and strength. It was considered a pity if the prisoner could not be kept alive in a state of agony for at least twenty-four hours before he or she died.)

Numerous other examples can be cited describing the failure of humans to include negative future circumstances in their deliberations. Again and again we see in the industrialized world the adoption of a technology that results in future problems. This is not to paint a dark portrait of technology or to suggest that we should abandon our technological ways. Our whole history as a nation has been one of progress and growth. It is merely a reminder that every new opportunity brings with it an obligation to consider the consequences of our actions, and this we seem rather reluctant to do.

COUNTERPOINT AND APPLICATION

If technologizing, that is, creating and using technology, is so natural to being human, then it would appear that it is always an ethical process, as it always works. It is not very useful for any species to go against its nature in the quest for survival, except as a part of the evolutionary process, and natural selection would seem to be quite adequate to this end. Why all the fuss about the ethical nature of technology? It is neutral. It is what it is. Talking about the ethical nature of technology is like talking about the ethical nature of a stick. Isn't that true?

Of course, that is not true at all. It must be remembered that the drive to create technology and thus evolve externally to our bodies is indeed a natural process, yet it still entails free will, or choice, on our part. An almost infinite array of technological possibilities is available to us, depending on how we choose to apply the basic principles that constitute our understanding of the physical universe (physical laws). It is because of that choice that we must consider ethical content.

Surely, the homeostatic tendencies of the species goes a long way toward allowing us to adjust if we make mistakes in our choices in technological design and creation. But considering the speed at which the world changes and the far-reaching effects of even the seemingly most insignificant changes in methodology, it becomes critical to consider the usefulness of technological change in the broadest of terms, and that is a matter of what is the ethical thing to do. Technology has ethical content by virtue of the free will with which we create it. What do we choose to do and not do? We make those choices in a desire to improve our position in life, either individually, collectively, or both. Do we know that our choices are sound ones, and do they truly work to achieve the goals that they are designed to achieve? Therein lies the ethical issue.

When looking at technology and creating technological change through the modification, production, or application of technology, it is wise to think in a broader context. It is best to consider why exactly we are doing whatever it is we are doing, what our goals are, and whether the process undertaken actually achieves those goals. Additionally, we must consider what other goals or conditions are affected by the new creation or application and how that affects our overall goals in life. In other words, what would happen if we all behaved like the Native American confederacy mentioned earlier and considered the consequences of our actions for the next seven generations, or 150 years. How would we behave differently?

EXERCISES

1. Choose a technology with which you are familiar and that is generally used in your society on an ongoing basis. It could be anything from computers to airplanes to rubber gloves. Now begin to consider the positive reasons for the existence of this technology. Make a list of its benefits. Note how each is of value to the culture and why the technology exists as a method of doing what works.

   Next, consider the other side of the same technology. How does this technology impact you and others negatively? If you are having trouble finding anything negative about the technology you have chosen, think again. There is no such thing as a technology without a cost, and not just a monetary cost. How does the technology change your job and the jobs of others? How does it affect relationships and values? How can it be used for harmful purposes? Now compare the lists and determine whether the benefits outweigh the costs.

2. Consider ways in which the development of technology could be approached to create a more positive outcome for a society as a whole. If you were in charge, how would you change licensing and manufacturing regulations to better steer technology away from negative consequences? Is this a practical idea? Should we control the development and dissemination of technology because of possible negative effects? Is this concept in opposition to the concept of creative freedom, stifling potential new ideas for fear of censure? Considering that new ideas can be controlled and kept secret through governmental and industrial forces, is this done to our benefit or our detriment?

3. It can be argued that every technology has both positive and negative possible consequences and that the technology itself is neutral. The truth may be that it is in the application that the ethical or unethical nature of a technology exists and nowhere else. To explore this, consider one of the following technologies and make a list of both the benefits and costs to us all as a result of that technology's use. Make the lists equally long, even if you have to cite potential but not realized uses or consequences of the technology.
   a. Nano-technology
   b. Urban development
   c. Nuclear weapons
   d. Dynamite
   e. Helicopters
   f. Mono-cropping agriculture
   g. Printing presses
   h. Computers

Ethical Decision-Making Frameworks

O. C. FERRELL

To establish policies and rules that encourage employees to behave ethically and in accordance with organizational objectives, managers need to understand how and why people make ethical decisions. They need to consider how employees acquire their moral philosophies and how these philosophies interact with other components of the ethical decision-making process, such as opportunity, organizational culture, codes of ethics, and the influence of other people. Philosophers, social scientists, and academics have developed various models of this process.

In this chapter, we introduce several frameworks and models that attempt to describe how people make ethical decisions in business. We discuss Kohlberg's model of cognitive moral development, Trevino's interactionist model, Ferrell and Gresham's contingency framework, and the synthesis model. Although there are many models of ethical decision making, we believe these are among the most descriptive and most applicable to the business world. See Table 3-1 for an overview of this chapter.

KOHLBERG'S MODEL OF COGNITIVE MORAL DEVELOPMENT

Psychologist Lawrence Kohlberg developed a model that describes the cognitive moral development process—that is, the stages through which people progress in their devel-

O. C. Ferrell and John Fraedrich, *Business Ethics: Ethical Decision Making and Cases,* First Edition. Copyright © 1991 by Houghton Mifflin Company. Reprinted with permission.

opment of intellectual thought.[1] Although Kohlberg's model was not developed specifically for business, it has been used to justify ethical decisions within organizations. According to **Kohlberg's model of cognitive moral development,** different people make different decisions in similar ethical situations because they are in different stages of cognitive moral development. Kohlberg proposed that individuals develop through six stages:

1. *The stage of punishment and obedience.* An individual in this stage defines right as literal obedience to rules and authority. For example, some companies forbid their buyers to accept gifts from salespeople; a buyer in this stage might justify his refusal to accept gifts from salespeople by referring to the company rule.

2. *The stage of individual instrumental purpose and exchange.* An individual in this stage defines right as serving one's own or another's needs and making fair deals. In this stage, the individual no longer looks just at specific rules or authority figures, but evaluates behavior on the basis of its fairness to all. For example, a sales representative doing business for the first time in a foreign country may be expected to give customers "gifts." Although gift giving may be against company policy in the United States, the salesperson may decide that certain company rules designed for operating in the United States do not apply overseas. In some foreign countries' cultures, gifts may be considered part of a person's pay. In addition, standard practice may designate acceptable levels of gift giving. So, in this instance, not giving a

Table 3.1
An Overview of This Chapter

Kohlberg's Model of Cognitive Moral Development

Trevino's Interactionist Model
    Cognitive Development
    Individual Variables
    Situational Variables
Ferrell and Gresham's Contingency Framework
    Individual Factors
    Significant Others
    Opportunity
A Synthesis Model of Ethical Decision Making

---

gift might represent an unfair deal or failure to take into account another person's needs.

3. *The stage of mutual interpersonal expectations, relationships, and conformity.* An individual in this stage emphasizes others rather than himself or herself. Again, motivation is derived from obedience to rules. A production manager in this stage might obey upper management's order to speed up an assembly line if she believed that this action would generate more profit for the company and thus maintain her employees' jobs.

4. *The stage of social system and conscience maintenance.* An individual in the fourth stage determines what is right by considering his or her duty to society. For example, a number of years ago an employee of Brown and Root, Inc., discovered that Peruvian safety standards for building highways were inadequate. The standards prescribed no special precautions to be taken when cutting channels through unstable rock formations; as a result, rock slides were likely to result. The employee felt it was his duty to complain because the rock slides might endanger construction workers or travelers using the highways.[2]

5. *The stage of prior rights, social contract, or utility.* In this stage, an individual is concerned with upholding the basic rights, values, and legal contracts of society. He or she believes that legal contracts define what is right or wrong. For example, a pivotal point in the rejection of Judge Robert H. Bork's nomination to the Supreme Court in 1987 was his view about individual rights and the Constitution. Bork believed that the Constitution defines the rights of citizens. Hence, if a certain right is not explicitly or implicitly stated in the Constitution, Bork concluded that a citizen does not

have that right. His view of explicit rights and social contracts illustrates this moral developmental stage.

6. *The stage of universal ethical principles.* A person in the final stage believes that right is determined by universal ethical principles that all should follow. These individuals believe that there are indeed inalienable rights, which are universal in nature and consequence. For example, a businessperson at this stage might argue for discontinuing a product that has caused death and injury because the inalienable right to life makes killing wrong, regardless of the reason.

An individual's perceptions when first confronted with an ethical decision depend on the stage of cognitive moral development that he or she has reached.

Kohlberg's six stages can be reduced to three levels of ethical concern. Initially, a person is concerned with his or her own immediate interests and with external rewards and punishments. At the second level, an individual defines right as conforming to the expectations of good behavior of the larger society or some significant reference group. Finally, at the third, or "principled" level, an individual sees beyond the norms, laws, and authority of groups or individuals. Kohlberg's model therefore implies that a person's level of moral development influences his or her perception of and response to an ethical issue.

Kohlberg's cognitive moral development model is important in understanding ethical decision making in business because it helps explain why some people may change their moral beliefs or values. According to his model, as people progress through stages of moral development and with time and experience, they may change their values and ethical behavior. Kohlberg's theory set the stage for the formulation of many models of ethical decision making in organizations, including those described in this chapter.

TREVINO'S INTERACTIONIST MODEL

Trevino's interactionist model sees the ethical decision-making process as an interaction of three factors: (1) cognitive moral development, (2) individual variables, and (3) situational variables (see Figure 3.1).[3]

Cognitive Moral Development

Management professor Linda K. Trevino adapted Kohlberg's model of cognitive moral development in constructing her own model. In Trevino's view, a person's perception of an ethical situation is determined by his or her

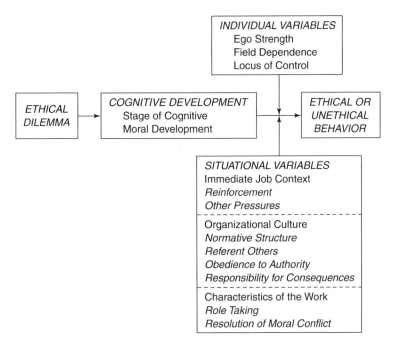

Figure 3.1

Trevino's Interactionist Model of Ethical Decision Making in Organizations

Source: Adapted from Linda K. Trevino, "Ethical Decision Making in Organizations: A Person-Situation Interactionist Model," *Academy of Management Review,* 11 (July 1986), p. 603. Reprinted by permission.

stage of cognitive moral development. Behavior, however, depends on the interaction of this perception with additional individual and situational variables.

Individual Variables

In Trevino's model, individual variables are factors within the individual that contribute to the ethical decision process. Ego strength, field dependence, and locus of control significantly influence decision making in business.

Ego Strength   Ego strength refers to the strength of an individual's personal convictions. For example, a manager who has high ego strength will probably act in accordance with her personal beliefs about what is right and wrong. If the manager has been taught not to give or receive bribes and has a strong ego, then the probability of her succumbing to bribes is low.

In contrast, a manager with weak ego strength would more likely go against his or her convictions. For instance,

in a situation in which lying might benefit a manager, he might lie despite a personal belief that honesty is important.

Field Dependence   Field dependence refers to the tendency to rely on external sources in ambiguous situations. Managers who are highly field dependent will look to those around them for help in making decisions in unfamiliar situations. Manager trainees, for example, often have mentors who help them understand the "company way" of doing things. Field-dependent trainees change their attitudes and values to mirror those of their superiors. Many Wall Street newcomers appear to have been very field dependent in that they picked up their cues on insider trading from superiors.

By contrast, managers with low field dependence rely on themselves and their own values when making decisions in ambiguous situations. Managers who are labeled stubborn may simply be very field independent.

Locus of Control   Locus of control refers to a person's perception of how much control he or she exerts over events in his or her life. A person with an internal locus of control believes that his or her own efforts determine outcomes. Such individuals believe they are masters of their own destiny, and that they make things happen rather than react to events. For example, a manager who consistently accepts responsibility for his or her decisions has an internal locus of control.

Conversely, a person with an external locus of control perceives events as beyond his or her control. Such people attribute much of their success or failure to providence, luck, or destiny. They are somewhat fatalistic in their attitudes toward business and life in general. Managers who continually refuse to take personal responsibility for the consequences of their behavior, and attribute the results to bad luck, probably have an external locus of control.

Ego strength, field dependence, and locus of control provide some insight into why people maintain or modify their value structures when making ethical decisions. For example, a manager with low ego strength, high field dependence, and an external locus of control might well go against her personal value structure, and might explain the result of a decision, whether positive or negative, as due to luck or pressures to conform. Conversely, a manager with high ego strength, low field dependence, and an internal locus of control will probably rely on his own values and, whether right or wrong, takes responsibility for decisions. This person would be less likely to depend on others in resolving ethical issues.

Situational Variables

Situational variables relate to factors in the environment. In a business situation, these factors include coworkers and organizational culture. Many decisions made by managers have no counterpart outside the business world. As a result, their other experiences provide very few norms to guide ethical decision making. Business students entering the job market for the first time have probably never experienced difficult situations such as plant closings and layoffs, employee discipline, antitrust issues, bribes, product liability and safety issues, environmental issues, or conflicts of interest. When managers must make ethical decisions that arise only in a business context, situational variables become very important. Trevino's adapted model includes situational variables such as the person's immediate job context, organizational culture, and work characteristics.

Immediate Job Context   The immediate job context includes the motivational "sticks and carrots" that superi-

ors use to influence employee behavior. Pay raises, bonuses, and public recognition are positive reinforcers, while demotions, firings, reprimands, and pay penalties are negative reinforcers. For example, if a salesperson is given public recognition and a large bonus for making a valuable sale that was obtained through aggressive tactics, that person will probably be motivated to use aggressive sales tactics in the future, even if such behavior goes against his or her personal value system.

Organizational Culture   The culture of an organization can be defined as a set of values, beliefs, goals, norms, and rituals shared by members (employees) of an organization. As time passes, a company or organization comes to be seen as a living organism with a mind and will of its own. For example, The Walt Disney Company requires that all new employees take a course in the traditions and history of Disneyland and Walt Disney. The corporate culture at American Express Company stresses that employees help customers out of difficult situations whenever possible. This attitude is reinforced through numerous company "legends" of employees who have gone above and beyond the call of duty to help customers. This strong tradition of customer loyalty might encourage an American Express employee to take unorthodox steps to help a customer who encounters a problem while traveling overseas. Such strong traditions and values have become a driving force in many companies including McDonald's Corp., IBM, Procter & Gamble Co., and Coors.

Organizational culture includes norms for seeking advice in ethical decision making. Trevino points out that decision makers turn to "referent others"—people, usually peers, within the organization who can provide information to help resolve an ethical dilemma. A referent other is one who cannot directly exert any type of authority over the decision maker in a business situation. Referent others offer help in the form of advice and information, perhaps at the water cooler or over coffee.

Obedience to authority is another aspect of organizational culture. For example, if respect for superiors is emphasized, employees may feel they are expected to carry out the orders of superiors, even if those orders are contrary to the employee's feelings of right and wrong. Later, if a decision is judged to have been wrong, they are likely to say, "I was only carrying out orders," or "My boss told me to do it this way."

Characteristics of Work   Kohlberg's cognitive moral development model suggests that people continue to change their decision priorities beyond their formative years, as a rule. The more a person considers differing perspectives on an ethical situation, the farther he or she has progressed in the moral development process. In the context of business, an individual's moral development can be influenced by the nature of his or her work.

Experience in resolving moral conflicts accelerates progress in moral development. A manager relying on a specific set of values or rules may eventually come across a situation to which the rules do not apply. For example, suppose a manager has a policy of firing any employee whose productivity declines for four consecutive months. Now suppose this happens with an employee who is going through a difficult divorce, but who will probably be a top performer again within one or two months. Because of the circumstances and the perceived value of the employee, the manager may bend the rule. He or she has changed because of on-the-job moral development. Managers in the highest stages of the moral development process seem to be more democratic than autocratic. They are likely to be more aware of the ethical views of others involved in an ethical decision-making situation.

The several variables in Trevino's model combine to determine how an employee makes an ethical or unethical decision. The variables described are not all-inclusive, yet they are consistent with social science theory and research.

FERRELL AND GRESHAM'S CONTINGENCY FRAMEWORK

Professors O. C. Ferrell (one of the authors of this book) and Larry Gresham developed **Ferrell and Gresham's contingency framework** (Figure 3.2), which describes variables that affect the individual, the ethical issue, and the evaluation of behavior within an ethical situation.[4] Like Trevino's model, this framework attempts to identify the factors that influence ethical decision making. It focuses on three main variables that influence a decision maker's behavior: individual factors, significant others, and opportunity.

Individual Factors

The individual factors identified by Ferrell and Gresham—knowledge, values, attitudes, and intentions—can be represented by moral philosophies. Moral philosophies are principles or rules that people use to decide what is right or wrong. Ferrell and Gresham assume that people are guided by different moral philosophies in making decisions in an ethical situation. Many businesspeople follow teleological philosophies, assessing the moral worth of a behavior by examining its likely consequences; an executive might assume, for example, that the firm's economic success should benefit employees, management, stockholders, consumers, and society. Others, however, follow deontological philosophies and stress the individual, often through formal rules that dictate proper behavior. From this perspective, for example, if lying is considered wrong, then lying to cover up a product defect or to increase sales would always be considered wrong.

Significant Others

Ferrell and Gresham, like Trevino, incorporate the influence of other people, such as peers, managers, and subordinates, in their model. (They refer to these influences as *significant others*.) Ferrell and Gresham include two other variables in their model: differential association and role-set configuration.

Differential Association   Differential association assumes that a person learns ethical or unethical behavior through interactions with people who are part of his or her intimate personal groups, or role sets. In other words, a decision maker who associates with others who behave unethically will probably behave unethically too. For example, if an office clerk associates with workers who steal supplies, such as notepads, pens, and computer paper, from their employer, he or she may come to believe that these activities are acceptable and begin stealing too. In fact, many observers believe that peers can change a person's original value system.[5] This value change, whether temporary or not, appears to be greater when the significant other is the decision maker's superior, especially if the decision maker is new to the firm.

Role-Set Configuration   Role-set configuration refers to the role relationships that people have because of their social status in an organization. A role-set configuration is a mixture of characteristics of significant others who form the role set, and may include their location and authority as well as their perceived beliefs and behaviors. For example, an assembly-line worker may not have any direct legitimate power (i.e., the person is not a foreman or supervisor) but may influence other people's decisions because of other characteristics, such as age, time with company, or expertise. Other employees might ask this person for advice on how to ask a supervisor for a raise or on whether to report an incident of unethical activity by a coworker.

One important dimension of role-set configuration appears to be the organizational distance—the number of layers of personnel—between the person making the

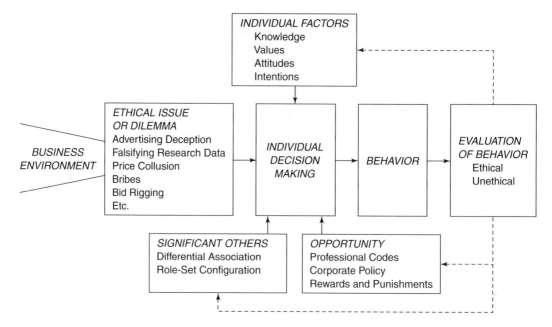

Figure 3.2

Ferrell and Gresham's Contingency Framework

Source: Adapted from O. C. Ferrell and Larry Gresham, "A Contingency Framework for Understanding Ethical Decision Making in Marketing," *Journal of Marketing,* 49 (Summer 1985), pp. 87–96, published by the American Marketing Association. Used by permission.

decision and significant others. For example, if there are four layers of management between a decision maker and the regional vice president, the vice president may have only marginal influence on the decision maker. With fewer layers between them, the influence would be greater.

Opportunity

Opportunity results from conditions that either limit barriers or provide rewards, whether internal or external. Examples of internal rewards include feelings of goodness and personal worth generated by performing altruistic activities. External rewards refer to what an individual expects to receive from others in the social environment. Rewards are external to the individual to the degree that they bring social approval, status, and esteem. Certainly, the absence of punishment provides an opportunity for unethical behavior because some individuals may act without regard for consequences.

Individual factors, significant others, and opportunity interact to influence a person's perception of and response

to an ethical issue. According to Ferrell and Gresham's framework, after making a decision and acting on it, the person evaluates the rightness or wrongness of his or her behavior by considering its consequences. The evaluation of the decision then influences future decisions in similar situations. For example, if an advertising manager realizes after the fact that a particular campaign was unethical, he or she may change future advertising decisions.

A SYNTHESIS MODEL OF ETHICAL DECISION MAKING

In 1989 O. C. Ferrell, John Fraedrich (the authors of the text), and Larry Gresham synthesized several significant ethical decision-making models, including the ones described in this chapter, to provide a more understandable description of the process of ethical decision making.[6] The **synthesis model,** shown in Figure 3.3, integrates components from all the earlier models to describe how people make decisions about ethical issues.

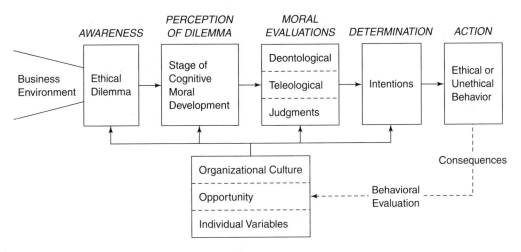

Figure 3.3
A Synthesis Model of Ethical Decision Making
Source: Adapted from O. C. Ferrell, Larry G. Gresham, and John Fraedrich, "A Synthesis of Ethical
Decision Models for Marketing," *Journal of Macromarketing,* 9 (Fall 1989), p. 60. Reprinted by permission.

In the first step of the synthesized model, an individual recognizes an ethical issue or dilemma resulting from forces in the business environment. Whether the person perceives the situation as an ethical dilemma depends on his or her stage of cognitive moral development, which also affects how he or she will deal with the ethical issue. For example, if a sales clerk gives too much change to a person at the lowest level of moral development, who is concerned only with immediate personal rewards, he or she might keep the change without saying anything. A person in one of Kohlberg's higher levels of moral development, however, would probably inform the clerk of the error and return the extra change.

Having recognized the need for ethical choice, the person then develops a set of possible solutions. Next, he or she evaluates each of the alternative solutions. The moral evaluation has three components: deontological, teleological, and judgmental (shown in the middle box in Figure 3.3). In the deontological evaluation, the person assesses the rightness or wrongness of each alternative, using rules or principles to preserve the inalienable rights of the individual. The teleological evaluation assesses the likely consequences of each alternative for each of the groups involved (employees, customers, stockholders, and others). The decision maker also evaluates the probability that each consequence will happen to each group, the

(un)desirability of each consequence, the importance of each consequence, and the importance of each group of people affected. Then, in the judgmental phase, the individual judges the alternatives to determine which is most "ethical" as well as which will result in the most desirable consequences.[7] Researchers have not determined exactly how this process works.

The next component of the synthesized model is a determination of the individual's intentions—that is, the final decision as to what action to take. Then follows the behavior or action itself.[8] When intentions and behavior are inconsistent with ethical judgments, the person may feel guilt. For example, an advertising account executive asked by her client to create an advertisement she perceives as misleading has two alternatives: to comply or to refuse. In the latter case she stands to lose business from that client, and possibly her job. Other factors—such as pressure from the client, the need to keep her job to pay her debts and living expenses, and the possibility of a raise if she does a good job developing the advertisement—influence her resolution of this ethical dilemma. Because of these other factors, she may decide to act unethically—to develop the advertisement even though she believes it to be inaccurate. Because her actions are inconsistent with her ethical judgment, however, she will probably feel guilty about her decision.

The final components of the decision process—organizational culture, opportunity, and individual variables—are shown at the bottom of Figure 3.3. They influence the entire process of ethical decision making, as the arrows indicate. Organizational culture includes factors such as organizational values and shared beliefs, significant others (differential association and role-set configuration), and obedience to authority. Opportunity comprises the conditions that limit or permit ethical or unethical behavior. Individual variables include factors such as ego strength, locus of control, field dependence, personal experiences, knowledge, and attitudes. These variables were described in detail in the earlier discussions of the Ferrell and Gresham and Trevino models.

On the basis of all these components, the individual chooses an action or behavior to resolve the ethical dilemma. After acting, he or she evaluates the consequences of the behavior. As a result of this final evaluation, the person may modify his or her behavior in similar situations in the future. For instance, after being criticized by supervisors and peers, the individual may decide the behavior was unacceptable and resolve not to act that way in the future. Similarly, a loss in self-esteem or feelings of guilt may lead the individual to modify his or her behavior in the future.

To illustrate the synthesized model, consider the case of accountant Jose Gomez, who was convicted of approving false financial statements of E.S.M. Government Securities, Inc., a firm that later went out of business. In 1979 E.S.M. executives asked Gomez to approve financial statements that hid millions of dollars in losses over a period of several years. The young accountant recognized an ethical dilemma arising from the business environment. He wrestled with the decision for several days, considering the consequences to himself and his family, his employer (Grant Thornton, a Chicago accounting firm), and his client. Gomez says he never *intended* to do anything wrong. However, because of personal financial difficulties (individual variables), pressure from the accounting firm to produce new clients (organizational culture), and pressure from the client to verify the false statements (situational factors, opportunity), Gomez agreed to sign off on the financial statements even though he knew they were false (unethical behavior). Thus, although Gomez's intentions were ethical, other factors influenced his decision to behave unethically and illegally. When E.S.M. Government Securities failed in 1985, investors lost nearly $320 million. Gomez was convicted of fraud in 1986 and sentenced to twelve years in prison.[9]

SUMMARY

Various models have been developed to describe how personal moral philosophies and other factors contribute to individual ethical decision making in the business environment. According to Kohlberg's model of cognitive moral development, individuals make different decisions in similar ethical situations because they are in different stages of moral development. Kohlberg proposed that everyone is in one of six stages of moral development: (1) the stage of punishment and obedience; (2) the stage of individual instrumental purpose and exchange; (3) the stage of mutual interpersonal expectations, relationships, and conformity; (4) the stage of social system and conscience maintenance; (5) the stage of prior rights, social contract, or utility; or (6) the stage of universal ethical principles. Kohlberg's six stages can be further reduced to three levels of ethical concern, corresponding to immediate self-interest, social expectations, and general principles. This model helps us understand ethical decision making in business because it explains why some people may change their beliefs or moral values.

Trevino's interactionist model states that the ethical decision-making process is built around cognitive moral development, individual variables, and situational variables. The cognitive moral development aspect of her model is derived from Kohlberg's cognitive moral development model. Individual variables that influence ethical decision making include ego strength, field dependence, and locus of control within the individual. Trevino's situational variables relate to factors in the work situation, such as a person's immediate job context, organizational culture (corporate culture, referent others, and obedience to authority), and work characteristics.

Ferrell and Gresham's contingency framework describes variables that affect the individual, the ethical issue, and the evaluation of behavior within an ethical situation. Their framework focuses on three variables: individual factors, significant others, and opportunity. Individual factors such as knowledge, values, attitudes, and intentions can be represented by moral philosophies. Significant others include peers, superiors, subordinates, and two more general factors: differential association and role-set configuration. Opportunity results from conditions that either limit barriers or provide rewards. After making a decision and acting on it, the decision maker evaluates the outcomes of his or her behavior and judges its rightness.

Finally, the synthesis model integrates components from other models. According to this model, a person first recognizes an ethical issue or dilemma resulting from forces in the business environment. Whether the person perceives the situation as an ethical dilemma depends on his or her stage of cognitive moral development. If the person perceives that the dilemma is an ethical one, he or she then develops a set of possible solutions and evaluates them deontologically, teleologically, and judgmentally. Next, the individual determines his or her intentions. When intentions, actual behavior, and ethical judgments are inconsistent, the person may feel guilt. Organizational culture, opportunity, and individual variables influence the other stages of the process in the synthesis model. After

acting, the individual evaluates the consequences of the behavior, and may modify his or her behavior in similar situations in the future.

IMPORTANT TERMS FOR REVIEW

Kohlberg's model of cognitive development

Trevino's interactionist model

Ferrell and Gresham's contingency framework
    synthesis model

AN ETHICAL DILEMMA*

Bill Conte was in a bind. A recent graduate of a prestigious business school, he had taken a job in the auditing division of Greenspan & Co., a fast-growing leader in the accounting industry. Greenspan relocated Bill, his wife, and their one-year-old daughter from the Midwest to the East Coast. On arriving they bought their first home and a second car. Bill was told that the company had big plans for him. Thus he did not worry about being financially overextended.

Several months into the job, Bill found he was working late into the night to complete his auditing assignments. He realized that the company did not want its clients billed for excessive hours, and that he needed to become more efficient if he wanted to move up in the company. He asked one of his friends, Ann, how she managed to be so efficient in auditing client records.

Ann quietly explained: "Bill, there are times when being efficient isn't enough. You need to do what is required to get ahead. The partners just want results—they don't care how you get them."

"I don't understand," said Bill.

"Look," Ann explained, "I had the same problem you have a few years ago, but Mr. Reed [the manager of the auditing department] explained that everyone eats time so that the group shows top results and looks good. And, when the group looks good, everyone in it looks good. No one cares if a little time gets lost in the shuffle."

Bill realized that "eating time" meant not reporting all the hours required to complete a project. He also remembered one of Mr. Reed's classic phrases, "results, results, results." He thanked Ann for her input and went back to work. Bill thought of going over Mr. Reed's head and asking for advice from the division manager, but he had met her only once and did not know anything about her.

QUESTIONS

1. What should Bill do?

2. If anyone is hurt by this action, it would seem to be young accountants, who are expected to work long hours. Why is this an ethical problem?

3. Using the synthesis model, describe the process through which Bill attempts to resolve his dilemma.

*This case is strictly hypothetical; any resemblance to real persons, companies, or situations is coincidental.

NOTES

1. Lawrence Kohlberg, "Stage and Sequence: The Cognitive Developmental Approach to Socialization," in *Handbook of Socialization Theory and Research,* ed. D. A. Goslin (Chicago: Rand McNally, 1969), pp. 347–480.
2. Charles Peters and Taylor Branch, *Blowing the Whistle: Dissent in the Public Interest* (New York: Praeger Publishers, 1972), pp. 182–185.
3. Linda K. Trevino, "Ethical Decision Making in Organizations A Person-Situation Interactionist Model," *Academy of Management Review,* 11 (July 1986), pp. 601–618.
4. O. C. Ferrell and Larry Gresham, "A Contingency Framework for Understanding Ethical Decision Making in Marketing," *Journal of Marketing,* 49 (Summer 1985), pp. 87–95.
5. Mary Zey-Ferrell, Mark Weaver, and O. C. Ferrell, "Predicting Unethical Behavior Among Marketing Practitioners," *Human Relations,* 32 (July 1979), pp. 557–569.
6. O. C. Ferrell, Larry G. Gresham, and John Fraedrich, "A Synthesis of Ethical Decision Models for Marketing," *Journal of Macromarketing,* 9 (Fall 1989), pp. 55–64.
7. Shelby D. Hunt and Scott Vitell, "A General Theory of Marketing Ethics," *Journal of Macromarketing,* 6 (Spring 1986), pp. 5–16.
8. Martin Fishbein and Icek Ajzen, *Belief, Attitude, Intention and Behavior: An Introduction to Theory and Research* (Reading, Mass.: Addison-Wesley, 1975).
9. Martha Brannigan, "Auditor's Downfall Shows a Man Caught in Trap of His Own Making," *The Wall Street Journal,* March 4, 1987, p. 31; and Martha Brannigan, "Aftermath of Huge Fraud Prompts Claims of Regret," *The Wall Street Journal,* March 4, 1987, p. 31.

QUESTIONS

1. Why is it important to implement ethical decision making in personal, professional and academic settings? Provide an example of a business situation where ethics must be considered before a manager makes her final decision.

2. How does Kohlberg's model correlate to ethical decision making within organizations? Provide one example for each of the six stages.

3. Think about a technological disaster that you are familiar with (The Challenger, Chernobyl, Three Mile Island, the Exxon Valdez oil spill, etc.) Pick one disaster and correlate the decisions that were made that caused the disaster to Trevino's Interactionist Model.

4. List and describe three specific, business related, situational variables that are not mentioned in the chapter. Why are they important? How do the variables that you listed correlate to ethical issues?

5. Explain Ferrell and Gresham's Contingency Framework. How does this theory correlate to social issues in technology?

6. What is a synthesis model?

Doing Well by Doing Good

Anti-*globalization* protesters see companies as unethical as well as exploitative. Firms demur, of course, but face an awkward question: Does virtue pay?

To many people the very concepts of "business" and "ethics" sit uneasily together. Business ethics, to them, is an oxymoron—or, as an American journalist once put it, "a contradiction in terms, like jumbo shrimp." And yet, in America and other western countries, companies increasingly wonder what constitutes ethical corporate behavior, and how to get their employees to observe it. Management schools teach courses on the subject to their students. Business ethics is suddenly all the rage.

Fashionable perhaps—but also vague. Protesters in Washington, DC, were this week railing against corporate immorality as well as the IMF. But plenty of people retort that companies should not be in the business of ethics at all—let alone worrying about social responsibility, morals or the environment. If society wants companies to put any of these ahead of the pursuit of shareholder value, then governments should regulate them accordingly. Thirty years ago Milton Friedman, doyen of market economics, summed up this view by arguing that "there is one and only one social responsibility of business—to use its resources and engage in activities designed to increase its profits."

Even those who think companies do have wider responsibilities argue about the best way to pursue them. Ulrich Steger, who teaches environmental management at

© 2000 The Economist Newspaper Group, Inc. Reprinted with permission. Further reproduction prohibited. www.economist.com

the International Institute for Management Development in Lausanne, says that companies cannot possibly hope to pursue a single abstract set of ethical principles and should not try. No universal set of ethical principles exists; most are too woolly to be helpful; and the decisions that companies face every day rarely present themselves as ethics versus economics in any case. He says that companies should aim instead for "responsible shareholder-value optimization": their first priority should be shareholders' long-term interests, but, within that constraint, they should seek to meet whatever social or environmental goals the public expects of them.

Certainly companies, which increasingly try to include their ethical principles in corporate codes, stumble over how to write in something about the need for profitability. Or, to put the dilemma more crudely: when money and morality clash, what should a company do? Most firms try to resolve this with the consoling belief that such clashes are more imagined than real, and that virtue will pay in the end. Yet they cannot always be right.

Indeed, companies face more ethical quandaries than ever before. Technological change brings new debates, on issues ranging from genetically modified organisms to privacy on the Internet. Globalization brings companies into contact with other countries that do business by different rules. Competitive pressures force firms to treat their staff in ways that depart from past practice. Add unprecedented scrutiny from outside, led by non-governmental organizations (NGOs), and it is not surprising that dealing with ethical issues has become part of every manager's job.

DON'T LIE, DON'T CHEAT, DON'T STEAL

In America, companies have a special incentive to pursue virtue: the desire to avoid legal penalties. The first attempts to build ethical principles into the corporate bureaucracy began in the defense industry in the mid-1980s, a time when the business was awash with kickbacks and $500 screwdrivers. The first corporate-ethics office was created in 1985 by General Dynamics, which was being investigated by the government for pricing scams. Under pressure from the Defense Department, a group of 60 or so defense companies then launched an initiative to set up guidelines and compliance programs. In 1991, federal sentencing rules extended the incentive to other industries: judges were empowered to reduce fines in cases involving companies that had rules in place to promote ethical behavior, and to increase them for those that did not.

But the law is not the only motivator. Fear of embarrassment at the hands of NGOs and the media has given business ethics an even bigger push. Companies have learned the hard way that they live in a CNN world, in which bad behavior in one country can be seized on by local campaigners and beamed on the evening television news to customers back home. As non-governmental groups vie with each other for publicity and membership, big companies are especially vulnerable to hostile campaigns.

One victim was Shell, which in 1995 suffered two blows to its reputation: one from its attempted disposal of the Brent Spar oil rig in the North Sea, and the other over the company's failure to oppose the Nigerian government's execution of Ken Saro-Wiwa, a human-rights activist in a part of Nigeria where Shell had extensive operations. Since then, Shell has rewritten its business principles, created an elaborate mechanism to implement them, and worked harder to improve its relations with NGOs.

Remarkably, Shell's efforts had no clear legal or financial pressure behind them. Neither of the 1995 rows, says Robin Aram, the man in charge of Shell's policy development, did lasting damage to the company's share price or sales—although the Brent Spar spat brought a brief dip in its market share in Germany, thanks to a consumer boycott. But, he adds, "we weren't confident that there would be no long-term impact, given the growing interest of the investment community in these softer issues." And he also concedes that there was "a sense of deep discomfort from our own people." People seem happier working for organizations they regard as ethical. In a booming jobs market, that can become a powerful incentive to do the right thing.

THE QUEST FOR VIRTUE

In America there is now a veritable ethics industry, complete with consultancies, conferences, journals and "corporate conscience" awards. Accountancy firms such as Pricewaterhouse Coopers offer to "audit" the ethical performance of companies. Corporate-ethics officers, who barely existed a decade ago, have become *de rigueur,* at least for big companies. The Ethics Officer Association, which began with a dozen members in 1992, has 650 today. As many as one in five big firms has a full-time office devoted to the subject. Some are mighty empires: at United Technologies, for example, Pat Gnazzo presides over an international network of 160 business-ethics officers who distribute a code of ethics, in 24 languages, to people who work for this defense and engineering giant all round the world.

For academic philosophers, once lonely and contemplative creatures, the business ethics boom has been a bonanza. They are employed by companies to run "ethics workshops" and are consulted on thorny moral questions. They also act as expert witnesses in civil lawsuits "where lawyers usually want to be able to tell the judge that their client's behavior was reasonable. So you are usually working for the defendants. They want absolution," says Kirk Hanson, a professor at Stanford Business School.

Outside America, few companies have an ethics bureaucracy. To some extent, observes IMD's Mr. Steger, this reflects the fact that the state and organized labor both still play a bigger part in corporate life. In Germany, for example, workers' councils often deal with issues such as sexual equality, race relations and workers' rights, all of which might be seen as ethical issues in America.

In developing a formal ethics policy, companies usually begin by trying to sum up their philosophy in a code. That alone can raise awkward questions. The chairman of a large British firm recalls how his company secretary (general counsel) decided to draft an ethics code with appropriately lofty standards. "You do realize," said the chairman, "that if we publish this, we will be expected to follow it. Otherwise our staff and customers may ask questions." Dismayed, the lawyer went off to produce something more closely attuned to reality.

Not surprisingly, codes are often too broad to capture the ethical issues that actually confront companies, which range from handling their own staff to big global questions of policy on the environment, bribery and human rights. Some companies use the Internet to try to add precision to general injunctions. Boeing, for instance, tries to guide staff through the whole gamut of moral quandaries, offering an online quiz (with answers) on how to deal with everything from staff who fiddle their expenses on business trips to suppliers who ask for kickbacks.

The best corporate codes, says Robert Solomon of the University of Texas, are those that describe the way everybody in the company already behaves and feels. The worst are those where senior executives mandate a list of principles—especially if they then fail to "walk the talk" themselves. However, he says, "companies debate their values for many months, but they always turn out to have similar lists." There is usually something about integrity; something about respect for the individual; and something about honoring the customer.

The ethical issues that actually create most problems in companies often seem rather mundane to outsiders. Such as? "When an individual who is a wonderful producer and brings in multiple dollars doesn't adhere to the company's values," suggests Mr. Gnazzo of United Technologies: in other words, when a company has to decide whether to sack an employee who is productive but naughty. "When an employee who you know is about to be let go is buying a new house, and you're honor-bound not to say anything," says Mr. Solomon. "Or, what do you do when your boss lies to you? That's a big one."

Issues such as trust and human relations become harder to handle as companies intrude into the lives of their employees. A company with thousands of employees in South-East Asia has been firing employees who have AIDS, but giving them no explanation. It now wonders whether this is ethical. Several companies in America scan their employees' e-mail for unpleasant or disloyal material, or test them to see if they have been taking drugs. Is that right?

Even more complicated are issues driven by conflicts of interest. Edward Petry, head of the Ethics Officer Association, says the most recent issue taxing his members comes from the fad for Internet flotations. If a company is spinning off a booming e-commerce division, which employees should be allowed on to the lucrative "friends-and-family" list of share buyers?

Indeed, the revolution in communications technologies has created all sorts of new ethical dilemmas—just as technological change in medicine spurred interest in medical ethics in the 1970s. Because it is mainly businesses that develop and spread new technologies, businesses also tend to face the first questions about how to use them. So companies stumble into such questions as data protection and customer privacy. They know more than ever before about

their customers' tastes, but few have a clear view on what uses of that knowledge are unethical.

FOREIGNERS ARE DIFFERENT

Some of the most publicized debates about corporate ethics have been driven by globalization. When companies operate abroad, they run up against all sorts of new moral issues. And one big problem is that ethical standards differ among countries.

Reams of research, says Denis Collins in an article for the *Journal of Business Ethics,* have been devoted to comparing ethical sensitivities of people from different countries. As most of this work has been North American, it is perhaps not surprising that it concluded that American business people are more "ethically sensitive" than their counterparts from Greece, Hong Kong, Taiwan, New Zealand, Ukraine and Britain. They were more sensitive than Australians about lavish entertainment and conflicts of interest; than French and Germans over corporate social responsibility; than Chinese in matters of bribery and confidential information; and than Singaporeans on software piracy. Given such moral superiority, it is surprising that American companies seem to turn up in ethical scandals at least as often as those from other rich countries.

Many companies first confronted the moral dilemmas of globalization when they had to decide whether to meet only local environmental standards, even if these were lower than ones back home. This debate came to public attention with the Bhopal disaster in 1984, when an explosion at a Union Carbide plant in India killed at least 8,000 people. Most large multinationals now have global minimum standards for health, safety and the environment.

These may, however, be hard to enforce. BP Amoco describes in a recent environmental and social report a huge joint venture in inland China. "Concerns remain around the cultural and regulatory differences in risk assessment and open reporting of safety incidents," the report admits. "For instance, deference to older and more senior members of staff has occasionally inhibited open challenging of unsafe practices." BP Amoco thinks it better to stay in the venture and try to raise standards. But Shell claims to have withdrawn from one joint venture because it was dissatisfied with its partner's approach. Most companies rarely talk about these cases, creating the suspicion that such withdrawals are rare.

Bribery and corruption have also been thorny issues. American companies have been bound since 1977 by the Foreign Corrupt Practices Act. Now all OECD countries have agreed to a convention to end bribery. But many companies turn a blind eye when intermediaries make such payments. Only a few, such as Motorola, have accounting systems that try to spot kickbacks by noting differences between what the customer pays and what a vendor receives.

Some corruption is inevitable, say companies such as Shell, which work in some of the world's nastiest places. "If someone sticks a Kalashnikov through the window of your car and asks for 20 naira, we don't say that you shouldn't pay," says Mr. Aram. "We say, it should be recorded." United Technologies' Mr. Gnazzo takes a similar view: "We say, employees must report a gift so that everybody can see it's a gift to the company, and we can choose to refuse it. Every year we write to vendors, saying that we don't want gifts we want good service."

RIGHTS AND WRONGS

Human rights are a newer and trickier problem. Shell has written a primer on the subject, in consultation with Amnesty International. It agonizes over such issues as what companies should do if they have a large investment in a country where human rights deteriorate; and whether companies should operate in countries that forbid outsiders to scrutinize their record on human rights (yes, but only if the company takes no advantage of such secrecy and is a "force for good").

The force-for-good argument also crops up when companies are accused of underpaying workers in poor countries, or of using suppliers who underpay. Such problems arise more often when there are lots of small suppliers. At Nike, a sporting-goods firm, Dusty Kidd, director of labor practices, has to deal with almost 600 supplier-factories around the world. The relationship is delicate: "They are independent businesses, but we take responsibility," says Mr Kidd. When, last year, Nike insisted on a rise in the minimum wage paid by its Indonesian suppliers, it claims to have absorbed much of the cost.

NGOs have berated firms such as Nike for failing to ensure that workers are paid a "living wage." But that can be hard, even in America. "I once asked a university president, do you pay a living wage on your campus?" recalls Mr. Kidd. "He said that was different. But it isn't." In developing countries, the dilemma may be even greater: "In Vietnam, our workers are paid more than doctors. What's the social cost if a doctor leaves his practice and goes to work for us? That's starting to happen."

Stung by attacks on their behavior in the past, companies such as Shell and Nike have begun to see it as part of their corporate mission to raise standards not just within their company, but in the countries where they work. Mr. Kidd, for instance, would like Nike's factories to be places where workers' health actually improves, through better education and care, and where the status of women is raised. Such ideals would have sounded familiar to some businessmen of the 19th century: Quaker companies such as Cadbury and Rowntree, for instance, were founded on the principle that a company should improve its workers' health and education. In today's more cynical and competitive world, though, corporate virtue no longer seems a goal in its own right.

When, in the late 1980s, companies devoted lots of effort to worrying about the environment, they told themselves that being clean and green was also a route to being profitable. In the same way, they now hope that virtue will bring financial, as well as spiritual, rewards. Environmental controls can, for instance, often be installed more cheaply than companies expect. Ed Freeman, who teaches ethics at the Darden Business School at the University of Virginia, recalls how the senior executive of a big chemical company announced that he wanted "zero pollution." His engineers were horrified. Three weeks later, they returned to admit that they could end pollution and save money. "The conflict between ethics and business may be a lot less than we think," he argues.

Most academic studies of the association between responsible corporate ethics and profitability suggest that the two will often go together. Researchers have managed to show that more ethically sensitive sales staff perform better (at least in America; the opposite appears to be the case in Taiwan); that share prices decline after reports of unethical conduct; and that companies which state an ethical commitment to stakeholders in their annual reports do better financially. But proving a causal link is well-nigh impossible.

What of the growing band of ethical investors? "I don't know of a single one of these funds which looks at the effectiveness of a company's internal ethics program," says the EOA's Mr. Petry, sadly. So a defense firm scores bad marks for being in a nasty industry, but no offsetting good marks for having an elaborate compliance program.

And then there is the impact on employees. It may be true that they like working for ethically responsible companies. But, says Stanford's Mr. Hanson, "I see a lot of my graduate students leaving jobs in not-for-profits to go and work for dot.coms." Few dot.coms would know a corporate ethics code if it fell on their heads. Small firms, in particular, pay far less attention than bigger rivals to normalizing ethical issues and to worrying about their social responsibilities. Yet employment is growing in small companies and falling in big ones.

There may still be two good reasons for companies to worry about their ethical reputation. One is anticipation: bad behavior, once it stirs up a public fuss, may provoke legislation that companies will find more irksome than self-restraint. The other, more crucial, is trust. A company that is not trusted by its employees, partners and customers will suffer. In an electronic world, where businesses are geographically far from their customers, a reputation for trust may become even more important. Ultimately, though, companies may have to accept that virtue is sometimes its own reward. One of the eternal truths of morality has been that the bad do not always do badly and the good do not always do well.

QUESTIONS

1. Why might many people believe that the terms business and ethics are contradictory? Provide two examples to explain your answer.

2. Do you agree or disagree with the author that "no universal set of ethical principles exist?" Support your answer with facts, examples or specific statements explaining your answer.

3. What do you think the correlation is between business and ethics? What part does the avoidance of litigation, and the embarrassment of media exploiting conflicts play in maintaining ethics in businesses?

4. What part does globalization play in new moral dilemmas?

5. Why should companies worry about their ethical reputation?

5 Whistle-Blowing

The economic and technological triumphs of the past few years have not solved as many problems as we thought they would, and, in fact, have brought us new problems we did not foresee.

HENRY FORD II

Whistle-blowing occurs when people believe that something is wrong with a system, process, or product and feel personally responsible or morally obligated to let officials know about the wrongdoing. In an article entitled "Blowing the Whistle: The Organizational and Legal Implications for Companies and Employees," published in *Psychology Today* (1996), it was pointed out that traditionally, whistle-blowers have long histories of successful employment and a firm belief in their organizations. When they perceive that something within their organization as potentially harmful, they become morally convicted to correcting, changing, or reporting the misdeed. Personal responsibility and concern about others is at the heart of their conviction. Whistle-blowers believe that by reporting the misconduct or hazard, their concerns will not be ignored and the problem will be solved. Whistle-blowers tend to have high ethical standards, strong religious beliefs, and a faith that superiors share the same desire as themselves to correct wrongs.

Whistle-blowing often occurs when an individual believes that decision making by a company or the government may be breaking the law, financially profitable but morally wrong, potentially dangerous, or a case of exploitation of management authority (VAOIG, 1998, p. 1). Some factors upon which people blow the whistle include the following: "misleading promotional efforts to secure public funding; cheap, unreliable designs; testing shortcuts; mis-

represented test results or experiments; shoddy manufacturing procedures; intermittently defective products; botched installations; careless or inadequate operational procedures; negligent waste disposal" (McGinn, 1991, p. 158). The reality is that whistle-blowers have strong ethical and moral beliefs that carry them through the rough times they are often subjected to after they report the wrongdoing. It is only through family and peer support systems or strong religious beliefs that they can come to terms with the isolation, demotion, intimidation, and threat of loss of work that they may experience after blowing the whistle (Rechtschaffen, 1996, p. 38).

In fact, "[W]histle-blowers are part and parcel of the corporate culture on which they blow the whistle. They are often rather senior because it is those issuing orders who usually have the most control over and the most knowledge about what is occurring within the corporation." (Koehn, 1998, p. 3). Even though whistle-blowing can be personally costly, it remains an important part of social and personal decisions in the face of precarious conflicts. Individuals have to go through soul-searching and heart-wrenching decisions when making the decision to blow the whistle and when dealing with the ramifications of this decision. The two examples of whistle-blowing described in the upcoming text are 1) the classic case of Roger Boisjoly, the chief engineer for the Challenger space shuttle, and 2) the case of the "nuclear warriors."

REFERENCES

Koehn, D. (1998). Whistleblowing and trust: Some lessons from the ADM scandal. Business Ethics Magazine Articles. Available: http://condor.depaul.edu/eumes/adm.nun.

McGinn, R. (1991). *Science, Technology and Society.* Upper Saddle River, NJ, Prentice Hall.

Presidential Commission on the Space Shuttle *Challenger* Accident, February 25, 1986.

Rechtschaffen, G., and Yardley, J. (1996, March). Whistleblowing and the law: Are you legally protected if you blow the whistle? *Management Accounting.* pp. 38–41.

VAOIG Hotline. *Whistleblowing information.* Available: http://www/va/gov/VAOIG/hotline/wmsue.nun.

Case Study: Roger Boisjoly, Chief Engineer at Morton Thiokol

"Whistleblowing arises from an unsuccessful attempt to achieve change through the chain of command. When issues are critical and management's response is unsatisfactory, employees have little recourse but to circumvent the chain of command. All too often they are punished in a modern version of blaming the messenger."

ROGER BOISJOLY, CHIEF ENGINEER AT MORTON THIOKOL

One of the saddest days in the history of America's space program was January 28, 1986, the day that the Challenger space shuttle exploded. Six astronauts and a school teacher, Christa McAuliffe, died when the shuttle exploded 73 seconds after liftoff. The nation, transfixed and confused, watched the disaster on their television sets, wondering how something that seemed so good could have pivoted so quickly into terror . . . How could it have happened?

On January 24, 1985, Roger Boisjoly, chief engineer at Morton Thiokol, watched the launch of Flight 51-C and noted that the temperature that day was much cooler than it had been during other recorded launches. When he inspected the solid rocket boosters, he found that "both the primary and secondary-ring seals on a field joint had been blackened" from what he thought might be severe hot gas blowby related to low temperatures (Boisjoly and Curtis, 1987, p. 9). Boisjoly presented his findings to NASA's Marshall Space Flight Center and received tough questioning from the Flight Readiness Review committee. He decided that he needed further evidence before he could positively link low temperatures with hot gas blowby. He further studied the effects of temperature on the seals of Flight 51-B and found that, in fact, there was a direct correlation between low temperatures and the chance for a catastrophe to result (Boisjoly and Curtis, 1987, pp. 9–10).

During the following months, Boisjoly felt frustration that management did not seem to be listening to him about the problem he had discovered. On January 27 (the day before the Challenger disaster), the predicted temperature for the launch was 18 degrees Fahrenheit. This low temperature prompted Morton Thiokol, the Marshall Space Flight Center, and the Kennedy Space Center to discuss their concerns about the safety of the launch, since no shuttle had ever been launched at a temperature lower than 53 degrees Fahrenheit. Boisjoly again stated his concern that the O-rings on the shuttle's solid rocket boosters would stiffen in the cold and lose their ability to act as a seal, potentially causing a fatal disaster (Schlager, 1994, p. 611).

After the meeting, Boisjoly returned to his office and created the following journal entry:

> I sincerely hope that this launch does not result in a catastrophe. I personally do not agree with some of the statements made . . . stating that SRM-25 is okay to fly tomorrow. (Boisjoly and Curtis, 1987, p. 13)

On January 28, 1986, as Boisjoly and his engineering team had predicted, the Challenger exploded, and all of those on board died. As the nation mourned, the Rogers Commission was formed to investigate the reason for the explosion. Research after the fact showed that there was "a failure in the joint between the two lower segments of the right Solid Rocket Motor . . . (i.e.) the destruction of the seals that are intended to prevent hot gases from leaking through the joint during the propellant burn. . ." (The Commission, 1986, Vol .4). It was a tragic technological disaster that could have been prevented. The seal had leaked because the booster was used at a temperature below its range of safe operation, something that

Roger Boisjoly had tried to warn management about (Westrum, 1991).

The experience of Roger Boisjoly is representative of what whistle-blowers often face when they make a moral decision to tell the truth. Their truths are based on their own perceptions, values, and morals; the question is why others do not choose to listen or act upon their recommendations.

REFERENCES

Boisjoly R., and Curtis, E. (1987, October). *Roger Boisjoly and the Challenger disaster: A case study in engineering management, corporate loyalty and ethics.* Lowell, MA, University of Lowell. (Reprinted from the Proceedings of the ASEM Eighth Annual Meeting, October 1987.)

Schlager, N. (1994). *When Technology Fails: Significant Technological Disasters, Accidents and Failures of the Twentieth Century.* Washington, DC, Gale Research.

Westrum, R. (1991). *Technologies and Society: The Shaping of People and Things.* Belmont, CA, Wadsworth Publishing Company. p. 132.

QUESTIONS

1. What was the main cause of the Challenger explosion? What do you think might be different about the respective answers of NASA, Morton Thiokol, and Roger Boisjoly to this question? Use the Internet or sources from the library to provide additional information to support your answer.

2. Pretend for a moment that you are Roger Boisjoly. Your recommendation that the launch be postponed has been overridden. What should you do? Provide a detailed explanation as to why you would or would not pick each of the following options.
   a. Resign your position in protest.
   b. Blow the whistle by calling a local radio or television station.
   c. Tell the astronauts and their families of your concerns.
   d. Do nothing.

3. What is your opinion of Roger Boisjoly's role as a whistle-blower? How would you have handled this situation differently?

ETHICAL ISSUES IN CYBERSPACE

"Technology is changing every 60 days, so we're playing catch-up. We're trying to legislate very difficult areas. Criminals can be somewhat anonymous, and the ways of concealing their identities make it easier [for them] to commit a fraud. When we read something on the Internet, we tend to take it as gospel, but it doesn't have to be right. You have to take it with a grain of salt."

MASTER SGT. JAMES MURRAY, ILLINOIS STATE POLICE

"A 1994 FBI report estimated that economic loss from Internet crime ranged from $164 million to $5 billion per year. It is difficult to get accurate calculations, because many people who are taken advantage of on the Internet don't report it; also, it is difficult to enforce because many people are known by a one-word screen name and leave few options for verification."

ERICA HARRINGTON, "AUTHORITIES PLAYING CATCH-UP AGAINST CYBERCRIME,"
CHICAGO TRIBUNE, DECEMBER 15, 1996

"'We face a unique, disturbing and urgent circumstance, because it's children who are the computer experts in most families,' said Indiana Republican Senator Dan Coats, who is interested in cleaning up the Internet. The result of Senator Coat's interest and concern was the creation of the Communications Decency Act. This act extends regulations written to govern the dial-a-porn industry into computer networks. The Communications Decency Act was designed to impose fines up to $100,000 and prison terms of two years on anyone who knowingly makes 'indecent' materials available to children under 18."

PHILLIP ELMER DEWITT, "CYBERPORN," TIME, JULY 3, 1997

"Another form of cybercrime is extortion. A computer hacker can obtain bits of information from the Internet and then transmit an extortion through e-mail demanding money. If they are not paid, they might threaten with a 'mailbomb,' which is the transmission of constant e-mail until the system overloads and becomes dysfunctional."

JONES INTERNATIONAL, LTD, "CRIME ON THE INTERNET," APRIL 1996,
HTTP://WWW.DIGITAL CENTURY.COM/ENCYCL/UPDATE/CRIME/HTML

"The Internet can also be used to fight crime. The FBI Web site receives 100,000 visits a day from people with information about the most wanted fugitives or missing persons posted there. The FBI also uses the Web page to request information about various crimes, like the Oklahoma City bombing. There are also private groups specializing in finding missing persons. The most effective tool in finding a child is a photograph. The National Center for Missing and Exploited Children receives between 25,000 and 30,000 visits a week about missing children. It is less expensive to put a picture of a missing child on the 'Net than it would be to do a mass mailing. Because of the Internet, pictures of missing children reach many quickly."

ERICA HARRINGTON, "AUTHORITIES PLAYING CATCH-UP AGAINST CYBERCRIME,"
CHICAGO TRIBUNE, DECEMBER 15, 1996

"President Clinton signed a crime bill which contains a two-page revision to the federal computer crime statute. The bill makes it illegal to transmit viruses and worms over the electronic network. An important change in the law is the creation of two levels of computer crime: Actions taken 'with reckless disregard' for the damage they could cause are classified as misdemeanors and intentionally harmful acts are classified as felonies."

M.B. BETTS, "STATUTE OUTLAWS VIRUSES," COMPUTERWORLD, OCTOBER 1994, P. 65

"Search and seizure laws still need clarification. Currently the law requires that a warrant be specific in what can be seized and the seizure must be the original document. This is difficult when dealing with computers, because obtaining a specific original file could mean that the entire hard drive would be confiscated. It is hard to get a warrant that allows the police to find the information without jeopardizing the privacy of the rest of the information in the hard drive."

L. REYNOLDS, "CONSTITUTIONAL LAW IN THE ELECTRONIC AGE,"
MANAGEMENT REVIEW, 1995, PP. 24–25

Conclusion

Technology allows us to transport people more quickly, communicate in a variety of formats (e.g. telephone, e-mails, chat boxes, etc.), warm up our food rapidly in the microwave, and access and store data efficiently. Technologies envelop our very existence. As technologies have become a part of everyday life, it is important to ask about their ethical ramifications. For example, what part does our own moral development (Kohlberg's theory) play in our adult ethical decision making?

Part I has provided theory and practical issues that correlate to moral interactions with technology and within business environments. It has defined technology, and explained Kohlberg's, Trevino's, Ferrell and Gresham's theories of ethical decision making. It has explained the part that managers, technicians and individuals play in creating ethical environments in science and technology and business. The goal of this part is to ask you to evaluate your role in utilizing technology, always keeping in mind the ethical decision making models.

The next part of the book, the history of technology, examines the timeline of technological development and its impact on humanity.

A Bridge to Your 21st Century Understanding

Complete and discuss the following flowchart.

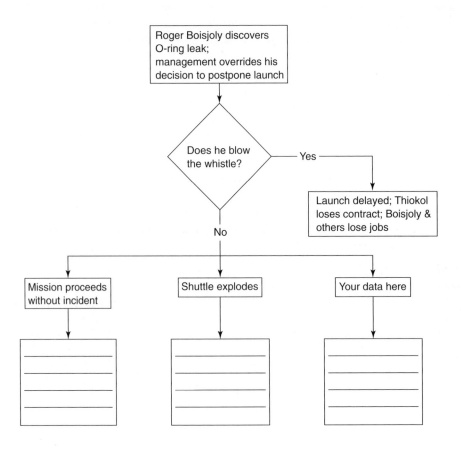

INTERNET EXERCISE

1. Use any of the Internet search engines (e.g., Alta Vista, Yahoo, Infoseek, etc.) to re-search the following topics:
    a. The definitions of whistle-blowing, cybercrime, and cyberporn.
    b. Additional factors about the tragedies of the Challenger, the Nuclear Warriors, the molasses spill, and Bhopal. Try to find new areas of interest that are not presented within the articles. What was the financial impact of each of these tragedies?

c. The amount of money spent in lawsuits nationally or within specific states on legal fees resulting from whistle-blowing.

d. Find a new article on ethics and technology and summarize it.

e. Find five ways that technology provides positive experiences for each of us daily.

Beyond the Bridge: Future Options . . .

Use any of the Internet search engines (e.g., Alta Vista, Yahoo, Infoseek, etc.) to research the following topics:

1. Roger Boisjoly is an example of a whistle-blower. Find two more examples using the Internet. Create a summary of each whistle-blowing experience and cite the Internet address(es) at which you found your information.

2. Search for information or opinions on psychological, social, or moral reasons for blowing the whistle.

3. Search the Internet for a technological disaster (Chernobyl, Three Mile Island, Bhopal) and identify how it relates to deontological and consequentialest theory.

USEFUL WEB SITES

http://ethics.acusd.edu//	Provides updates in current ethics related literature
http://www.iit.edu/department/csep/public/www/codes	Codes of ethics
http://roger.babson.edu/ethics/business.htm	Analyze ethical business cases
http://onlineethics.org/essays/practice/	Essays in business and engineering ethics
http://www.ethics.ubc.ca/resources	Center for Applied Ethics

History of Technology

A monument dedicated to the men and women workers of the Industrial Revolution in Birmingham, England. Photo courtesy of Ahmed S. Khan.

OBJECTIVES

Part II will help you to:

- See that technology and human development run hand in hand.
- Know that sometimes technology development and use isn't always altruistic; that progress sometimes causes pain.
- Appreciate the fact that technological development sometimes can threaten stability and peace.

INTRODUCTION

Have you ever heard the old saying, "Necessity is the mother of invention"? What do you think it means? Often it is interpreted to mean that things are created to take care of a need. Want to preserve an idea? Invent writing. Need to get someplace fast? Invent the car, train or plane. Got a neighbor who keeps trespassing on your property? A fence might be the answer, and, in extreme cases, a gun has been known to do the trick.

It seems as if human development and technological development have proceeded hand in hand. Indeed, as we survey all that we have done with the technology that we have developed, it would be hard to resist a moment of self-congratulation. But before we break our arms patting ourselves on the back, as some might say, we should recognize that there is a growing chorus of voices asking whether, in fact, technology has become humanity's master and not the other way around. Maybe the old saying, which began this introduction, ought to read, "Invention is the mother of necessity." The inference here is that those things people never dreamed of having a few hundred years ago have become the things that people today couldn't dream of *not* having. Some equally disturbing questions brought about by the use of technology are "Has technology helped everyone, or just the few who can understand and afford it?" and "Have we become so dependent on technology that we have attached our spiritual as well as material happiness to it?"

This part of the text is designed to give you a brief introduction to the role that technology has played in human development. It will invite you to look at a technology time line. As you do, imagine the impact that each new technology had on humans and how the technology changed individual and societal lifestyles forever. Then you'll get to read two stories about the same subject. In "George," you'll meet someone who lived millions of years ago, literally hanging on to life by his fingernails. His understanding of technology is next to zero, and so are his chances of survival. "George" gets by with some luck and the few wits he has at his disposal. But what happens when "George" discovers technology? The answer is a "'New' George." Is his life any better with his new tools? You decide.

900,000 B.C. Stone tools found in Kenya

13,000 B.C. Sledge (a load fastened to two long, flat pieces of wood)

6000 B.C. Iron (boat made of simple hollowed-out logs found in Holland)

3000 B.C. Bronze, candle, clock, dam, irrigation, mining, oar, rope, silver, sewers used in Nineveh, Assyria

2500 B.C. Brick (clay) at Mesopotamia. Ink used in Egypt and China (carbon and ink in water base and resin)

2000 B.C. Coal mining, copper mining, spoked wheel, wooden lock used for security in Egypt

1500 B.C. Air compressor, mercury, paint, simple pulleys used by Assyrians

1100 Gunpowder developed in China

330 B.C. Aristotle, Greek philosopher, was first person to write about ecological concepts

1455 The Gutenberg Bible is created using movable-type printing press

1550 Robot; Hans Bullman, Germany, built spring-wound people-like figures that could walk and play instruments

1596 Toilet; John Harington invented the first "water closet"

1600 Electric insulator, magnetism, glass eye; William Gilbert introduced the term *electricity* into language

1609 Johannes Kepler discovered that planets moved in eliptical orbits

1610 Microscope invented by Galileo Galilei, Italian astronomer

2700 B.C. Concrete and cement (made of lime and gypsum and used to create the pyramids; Roman builders used volcanic ash and water)

5500 B.C. Sickle found in caves in Mount Carmel in Palestine (flint teeth mounted on antler bone)

7000 B.C. Metal (the first metals used were those that could be separated from ore—e.g., gold and copper)

20,000 B.C. Oil lamps, bow and arrow (wooden shafts with a piece of flint tied to the shaft with resin or pitch)

1 A.D. Wheelbarrow in China, waterwheel

220 B.C. Explosives (Chinese used potassium nitrate for fireworks)

250 B.C. Automation begins with the water clock built by Ctesibius in Egypt

450 B.C. Democritus posited (but did not prove experimentally) the existence of atoms

800 B.C. Iron hand tools used by Assyrians; Chinese bronze; Etruscan dentists made false teeth

1656 Pendulum clock invented by the Dutch physicist and astronomer Christian Huygens

1623 Mechanical calculator invented by Wilhelm Schickard, German mathematician

1608 Telescope invented by Dutch optician named Hans Lippershey

800 Acetic acid, distillation, horse collar, nitric acid

200 Glass, paper manufacture in China

50 Gear, lever, mirror, tunnel, iron horseshoe, metal stirrup

Source: Table represents information adapted from *Milestones in Science and Technology: The Ready Reference Guide to Discoveries, Inventions, and Facts, Second Edition* by Ellis Mount and Barbara A. List. © 1994 by The Oryx Press. Used with permission from The OryxPress, 4041 N. Central Ave., Suite 700, Phoenix, AZ 85012.

1743 First passenger elevator invented for use by Louis XV of France at Versailles palace

1756 Dentures (Prussian dentist Philip Pfaff described steps to make wax impressions and cast models for missing teeth)

1666 Isaac Newton discovered gravity

1745 Electric capacitor (The first storage device for an electric charge was invented by Edwald von Kliest of Germany); first working electric capacitor (Leyden jar invented by Peter van Musschenbroek at University of Leyden)

1757 Achromatic lens (John Dolland of England invented lenses made of two kinds of glass for the telescope)

1676 Andre Felibien invented the screwdriver

1698 Steam pump invented by Thomas Savery, an English military engineer (pump used to drain water from mines)

1718 Machine gun invented by James Puckle in England; fired 63 rounds in seven minutes

1758 First railroad passenger car made of wood (Great Britain)

1749 Electricity (Benjamin Franklin discovered that electricity has two states: positive and negative)

1760 First bifocals were designed by Benjamin Franklin (U.S.)

1725 Punched card (Basile Bouchon of France used punched paper to control a loom)

1752 Diving suit (Englishman John Smeaton invented an air pump that fit into a diving bell)

1776 David Bushnell (U.S.) built a hand-cranked submarine

1714 Thermometer invented by Gabriel Danile Fahrenheit in Poland

1748 Electroscope, an instrument that detects electric charges, invented by Jean Antoine Nollet of France; steel-nibbed pen invented by Johann Janssen in france

1769 First automobiles had three wheels, was powered by a steam engine, and had a top speed of 3 mph; they were bulky and inefficient

1681 Barograph, a device for recording atmospheric pressure, invented by English physicist Robert Hooke

1661 Pollution control (John Evelyn of England) proposed methods for preventing smoke from polluting the air

1744 "Franklin stove" designed by Benjamin Franklin in United States; widely adopted in Europe decades later

1764 Spinning jenny (Invented by English weaver Hames Hargreaves, the machine could do the work of 30 spinning wheels)

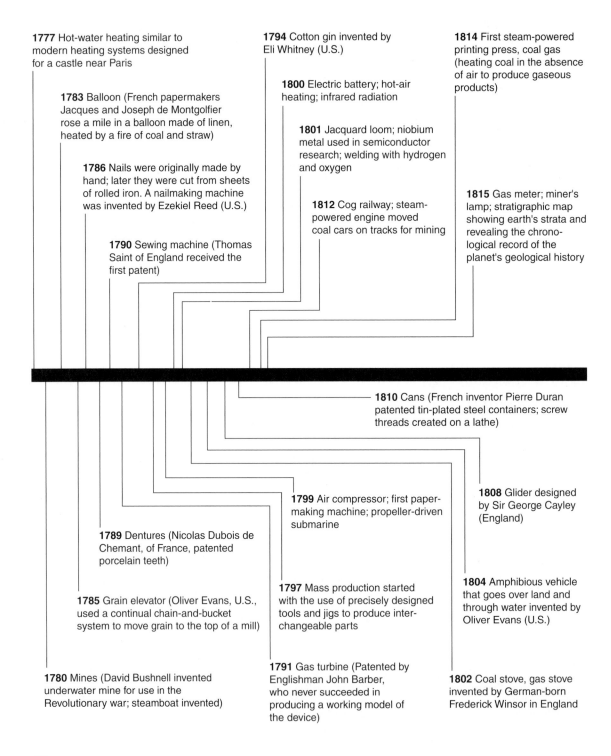

1777 Hot-water heating similar to modern heating systems designed for a castle near Paris

1783 Balloon (French papermakers Jacques and Joseph de Montgolfier rose a mile in a balloon made of linen, heated by a fire of coal and straw)

1786 Nails were originally made by hand; later they were cut from sheets of rolled iron. A nailmaking machine was invented by Ezekiel Reed (U.S.)

1790 Sewing machine (Thomas Saint of England received the first patent)

1794 Cotton gin invented by Eli Whitney (U.S.)

1800 Electric battery; hot-air heating; infrared radiation

1801 Jacquard loom; niobium metal used in semiconductor research; welding with hydrogen and oxygen

1812 Cog railway; steam-powered engine moved coal cars on tracks for mining

1814 First steam-powered printing press, coal gas (heating coal in the absence of air to produce gaseous products)

1815 Gas meter; miner's lamp; stratigraphic map showing earth's strata and revealing the chrono-logical record of the planet's geological history

1810 Cans (French inventor Pierre Duran patented tin-plated steel containers; screw threads created on a lathe)

1799 Air compressor; first paper-making machine; propeller-driven submarine

1808 Glider designed by Sir George Cayley (England)

1789 Dentures (Nicolas Dubois de Chemant, of France, patented porcelain teeth)

1797 Mass production started with the use of precisely designed tools and jigs to produce inter-changeable parts

1804 Amphibious vehicle that goes over land and through water invented by Oliver Evans (U.S.)

1785 Grain elevator (Oliver Evans, U.S., used a continual chain-and-bucket system to move grain to the top of a mill)

1780 Mines (David Bushnell invented underwater mine for use in the Revolutionary war; steamboat invented)

1791 Gas turbine (Patented by Englishman John Barber, who never succeeded in producing a working model of the device)

1802 Coal stove, gas stove invented by German-born Frederick Winsor in England

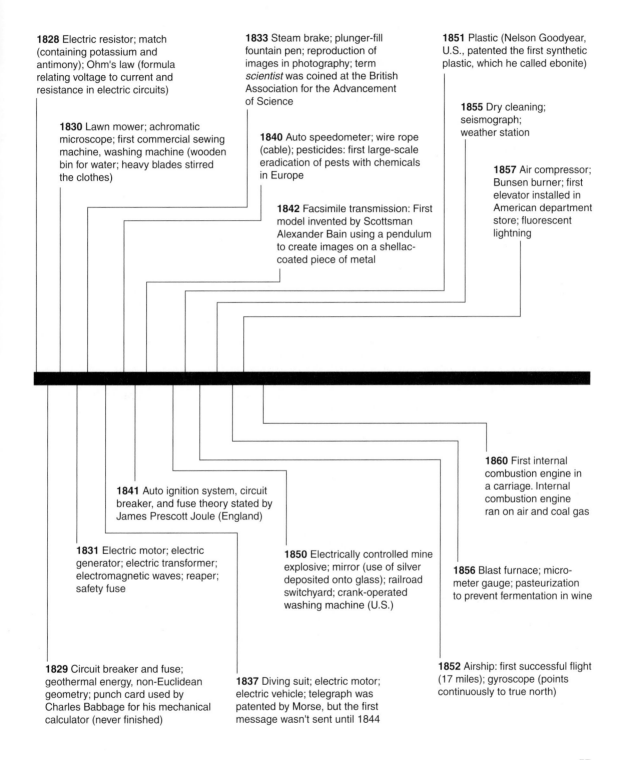

1828 Electric resistor; match (containing potassium and antimony); Ohm's law (formula relating voltage to current and resistance in electric circuits)

1830 Lawn mower; achromatic microscope; first commercial sewing machine, washing machine (wooden bin for water; heavy blades stirred the clothes)

1833 Steam brake; plunger-fill fountain pen; reproduction of images in photography; term *scientist* was coined at the British Association for the Advancement of Science

1840 Auto speedometer; wire rope (cable); pesticides: first large-scale eradication of pests with chemicals in Europe

1842 Facsimile transmission: First model invented by Scottsman Alexander Bain using a pendulum to create images on a shellac-coated piece of metal

1851 Plastic (Nelson Goodyear, U.S., patented the first synthetic plastic, which he called ebonite)

1855 Dry cleaning; seismograph; weather station

1857 Air compressor; Bunsen burner; first elevator installed in American department store; fluorescent lightning

1841 Auto ignition system, circuit breaker, and fuse theory stated by James Prescott Joule (England)

1860 First internal combustion engine in a carriage. Internal combustion engine ran on air and coal gas

1831 Electric motor; electric generator; electric transformer; electromagnetic waves; reaper; safety fuse

1850 Electrically controlled mine explosive; mirror (use of silver deposited onto glass); railroad switchyard; crank-operated washing machine (U.S.)

1856 Blast furnace; micro-meter gauge; pasteurization to prevent fermentation in wine

1829 Circuit breaker and fuse; geothermal energy, non-Euclidean geometry; punch card used by Charles Babbage for his mechanical calculator (never finished)

1837 Diving suit; electric motor; electric vehicle; telegraph was patented by Morse, but the first message wasn't sent until 1844

1852 Airship: first successful flight (17 miles); gyroscope (points continuously to true north)

1868 First tilting dental chair invented. First electric dental drill invented. (The drill wasn't put on the market until 1872 and wasn't used until the 1890s due to the lack of electricity)

1862 Rapid-firing machine gun; acoustics of music

1865 Genetics (Gregor Mendel of Austria conducted experiments crossbreeding various strains of garden peas.) Electrolytic refining of copper

1876 Telephone patent was granted to Alexander Graham Bell by the U.S. Supreme Court. Elisha Gray filed for, and failed to receive, the patent two hours after Bell

1877 Duplicating was invented by Thomas Alva Edison. Carbon microphone created by Emile Berliner and Thomas Alva Edison. Moving-coil microphone invented by Charles Cuttris and Werner Siemens;

phonograph was patented by Thomas Alva Edison. Emile Berliner invented a wax phonograph record that produced a fuzzy sound

1869 Sir William Herschel (England) devised fingerprint identification system; steam-powered monorail system invented in Syria; periodic table of elements created; stock ticker (printer of stock prices) invented by Thomas Alva Edison; DNA discovered by Johann Friedrich Miescher; automatic air brake invented for railroad cars

1864 Nitroglycerine used as a detonator by Alfred Nobel. James Clerk Maxwell created a set of four equations that described the relation between electricity and magnetism

1861 Color photography (Thomas Sutton and James Clerk Maxwell took pictures, made positive transparencies, and projected them on a screen); tanker named *Elizabeth Watts* was the first ship built in the United States to carry oil from Pennsylvania to London

1876 Carpet sweeper invented by Melville Bissell (U.S.); J. Willard Gibbs, American physicist, wrote about the general principles of physical chemistry

1867 Barbed wire patented. Dynamite patented. First successful torpedo invented. First commercial typewriter invented

1866 Ernst Heinrich Haeckle was the first to use the term *ecology*

1870 Refrigerator invented and manufactured using ammonia

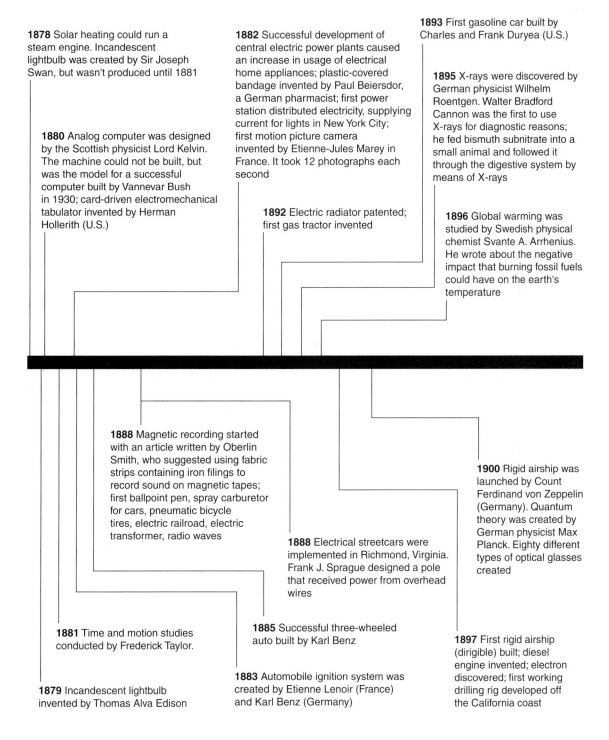

1878 Solar heating could run a steam engine. Incandescent lightbulb was created by Sir Joseph Swan, but wasn't produced until 1881

1882 Successful development of central electric power plants caused an increase in usage of electrical home appliances; plastic-covered bandage invented by Paul Beiersdor, a German pharmacist; first power station distributed electricity, supplying current for lights in New York City; first motion picture camera invented by Etienne-Jules Marey in France. It took 12 photographs each second

1893 First gasoline car built by Charles and Frank Duryea (U.S.)

1895 X-rays were discovered by German physicist Wilhelm Roentgen. Walter Bradford Cannon was the first to use X-rays for diagnostic reasons; he fed bismuth subnitrate into a small animal and followed it through the digestive system by means of X-rays

1880 Analog computer was designed by the Scottish physicist Lord Kelvin. The machine could not be built, but was the model for a successful computer built by Vannevar Bush in 1930; card-driven electromechanical tabulator invented by Herman Hollerith (U.S.)

1892 Electric radiator patented; first gas tractor invented

1896 Global warming was studied by Swedish physical chemist Svante A. Arrhenius. He wrote about the negative impact that burning fossil fuels could have on the earth's temperature

1888 Magnetic recording started with an article written by Oberlin Smith, who suggested using fabric strips containing iron filings to record sound on magnetic tapes; first ballpoint pen, spray carburetor for cars, pneumatic bicycle tires, electric railroad, electric transformer, radio waves

1888 Electrical streetcars were implemented in Richmond, Virginia. Frank J. Sprague designed a pole that received power from overhead wires

1900 Rigid airship was launched by Count Ferdinand von Zeppelin (Germany). Quantum theory was created by German physicist Max Planck. Eighty different types of optical glasses created

1881 Time and motion studies conducted by Frederick Taylor.

1885 Successful three-wheeled auto built by Karl Benz

1879 Incandescent lightbulb invented by Thomas Alva Edison

1883 Automobile ignition system was created by Etienne Lenoir (France) and Karl Benz (Germany)

1897 First rigid airship (dirigible) built; diesel engine invented; electron discovered; first working drilling rig developed off the California coast

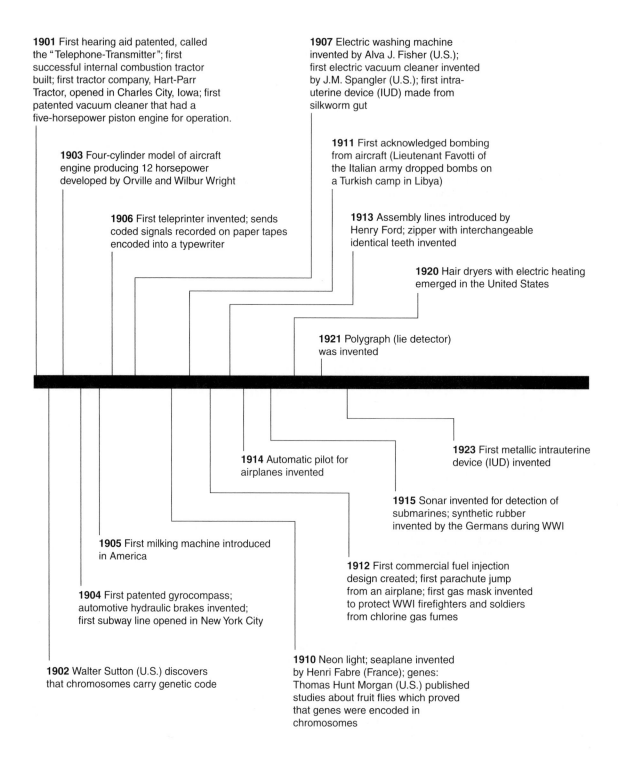

1901 First hearing aid patented, called the "Telephone-Transmitter"; first successful internal combustion tractor built; first tractor company, Hart-Parr Tractor, opened in Charles City, Iowa; first patented vacuum cleaner that had a five-horsepower piston engine for operation.

1907 Electric washing machine invented by Alva J. Fisher (U.S.); first electric vacuum cleaner invented by J.M. Spangler (U.S.); first intra-uterine device (IUD) made from silkworm gut

1903 Four-cylinder model of aircraft engine producing 12 horsepower developed by Orville and Wilbur Wright

1911 First acknowledged bombing from aircraft (Lieutenant Favotti of the Italian army dropped bombs on a Turkish camp in Libya)

1906 First teleprinter invented; sends coded signals recorded on paper tapes encoded into a typewriter

1913 Assembly lines introduced by Henry Ford; zipper with interchangeable identical teeth invented

1920 Hair dryers with electric heating emerged in the United States

1921 Polygraph (lie detector) was invented

1923 First metallic intrauterine device (IUD) invented

1914 Automatic pilot for airplanes invented

1915 Sonar invented for detection of submarines; synthetic rubber invented by the Germans during WWI

1905 First milking machine introduced in America

1912 First commercial fuel injection design created; first parachute jump from an airplane; first gas mask invented to protect WWI firefighters and soldiers from chlorine gas fumes

1904 First patented gyrocompass; automotive hydraulic brakes invented; first subway line opened in New York City

1902 Walter Sutton (U.S.) discovers that chromosomes carry genetic code

1910 Neon light; seaplane invented by Henri Fabre (France); genes: Thomas Hunt Morgan (U.S.) published studies about fruit flies which proved that genes were encoded in chromosomes

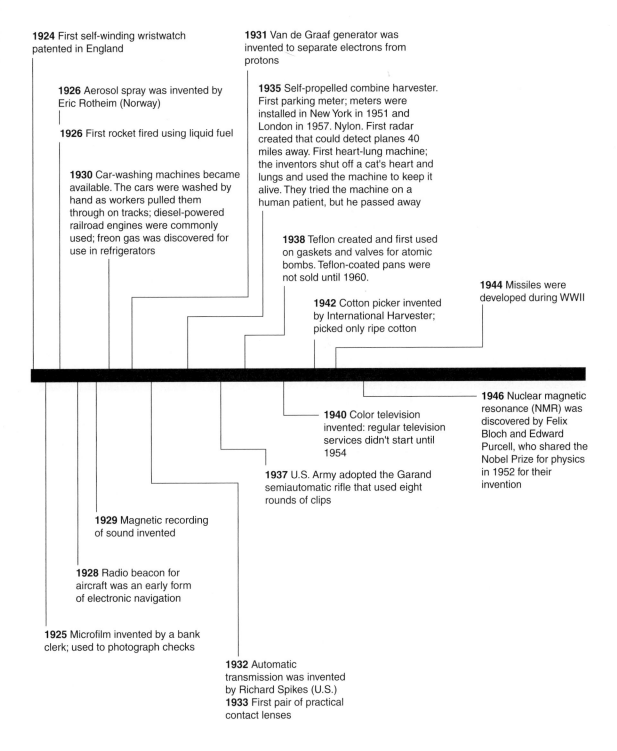

1924 First self-winding wristwatch patented in England

1931 Van de Graaf generator was invented to separate electrons from protons

1926 Aerosol spray was invented by Eric Rotheim (Norway)

1926 First rocket fired using liquid fuel

1935 Self-propelled combine harvester. First parking meter; meters were installed in New York in 1951 and London in 1957. Nylon. First radar created that could detect planes 40 miles away. First heart-lung machine; the inventors shut off a cat's heart and lungs and used the machine to keep it alive. They tried the machine on a human patient, but he passed away

1930 Car-washing machines became available. The cars were washed by hand as workers pulled them through on tracks; diesel-powered railroad engines were commonly used; freon gas was discovered for use in refrigerators

1938 Teflon created and first used on gaskets and valves for atomic bombs. Teflon-coated pans were not sold until 1960.

1944 Missiles were developed during WWII

1942 Cotton picker invented by International Harvester; picked only ripe cotton

1946 Nuclear magnetic resonance (NMR) was discovered by Felix Bloch and Edward Purcell, who shared the Nobel Prize for physics in 1952 for their invention

1940 Color television invented: regular television services didn't start until 1954

1937 U.S. Army adopted the Garand semiautomatic rifle that used eight rounds of clips

1929 Magnetic recording of sound invented

1928 Radio beacon for aircraft was an early form of electronic navigation

1925 Microfilm invented by a bank clerk; used to photograph checks

1932 Automatic transmission was invented by Richard Spikes (U.S.)
1933 First pair of practical contact lenses

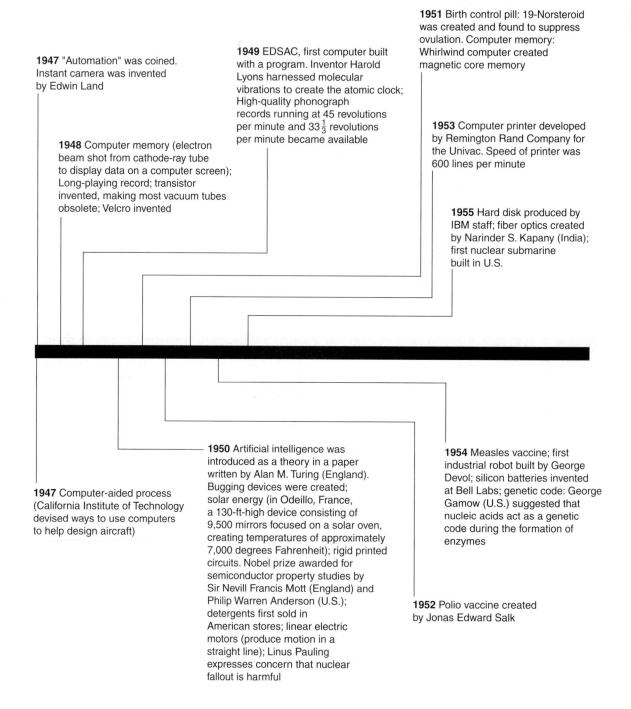

1947 "Automation" was coined. Instant camera was invented by Edwin Land

1948 Computer memory (electron beam shot from cathode-ray tube to display data on a computer screen); Long-playing record; transistor invented, making most vacuum tubes obsolete; Velcro invented

1949 EDSAC, first computer built with a program. Inventor Harold Lyons harnessed molecular vibrations to create the atomic clock; High-quality phonograph records running at 45 revolutions per minute and $33\frac{1}{3}$ revolutions per minute became available

1951 Birth control pill: 19-Norsteroid was created and found to suppress ovulation. Computer memory: Whirlwind computer created magnetic core memory

1953 Computer printer developed by Remington Rand Company for the Univac. Speed of printer was 600 lines per minute

1955 Hard disk produced by IBM staff; fiber optics created by Narinder S. Kapany (India); first nuclear submarine built in U.S.

1947 Computer-aided process (California Institute of Technology devised ways to use computers to help design aircraft)

1950 Artificial intelligence was introduced as a theory in a paper written by Alan M. Turing (England). Bugging devices were created; solar energy (in Odeillo, France, a 130-ft-high device consisting of 9,500 mirrors focused on a solar oven, creating temperatures of approximately 7,000 degrees Fahrenheit); rigid printed circuits. Nobel prize awarded for semiconductor property studies by Sir Nevill Francis Mott (England) and Philip Warren Anderson (U.S.); detergents first sold in American stores; linear electric motors (produce motion in a straight line); Linus Pauling expresses concern that nuclear fallout is harmful

1954 Measles vaccine; first industrial robot built by George Devol; silicon batteries invented at Bell Labs; genetic code: George Gamow (U.S.) suggested that nucleic acids act as a genetic code during the formation of enzymes

1952 Polio vaccine created by Jonas Edward Salk

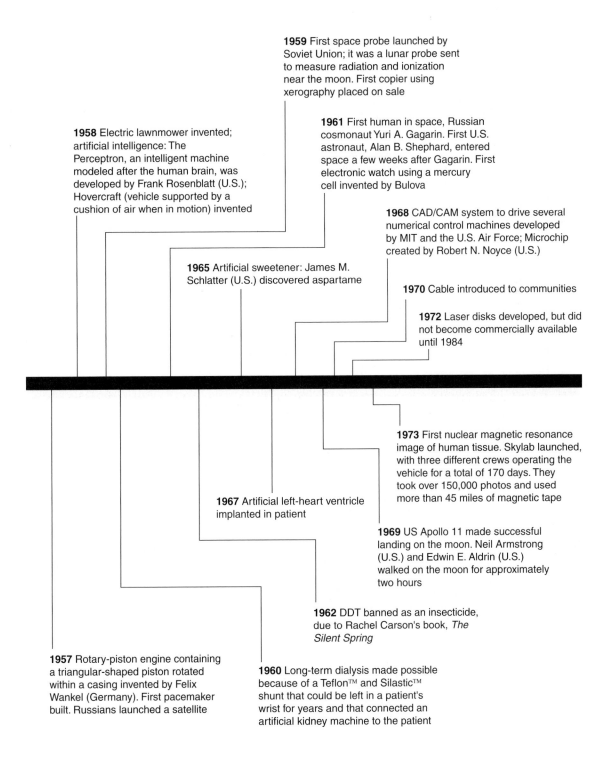

1959 First space probe launched by Soviet Union; it was a lunar probe sent to measure radiation and ionization near the moon. First copier using xerography placed on sale

1958 Electric lawnmower invented; artificial intelligence: The Perceptron, an intelligent machine modeled after the human brain, was developed by Frank Rosenblatt (U.S.); Hovercraft (vehicle supported by a cushion of air when in motion) invented

1961 First human in space, Russian cosmonaut Yuri A. Gagarin. First U.S. astronaut, Alan B. Shephard, entered space a few weeks after Gagarin. First electronic watch using a mercury cell invented by Bulova

1968 CAD/CAM system to drive several numerical control machines developed by MIT and the U.S. Air Force; Microchip created by Robert N. Noyce (U.S.)

1965 Artificial sweetener: James M. Schlatter (U.S.) discovered aspartame

1970 Cable introduced to communities

1972 Laser disks developed, but did not become commercially available until 1984

1973 First nuclear magnetic resonance image of human tissue. Skylab launched, with three different crews operating the vehicle for a total of 170 days. They took over 150,000 photos and used more than 45 miles of magnetic tape

1967 Artificial left-heart ventricle implanted in patient

1969 US Apollo 11 made successful landing on the moon. Neil Armstrong (U.S.) and Edwin E. Aldrin (U.S.) walked on the moon for approximately two hours

1962 DDT banned as an insecticide, due to Rachel Carson's book, *The Silent Spring*

1957 Rotary-piston engine containing a triangular-shaped piston rotated within a casing invented by Felix Wankel (Germany). First pacemaker built. Russians launched a satellite

1960 Long-term dialysis made possible because of a Teflon™ and Silastic™ shunt that could be left in a patient's wrist for years and that connected an artificial kidney machine to the patient

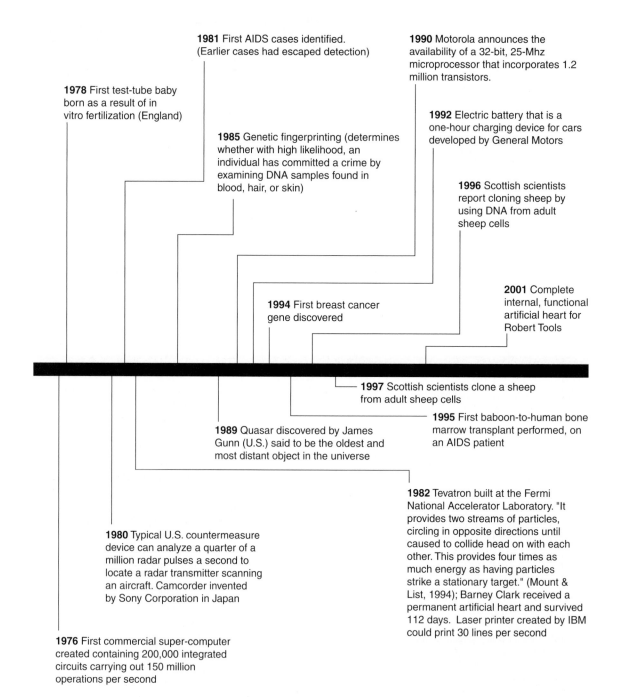

1981 First AIDS cases identified. (Earlier cases had escaped detection)

1990 Motorola announces the availability of a 32-bit, 25-Mhz microprocessor that incorporates 1.2 million transistors.

1978 First test-tube baby born as a result of in vitro fertilization (England)

1992 Electric battery that is a one-hour charging device for cars developed by General Motors

1985 Genetic fingerprinting (determines whether with high likelihood, an individual has committed a crime by examining DNA samples found in blood, hair, or skin)

1996 Scottish scientists report cloning sheep by using DNA from adult sheep cells

2001 Complete internal, functional artificial heart for Robert Tools

1994 First breast cancer gene discovered

1997 Scottish scientists clone a sheep from adult sheep cells

1995 First baboon-to-human bone marrow transplant performed, on an AIDS patient

1989 Quasar discovered by James Gunn (U.S.) said to be the oldest and most distant object in the universe

1982 Tevatron built at the Fermi National Accelerator Laboratory. "It provides two streams of particles, circling in opposite directions until caused to collide head on with each other. This provides four times as much energy as having particles strike a stationary target." (Mount & List, 1994); Barney Clark received a permanent artificial heart and survived 112 days. Laser printer created by IBM could print 30 lines per second

1980 Typical U.S. countermeasure device can analyze a quarter of a million radar pulses a second to locate a radar transmitter scanning an aircraft. Camcorder invented by Sony Corporation in Japan

1976 First commercial super-computer created containing 200,000 integrated circuits carrying out 150 million operations per second

"George"
PAUL ALCORN

He was a small creature, no more than forty inches in height, and he was, as usual, hungry. His brain was about half the size of modern Homo sapiens, about 650 cubic centimeters, and he lived on a wide, flat plain in Africa, where the tall grass and clutches of low-branched trees made a hunter's paradise for him and his fellow creatures. His kind would one day be known as Homo habilis, but that was nearly two million years in the future. For our purposes, we will call him George.

As I have said, he was hungry. It was a characteristic of his species. Warm blooded and a hunter, George and his companions spent much of their time ranging out across the plain near their most recent campsite in search of food. They were omnivorous, as likely to devour the tough nutty fruit of a nearby berry bush as the raw flesh of some small reptile or insect that failed to escape their notice in time. From day to day, George and his fellows satisfied their internal furnace with the fuel of whatever they could find, always searching for the great kill that would allow them to gorge themselves and replenish the dwindling supply of protein gathered from the last great kill some days or even weeks earlier.

Today they were near the high rock carapace to the east, though they had no concept of direction in those terms. It was merely the "high place over there, where the sun rises." George was scouting ahead of the pack, a chore he seemed to relish. A loner, he would often run ahead, somehow

Excerpted from *Social Issues in Technology: A Format for Investigation,* 2d ed., by Paul A. Alcorn, © 1997. Reprinted by permission of Prentice-Hall, Inc., Upper Saddle River, NJ.

enjoying the prospect of being the first to sight a potential prey, hoping to be the first to wrestle it to the ground, to pound it to death with his clenched hands, or to tear its throat with his teeth.

The band was in the narrow passage that led into the center of the mound of rocks, close to where a fellow hunter had perished only days before at the hands of another predator, a huge cat creature with claws to tear at the throat and jaws to sink deeply into the flesh and break the victim's neck. George had seen it happen. He remembered it all too well. He was cautious, listening and sniffing the air, remembering what had happened to . . . who was it? His simple consciousness forgot those things easily, but the memory of the danger remained solidly in his mind.

The others were far behind him and out of sight as he turned into the natural bowl formed by the circle of high, flat rocks near the center of the carapace. He could feel the eyes on him, almost smell death in the air. Instinctively, he knew he was not alone. Turning quickly, he scanned the rocks above, seeking any telltale clue of whatever was lurking there. He spun so quickly and jerked his head about so violently in his panic that he nearly missed the low, flat, black furred head, the huge yellowish eyes that stared back at him.

Above George and a little to his left was the same sleek creature that had made a meal of his fellow hunter only days before. George panicked. He turned and leaped to the side of a sheer rock, clinging with his toes and fingertips as the cat made a lunging pass at him. The panther missed the small ape-man by inches. George scrambled toward the summit of the rock, churning his legs wildly in search of some foothold, reaching out blindly with his hands for any purchase further up the rock face. Lacerations appeared on his knees and thighs as he slid against the sharp black obsidian. His fingers numbed as they bit again and again into the narrow, knifelike crevices above. But he was making progress. Below him, the cat yowled and paced, panting heavily and leaping toward the fleeing figure.

Springing with all its might, the huge panther nearly reached George, who pulled forward with a last great effort and reached a wide ledge nearly halfway up the rock face. As he slid himself onto the strip of rock and flattened himself against the wall, a single stone slipped over the edge and fell, striking the huge cat squarely on the nose. With a howl, the panther retreated. George kicked another rock toward the beast, then another, missing both times. In panic, he grabbed several more and hurled them toward the beast, striking him again, this time dead center at the skull. The panther slumped to the ground, stunned by the blow. George grabbed for another loose rock and another, improving his aim with each throw, grasping larger and larger rocks until at last he found himself holding heavy slabs of obsidian over his head with both hands and hurling them down on the lifeless victim. He struck the creature again and again and again.

In the night, belly taut, legs splayed out before him, George lay with the other hunters in the natural bowl of the rocks stuffed with the meat of the dead panther. He was smiling, staring up into the night sky at the bright starlit veil, a swath of white that spread across the sky like a river in the firmament. Absently, he licked his hand and passed the thick saliva over the crusted scratches on his belly and legs. Around him, the sound of night creatures echoed off the surrounding walls as predator and prey continued the struggle for survival.

In his right hand, George clutched at a single round stone, about three inches long across its short axis and weighing nearly half a pound. He felt safe now. He knew that he could fend off any attack. Tomorrow he would try his luck again with his newfound weapon. Tomorrow he would try it against one of the doglike scavengers of the plain or use it to bring down a bird near the river. Perhaps he would never need to be hungry again. Had he not slain the mighty panther singlehandedly? Had he not proven himself the greatest hunter of them all? Who knew what he might be able to do the next time? Who could really know?

QUESTIONS

1. Who was George? Describe him in detail.

2. Reflect on George's life; how was his life different from that of humans today? Write a minimum of three paragraphs on the differences.

3. How does the last paragraph of this article correlate to the importance of technology in the life of *homo habilis?*

The "New" George

JOHN MORELLO

8

In a previous article you met "George," who lived millions of years ago. Paul Alcorn captured George's life in vivid detail. He depended on plants and animals for his survival, picking what he could from the trees around him, and hunting animals that sometimes were in the process of hunting him. He seemed to be living on the edge of extinction and in the constant grip of fear. His less-than-modest mental capabilities, braced by his even more breathtaking lack of tools and the skills to use them, made George's odds of survival long. But one day, while fighting for his life, he discovered the power of a rock and the skills and technique to use it. The blending of technology and human need changed George's life. But while technology seemingly made George's life a little easier, what lurked below the surface of that discovery was something most of us already know: technology isn't always a blessing. Consider for a moment what would happen if we provided George with a few of the technological marvels he was forced to do without so many years ago. Many of the technologies George will use can be found in the timeline in Part II. As you'll see, each one comes with implications and obligations that George, or the rest of us, for that matter, probably didn't count on when they first entered our lives.

For starters, let's go back to when we first met George. Alcorn says George and his companions spent much of their time looking for food. It was necessary in order to survive. Without proper storage, Monday's dinner was Tuesday's carrion, so every day was a struggle. But suppose George had refrigeration? The first refrigerator was invented around 1870. It used ammonia to help keep food fresh. Later, ammonia would be replaced with freon. Now George could kill his prey, drag it back to his lair, eat what he wanted that day, and store the rest. George's life would now be less tenuous. However, like the rest of us, George would learn that this particular technology is not without its headaches. Refrigeration depends on chemicals, specifically ammonia and freon, to keep things cold. These can be nasty if they come into contact with human flesh or the respiratory system. Maintaining such a device can be tricky. There are laws and regulations governing the recharging and maintenance of refrigerators, and George could find himself in trouble with the law regarding the proper disposal of hazardous materials and, if the electricity goes off, he's got a brand new problem to deal with. Equally problematic could be the way refrigeration affects food. Some foods become dry and lose their flavor if not packaged properly. George will need Tupperware or disposable containers to keep foods fresh. And, if he opts for disposable containers, where is he going to put them when he wants to dispose of them? George's companions might not appreciate the mountain of disposable containers in the back of the cave. Last, but not least, how about cleanliness? Like everybody else, George is going to have to clean out the refrigerator. And, when he reaches all the way to the back and finds something that was left there a couple of millennia ago, he may decide that the daily grind of looking for food wasn't so bad after all. Of course, he could wait until the 1950s when the first McDonald's opened.

George also liked plants. Alcorn said he was omnivorous. But the berries George feasted on may have been spoiled by insects that ate their way into the fruit, and who either left or decided to make that blueberry a weekend retreat. Pesticides could have helped. The first pesticides began to appear in large quantity in Europe around 1840. They would have helped George handle his bug problems, and made sure other plants, namely weeds, didn't grow up around his berries, blocking the sun and choking off their water supply. But George would have to remember which pesticides he was using. DDT would be pulled off the market by 1962, and others containing dioxin would be linked with all sorts of diseases. He would have to wash his fruits and vegetables carefully in order to avoid ingesting any pesticides directly. He would also have to make sure he knew how much pesticide to use. Too much and he might harm the soil. Add way too much, and the first heavy rain will wash the excess into your drinking water. These would prove to be pretty strict limitations for someone like George, who couldn't read. They're still pretty tough for those of us who can. Maybe George is better off taking his chances without it.

Alcorn says George was also a hunter. He enjoyed the prospect of being the first to sight a potential prey, wrestling it to the ground, pounding it to death with his fists or using his teeth to tear its throat. If George made a practice of using his teeth as a weapon, and it sounds like he did, pretty soon those teeth would be worn to the gums, making him an excellent candidate for dentures. The Etruscans produced the first set around 800 B.C.; more fashionable and reliable models came out in the 18th century. Still, dentures require some getting used to; no heat-of-the-moment throat ripping for someone who keeps his teeth in a jar at night, or, in George's case, on the nearest rock. Chemicals are needed to keep them clean, adhesives necessary to keep them firmly in place, insurance companies to decide if they're covered, and regular visits to the dentist to make sure everything is opening and shutting the way it should. George would soon learn what many of us already know: technological substitutions are nice, but nothing beats the real thing.

During his encounter with the panther, George discovered that sometimes the hunter can become the hunted. In his desperation, he reached for a rock, and then another, and then another after that. Giving his hand–eye coordination on-the-job training, George discovered that a well-placed rock was better than a fist. In time, George would come to depend on more sophisticated weapons. Gunpowder will come on the scene about A.D. 1100. By the 1700s, machine guns will be introduced and will be perfected in the 19th century. But in George's first battle for dominance among species, the rock allowed him to be the dominator. That night, as he and his companions gorged themselves on panther meat, George held a rock in his hands. It was now a symbol of power in man's confrontation with the dangerous world around him. At first, George might only use his rock to kill for food. But would he stop there? What would stop him from using his new technology to establish and defend exclusive hunting rights in areas he shared with other humans? Would he resist the notion of using his new technology to settle disputes? Or would he use it to acquire property belonging to another? And how long would it be before others learned to use a rock, and then a stick, and so on?

Yes, the night of George's first panther kill held out unlimited possibilities. Who knew what he or someone else might do with his new tool the next time? Who could really know?

QUESTIONS

1. Describe how George might use refrigeration, and how the use of refrigeration might affect his life?

2. Make a list of technologies on which you depend. How many of them have made your life significantly easier? How many of them have made your life more complicated?

3. Suppose for a moment that you were George. What other problems might you encounter in living his life with some of the technologies he's suddenly confronted with?

Conclusion

As the timeline in Part II shows, human and technological development seemed to have traveled a parallel course. Some of the developments may have surprised you. But the real surprise must have come when those new technologies first burst upon the scene and into the consciousness of our ancestors. These new devices now required would-be users to figure out how, when and if they should be used. Sometimes the failure to understand how a technology operated resulted in disaster. Other times, people—knowing full well the implications of technology use—went ahead anyway and provoked disaster.

Part II should have gotten you thinking about our relationship with the technology all around us. Are we really living in an age when technology runs us, and not the other way around? Or are only those people who can afford and operate technology destined to rule?

Part III of this book deals with the role energy plays in society. Technology has provided ways to get oil, gas and nuclear power into our hands, but finding ways to use them efficiently and safely hasn't been as easy. How do we balance the need to develop energy sources to keep society going? How do we find ways to conserve energy by developing technologies that don't require so much? And, what have been the economic, social and environmental costs of this struggle?

INTERNET EXERCISE

1. Use any Internet search engine you choose to find answers to the following questions
   a. In 1900, what was the estimated life expectancy of males living in the U.S.? Females?
   b. What was the infant mortality rate in the U.S. in 1900? 1999?
   c. How have work schedules been affected by developments in technology since the beginning of the twentieth century?
   d. How much leisure time do Americans enjoy today as opposed to 1900? How has technology affected this statistic?

USEFUL WEB SITES

http://www.colby.edu/sci.tech/STS.html	Science, Technology and Society
http://echo.gmu/edu.center/index.html	Virtual Center for History and Technology
http://www.englib.cornell.edu/ice/lists/historytechnology/historytechnology.html	History of Technology resources available on the Internet.

Energy

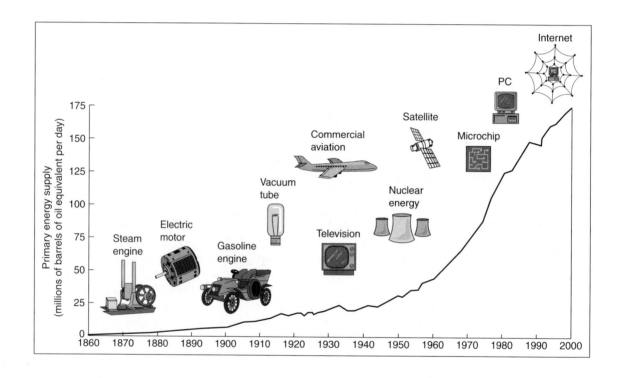

OBJECTIVES

Part III will help you to

- Realize the historical background and major and minor implications of energy sources
- Discover various problems, urgencies and implications of increased use of fossil fuel
- Distinguish between developed and developing world's energy consumption issues and potential problems
- Analyze and use tables and graphs to promote adoption of appropriate energy policies at individual, national and international level for achieving a balance between economic growth and sustainable environment
- Understand the impact of renewable and non-renewable energy sources for a sustainable environment
- Determine ethical considerations of increased fossil fuel usage
- Realize the importance of the development of new energy sources for a sustainable environment
- Appreciate the importance and potential contributions of individual awareness, ethical action to the planning of the future

INTRODUCTION

The Energy is the go of things.
JAMES CLARK MAXWELL

Energy is the most basic element of progress for all economies—the pivotal force that sustains life and ensures a standard of living and, ultimately, the standard of life. Energy technologies are society's most basic infrastructure that enables economic growth.

In the pre-Industrial Revolution era, the population of planet Earth was small, and energy needs were limited to cooking and heating. Energy could be exploited without serious damage to the atmosphere, hydrosphere, or geosphere. But the dawn of the Industrial

Figure P3.1
World Fossil Fuel Use, 1950–96
Source: *Vital Sign 1997,* Worldwatch Institute.

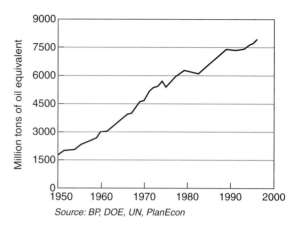

Source: *BP, DOE, UN, PlanEcon*

Revolution created an energy-hungry genie whose appetite for hydrocarbons has grown to such a degree that it is jeopardizing future prospects for a sustainable environment (Figures P3.1, P3.2).

Since the Industrial Revolution, atmospheric carbon dioxide levels have risen from an estimated 280 parts per million to 362 parts per million, the highest in 150 years (Figure P3.3). The mainstream scientific community, including the Intergovernmental Panel on Climatic Change, with 2,500 of the world's leading atmospheric scientists, now finds evidence that human activity is indeed altering the earth's climate (Figure P3.4). In 1996, worldwide carbon emissions from the burning of fossil fuels climbed to 6.25 billion tons, reaching a new high for the second year in a row (Figures P3.5, P3.6).

The world's output of goods and service is growing fast. In 1996 it expanded by 3.8 percent, up modestly from 3.5-percent growth in 1995. The gross world product (GWP) climbed from $26.9 trillion to $28.0 trillion (1995 dollars) over the same period, and GWP per person increased from $4,733 to $4,846. If the world economy keeps a similar growth trend into the 21st century, the increased use of fossil fuels would be catastrophic for the environment of the planet. According to some estimates, the entire biosphere (i.e., the

Figure P3.2
Fossil Fuel Use by Type of Fuel, 1950–96
Source: *Vital Sign 1997,* Worldwatch Institute.

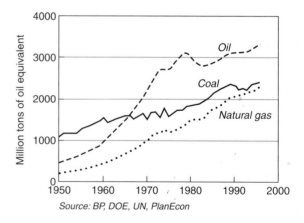

Source: *BP, DOE, UN, PlanEcon*

Figure P3.3
Atmospheric Concentration of
Carbon Dioxide, 1764–1996
Source: *Vital Sign 1997,* Worldwatch
Institute.

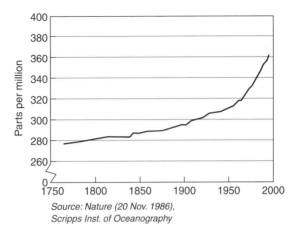

Source: Nature (20 Nov. 1986),
Scripps Inst. of Oceanography

earth) provides at least $33 trillion worth of free materials every year, in comparison to the
GWP of $28 trillion. The economies of the earth would grind to a halt without the services
of ecological life-support systems.

Since the Industrial Revolution, the world has become increasingly dependent on fossil fuels. Modern civilization is actually based on nonrenewable resources, putting a finite
limit on the length of time our civilization can exist. The following facts illustrate the magnitude of the energy problems the world faces today:

- The United States is responsible for almost 25 percent of the world's total energy consumption. Americans use 1 million gallons of oil every two minutes.

- The United States, the largest single source of carbon emissions, is responsible for
  23 percent of the emissions of this climate-changing gas.

- The energy currently wasted by U.S. cars, homes, and appliances equals more than
  twice the known energy reserves in Alaska and the U.S. outer continental shelf.

Figure P3.4
Average Temperature at the
Earth's Surface, 1866–1996
Source: *Vital Sign 1997,* Worldwatch
Institute.

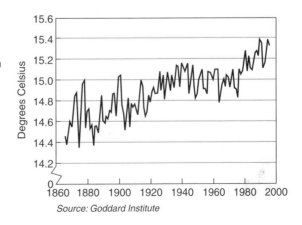

Source: Goddard Institute

Figure P3.5
World Carbon Emissions from
Burning of Fossil Fuel, 1950–96
Source: *Vital Sign 1997,* Worldwatch
Institute.

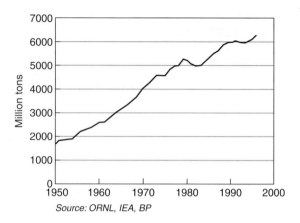

Source: ORNL, IEA, BP

- Americans could cut energy consumption in half by the year 2030 simply by using energy more efficiently and by using more renewable energy sources.

- A human being living at survival level needs about 2,000 kilocalories of energy per day. Americans consume 230,000 kilocalories of energy a day, 115 times the level needed for survival.

- If just 1 percent of America's 140 million car owners were to tune up their cars, nearly a billion pounds of carbon dioxide could be eliminated.

- One-fifth of the world's population now accounts for 70 percent of the globe's energy use.

- China, the world's fastest growing economy, accounts for 14 percent of carbon emissions, mainly because of its heavy dependence on coal.

- One person in Western Europe uses as much energy as 80 people in sub-Sahara Africa.

- A U.S. citizen uses as much energy as 330 citizens of Bangladesh.

Figure P3.6
Carbon Emissions from Burning
of Fossil Fuel, by Economic
Region, 1950–96
Source: *Vital Sign 1997,* Worldwatch
Institute.

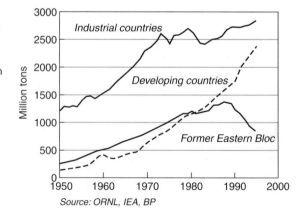

Source: ORNL, IEA, BP

Energy (noun)

1. The capacity for vigorous activity.

2. The ability of matter or radiation to do work either because of its motion (kinetic energy), or because of its mass (released in nuclear fission, etc.), or because of its electric charge, etc.

3. Fuel and other resources used for the operation of machinery, etc.

> *"Energy is the go of things"*
> JAMES CLARK MAXWELL

> $E = mc^2$
> ALBERT EINSTEIN

The capacity of doing work.
Work is done when a body is moved by a force. The rate of doing work is known as power. Power is energy spent per time unit. The Units of energy are as follows:

- ergs
- joules
- calories
- watts
- kilowatt hours (kwh)

- BTUs
- horsepower
- foot-pounds
- barrels of oil
- metric tons of coal

Energy is interchangeable with matter and takes many forms (Figure P3.7):

kinetic
potential
electrical
magnetic
heat
chemical
nuclear
sound
light
mass

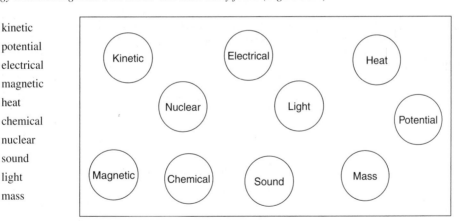

Figure P3.7
Universal Set of Energy

The approximate amount of total energy as mass and radiation in the universe is equal to 4.5×10^{58} kwh, which is a small fraction of the amount of energy as gravitation in the universe.
The sun's daily output energy is 8.33×10^{25} kwh.
The earth receives 4.14×10^{15} kwh of energy daily.

- Coal has been mined for 800 years, but over half of it has been extracted in the past 37 years.

- Petroleum has been pumped out of the ground for about 100 years, but half of it has been consumed during the past 25 years.

- If the rest of the world used energy at the same rate as the United States, global energy use would be four times as large as it is, and its impact on environment would be increased manyfold.

The world as a whole could not sustain the rate of energy consumption now enjoyed by the industrial nations. The Third World is trying to industrialize, but the high energy costs are adding to staggering debt loads. A reduction in the use of oil by the industrial nations is essential for world peace and global justice and also for environmental protection and long-term sustainability of the planet.

According to some estimates, most of the increases in energy demand will probably occur in the developing world, where population growth rates are high and industrialization and urbanization are under way. In contrast, demand is expected to remain stable or

Table P3.1
Units of Conversion between Energy and Power

1 watt (w) = 1 joule (j) for 1 second
1 kilowatt (kw) = 1,000 watts
1 megawatt (mw) = 1,000 kilowatts (10^6 watts)
1 gigawatt (gw) = 1,000 megawatts (10^9 watts)
1 terawatt (tw) = 1,000 gigawatts (10^{12} watts)
1 twh = 1 billion kwh
1 kw (capacity) = 8,760 kilowatt-hours (kwh; maximum annual production)
1 kwh = 1.34 horsepower-hours (hph)
1 kwh = 3,415 BTU
1 hph = 2,547 BTU
1 hph = 18 ft^3 CH_4 (methane) using internal combustion
1 metric ton coal equivalent = 8,000 kwh (heat equivalent)
1 ton = 2,000 pounds
1 metric ton = 2205 pounds
1 short ton = .9072 metric tons
1 million metric tons of crude oil of petroleum products = 4×10^{13} BTU
1 million metric tons of coal = 2.8×10^{13} BTU
1000 million m^3 of natural gas = 3.6×10^{13} BTU
1 barrel of crude oil = 5.8×10^6 BTU
1 watt-hour = 3.6×10^3 joules
1 BTU = 1.055107×10^3 joules
1 calorie = 4.184 joules
1 foot-pound = 1.356 joules
1 metric ton coal equivalent (MTCE) = 8,000 kwh (1,000 kwh = .125 MTCE)
1,000 ft hydrogen = 79 kwh
1 langley = .21622 kwh per square foot

drop in the industrialized countries, where population growth rates are low. Demand could stabilize or decline in eastern Europe and former Soviet republics, depending on the success of economic reforms. Much hinges on consensus and whether sustainable policies are enacted.

A report by the World Resources Institute concludes that if efficiency is strenuously pursued in all countries, global energy demand in 2020 will be only 10 percent higher than in 1980, even with the expected growth in population. Energy use in the Northern Hemisphere would have to be cut in half, while that in the Southern Hemisphere would grow, and living standards in the Third World in 2020 could then be comparable to those of Western Europe in the mid-seventies. The use of renewable energy sources (hydroelectric, biomass, wind power, solar thermal, and photovoltaic sources) will contribute to sustainability and justice, and in most cases the environmental impact will be lower than that of fossil fuels.

As we discuss the energy issue in relation to our future, our challenge will be to support world economic growth and unprecedented energy needs in a sustainable approach, a formula that demands both economic feasibility and preservation of the environment. This part of the text investigates present energy technologies and concerns and future strategies that will lead the reader to become aware of both future challenges and their potential solutions.

> *Man strides over the earth, and deserts follow in his footsteps*
> ANCIENT PROVERB

9

Fossil Fuel Fundamentals

Ahmed S. Khan

Barbara A. Eichler

Energy is the lifeblood of the global economy. Despite tremendous advances in the various technological fields, fossil fuels still remain the number one source of energy, as they were a century ago. The exponential growth of population during the past fifty years, coupled with inadequate research efforts in developing efficient and environment-friendly energy technologies, has led to an increased demand for the fossil fuels. The world consumes more than 65 million barrels of petroleum each day. By 2015, the consumption will increase to 99 million barrels per day. Table 9.1 presents a summary of world crude oil, coal and natural gas reserves.

Coal, crude oil and natural gas are fossil fuels. These fuels are very precious resources because they are non-renewable. Fossil fuels were formed millions of years ago due to decomposition of plants and animal matter. The fossil fuels are burnt with oxygen from air to produce heat, which in turn is used in a heat engine to produce mechanical power. The increased use of fossil fuels has led to increased carbon emissions, which are not only endangering the flora and fauna, but are also posing serious health, social and economic problems.

Due to lack of technological advances in the renewable energy sources, the use of fossil fuels will continue to dominate the world's energy supply in the foreseeable future. Political leaders of the developed world are still more concerned about economic growth without giving serious thought to addressing the issues of acid rain, smog, and global warming. The 2001 United Nations report on global climate change warns that the world's poorest countries would "bear the brunt of devastating changes" from global warming.

ENERGY PRODUCTION AND CONSUMPTION

Most commercial energy produced is from fossil fuels. Developed countries consume high amounts of oil, coal and natural gas. The United States leads the world in total energy consumption, but Canada and Norway are the highest per capita users. With only 5% of the world's population, the United States consumes about one-quarter of the world's energy. The developing countries also use one-quarter of the world's energy but they represent 77% of world's population. Most of the least developed countries (LDCs) receive a meager energy ration, well below the moderate threshold levels necessary for economic development. They lack resources to buy energy rations to promote development. Poor rural people in the developing countries are forced to use wood for their energy needs, and thus contribute to the severe problems of deforestation, serious environmental degradation and pollution.

FOSSIL FUEL ADVANTAGES AND DISADVANTAGES

Fossil fuels have literally fueled highly accelerated growth since the early 19th century. Coal, charcoal, oil, and gas have provided efficient and powerful energy sources that have been fairly easily accessible from the earth's resources, which have changed world societies' technologies and growth patterns rapidly. With these two advantages of accessibility and power efficiency, the world has

Table 9.1
World Crude Oil, Coal and Natural Gas Reserves, January 1, 2000 (Oil and Gas Journal)

Region/Country	Crude Oil (Billion Barrels)	Recoverable Coal (Million Short Tons)	Natural Gas (Trillion Cubic Feet)
North America	55.1	286,614	261.3
Central & South America	89.5	23,781	222.7
Western Europe	18.8	99,658	159.5
Eastern Europe & Former U.S.S.R.	58.9	288,386	1,999.2
Middle East	675.6	213	1,749.2
Africa	74.9	67,695	394.2
Far East & Oceania	44.0	322,255	363.5
World Total	1,016.8	1,088,602	5,149.6

Source: http://www.eia.doe.gov/emeu/iea/

developed an increasing reliance on fossil fuels, despite looming environmental, resource, and political concerns.

In 1850, wood accounted for nearly 90 percent of world energy use. Coal then became more dominant in the 1890s, where wood and coal each accounted for approximately half of the total. In 1910, coal use expanded to about 60 percent and then shrank as oil and natural gas use increased and assumed significant amounts of world energy production and use (Dunn, 2001, p. 88). Today, crude oil is the most heavily used type of energy, accounting for 43 percent of world oil energy production. Coal production is 25 percent, natural gas 22 percent, with hydroelectric, nuclear power and renewables accounting for the remaining percentages (Blair, 1997, p. 12).

Despite the enormous economic and technological growth of the past 200 years—made possible by the use of fossil fuels—most energy experts caution that this fossil fuel mentality has to be rethought and changed in the early 21st century. There are urgent concerns regarding supply, rapid world economic growth and energy demands (especially in the LDCs) and heavy air pollution effects with possible global warming.

Although world estimates of fossil fuel supplies differ greatly, Barbour states "global reserves of oil will last for 44 years, natural gas for 60 years and coal for three centuries at current depletion rates" (Barbour, 1993, p. 117). Not only are supplies a problem, but the distributions are very unequal, with the Middle East holding approximately 2/3 of the world's oil supply, the former Soviet Union and the U.S. possessing most of the coal, and the U.S., the former Soviet Union and the Middle East possessing most of the natural gas supplies. Such unequal distribution of

sources leads to pricing monopolies, fewer opportunities for the LDCs and, most especially, political tensions and power struggles when supplies become costly or threatened. The central problem, then, is that these fossil fuels are not equally distributed, are not renewable, and at present rates of use cannot support 21st century energy growth projections. According to a recent study by a national energy policy research group, U.S. energy consumption alone is projected to increase by about 32 percent by 2020.

Another key problem is environmental degradation. Fossil fuels are dirty with large quantities of polluting emissions. These cause acidification of rain, lakes, rivers and soils and significant air pollution and air particulants. Coal is the worst polluter, then oil, then gas. Refer to chart below:

Emissions from a 1000 Megawatt Generating Plant (in thousands tons per year)

Fuel	Sulfur Oxide	Nitrogen Oxide	Carbon Dioxide
Coal	70	25	1600
Oil	30	12	1400
Gas	—	15	1100

Source: EPA.

Barbour states that carbon dioxide(CO_2) emissions changes in the atmosphere have been very significant. The burning of all fossil fuels results in the formation of CO_2 and "the enormous quantities released into the air by fossil fuels have increased the CO_2 content of the air by 25 percent in the past hundred years. At the present emission

Storm Brewing Over Energy
Resources & Environmental
Effects. Photo Courtesy of
Ahmed S. Khan

rate, it will have doubled by 2030. Coal is the worst offender, releasing 24 kilograms of CO_2 per billion joules of heat produced, oil is the next (20 kg.) and natural gas produces least (14 kg.). In the United States, coal-burning utilities account for a third of the CO_2 emitted, oil-burning vehicles another third, and fuels burned in homes and industry, the final third" (1993, p. 120). The United States alone is responsible for emitting ¼ of the world's annual emissions of CO_2. Refer to the chart below:

World CO_2 Emissions from Consumption and Flaring of Fossil Fuels, 1999

Country, Rank	Millions of Metric Tons (2,204.62 pounds equal a metric ton)
1. U.S.	1,519.89
2. China	668.73
3. Russia	400.09
4. Japan	306.65
5. India	243.28
6. Germany	229.93
7. United Kingdom	152.39
8. Canada	150.90
9. Italy	121.28
10. France	108.59

Source: U.S. Dept of Energy, Energy Information Administration, International Energy Annual, 1999.

The world adds about 6 billion tons of CO_2 to the atmosphere from fossil fuel combustion each year where sustainable, stable amounts would be about 1 billion. Concern is that such continual CO_2 concentrations added to the environment will cause severe atmospheric changes and raise global warming temperatures, resulting in problems of climate change, ice melts, and other drastic ecosystem patterns.

As can be seen, these are some of the major disadvantages for a continuing global policy of fossil fuel reliance. There are many more arguments to support this position, but these are some of the fundamental and urgent issues surrounding the continued reliance on a fossil fuel energy and economy framework. A report by the National Academy of Science estimates that the U.S., by use of conservation and of more renewable alternative energies, could reduce fossil fuel reliance and carbon emissions by more than half with no reduction of lifestyle.

THE UNITED STATES DEPENDENCE ON FOSSIL FUEL

In 1993, close to 90% of the United States' energy supply came from coal, oil and natural gas. As the developing countries continue to industrialize, competition for the world's finite oil supply will increase, resulting in higher prices. The United States' dependence on fossil fuel is evident by the following facts. (See Table 9.2).

Increase of Fossil Fuel Use Changed the Landscape of the 20th Century. Photo Courtesy of Ahmed S. Khan

- Since 1983, the U.S. has experienced a steady increase in petroleum consumption. The present-day consumption level is equal to the 1978 all-time high petroleum consumption level (18.9 million barrels per day).

- The average horsepower of new vehicles has increased steadily by 63% [From 102 horsepower (1975) to 166 horsepower (1996)].

- U.S. dependence on imported oil has increased to record levels during last 25 years. Net imports provide 48 percent of U.S. oil consumption.

Table 9.2

Energy Consumption in the United States 1999 (Quadrillion Btu)

Energy Type	Consumption Percentage
Wood	3.0%
Hydroelectric	3.5%
Nuclear	8.1%
Coal	22.7%
Natural Gas	23.2%
Petroleum	39.5%

Source: http://www.ec.gc.ca/ind/English/Energy/Tech_Sup/ecsup4_e.cfm

- In 1972, the American Petroleum Institute (API) estimates of crude oil reserves were 36.3 billion barrels.

- In 1996, the Energy Information Administration (EIA) estimates of crude oil reserves were 22 billion barrels.

- Renewable energy was approximately 8.4% of U.S total consumption in 1972. It declined slightly to 7.6% in 1997.

- A new barrel of oil or gas reserves that cost about $15 to find in 1977 cost less than $5 to find today.

- Carbon dioxide emissions grew in all sectors during the past decade. The carbon emissions range between 5 and 6 metric tons per person per year.

- In 2001, for the first time in U.S. car sales history, trucks—pickups, sport-utility vehicles and minivans—outsold cars, taking 50.9% of the market.

- Automakers sold 17.2 million cars and trucks in 2001, compared with 17.4 million in 2000.

COAL

Coal is an organic rock present in the earth's crust. Coal has a very complex molecular structure. It contains carbon, hydrogen, sulfur and nitrogen. Based on the presence of these elements, coal is classified into two types: low-rank coal (lignite and subbituminos have low heating value) and high-rank coal (bituminous and anthracite have high heating value). Low-rank coals are roughly 50–100 million years old and high-rank coals were formed around 350 million years ago. The typical elemental composition of bituminous coal is as follows: 82% carbon, 9% oxygen, 5% hydrogen, 3% sulfur, 1% nitrogen. Since bituminous coal has higher heating value (13,000 Btu-15, 000 Btu per pound), most electric power plants prefer it to the low-rank coals. But the bituminous coal contains large amount of sulfur (up to 4%), which is converted to sulfur dioxide on combustion, and therefore contributes to the problem of acid rain.

U.S. coal reserves are estimated at more than 1,600 billion tons, the equivalent of more than six trillion barrels of oil. The United States is estimated to hold 29% of the world's coal reserves, the highest percentage of any county. Table 9.3 lists the World Coal Supply and Disposition.

The use of recent advances in ultra-clean fuel technology (coal gasification/liquefaction) can lower the overall amount of greenhouse gases introduced into the atmosphere and reduce the dependence on foreign oil. The coal gasification/liquefaction processes utilize carbonaceous matter (coal, coal waste, biomass, refinery waste and other materials) to produce liquid products that are environmentally friendly, known as ultra-clean fuels.

Advantages

- Coal is one of the most abundant energy resources.
- Coal is versatile; it can be burned directly or transformed into liquid, gas or feedstock.
- Inexpensive compared to other energy sources.

Drawbacks

- Source of pollution. Coal burning emits sulfur dioxide, nitrogen oxide and particles to the atmosphere, and leaves a residue of solid waste.
- Coal mining, especially strip-mining, can be very unsightly, and abandoned mines have marred the landscape.
- Coal liquefaction and gasification require large amount of water.
- Coal is bulky, so it is more difficult to transport and burn than liquid or gases.
- Coal is porous. It has water trapped in the pores (as much as 30% by weight), which reduces the heating value (for low-rank coals the heating value is less than 8,000 Btu per pound).
- Processes for making liquids and gas from coal are not fully developed.

PETROLEUM (CRUDE OIL)

Petroleum is a complex liquid mixture which contains hundreds of compounds. The majority of these compounds contain carbon and hydrogen, and thus have high heating value. Typical elemental analysis of crude oils are as follows: 83–87% carbon, 11–16% hydrogen, 0–7% + nitrogen, 0–4% sulfur. The yield of an oil well depends upon the type and age of the oil field. For example, a typical output of an oil well in Saudi Arabia is about 10,000 barrels per day whereas on average an oil well in the United States produces 15 barrels per day.

Table 9.3
World Coal Supply and Disposition, 1998 (Trillion Btu)

Region/Country	Production	Apparent Consumption
North America	25,918	23,461
Central & South America	1,223	950
Western Europe	5,614	10,263
Eastern Europe & Former U.S.S.R.	11,318	10,834
Middle East	24	289
Africa	5,447	3,723
Far East & Oceania	39,819	39,738
World Total	89,363	89,258

Source: http://www.eia.doe.gov/emeu/iea/

Crude oil, after its recovery from reservoirs in the earth, is refined by fractional distillation into various fuel products: gasoline (fuel for spark-ignition engines), diesel fuel (fuel for compression-ignition engines), kerosene (fuel for jet engines), fuel oils (fuel for industrial and residential furnaces). A very small remaining fraction is used to produce petrochemicals (used in pharmaceuticals, cosmetics, plastics, detergents, textiles, etc.). Table 9.4 presents a summary of World Petroleum Supply and Disposition.

Advantages

- Oil is one of the most abundant energy resources.
- Its liquid form makes it easy to transport and use.
- Oil has high heating value.

Drawbacks

- Oil burning leads to carbon emissions.
- Finite sources (some experts disagree).
- Oil recovery processes need to be developed to provide better yields.
- Oil drilling endangers the environment and ecosystems.
- Oil transportation (by ship) can lead to spills, causing environmental and ecological damage.

NATURAL GAS

Natural gas, also known as methane, is a colorless, odorless, fuel. It is one of the most commonly used sources of energy today. Natural gas is produced by drilling into the earth's crust where pockets of gas were trapped hundreds of thousands of years ago. One of the biggest advantages of natural gas is that it burns cleaner than many other forms of fossil fuels (coal and oil). It can help improve the quality of air and water. Natural gas combustion results in virtually no atmospheric emissions of sulfur dioxide or small particulate matter and far lower emissions of carbon monoxide, reactive hydrocarbons, nitrogen oxides and carbon dioxide than combustion of other fossil fuels (coal and oil). When natural gas (methane, a molecule made of one carbon atom and four hydrogen atoms) is burned, the principal products of combustion are carbon dioxide and water vapor. The burning of other fossil fuels (oil and coal) produce ash particles, which do not burn and can be carried into the atmosphere. Since burning of natural gas does not produce any reactive hydrocarbons, it can help improve air quality by not producing acid rain and damaging the ozone levels. Table 9.5 presents a summary of World Dry Natural Gas Supply and Disposition.

The burning of fossil fuels accounts for 75–80 percent of carbon dioxide emissions, and 20–30 percent of methane emissions into the environment. Burning of natural gas emits up to 45 percent less carbon dioxide than other fuels. If newer high-efficiency burning techniques are employed, the reduction in carbon dioxide emission can be nearly 70%.

Natural gas can make a significant contribution towards improving air quality if it is employed in the transportation sector. Advanced natural gas-fueled vehicles have the potential to reduce carbon monoxide by as much as 90 percent and reactive hydrocarbon emissions by as much as 85 percent, compared with gasoline vehicles. About 40,000 natural gas vehicles are in operation in the United States, and more than 500,000 around the world.

Table 9.4

World Petroleum Supply and Disposition, 1998 (Thousand Barrels per Day)

Region/Country	Oil Production	Apparent Consumption
North America	15,494	22,715
Central & South America	6,974	4,859
Western Europe	6,999	14,985
Eastern Europe & Former U.S.S.R.	7,458	4,968
Middle East	22,454	4,339
Africa	7,823	2,463
Far East & Oceania	7,922	19,312
World Total	75,124	73,642

Source: http://www.eia.doe.gov/emeu/iea/

Table 9.5
World Dry Natural Gas Supply and Disposition, 1998 (Billion Cubic Feet)

Region/Country	Dry Gas Production	Apparent Consumption
North America	26,021	25,414
Central & South America	3,120	3,120
Western Europe	9,638	13,999
Eastern Europe & Former U.S.S.R.	25,163	22,208
Middle East	6,596	6,237
Africa	3,699	1,836
Far East & Oceania	8,551	9,083
World Total	82,788	81,896

Source: http://www.eia.doe.gov/emeu/iea/

A "real-world" test of a natural gas vehicle (NGV), conducted by the Gas Research Institute, showed that, depending on the outside air temperature, emission of non-methane organic gases can be lowered to as low as 95% of that of gasoline or even reformulated-gasoline, and the emissions of nitrogen oxides can be lowered as much as 60%.

Fuel cell technology is one of the most environment-friendly advances in natural gas technology. NASA first used these cells in the 1960s to generate power in the space capsules. Fuel cells rely on the chemical interaction of natural gas and certain other metals like platinum, gold and other electrolytes to produce electricity. The only byproduct is water. The high cost of fuel cell technology has impeded the growth of its implementation. Fuel cells are being used in hospitals to generate power and are also being considered for use in vehicles.

Advantages

- Inexpensive compared to oil.
- Clean to burn, less polluting than other fossil fuels.
- Burning does not produce any ash particles.
- Has high heating value (about 24,000 Btu per pound).

Drawbacks

- Not a renewable source.
- Finite resource trapped in earth (some experts disagree).
- Inability to recover all in-place gas from a producible deposit because of unfavorable economics and lack of technology.

FUTURE STRATEGIES FOR A SUSTAINABLE ENVIRONMENT

As we approach the end of the cheap fossil fuel age, we must transform our lifestyles from a growth-oriented (infinite resources and energy, local and national outlook) to a balance-oriented lifestyle (finite resources and infinite energy with developing efficient renewable energy technologies, global outlook) by exploring the following strategies.

Conservation   Energy conservation should be the central focus of all energy strategies for meeting the future energy needs.

Alternate fuels   Allocation of funds to improve efficiency of renewable energy technologies (solar, wind, etc).

Development of new technologies   Development of high-temperature superconductors could reduce a large amount of losses in the transmission grid and thus could provide economical electricity.

Education   Teaching the importance of responsible use of energy at personal, national and international levels.

Environment-friendly policies   Public policies, which promote conservation and encourage use of alternate fuels, need to be developed and implemented at the national and international level.

An increased consumption of oil in the developed world, coupled with the uneven global distribution of reservoirs, has led to various wars during the 20th century. Unless serious strategies and plans are implemented to develop efficient alternate energy technologies, the deterioration of the environment will continue, and the peril of future wars over oil will loom.

In numerous ways, energy has improved our quality of life, but we have paid a heavy price in the form of irreversible damage to the environmental and ecological systems of our biosphere. Earth's atmosphere is a precious environment. In the infinite cosmic ocean, the earth remains the only planet which sustains life—let's keep its uniqueness.

QUESTIONS

1. Use any of the Internet search engines (e.g. Alta Vista, Yahoo, Infoseek, etc.) to research the following:
   (i) Define the following terms:
       a. Bcf
       b. Brine
       c. Btu
       d. CFCs
       e. Fuel cell
       f. Hydrocarbon
       g. Lithography
       h. Mcf
       i. methane
       j. NGV
       k. NES
       l. Lignite
       m. Subbituminos
       n. Bituminous
       o. Anthracite
   (ii) Statistics for carbon dioxide emissions, from the consumption and flaring of fossil fuels, for the following countries:
       a. Canada
       b. United States
       c. Mexico
       d. Chile
       e. Cuba
       f. France
       g. Germany
       h. Ireland
       i. United Kingdom
       j. Turkey
       k. Former U.S.S.R
       l. Oman
       m. Gabon
       n. Zimbabwe
       o. Pakistan
       p. India
       q. Bangladesh
       r. Vietnam

2. What effect does burning fossil fuels have on the environment? What are various strategies to reduce global warming?

3. Refer to Figure 9.1. Estimate how much you could realistically reduce your personal use of electricity and hence fossil fuel emissions on a yearly basis. Understand the urgency to help to reduce emissions. and try to apply that estimate.

4. List the advantages ($+$) and disadvantages ($-$) of coal:

Advantages ($+$)
_____
_____
_____
_____

Disadvantages ($-$)
_____
_____
_____
_____

5. List the advantages ($+$) and disadvantages ($-$) of natural gas.

Advantages ($+$)
_____
_____
_____
_____

Disadvantages ($-$)
_____
_____
_____
_____

Figure 9.1

Ten Steps You Can Take To Reduce Global Warming

As stated above, the U.S. with 5% of the world's population generates about 25 percent of the "greenhouse gases" that are polluting the atmosphere. This is about 40,000 pounds of CO_2 per person each year. These are some measures you can take to reduce your consumption of fossil fuels, along with the amount of CO_2 you'll avoid releasing into the atmosphere.

What You Can Do	Estimated Pounds of CO_2 Saved Per Year
1. Run your dishwasher with a full load and use the air option rather than heat to dry the dishes.	200
2. Wash clothes in warm or cold water, not hot.	Up to 500
3. Ask your utility company for a home energy audit to find out where your home is poorly insulated or energy-inefficient.	Potentially could cut thousands of pounds per year
4. Plant trees next to your home and paint your home a light color if you live in a warm climate and a dark color if you live in a cold climate.	Up to 10,000 pounds per year
5. As you replace home appliances, select the most energy efficient.	Potentially could cut thousands of pounds per year
6. Buy and replace your incandescent light bulbs with compact fluorescent bulbs.	500 pounds per year for each bulb
7. Choose a car with good gas mileage.	2,500 for each 10 mpg improvement
8. Buy minimally packaged goods; choose reusable products over disposable ones; recycle.	1000 if waste reduced by 25%
9. Leave your car at home twice a week (walk, bike, or use mass transit instead)	Up to 1590 pounds per year
10. Turn down your water heater thermostat; 120 degrees is usually hot enough.	For each 10-degree adjustment, you cut 500 pounds per year.

Source: Environmental Defense Fund.

6. List the advantages (+) and disadvantages (−) of petroleum (crude oil).

7. How can we avoid a future oil crisis and oil wars? List and explain three ways.

Advantages (+)
_____
_____
_____
_____

Disadvantages (−)
_____
_____
_____
_____

REFERENCES

Barbour, Ian. (1993). *Ethics in an Age of Technology.* New York: Harper Collins Publishers.

Blair, Cornelia, Landes, Alison, and Jacobs, Nancy, (eds). (1997). *Energy: An Issue of the 90's.* Wylie, Texas: Information Plus.

COAL: The Fuel of America's Industrialization, The Fuel of America's Future. [Online]. Retrieved February 9, 2002. Available at http://www.ultracleanfuels.com/html/about.html

de Souza, Anthony F. (1990). Resources. *Geography of World Economy.* Columbus, Ohio: Merrill Publishing.

Dunn, Seth. (2001) "Decarbonizing the Energy Economy." In Lester R. Brown, Christopher Flavin and Hilary French, et.al, *State of the World, 2001.* New York: Norton.

The Environmental Impact of Natural Gas [Online]. Retrieved February 9, 2002. Available at http://www.naturalgas.org/ENVIRON2.HTM

Eldridge, Earle. (2002) *"2001 car sales rank second best ever."* USA TODAY (01/04/2002) [Online]. Retrieved February 12, 2002. Available at http://www.usatoday.com/money/autos/2002/01/04/autos.htm

Fossil Fuels. [Online]. Retrieved February 9, 2002. Available at http://www.ems.psu.edu/~radovic/fossil_fuels.html

Overview of Natural Gas [Online]. Retrieved February 9, 2002. Available at http://www.naturalgas.org/ENVIRON2.HTM

UN Issues Global Warning: Rich Countries Must Cut Fossil Fuels. [February 19, 2001]. [Online]. Retrieved February 9, 2002. Available at http://www.foe.org.au/pr/190201.htm

World Resources 1998–89, A Report by Resources Institute. New York, Basic Books.

10

Energy for a New Century

CHRISTOPHER FLAVIN

The stone age did not end because the world ran out of stones, and the oil age will not end because we run out of oil.

DON HUBERTS, SHELL HYDROGEN
(DIVISION OF ROYAL DUTCH SHELL)

The age of oil has so dominated social and economic trends for the last 100 years that most of us have a hard time imagining a world without it. Oil is cheap, abundant, and convenient—easy to carry halfway around the world in a supertanker or across town in the tank of a family sport utility vehicle. From Joe Sixpack to the PhD energy economists employed by governments and corporations, we tend to assume that we will burn fossil fuels until they're gone, and that the eventual transition will be painful and expensive.

But if you turn the problem around, our current energy situation looks rather different: from an ecological perspective, continuing to depend on fossil fuels for even another 50 years—let alone the century or two it might take to use them up—is preposterous. As the new century begins, the world's 6 billion people already live with the dark legacy of the heavily polluting energy system that powered the last century. It is a legacy that includes impoverished lakes and estuaries, degraded forests, and millions of damaged human lungs.

Fossil-fuel combustion is at the same time adding billions of tons of carbon dioxide to the atmosphere each year, an inexorable escalation that must end soon if we are not to disrupt virtually every ecosystem and economy on the planet.

An energy transition in the new century is therefore ecologically necessary, but it is also economically logical. The

same technological revolution that has created the Internet and so many other 21st century wonders can be used to efficiently harness and store the world's vast supplies of wind, biomass, and other forms of solar energy—which is 6,000 times as abundant on an annual basis as the fuels we now use. A series of revolutionary technologies, including solar cells, wind turbines, and fuel cells can turn the enormously abundant but diffuse flows of renewable energy into concentrated electricity and hydrogen that can be used to power factories, homes, automobiles, and aircraft.

These new energy conversion devices occupy about the same position in the economy today that the internal combustion engine and electromagnetic generator held in the 1890s. The key enabling technologies have already been developed and commercialized, but they only occupy small niche markets—and their potential future importance is not yet widely appreciated. As with the automobile and incandescent lightbulb before them, the solar cell and hydrogen-electric car are steadily gaining market share—and may soon be ready to contribute to a third energy revolution. They could foster a new generation of mass-produced machines that efficiently and cleanly provide energy needed to take a hot shower, sip a cold beer, or surf the Internet.

Thanks to a potent combination of advancing technology and government incentives, motivated in large measure by environmental concerns, the once glacial energy markets are now shifting. During the 1990s, wind power has grown at a rate of 26 percent per year, while solar energy has grown at 17 percent per year. During the same

Reprinted with permission from Worldwatch magazine (www. worldwatch.org) Worldwatch Institute, Washington, D.C.

period, the world's dominant energy source—oil—has grown at just 1.4 percent per year.

Wind and solar energy currently produce less than 1 percent of the world's energy, but as the computer industry long ago discovered, double-digit growth rates can rapidly turn a tiny sector into a giant. In the past two years, perhaps a dozen major companies have joined Royal Dutch Shell in announcing major new investments in giant wind farms, solar manufacturing plants, and fuel cell development. The "alternative" energy industry is beginning to take on the same kind of buzz that surrounded John D. Rockefeller's feverish expansion of the oil industry in the 1880s—or Bill Gates's early moves in the software business in the 1980s. This January, stocks of solar and fuel cell companies suddenly jumped several-fold in a month, following the pattern of Internet stocks.

The 21st century may be as profoundly reshaped by the move away from fossil fuels as the 20th century was shaped by them. Energy markets, for example, could shift abruptly, drying up sales of conventional power plants and cars in a matter of years, and influencing the share prices of scores of companies. The economic health—and political power—of whole nations could be boosted, or in the case of the Middle East, sharply diminished. And our economies and lifestyles are likely to become more decentralized with the advent of new energy sources that provide their own transportation network—for example, the sunshine that already falls on our rooftops.

How quickly the world's energy economy is transformed will depend in part on whether fossil-fuel prices remain low and whether the opposition of many oil and electric power companies to a new system can be overcome. The pace of change will be heavily influenced by the pace of international negotiations on climate change and of the national implementation plans that follow. In the 1980s, California provided tax incentives and access to the power grid for new energy sources, which enabled the state to dominate renewable-energy markets worldwide. Similar incentives and access have spurred rapid market growth in several European countries in the 1990s. Such measures have begun to overcome the momentum of a century's investment in fossil fuels.

Earth Day 2000—with its central theme, "Clean Energy Now!"—provides a timely opportunity for citizens to express their desire for a new energy system, and to insist that their elected officials implement the needed policy changes. If they do so, smokestacks and cars may soon look as antiquated as manual typewriters and horse drawn carriages do.

QUESTIONS

1. Comment on the author's proposed new energy system. What key energy technologies does the author recommend for a sustainable environment?

2. What percent of the world's total energy is produced by solar and wind energy technologies?

3. What needs to be done at a personal level, community level, national level and international level to create a new energy system?

11

The Immortal Waste

Ahmed S. Khan

Barbara A. Eichler

If there ever was an element that deserved a name associated with hell, it is plutonium. This is not only because of its use in atomic bombs—which certainly would amply qualify it—but also because of its fiendishly toxic properties, even in small amounts.

Robert E. Wilson

Since December 1942—when Enrico Fermi successfully initiated the first nuclear fission reaction (Table 11.1)—millions of tons of nuclear waste have been accumulated as a result of a global race for power generation and weapon production. Today, in thirty nations 426 nuclear power reactors generate one sixth of the world's electricity production. Ninety-seven power reactors are under construction. A typical 1000 MW power reactor generates a large amount of nuclear waste. However, the significant part of the nuclear waste generated world wide is contributed by the nuclear weapons production. The International Atomic Agency (IAEA) has classified nuclear waste into three categories: high, intermediate and low level nuclear waste. (Table 11.2). The classification is based on waste's source, temperature and half-life.

For the last fifty years scientists have proposed various methods and techniques (Tables 11.3–11.4) for the permanent storage of nuclear waste but various governments around the world have not given serious consideration to this problem. As a result of this neglect, 80,000 tons of irradiated fuel and hundreds of thousands of tons of radiated waste are sitting in temporary storage sites.

Unlike chemical waste, whose toxicity can be neutralized or reduced by various techniques in order to minimize the adverse effects on environment and humans, the nuclear waste is radioactive (Table 11.5) and the potency of radioactivity can only be eliminated by natural decay. Depending on the type of element, it can take hundreds and thousands or even millions of years (Tables 11.6–11.8) for material decay to occur.

Accidental or intentional released radiation can rapidly spread through air and water to contaminate the environment. Radioactive waste from Soviet and U.S. weapons facilities has spread thousands of kilometers from the source contaminating the environment and people.

According to a 1991 study commissioned by International Physicians of Nuclear War "the fallout from the atmospheric atomic bomb testing has spread around the globe and will eventually cause an estimated 2.4 million cancer deaths." The accident at Chernobyl released 50 million curies into the environment, whereas the bombing of Hiroshima and Nagasaki released an estimated 1 million curies. The radiation released from Chernobyl will be responsible for an estimated 14,000 to 475,000 cancer deaths.

Nuclear energy provides over 17 percent of the world's electricity and displaces approximately 6 million barrels of oil per day. Today 426 nuclear power plants in 30 countries generate 318,271 MW. By 2000, in 32 countries 511 nuclear plants are estimated to generate 406,226 MW. In the U.S. 110 nuclear plants provide 19% of electricity. By 2000 more than 20% of electricity will be provided by 119 nuclear power plants. At first look the nuclear energy appears to be clean (i.e., it does not contribute to atmospheric emissions, but it generates the immortal nuclear waste, which possesses the greatest threat to the environment and people).

At present there is no permanent depository for high level nuclear wastes in the world. Yet there are about 426 nuclear power plants operating worldwide. Each nuclear power plant yields about 30 tons of high level waste. Therefore, in very simplistic terms not accounting

Table 11.1
Nuclear Energy: Chronology of Research and Development

1905	Albert Einstein developed the theory of the relationship between mass and energy. $E = mc^2$ i.e., energy is equal to the mass times the square of the speed of light.
1932	Preliminary work done by Frederic and Joliot-Curie led to the discovery of the neutron by James Chadwick of England.
1934	Enrico Fermi carried out a series of experiments in Rome that showed neutrons could cause the fission of many kinds of elements including uranium atoms.
1938	German scientists Otto Hahn and Fritz Strassman bombarded uranium with neutrons from a radium–beryllium source, and discovered a radioactive barium isotope among residual material. This indicated a new type of reaction–*fission*–took place. Some mass of uranium was converted to energy, thereby verifying Einstein's theory.
Dec. 2, 1941	A group of scientists led by Fermi initiated the first self-sustaining nuclear chain reaction in a laboratory at the University of Chicago.
July 16, 1945	The first atomic bomb was tested at Alamogordo, New Mexico, by the U.S. army under the code name "Manhattan Project."
Aug. 1, 1946	The Atomic Energy Act of 1946 established the AEC to control nuclear energy development and explore the peaceful uses of nuclear energy.
Dec. 20, 1951	The experimental Breeder Reactor I at Arco, Idaho, produced the first electric power from nuclear energy.
Dec. 8, 1953	President Eisenhower delivered his "Atoms for Peace" speech before the United Nations.
Jan. 21, 1954	The U.S. Navy launched the first nuclear-powered submarine, the U.S.S. *Nautilus,* which was capable of cruising 115,000 km (62,500 nautical miles) without refueling.
Aug. 30, 1954	President Eisenhower signed the Atomic Energy Act, which permitted and encouraged the participation of private industry in the development and use of nuclear energy and permitted greater cooperation with U.S. allies.
Jan. 10, 1955	The Atomic Energy Commission (AEC) announced the Power Demonstration Reactor Program. The AEC would cooperate with industry to construct and operate experimental nuclear power reactors.
Aug. 8–20, 1955	The first United Nations International Conference on peaceful uses of atomic energy was held in Geneva, Switzerland.
Sept. 2, 1957	The Price-Anderson Act granted financial protection to the public and to AEC licensees and contractors in the occurrence of a major accident involving a nuclear power plant.
Oct. 1, 1957	The International Atomic Energy Agency (IAEA) is established in Vienna, Austria, by the United Nations to promote the peaceful use of nuclear agency.
Dec. 2, 1957	The world's first large-scale nuclear power plant began operations in Shippingport, Pennsylvania.
July 21, 1959	The world's first nuclear powered merchant ship N.S. Savannah was launched in Camden, New Jersey.
Nov. 25, 1961	The U.S. Navy commissioned the world's largest ship, the U.S.S. *Enterprise*—a nuclear powered aircraft carrier capable of cruising up to 20 knots for distances up to 740,800 km (4,000,000 nautical miles) without refueling.
April 3, 1965	The first nuclear reactor in space (SNAP-10A) was launched.
March 5, 1970	The Treaty for Non-proliferation of Nuclear Weapons was ratified by the United States, the United Kingdom, the Soviet Union and 45 other nations.
1971	A total of 22 commercial nuclear power plants were in full operation in the United States.
August 1974	The federal government released the results of a safety study by Dr. Norman Rasmussen of MIT, which concluded that a meltdown in a power reactor would be extremely unlikely.

(continued)

Table 11.1
Nuclear Energy: Chronology of Research and Development, *continued*

Oct. 11, 1974	The Energy Reorganization Act of 1974 divided AEC functions between two newly formed agencies—the Energy Research and Development Administration (ERDA) and Nuclear Regulatory Commission (NRC).
1976	A total of 61 nuclear power plants with an aggregate capacity of 42,699 megawatts were producing 8.3% of electricity generated in the U.S.
March 28, 1979	The worst accident in a commercial reactor occurred at Three Mile Island (TMI) nuclear power station near Harrisburg, Pennsylvania. The cause was the loss of coolant from the reactor core due to mechanical and human errors. Without the cooling water surrounding the fuel, its temperature exceeded 5,000 degrees Farenheit, causing melting and damage to the reactor core. Due to the accident, the radioactive material normally confined to the fuel was released into the reactor's cooling water system.
1979	After the Three Mile Island accident the Nuclear Regulatory Commission (NRC) imposed stricter safety regulation and more rigid inspection procedures in order to improve the safety of nuclear reactors.
	The 12 percent of electricity produced commercially in the United States was generated by 72 licensed nuclear reactors.
1981	The Shippingport power station was shut down after 25 years of service.
1984	A total of 83 nuclear power reactors generated 14 percent of the electricity produced in the United States.
1986	The worst accident in nuclear history—the Chernobyl disaster—took place in the former Soviet Union on April 26.

Table 11.2
Nuclear Waste Classification

Type	Characteristics
High Level	Produced in two ways:
	(a) By reprocessing spent fuel to recover isotopes that can be used again as fuel.
	(b) Reactor fuel rods which contain long lived isotopes (with half-life of 30 years or more) Contains transuranic (heavier than uranium) elements.
	Most are harmful, highly radioactive, and must be handled and transported with shielding. Must be isolated from humans and environment for thousands of years.
	By 1990, 26,400 cubic meters of high level waste was generated.
Intermediate Level	Produced as reactor by-products and other material such as equipment, tools, etc. that have become radioactive.
	Less harmful, but cannot be handled and transported without shielding.
	By 1990, 3,400 cubic meters of Intermediate level waste was generated. IAEA estimates that by 1995 the rate of generation will be 3,800 cubic meters per year.
Low Level	Produced due to contamination of metal, paper, rags, etc. by radioactive material at power reactors, medical equipment, and other non-military sources.
	Contains no transuranic elements.
	Least harmful, can be handled and transported without shielding.
	By 1990, 370,000 cubic meters of low level waste was generated.

Source: "Nuclear Waste: The Challenge is Global," *IEEE Spectrum,* July 1990.

Table 11.3
Technical Options for Dealing with Irradiated Fuel

Method	Process	Problems	Status
Antarctica Ice Burial	Bury waste in ice cap	Prohibited by international law; low recovery potential, and concern over catastrophic failure	Abandoned
Geologic Burial	Bury waste in mined repository hundreds of meters deep	Difficulty predicting geology, groundwater flows, and human intrusions over long time period	Under active study by all nuclear countries as favored strategy
Long-term Storage	Store waste indefinitely in specially constructed buildings	Dependent on human institutions to monitor and control access to waste for long time period	Not actively being studied by gov'ts, though proposed by nongovernmental groups
Reprocessing	Chemically separate uranium and plutonium from fission products in irradiated fuel; decreases radioactivity by 3 percent	Increases volume of waste by 160-fold; poor economics; increases risk of nuclear weapons proliferation	Commercially under way in four countries; total of 16 countries have reprocessed or plan to reprocess irradiated fuel
Seabed Burial	Bury waste in deep ocean sediments	Possibly prohibited by international law; transport concerns; nonretrievable	Under active study by consortium of 10 countries
Space Disposal	Send waste into solar orbit beyond earth's gravity	Potential launch failure could contaminate whole planet; very expensive	Abandoned
Transmutation	Convert waste to shorter-lived isotopes through neutron bombardment	Technically uncertain whether waste stream would be reduced; very expensive	Under active study by United States, Japan, Soviet Union, and France

Source: Worldwatch Institute, *State of the World,* 1992, Norton.

for various types of reactors, the world's high level nuclear waste approaches 15,000 tons annually.

In 1988, the United States proposed its only licensed permanent high level waste dumping site 1200 feet under Yucca Mountain, Nevada, for its 110 operating nuclear plants. Since the proposal of Yucca mountains, however, there remains little progress in this site becoming a reality since the plan is fraught with problems and technical difficulties. In the meantime, the U.S. Department of Energy says it cannot offer a permanent storage site until 2010 at the earliest and the wastes continue to pile up. Some of the problems associated with the Yucca mountain plan are as follows.

LOCATION PROBLEMS

- The Department of Energy did not follow Nevada procedures and did not receive proper state authorization for the site.

- Nevada does not want the repository—NIMBY (Not in My Back Yard)—and passed a resolution in 1990 against storing radioactive wastes anywhere in Nevada's borders.

- The dump site was based wholly on political considerations and science's role in this decision was minimal.

- There is a young active volcano within 7 miles of the site and 32 active faults on the site itself. This renders the site

Table 11.4
Selected Country Programs on High-Level Waste Burial

Country	Earliest Planned Year	Status of Program
Argentina	2040	Granite site at Gastre, Chubut, selected.
Belgium	2020	Underground laboratory in clay at Mol.
Canada	2020	Independent commission conducting four-year study of government plan to bury irradiated fuel in granite at yet-to-be-identified site.
China	none announced	Irradiated fuel to be reprocessed; Gobi desert sites under investigation.
Finland	2020	Field studies being conducted; final site selection due in 2000.
France	2010	Three sites to be selected and studied; final site not to be selected until 2006.
Germany	2008	Gorleben salt dome sole site to be studied.
India	2010	Irradiated fuel to be reprocessed; waste stored for 20 years, then buried in yet-to-be-identified site.
Italy	2040	Irradiated fuel to be reprocessed and waste stored for 50–60 years before burial in clay or granite.
Japan	2020	Limited site studies; cooperative program with China to build underground research facility.
Netherlands	2040	Interim storage of reprocessing waste for 50–100 years before eventual burial, possibly in the seabed or in another country.
Soviet Union	none announced	Eight sites being studied for deep geologic disposal.
Spain	2020	Burial in unidentified clay, granite, or salt formation.
Sweden	2020	Granite site to be selected in 1997; evaluation studies under way at Aspo site near Oskarshamn nuclear complex.
Switzerland	2020	Burial in granite or sedimentary formation at yet-to-be-identified site.
United States	2010	Yucca Mountain, Nevada, site to be studied and, if approved, receive 70,000 tons of waste.
United Kingdom	2030	Fifty-year storage approved in 1982; explore options including seabed burial.

Source: Worldwatch Institute, *State of the World,* 1992, Norton.

Table 11.5
Types of Radiation

Mode of Radiation	Sources	Penetrating Power	Approx. Distance Traveled in Air	Shielding Material
Alpha α	Fission and fission products	Very small	5 cm	Paper
Beta β	Fission, fission products, activation products	Small	300 cm at 1 MeV	Water, plastic, wood
Gamma γ	Fission, fission products, activation products	Very large		Lead, plastic, paraffin
Neutron n	Fission	Very large		Water, plastic, paraffin

Table 11.6
Radioactive Decay

Type of Radiation	Nuclide	Half-Life
α	uranium-238	4.47 billion years
β	thorium-234	24.1 days
β	protactinium-234	1.17 minutes
α	uranium-234	245,000 years
α	thorium-230	8,000 years
α	radium-226	1,600 years
α	radon-222	3.823 days
α	polonium-218	3.05 minutes
β	lead-214	26.8 minutes
β	bismuth-214	19.7 minutes
α	polonium-214	0.000164 second
β	lead-210	22.3 years
β	bismuth-210	5.01 days
α	polonium-210	138.4 days
	lead-206	stable

Source: *Radiation Doses, Effects and Risks,* United Nations Environment Programs, Nairobi, Kenya.

Table 11.7
Half-Lives of Radioactive Elements

Element (Symbol-Mass No)	Half-life (years)	Decay Mode
Uranium		
U-232	72	α, β
U-233	1.59×10^5	α, β
U-235	7.03×10^8	α, β
U-236	2.34×10^7	α, β
U-238	4.46×10^9	α, β
U-239	23.5 minutes	α, β
Plutonium		
Pu-239	2.41×10^4	α, β
Tellurium		
Te-130	2×10^{21}	β
Indium		
In-115	5.1×10^{14}	β

unstable and also suggests that through these faults, contaminated water could escape through a network of geologic cracks. (Federal requirements prohibit the construction of a nuclear waste repository where water can travel 5 km from the burial site in less than 100 years.)

- The site is probably among the most highly mineralized areas on the continent. Two of North America's largest gold mines are 15 to 20 miles away, and gold and silver have been found at Yucca Mountain, making the site vulnerable to prospectors.

- Two years of planning have progressed for this site.

TECHNICAL PROBLEM

- The site will accommodate 63,500 tons of high-level waste. Radioactive waste already exceeds 22,500 tons (which, according to the Department of Energy, should remain isolated from the environment for 10,000 years). It would take 28 years every workday to fill the site. By the end of 28 years there will be tens of thousands more tons of waste to deal with and there will be no room for the new waste. Additionally, trucks and other means of moving the waste would be arriving from all parts of the country at 90 minute or more frequent intervals, posing serious safety and transportation questions.

During the past half-century, a number of major nuclear accidents have taken place at various nuclear reactors in the Soviet Union and the United States (Table 11.9), resulting

Table 11.8
Radioactivity and Thermal Output Per Metric Ton of Irradiated Fuel from a Light-Water Reactor

Age (years)	Radioactivity (curies)	Thermal Output (watts)
At Discharge	177,242,000	1,595,375
1	693,000	12,509
10	405,600	1,268
100	41,960	299
1,000	1,752	55
10,000	470	14
100,000	56	1

Sources: Ronnie B. Lipschultz, *Radioactive Waste: Politics, Technology and Risk* (Cambridge, MA: Ballinger Publishing Company, 1980); J.O. Biomeke et al., Oak Ridge National Laboratory, *Projections of Radioactive Wastes to Be Generated by the U.S. Nuclear Power Industry,* National Technical Information Service, Springfield, VA, February 1974.

Table 11.9
Major Nuclear Accidents

United States	Soviet Union
1951, Detroit Accident in a research reactor. Overheating of fissionable material because permissible temperatures had been exceeded. Air contaminated with radioactive gases.	**September 1957** Accident at reactor near Chelyabinsk. A spontaneous nuclear reaction occurred in spent fuel, causing a substantial release of radioactivity. Radiation spread over a wide area. The contaminated zone was enclosed within a barbed wire fence, and ringed by a drainage channel. The population was evacuated and the topsoil removed; livestock was destroyed and buried in pits.
24 June 1959 Meltdown of part of fuel rods due to failure of cooling system at experimental power reactor in Santa Susanna, California.	
3 January 1961 Steam explosion at an experimental reactor near Idaho Falls, Idaho. Three people died.	**7 May 1966** Prompt neutron power surge at a nuclear power station with a boiling-water nuclear reactor in the town of Melekess. Two people were exposed to severe doses of radiation.
5 October 1966 Partial core melt due to failure of cooling system at the Enrico Fermi reactor, near Detroit.	
19 November 1971 Almost 53,000 gallons (200,000 liters) of water contaminated with radioactive substances from an overflowing waste storage tank at Monticello, Minnesota, flowed into the Mississippi River.	**7 January 1974** Explosion of reinforced concrete gasholder for the retention of radioactive gases in No. 1 reactor of Leningrad nuclear power station.
28 March 1979 Core melt due to loss of cooling at the Three Mile Island nuclear power station. Radioactive gases released into the atmosphere and 172,000 cubic feet of liquid radioactive waste was discharged into the Susquehanna River. Population evacuated from vicinity of disaster.	**6 February 1974** Rupture of intermediate loop in No. 1 reactor at the Leningrad nuclear power station due to boiling of water. Three people were killed. Highly radioactive water with pulp from filter powder discharged into the environment.
	October 1975 Partial destruction of the core at No. 1 reactor of the Leningrad nuclear power station. About one and a half million curies of highly radioactive gases were discharged into the environment.
7 August 1979 About one hundred people received a radiation dose six times higher than the normal permissible level due to the discharge of highly enriched uranium from a plant producing nuclear fuel near the town of Irving, Texas.	**31 December 1978** No. 2 unit at the Byeloyarsk nuclear power station was heavily damaged by a fire started when a roof panel in the turbine fell onto a fuel tank. The reactor was out of control. In the effort to supply emergency cooling water to the reactor, eight persons were exposed to severe doses of radiation.
25 January 1982 A broken tube in a steam generator in the R.E. Ginna nuclear power plant, near Rochester, New York. A breakdown in the cooling system caused a leak of radioactive substances into the atmosphere.	**October 1982** Explosion of generator in No. 1 reactor of the Armyanskaya nuclear power station. The turbine hall burned down.
30 January 1982 Near the town of Ontario, New York, a breakdown in the cooling system caused a leak of radioactive substances into the atmosphere.	**September 1982** Destruction of the central fuel assembly of No. 1 reactor at the Chernobyl nuclear power station due to errors by the operational staff. Radioactivity was released into the immediate vicinity of the plant and into the town of Pripyat, and staff doing repair work were exposed to severe doses of radiation.
28 February 1985 At the Virgil C. Summer nuclear power station, in Jenkinsville, South Carolina, the reactor became critical too soon, leading to an uncontrolled nuclear power surge.	
19 May 1985 At the Indian Point 2 nuclear power station, near New York City, there was a leakage of several hundred gallons of radioactive water, some of which entered the environment outside the facility.	**27 June 1985** Accident in No. 1 reactor of the Balakovo nuclear power station. During start-up activities, a relief valve burst. Fourteen people were killed. This accident was due to errors made in haste and nervousness by inexperienced operational staff.
1986 Webbers Falls, Oklahoma, explosion of a tank containing radioactive gas at a uranium enrichment plant. One person was killed, eight others injured.	

Table 11.10
List of Nuclear Reactors

	Reactors In Operation	Electricity Generated (megawatts)	Percent of Electricity	Reactors Under Construction
NORTH & CENTRAL AMERICA				
Canada	18	12,185	15.6	4
Cuba	0	0	0	2
Mexico	1	654	—	1
U.S.	110	98,331	19.1	4
SOUTH AMERICA				
Argentina	2	935	11.4	1
Brazil	1	628	.7	1
EUROPE				
Belgium	7	5,500	60.8	0
Bulgaria	5	2,585	32.9	2
Czechoslovakia	8	3,264	27.6	8
East Germany	6	2,102	10.9	5
Finland	4	2,310	35.4	0
France	55	52,588	74.6	9
Hungary	4	1,645	49.8	0
Italy	2	1,120	—	0
Netherlands	2	508	5.4	0
Romania	0	0	0	5
Spain	10	7,544	38.4	0
Sweden	12	9,817	45.1	0
Switzerland	5	2,952	41.6	0
U.K.	39	11,242	21.7	1
West Germany	24	22,716	34.3	1
Yugoslavia	1	632	5.9	0
ASIA				
China	0	0	0	3
India	7	1,374	1.6	7
Iran	0	0	0	2
Japan	39	29,300	27.8	12
Pakistan	1	125	0.2	1
South Korea	9	7,200	50.2	2
Taiwan	6	4,924	35.2	0
USSR	46	34,230	12.3	26
AFRICA				
South Africa	2	1,842	7.4	0
TOTALS	426	318,271	—	97

Source: International Atomic Energy Agency (IAEA), Vienna.

in the contamination of environment. But it was the accident at Chernobyl that finally destroyed the myth of nuclear energy being a clean energy source. At a little past 1:24 a.m. on April 26, 1986, two mammoth explosions blew apart Unit Four of the Chernobyl nuclear power plant. The plant is located about 70 miles north of Kiev, the capital of Ukraine, a republic of the former Soviet Union. The roof of the plant was blown off and radioactive gases and materials were released in the atmosphere, reaching up to eleven hundred meters. According to recent estimates, more than 50 million curies were released as a result. The cause of the accident was the flawed design of the RBMK reactor (large power boiling reactor). After the accident, Soviet scientists were reluctant to modify the design of the RBMK reactor, but eventually modified it to make it safer. Forty RBMK type reactors are still operating in Eastern Europe and former Soviet states. And in 25 countries, 426 atomic power plants (Table 11.10) continue to operate and generate enormous amounts of nuclear waste.

What appear to be a clean source of energy are a few megawatts of low-cost power for today's consumption. But in the long run, these will cause an immortal radio-

---

Scenario I

Los Angeles, July 12, 20XX

The powerful earthquake (7.7 on the Richter scale) that shook the city this morning has severely damaged the core of XXXXXX nuclear reactor. A radioactive gas cloud has escaped into the atmosphere. Efforts are being made to contain the radioactivity.

Response:

Discuss how this accident could have been avoided. Are nuclear power stations better or worse than fossil fuel power generating stations?

Scenario II

Dateline: January 30, 20XX

A ship containing 50,000 cubic meters of nuclear waste returned to New York after visiting South America, Africa, and Asia in search of potential dump sites. All Third-World countries have refused to accept the nuclear waste, despite lucrative offers.

Today, the U.S. nuclear waste volume has reached 10 million cubic meters, compared with 500,000 cubic meters in the 1990s. Due to the saturation of temporary storage sites for nuclear waste, and because of a lack of proper planning and development of permanent storage for nuclear waste in the last century, the world's nuclear waste volumes have reached alarming levels. The potential leakage from temporary storage sites poses the greatest threat to the environment.

Action Item/Response:

Going back to 1960, draw a time-line, and label it by proposing appropriate action taken in each decade that could have prevented the nuclear waste dilemma the world faces today.

Time line: _____

         1950   1960   1970   1980   1990   2000   2010   2020   2030   20XX

Scenario III

Dateline: February 19, 20XX

Today the U.N. Security Council passed a resolution demanding the African republic of BANGO to accept 500,000 tons of nuclear waste for permanent storage. All African and Third-World countries have protested against this resolution. The government of BANGO has announced that it will not comply with the U.N. resolution. The U.N. is also considering a proposal for establishing common storage sites in Third-World countries for the nuclear waste generated by developed countries.

Response:

Discuss the implications of the U.N. resolution against BANGO. Is the U.N. justified in asking Third-World countries to accept the nuclear waste generated by the developed countries?

Scenario IV

St. Louis, April 1, 20XX

A train carrying high-level nuclear waste from a nuclear power plant in Illinois collided with an east-bound freight train while crossing a bridge over the Mississippi River. The cars containing waste canisters were badly damaged. Due to the colossal impact of the accident, many canisters ripped open and some fell into the river.

Response:

Discuss the impact of this accident on the environment. Should the transfer of high-level waste by train be allowed to continue?

---

active contamination of the environment. With the growing amount of nuclear waste and limited technical options for its storage and disposal, it is just a matter of time until major accidents happen, resulting in contamination of the environment.

REFERENCES

Grossman, D., and Shulman, S. (1989). A nuclear dump: The experiment begins. *Discover.* March, pp. 49–56.

Lenssen, N. (1992). Confronting the nuclear waste. *State of the World.* New York, W. W. Norton & Company.

Raloff, J. (1990). Fallout over Nevada's nuclear destiny. *Science News*, Jan. 6, 1990, vol 13, pp. 11–12.

State of the World (1992). New York, W. W. Norton & Company.

Newton, D. (1994). Chernobyl Accident. *When Technology Fails.* Detroit, MI. Gale Research Inc.

QUESTIONS

1. Define the following:
   a. Fission
   b. Low-level nuclear waste
   c. Intermediate-level nuclear waste
   d. High-level nuclear waste
   e. Alpha (α) radiation
   f. Beta (β) radiation
   g. Gamma (γ) radiation
   h. Isotope

2. What methods are being employed for the interim storage of commercial spent fuel and high-level nuclear waste?

3. Do you think that Yucca Mountain is a safe place for the permanent storage of nuclear waste? Explain your answer by citing facts.

4. Do you think that nuclear energy is a "clean" source of energy?

5. With more than 426 operational nuclear plants worldwide and 97 plants under construction, is the chance for a nuclear accident growing larger or smaller? What might be the consequences of a major nuclear accident?

6. What alternatives are there to nuclear energy?

7. "Considering the potential risks involved with nuclear technology and the past experiences at Three Mile Island and Chernobyl, the worldwide development of new nuclear plants should be stopped." Do you agree with this statement? Explain your answer.

8. A number of developing countries are constructing nuclear power plants for their growing energy needs. Due to scarce resources, for the majority of the developing countries, nuclear energy appears to be the most viable solution for their growing demand for energy. The developed countries are trying to prevent the transfer of nuclear technology to Third-World countries. Should Third-World nations be denied access to nuclear technology? Support your answer. What alternative sources would you recommend?

Case Study: Nuclear Warriors

Eric Pooley

George Betancourt looked up from his desk as George Galatis burst into the office, a bundle of papers under his arm. On that morning in March 1992, the two men—both senior engineers at Northeast Utilities, which operates five nuclear plants in New England—were colleagues but not yet friends. Apart from their jobs and first names, they seemed to have little in common. Betancourt, 45, was extravagantly rebellious—beard, biker boots, ponytail sneaking out the back of his baseball cap—while Galatis, 42, was square-jawed and devout: Mr. Smith Goes Nuclear. But Galatis respected Betancourt's expertise and knew he could count on him for straight answers.

On this day, Galatis wanted to know about a routine refueling operation at the Millstone Unit 1 nuclear plant in Waterford, Connecticut. Every 18 months the reactor is shut down so the fuel rods that make up its core can be replaced; the old rods, radioactive and 250°F hot, are moved into a 40-ft.-deep body of water called the spent-fuel pool, where they are placed in racks alongside thousands of other, older rods. Because the federal government has never created a storage site for high-level radioactive waste, fuel pools in nuclear plants across the country have become de facto nuclear dumps—with many filled nearly to capacity. The pools weren't designed for this purpose, and risk is involved: the rods must be submerged at all times. A cooling system must dissipate the intense heat they give off. If the system failed, the pool could boil, turning the plant into a lethal sauna filled with clouds of radioactive steam. And if earthquake, human error or

© 1996 Time Inc., Reprinted by permission.

mechanical failure drained the pool, the result could be catastrophic: a meltdown of multiple cores taking place outside the reactor containment, releasing massive amounts of radiation and rendering hundreds of square miles uninhabitable.

To minimize the risk, federal guidelines require that some older plants like Millstone, without state-of-the-art cooling systems, move only one-third of the rods into the pool under normal conditions. But Galatis realized that Millstone was routinely performing "full-core off-loads," dumping all the hot fuel into the pool. His question for Betancourt was, "How long has this been going on?"

Betancourt thought for a minute. "We've been moving full cores since before I got here," he said, "since the early '70s."

"But it's an emergency procedure."

"I know," Betancourt said. "And we do it all the time." What's more, Millstone 1 was ignoring the mandated 250-hr. cool-down period before a full off-load, sometimes moving the fuel just 65 hrs. after shutdown, a violation that had melted the boots of a worker on the job. By sidestepping the safety requirements, Millstone saved about two weeks of downtime for each refueling—during which Northeast Utilities has to pay $500,000 a day for replacement power.

Galatis then flipped through a safety report in which Northeast was required to demonstrate to the Nuclear Regulatory Commission that the plant's network of cooling systems would function even if the most important one failed. Instead, the company had analyzed the loss of a far less critical system. The report was worthless, the NRC hadn't noticed, and the consequences could be dire. If Millstone lost

101

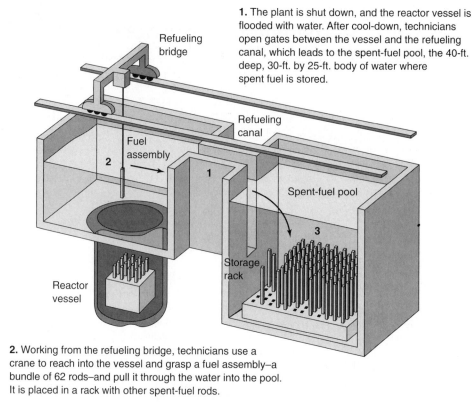

1. The plant is shut down, and the reactor vessel is flooded with water. After cool-down, technicians open gates between the vessel and the refueling canal, which leads to the spent-fuel pool, the 40-ft. deep, 30-ft. by 25-ft. body of water where spent fuel is stored.

Refueling bridge

Refueling canal

Fuel assembly

Spent-fuel pool

Storage rack

Reactor vessel

2. Working from the refueling bridge, technicians use a crane to reach into the vessel and grasp a fuel assembly–a bundle of 62 rods–and pull it through the water into the pool. It is placed in a rack with other spent-fuel rods.

3. The process is repeated for each of the 580 fuel assemblies. When the plant is ready to resume operation, two-thirds of the rods are returned to the core along with 190 fresh assemblies. The rest are stored in the pool for future disposal.

its primary cooling system while the full core was in the pool, Galatis told Betancourt, the backup systems might not handle the heat. "The pool could boil," he said. "We'd better report this to the NRC *now.*"

Betancourt saw that Galatis was right. "But you do that," he said, "and you're dogmeat."

Galatis knew what he meant. Once a leading nuclear utility, Northeast had earned a reputation as a rogue—cutting corners and, according to critics, harassing and firing employees who raised safety concerns. But if Galatis wanted to take on the issue, Betancourt told him, "I'll back you."

So began a three-year battle in which Galatis tried to fix what he considered an obvious safety problem at Millstone 1. For 18 months his supervisors denied the problem existed and refused to report it to the NRC, the federal agency charged with ensuring the safety of America's 110 commercial reactors. Northeast brought in outside consultants to prove Galatis wrong, but they ended up agreeing with him. Finally, he took the case to the NRC himself, only to discover that officials there had known about the procedure for a decade without moving to stop it. The NRC says the practice is common, and safe—if a plant's cooling system is designed to handle the heat load. But Millstone's wasn't. And when Galatis learned that plants in Delaware, Nebraska and New Jersey had similar fuel-pool troubles, he realized the NRC was sitting on a nationwide problem.

Ten years after the disastrous uncontained meltdown at Chernobyl, 17 years after the partial meltdown at Three Mile Island, most Americans probably give only passing thought to the issue of nuclear safety. But the story of George Galatis and Millstone suggests that the NRC itself may be giving only passing thought to the issue—that it may be more concerned with propping up an embattled, economically straitened industry than with ensuring public safety. When a nuclear plant violates safety standards and the federal watchdog turns a blind eye, the question arises, How safe are America's nuclear plants?

Though the NRC's mission statement promises full accountability—"nuclear regulation is the public's business," it says—the agency's top officials at first refused to be interviewed by TIME. After repeated requests, Chairwoman Shirley Ann Jackson, a physics professor who was appointed by President Clinton last summer, finally agreed to talk. But the veteran official in charge of the agency's day-to-day operations, executive director James M. Taylor, would provide only written answers to TIME's faxed questions.

"The responsibility for safety rests with the industry," Jackson told TIME. "Like any other regulatory body, NRC is essentially an auditing agency." Jackson argued that her agency is tough—"When we catch problems, it never makes the papers"—but added that with 3,000 employees and just four inspectors for every three plants, "we have to focus on the issues with the greatest safety significance. We can miss things."

In fact, Millstone is merely the latest in a long string of cases in which the NRC bungled its mandate and overlooked serious safety problems until whistle blowers came forward. The NRC's relationship with the industry has been suspect since 1974, when the agency rose from the ashes of the old Atomic Energy Commission, whose mandate was to promote nuclear power. The industry vetoes commission nominees it deems too hostile (two of five NRC seats are vacant), and agency officials enjoy a revolving door to good jobs at nuclear companies such as Northeast. "The fox is guarding the henhouse," says Delaware Senator Joseph Biden, who is pushing legislation to create an independent nuclear safety board outside the NRC. The Democrat, who is also calling for a federal investigation of NRC effectiveness, believes the agency "has failed the public."

It all comes back to money. "When a safety issue is too expensive for the industry, the NRC pencils it away," says Stephen Comley, executive director of a whistle-blower support group called We the People, which has brought many agency failures to light. "If the NRC enforced all its rules, some of the plants we've studied couldn't compete economically."

In a rare point of agreement with activists, the nuclear industry also says regulations threaten to drive some plants out of business, but it argues that many NRC rules boost costs without enhancing safety. "The regulatory system hasn't kept pace with advances in technology," says Steve Unglesbee, a spokesman for the Nuclear Energy Institute, the industry's p.r. unit. "Industry-wide, our safety record is improving. But NRC creates so many layers of regulation that every plant is virtually assured of being in noncompliance with something."

The NRC suggested as much in a 1985 agency directive on "enforcement discretion," which allowed the agency to set aside hundreds of its own safety regulations. Since 1990, Millstone has received 15 such waivers—more than any other nuclear station. In November, Jackson scaled back the policy, but she says this never endangered public safety. Others disagree.

"Discretionary enforcement was out of hand," says NRC acting Inspector General Leo Norton, who investigates agency wrongdoing but has no power to punish. "We shouldn't have regulations on the books and then ignore or wink at them."

Yet the tensions between cost and safety can only increase as deregulation of the nation's utilities ushers in a new era of rate-slashing competition. In some states, consumers will soon choose their electric company the way they now choose a long-distance telephone carrier. Companies with nuclear plants are at a disadvantage because nuclear-generated electricity can cost twice as much as fossil-generated power. No new plants have been ordered in 18 years, and a dozen have been mothballed in the past decade.

For now, however, nuclear power provides 20% of the electricity consumed in the U.S.; New England depends on nuclear plants for more than half its supply. Long-term, says Northeast senior vice president Donald Miller, Millstone and her sisters will survive only "if we start running them like a business [and] stop throwing money at issues." New England's largest power company, with $6.5 billion in assets and $3.7 billion in revenues last year, Northeast is slashing its nuclear work force of 3,000 employees by one-third over the next five years. Company CEO Bernard Fox says the move will not undermine safety.

George Galatis went to work at Northeast Utilities in June 1982 with a degree from Rensselaer Polytechnic Institute and experience with a top manufacturer of nuclear components. At Northeast, he started in the division that oversees the utility's 15 fossil-fuel plants, then moved to the nuclear group, specializing in performance and reliability. Eric DeBarba, Northeast's vice president of technical services, describes him as a solid engineer. "Nobody here ever questioned his honesty or motives," DeBarba says.

Galatis tells it differently. In March 1992 he began working on Millstone 1, one of three nuclear plants perched on a neck of land that juts into Long Island Sound from the shore of southeastern Connecticut. He was checking specifications for a replacement part for a heat exchanger in the spent-fuel cooling system. To order the proper part, he needed to know the heat load. So he pulled a safety report that should contain the relevant data.

But they weren't there.

"The report didn't contain the safety analysis for what we were doing," says Galatis. "No heat-load calculations." It was then he realized the plant had been routinely operating "beyond design basis," putting 23 million BTUs into a pool analyzed for 8 million, which is, he says, "a bit like running your car at 5,000 r.p.m."

Galatis raised the issue with members of Northeast's division of nuclear licensing. "They tried to convince me they had it analyzed," he says. He asked them to produce the documents, and they could not. Galatis sensed trouble when, in later talks, "they began denying that the first discussions had taken place." In June 1992 he spelled out the problem in a memo, calling the fuel pool a license violation and an "unreviewed safety question"—NRC lingo for a major regulatory headache—and adding other concerns he had found, such as the fact that some of the pool's cooling pipes weren't designed to withstand an earthquake, as they were required to do. Northeast sat on the memo for three months, until Galatis filed an internal notice-of-violation form, and Betancourt, a leader in the spent-fuel field for years, wrote a memo backing him up.

"When I started in the industry, 20 years ago," Betancourt says, "spent fuel was considered the ass end of the fuel cycle. No one wanted to touch it. Everyone wanted to be on the sexy side, inside the reactor vessel, where the action and danger were. No one noticed fuel pools until we started running out of room in them."

In 1982 Congress mandated that the Department of Energy begin to accept nuclear waste from commercial reactors in 1998. Consumers started paying into a federal fund meant to finance a storage site. Though the Energy Department has collected $8.3 billion, no facility has been completed; in a case of NIMBY writ large, no state wants such a site in its backyard. As the nation's stockpile of spent fuel reached 30,000 tons, activists seized the issue as a way to hobble the industry, and the Energy Department announced that a permanent facility planned for Yucca Mountain, Nevada, wouldn't be ready until 2010; Energy Secretary Hazel O'Leary now puts its chances of opening at no better than fifty-fifty. Bills to create temporary sites are stalled in both houses of Congress.

"Slowly, we woke up to this problem," says Betancourt. The NRC relaxed standards and granted license amendments that allowed plants to "rerack" their rods in ever more tightly packed pools. Sandwiched between the rods is a neutron-absorbing material called Boraflex that helps keep them from "going critical." After fuel pools across the country were filled in this way, the industry discovered that radiation causes Boraflex to shrink and crack. The NRC is studying the problem, but at times its officials haven't bothered to analyze a pool's cooling capacity before granting a reracking amendment. "It didn't receive the attention that more obvious safety concerns got," says Inspector General Norton.

Then, in late 1992, David Lochbaum and Don Prevatte, consultants working at Pennsylvania Power & Light's Susquehanna plant, began to analyze deficiencies in spent-fuel cooling systems. They realized that a problem had been sneaking up on the industry: half a dozen serious accidents at different plants had caused some water to drain from the pools. In the worst of them, at Northeast's Haddam Neck plant in 1984, a seal failure caused 200,000 gal. to drain in just 20 min. from a water channel next to the fuel pool. If the gate between the channel and the pool had been open, the pool could have drained, exposing the rods and causing a meltdown. Says Lochbaum: "It was a near miss."

The NRC insists that the chance of such an accident is infinitesimal. But the agency's risk-assessment methods have been called overly optimistic by activists, engineers and at least one NRC commissioner. The agency's analysis for a fuel-pool drainage accident assumes that at most one-third of a core is in the pool, even though plants across the country routinely move full cores into pools crowded with older cores. If the NRC based its calculations on that scenario, says Lochbaum, "it would exceed the radiation-dose limits set by Congress and scare people to death. But the NRC won't do it." The NRC's Taylor told TIME that the agency analyzes dose rates at the time a plant opens—when its pool is empty. The law, he said, "does not contain a provision for rereview."

Lochbaum and Prevatte reported Susquehanna to the NRC and suggested improvements to its cooling system. The NRC, Lochbaum says, didn't read the full report. He and Prevatte called Congress members, pushed for a public hearing and presented their concerns to NRC staff. Conceding that Lochbaum and Prevatte "had some valid points," the agency launched a task force and in 1993 issued an informational notice to the 35 U.S. reactors that share Susquehanna's design, alerting them to the problem but requiring no action. One of the plants was Millstone 1.

In 1992, Galatis didn't know about Lochbaum's struggle to get fuel-pool problems taken seriously. He did know he would face resistance from Northeast, where the bonus system is set up to reward employees who don't raise safety issues that incur costs and those who compromise productivity see their bonuses reduced. (Northeast says it has a second set of bonuses to reward those who raise safety issues. Galatis never got one.)

"Management tells you to come forward with problems," says Millstone engineer Al Cizek, "but actions speak louder than words." A Northeast official has been quoted in an NRC report saying the company didn't have to resolve a safety problem because he could "blow it by" the regulators. An NRC study says the number of safety and harassment allegations filed by workers at Northeast is three times the industry average. A disturbing internal Millstone report, presented to CEO FOX in 1991 and obtained by TIME, warns of a "cultural problem" typified by chronic failure to follow procedures, hardware problems that were not resolved or were forgotten, and a management tolerant of "willful [regulatory] noncompliance without justification." The report, written by director of engineering Mario Bonaca, changed nothing. "We've been working at this," says Fox, "but making fundamental change in a complex, technical environment is really hard."

A 1996 Northeast internal document reports that 38% of employees "do not trust their management enough to willingly raise concerns [because of] a 'shoot the messenger' attitude" at the company. In recent years, two dozen Millstone employees have claimed they were fired or demoted for raising safety concerns; in two cases, the NRC fined Northeast. In one, Paul Blanch, who had only recently been named engineer of the year by a leading industry journal, was subjected to company-wide harassment after he discovered that some of Millstone Unit 3's safety instrumentation didn't work properly.

Galatis had watched that case unfold. "George knew what he was getting into," says Blanch. "He knew Northeast would come after him. He knew the NRC wouldn't protect him. And he did it anyway."

In January 1993, Galatis pushed for a meeting with Richard Kacich, Northeast's director of nuclear licensing. Galatis outlined the pool's problems and asked for a consultant, Holtec International, to be brought in. Holtec agreed with Galatis that the pool was an unanalyzed safety question; later the consultant warned that a loss of primary cooling could result in the pool's heating up to 216°F—a nice slow boil.

Galatis sent a memo to DeBarba, then vice president of nuclear engineering, in May 1993. Galatis was threatening

to go to the NRC, so DeBarba created a task force to address "George's issues," as they were becoming known. The aim seems to have been to appease Galatis and keep him from going public. DeBarba says the calculations that Holtec and Galatis used were overly conservative and that experience told him there was no problem. The pool hadn't boiled, so it wouldn't boil. If a problem ever developed, there would be plenty of time to correct it before it reached the crisis stage. "We live and work here. Why would we want an unsafe plant? We had internal debate on this topic," DeBarba told TIME. "Legitimate professional differences of opinion." In 1977, he says, the NRC stated, "We could make the choice [of a full-core off-load] if it's 'necessary or desirable for operational considerations.' But that does not mean that what George raised was not an issue. We have rules on this, and we want to get it right."

By October 1993, Galatis was writing to the chief of Northeast's nuclear group, John Opeka, and to Fox, who was then company president. Galatis mentioned the criminal penalties for "intentional misconduct" in dealings with the NRC. Opeka objected to Galatis' abrasive tone but hired another consulting firm, which also agreed with Galatis. Northeast moved on to yet another consultant, a retired NRC official named Jim Partlow.

In December, during a four-hour interview that Galatis calls his "rape case"—because the prosecutor, he says, put the victim on trial—Partlow grilled Galatis about his "agenda" and "motives." After Galatis showed him the technical reports, Partlow changed his mind about Galatis and began questioning Kacich about the apparent violations. In two March 1994 memos to Kacich, Partlow backed Galatis, scolded the utility for taking so long to respond to him and suggested that they should reward Galatis "for his willingness to work within the NU system . . . Let him know that his concern for safety . . . . is appreciated."

DeBarba and Kacich created another task force but did not modify the cooling system. Kacich began having conversations with Jim Andersen, the NRC's project manager for Millstone 1, about Galatis' concerns and how to get through the spring 1994 off-load. Andersen, who works at NRC headquarters in Washington, has told the inspector general that he knew all along Millstone was off-loading its full core but didn't know until June 1993 that it was a problem. Even then he did not inform his superiors. In a bow to Galatis, Millstone modified its off-load procedure, moving all the rods but doing so in stages. Before the off-load, Northeast formally reported to Andersen what he'd known for months: that Millstone might have been operating outside its design basis, a condition that must be reported within 30 days.

During the spring outage, a valve was accidentally left open, spilling 12,000 gal. of reactor-coolant water—a blunder that further shook Galatis' faith. He began to see problems almost everywhere he looked and proposed the creation of a global-issues task force to find out whether Millstone was safe enough to go back online. His bosses agreed. But when the head of the task force left for a golf vacation a few weeks before the plant was scheduled to start up, Galatis says, he knew it wasn't a serious effort. So he made a call to Ernest Hadley, the lawyer who had defended whistle blower Blanch against Northeast two years before.

An employment and wrongful-termination lawyer, Hadley has made a career of representing whistle blowers, many of them from Millstone. For 10 years he has also worked with Stephen Comley and We the People. Comley, a Massachusetts nursing-home operator, is a classic New England character, solid and brusque. He founded We the People in 1986 when he realized the evacuation plans for Seabrook Station, a plant 12 miles from his nursing home, included doses of iodine for those too old and frail to evacuate.

"Some of us were expendable," says Comley. "That got me going." For years he was known for publicity stunts— hiring planes to trail banners above the U.S. Capitol—and emotional outbursts at the press conferences of politicians. The NRC barred him from its public meetings until a judge ordered the ban lifted. But Comley's game evolved: instead of demanding that plants be shut down, he began insisting they be run safely. He teamed up with the sharp-witted Hadley to aid and abet whistle blowers and sank his life savings into We the People before taking a dime in donations. Comley, says the NRC's Norton, "has been useful in bringing important issues to our attention. Steve can be a very intense guy. I don't think it's good for his health. But people who seem—not fanatical, but overly intense— help democracy work."

In April, 1994, two years after he discovered the problems with Millstone's cooling system, Galatis reported the matter to the NRC. He spoke to a "senior allegations coordinator," waited months, then refiled his charges in a letter describing 16 problems, including the cooling system, the pipes that couldn't withstand seismic shock, the corporate culture. "At Northeast, people are the biggest safety problem," Galatis says. "Not the guys in the engine room. The guys who drive the boat."

Galatis told DeBarba and Kacich that he was going to the NRC. He continued to experience what he calls "subtle forms of harassment, retaliation and intimidation." His performance evaluation was down-graded, his personnel file forwarded to Northeast's lawyers. DeBarba "offered" to move him out of the nuclear group. He would walk into a meeting, and the room would go suddenly silent. DeBarba says he is unaware of any such harassment.

With missionary zeal, Galatis continued to forward allegations to the NRC. Yet four months passed before Galatis finally heard from Donald Driskill, an agent with the NRC's Office of Investigations (the second watchdog unit inside the NRC, this one tracks wrongdoing by utilities). Galatis felt that Driskill was too relaxed about the case. Driskill talked to Northeast about Galatis' charges— a breach of confidentiality that the NRC calls "inadvertent." When Hadley complained to him about Northeast's alleged harassment of Galatis, Driskill suggested he talk to Northeast's lawyer: "He's a really nice guy."

While playing detective—sniffing through file drawers and computer directories—Galitis found items that he felt suggested collusion between the utility and its regulator. Safety reports made it clear that both on-site inspectors and officials from the NRC's Office of Nuclear Reactor Regulation had known about the full-core off-loads since at least 1987 but had never done anything about them. Now, to clear the way for the fall 1995 off-load, NRC officials were apparently offering Northeast what Galatis calls "quiet coaching." One sign of this was a draft version of an NRC inspection report about the spent-fuel pool that had been E-mailed from the NRC to Kacich's licensing department. "What was that doing in Northeast's files?" asks Inspector General Norton.

On June 10, 1995, Jim Andersen visited the site to discuss Galatis' concerns with Kacich's staff. Andersen wouldn't meet with Galatis but huddled with Kacich's team, trying to decide how to bring Millstone's habits into compliance with NRC regulations, either by requesting a license amendment— a cumbersome process that requires NRC review and public comment—or by filing an internal form updating the plant's safety reports. This was the easier path, but it could be used only if the issue didn't constitute an unreviewed safety question. Andersen told DeBarba and Kacich that the license amendment "is the cleaner way to go," but they weren't sure there was enough time to get an amendment approved before the next off-load, scheduled for October 1995.

On July 10, Betancourt met with Ken Jenison, an inspector from the NRC's Region 1 office, and gave testimony in support of Galatis' safety allegations. Less than a week later, Betancourt was called to the office of a good-natured human-resources officer named Janice Roncaioli. She complained that he wasn't a "team player," Betancourt says, and ran through the company's termination policies. Rancaioli called Betancourt's account of the meeting

NEAR MISSES

The Nuclear Regulatory Commission's Office of the Inspector General—a watchdog that can investigate but not punish—has looked into an array of cases in which safety problems were ignored by NRC staff. Some highlights:

- After a 1975 fire knocked out equipment at the Browns Ferry plant in Alabama, the NRC approved a material called Thermo-Lag as a "fire barrier" to protect electrical systems. Between 1982 and 1991, however, the NRC ignored seven complaints about Thermo-Lag; when an engineer testified that fire caused it to melt and give off lethal gases, the NRC closed the case without action. After more complaints and an inspector general's investigation, the NRC "reassessed." Now, it says, "corrective action is ongoing."

- In 1980 workers at Watts Bar 1, a plant then under construction by the Tennessee Valley Authority, flooded the NRC with some 6,000 allegations of shoddy workmanship and safety lapses—enough to halt construction for five years. The NRC breached confidentiality and identified whistle blowers such as electrical supervisor Ann Harris to the TVA; several were fired. After 23 years and $7 billion, Watts Bar 1 was completed last fall. Though workers say the TVA has abandoned thorough safety inspections in favor of a "random sampling" program, the NRC in February granted an operating license to Watts Bar, the last U.S. nuclear plant scheduled for start-up.

- In the early 1980s, when Northeast Utilities Seabrook Station in New Hampshire was under construction, Joseph Wampler warned the NRC that many welds were faulty. His complaints went unanswered, and he was eventually fired. Blacklisted, he says, Wampler moved to California and revived his career. But in 1991 the NRC sent a letter summarizing Wampler's allegations—and providing his full name and new address—to several dozen nuclear companies. His career was destroyed a second time; he now works as a carpenter. The NRC fined Northeast $100,000 for problems with the welds.

- In 1990 Northeast engineer Paul Blanch discovered that the instruments that measure the coolant level inside the reactor at Millstone 3 were failing. Blanch was forced out, and the problem went uncorrected. In 1993 the NRC's William Russell told the inspector general that the agency had exercised "enforcement discretion," a policy that allows it to waive regulations. Later Russell said the remark had been taken out of context.

- Last December a worker at the Maine Yankee plant in Bath charged that management had deliberately falsified computer calculations to avoid disclosing that the plant's cooling systems were inadequate. The NRC didn't discover this, the Union of Concerned Scientists told reporters, because it didn't notice that Maine Yankee had failed to submit the calculations for review—though they were due in January 1990.

- In 1988 a technician at the Nine Mile Point plant near Oswego, New York, called the NRC with allegations of drug use and safety violations at the plant. The NRC executive director at the time, Victor Stello Jr., took a personal interest in the matter, but his chief aim seemed to be building a case against Roger Fortuna, the deputy director of the NRC's Office of Investigation, for leaking secrets to the watchdog group We the People. The NRC demanded that We the People head Steve Comley turn over tapes he had allegedly made of conversations with Fortuna. When Comley refused, he was ruled in contempt and fined $350,000 (he still has not paid). The charges against Fortuna were found to be without merit, and when the case came to light—during hearings to confirm Stello as Assistant Secretary of Energy—Stello withdrew his name. "The tension between enforcement and appeasement," a ranking NRC official says, "tugs at this agency every day."

"slanted" but would not comment further, citing employee-confidentiality rules.

In a July 14 meeting, Jenison, one official who wasn't going to stand for any regulatory sleight-of-hand, told DeBarba and Kacich that if Northeast tried to resolve its licensing problems through internal paperwork alone, he would oppose it. Northeast had to get a license amendment approved before it could off-load another full core, and time was running out. DeBarba and Kacich called on Galatis and Betancourt to help them write the amendment request. The plan included, for the first time, the cooling-system improvements Galatis had been demanding for three years. It was a kind of victory, but he felt disgusted. "The organizational ethics were appalling," he says. "There's no reason I should have had to hire a lawyer and spend years taking care of something this simple."

So Galatis helped Kacich with the amendment request, which was filed July 28. Then he and Hadley drew up another document: a petition that asked the NRC to deny Northeast's amendment request and suspend Millstone's license for 60 days. The petition, filed on behalf of Galatis and We the People, charged that Northeast had "knowingly, willingly, and flagrantly" violated Millstone 1's license for 20 years, that it had made "material false statements" to the NRC and that it would, if not punished, continue to operate unsafely.

On Aug. 1, Betancourt was called into DeBarba's office; Roncaioli was present, and DeBarba told Betancourt he was being reassigned. "We want to help you, George," Betancourt recalls DeBarba saying, "but you've got to start thinking 'company.'" It was all very vague and, Betancourt thought, very intimidating. On Aug. 3—the day Betancourt was scheduled to meet with the Office of Investigations—Roncaioli called him to her office again. According to Betancourt, she said she wanted to "reaffirm the meaning" of the DeBarba meeting. Betancourt's wife and children began to be worried that he would be fired. "Why don't you just do what they want you to?" his eldest girl asked. Betancourt didn't know quite how to answer. "Your own daughter telling you to roll over," he says.

After Galatis filed his petition, on Aug. 21, he found himself in many of New England's newspapers. As citizen's groups called meetings, Northeast and the NRC assured everyone that the full-core off-load was a common practice that enhanced safety for maintenance workers inside the empty reactor vessel. "We've been aware of how they off-loaded the full core," NRC spokeswoman Diane Screnci told one paper. "We could have stopped them earlier."

At a citizens' group meeting, Galatis met a mechanic named Pete Reynolds, who had left Millstone in a labor dispute two years before. Reynolds shared some hair-raising stories about his days off-loading fuel. He told Galatis—and has since repeated the account to TIME—that he saw work crews racing to see who could move fuel rods the fastest. The competition, he said, tripped radiation alarms and overheated the fuel pool. Reynolds' job was to remove the big bolts that hold the reactor head in place. Sometimes, he said, he was told to remove them so soon after shutdown that the heat melted his protective plastic booties.

Galatis knew that if such things had happened, they would be reflected in operator's logs filed in Northeast's document room. So, on Oct. 6, he appeared in the room and asked for the appropriate rolls of microfiche. The logs backed up what Reynolds had said: Millstone had moved fuel as soon as 65 hrs. after shutdown—a quarter of the required time. The logs noted the sounding of alarms. Galatis wondered where the resident inspector had been.

The deadline came for Millstone's off-load, but the amendment still had not been granted. Connecticut's Senator Chris Dodd, Representative Sam Gejdenson and a host of local officials were asking about the plant's safety, and Millstone scheduled a public meeting for late October. Senior vice president Don Miller sent a memo to his employees warning them that "experienced antinuclear activists" had "the intention of shutting the station down and eliminating 2,500 jobs." The memo stirred up some of Galatis' colleagues. "You're taking food out of my girl's mouth," one of them told him.

DeBarba assembled a task force to assess what had to be done to get the pool ready for the overdue off-load, but he kept Galatis and Betancourt off the team. The task force came up with six serious problems, most already raised by Galatis. Scrambling to fix the pool in a few weeks, DeBarba hired extra people. The plant shut down anticipating permission to move fuel.

Galatis and Hadley had been waiting two months for a reply to their petition to deny Northeast's amendment. Finally, on Oct. 26, a letter from William Russell, director of the NRC's Office of Nuclear Reactor Regulation, informed them that their petition was "outside of the scope" of the applicable regulatory subchapter. Two weeks later, the NRC granted Northeast's amendment. Millstone started moving fuel the next morning.

Because of Galatis, the plant is still shut down. "What's especially galling," says Hadley, "is that the NRC ignored my client and denied his motion, then validated his concerns after the fact." In late December, Inspector General Norton released his preliminary report. He found that Northeast had conducted improper full-core off-loads for 20 years. Both the NRC's on-site inspectors and headquarters staff, the report said, "were aware" of the practice but somehow "did not realize" that this was a violation. In other words, the NRC's double-barreled oversight system shot blanks from both barrels. Norton blamed bad training and found no evidence of a conspiracy between Northeast and the NRC to violate the license. He is still investigating possible collusion by the NRC after Galatis came forward. What troubled him most, Norton told TIME, is that agency officials all the way up to Russell knew about the off-loads and saw nothing wrong with them. "The agency completely failed," says Norton. "We did shoddy work. And we're concerned that similar lapses might be occurring at other plants around the country."

In a second investigation, the Office of Investigations is looking into Northeast's license violations and the alleged harassment of Galatis and Betancourt. The intense public scrutiny their case has received will, Galatis says, "make it harder for them to sweep this one under the rug."

On Dec. 12, Russell sent a letter informing Northeast that because "certain of your activities may have been conducted in violation of license requirements," the NRC was considering penalties. In an extraordinary move, Russell demanded a complete review of every system at Millstone 1, with the results "submitted under oath," to prove that every part of the plant is safe—the global examination Galatis asked for two years ago. The results, Russell wrote, "will be used to decide whether or not the license of Millstone Unit 1 should be suspended, modified or revoked."

Now the pressure is on NRC Chairwoman Jackson to prove her commitment to nuclear safety—and her ability to reform an inert bureaucracy. "I will not make a sweeping indictment of NRC staff," Jackson, a straight-talking physicist who in July 1995 became both the first female and the first African American to run the NRC, told TIME. "Does that mean everybody does things perfectly? Obviously not. We haven't always been on top of things. The ball got dropped. Here's what I'm saying now: The ball will not get dropped again."

In response to the problems Galatis exposed, Jackson launched a series of policies designed to improve training, accountability and vigilance among inspectors and NRC staff. She ordered the agency's second whistle-blower study in two years and a nationwide review of all 110 nuclear plants, to find out how many have been moving fuel in violation of NRC standards. The results will be in by April, along with a menu of fuel-pool safety recommendations. (By using a technique called dry-cask storage, utilities could empty their pools and warehouse rods in airtight concrete containers, reducing risk. In the past, the NRC has ruled that the process isn't cost effective.)

Jackson still refuses to meet with Galatis or even take his phone call. "Mr. Galatis is part of an adjudicatory process," she explains. But in a letter turning down Stephen Comley's request that she meet with him and Galatis, Jackson wrote, "The avenues you have been using to raise issues are the most effective and efficient ways. I see no additional benefit to the meeting."

Asked by TIME if she considered three years and two wrecked careers "the most efficient" way to raise the fuel-pool issue, Jackson offered a thin smile. "I'm changing the process," she said. "When all is said and done, then Mr. Galatis and I can sit and talk."

For Galatis, the endgame should have been sweet. On Dec. 20, a Millstone technical manager fired off a frank piece of E-mail warning his colleagues that "the acceptance criteria are changing. Being outside the proper regulatory framework, even if technically justifiable, will be met with resistance by the NRC. Expect no regulatory relief." DeBarba put 100 engineers on a global evaluation of the plant, and they turned up more than 5,000 "items" to be addressed before the plant could go back online. The company announced a reorganization of its nuclear division in which DeBarba and Miller were both promoted. Miller, who told TIME that "complacency" was to blame for the utility's troubles, was put in charge of safety at Northeast's five nuclear plants. On Jan. 29, the NRC, citing chronic safety concerns, employee harassment "and historic emphasis on cost savings vs. performance," enshrined all three Millstone plants in the agency's hall of shame: the high-scrutiny "watch list" of troublesome reactors. Northeast announced that Millstone would stay down at least through June, at a cost of $75 million. And Standard & Poor's downgraded Northeast's debt rating from stable to negative.

"A hell of an impact," says Betancourt, but "I'm going to lose my job."

"If I had it to do over again," says Galatis, "I wouldn't." He believes his nuclear career is over. (Though still employed by Northeast, he knows that whistle blowers are routinely shut out by the industry.) He's thinking about entering divinity school.

In January, Northeast laid off 100 employees. To qualify for their severance money, the workers had to sign elaborate release forms pledging not to sue the utility for harassment. Four engineers say they were fired in retaliation for their testimony to the NRC four years ago on behalf of whistle blower Blanch. The company denies any connection between the layoffs and Blanch's case. That makes Blanch chuckle. "The two Georges had better watch their backs," he says. "Up at Northeast, they've got long memories."

In the end, Galatis believes, the NRC's recent flurry of activity is little more than window dressing. "If they wanted to enforce the law," he says, "they could have acted when it counted—before granting the license amendment. Whatever wrist slap they serve up now is beside the point."

"I believe in nuclear power," he says, "but after seeing the NRC in action, I'm convinced a serious accident is not just likely but inevitable. This is a dangerous road. They're asleep at the wheel. And I'm road-kill."

QUESTIONS

1. What is the solution to the problem of storing nuclear reactive waste?

2. Why was Millstone sidestepping the safety requirements? What would have happened if Millstone lost its primary cooling system while the full core was in the pool?

3. What would you have done if you had been in Galatis's and Betancourt's position? Defend your reasoning for your actions.

4. What part does the NRC play in monitoring the lethality of this situation?

A Bridge to Your 21st Century Understanding

Complete and discuss the following flowchart.

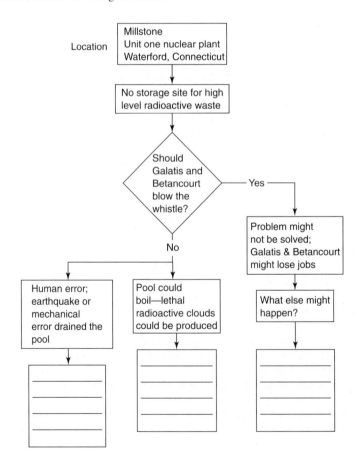

The Hydrogen Experiment

Seth Dunn

In Reykjavík, Iceland, scientists, politicians, and business leaders have conspired to put-into motion a grand experiment that may end the country's—and the world's—reliance on fossil fuels forever. The island has committed to becoming the world's first hydrogen economy over the next 30 years.

Riding from the airport to Iceland's capital, Reykjavík, gives one the sensation of having landed on the moon. Black lava rocks cover the mostly barren landscape, which is articulated by craters, hills, and mountains. Other parts of the island are covered by a thin layer of green moss. American astronauts traveled here in the 1960s to practice walking the lunar surface, defining rock types, and taking specimens.

I, too, have traveled here on a journey of sorts to a new world—a world that is powered not by oil, coal, and other polluting fossil fuels, but one that relies primarily on renewable resources for energy and on hydrogen as an energy carrier, producing electricity with only water and heat as byproducts. My quest has brought me to the cluttered office of Bragi Arnason, a chemistry professor at the University of Iceland whose 30 year old plan to run his country on hydrogen energy has recently become an official objective of his government, to be achieved over the next 30 years. "I think we could be a pilot country, giving a vision of the world to come," he says to me with a quiet conviction and a deep, blue-eyed stare that reminds me of this country's hardy Viking past.

When he first proposed this hydrogen economy decades ago, many thought he was crazy. But today, "Professor Hydrogen," as he has been nicknamed, is something of a national hero. And Iceland is now his 39,000-square-mile lab space for at long last conducting his ambitious experiment. Already, his scientific research has led to a multi-

million-dollar hydrogen venture between his university, his government, other Iceland institutions, and a number of major multinational corporations.

I am not alone in my expedition to ground zero of the hydrogen economy: hundreds of scientists, politicians, investors, and journalists have visited over the past year to learn more about Iceland's plans. My journey is also an echo of what happened in the 18th century, when merchants and officials flocked to another North Atlantic island—Great Britain—to witness the harnessing of coal.

Today, many experts are watching Iceland closely as a "planetary laboratory" for the anticipated global energy transition from an economy based predominantly on finite fossil fuels to one fueled by virtually unlimited renewable resources and hydrogen, the most abundant element in the universe. The way this energy transition unfolds over the coming decades will be greatly influenced by choices made today. How will the hydrogen be produced? How will it be transported? How will it be stored and used? Iceland is facing these choices right now, and in plotting its course has reached a fork in the road. It must choose between developing an interim system that produces and delivers methanol, from which hydrogen can be later extracted, or developing a full infrastructure for directly transporting and using hydrogen. Whether the country tests incremental improvements or more ambitious steps will have important economic and environmental implications, not only for Iceland but for other countries hoping to draw conclusions from its experiment.

Iceland is not undertaking this experiment in isolation. Its hydrogen strategy is tied to three major global trends.

Reprinted with permission from Worldwatch magazine (www.worldwatch.org) Worldwatch Institute, Washington, D.C.

The first of these is growing concern over the future supply and price of oil—already a heavy burden on the Icelandic economy. The second is the recent revolution in bringing hydrogen-powered fuel cells—used for decades in space travel—down to earth, making Árnason's vision far more economically feasible than it was just ten years ago. The third is the accelerating worldwide movement to combat climate change by reducing carbon emissions from fossil fuel burning, which in its current configuration places constraints on Iceland that make a hydrogen transition particularly palatable. How the island's plans proceed will both help to shape and be shaped by these broader international developments.

A HEAD START

Straddling the Mid-Atlantic Ridge, Iceland is a geologist's dream. Providing inspiration for Jules Verne's *Journey to the Center of the Earth,* the island's volcanoes have accounted for an estimated one-third of Earth's lava output since A.D. 1500. Eruptions have featured prominently in Icelandic religion and history, at times wiping out large parts of the population. Reykjavík is the only city I know that has a museum devoted solely to volcanoes. There, one can find out the latest about the 150 volcanoes that remain active today.

Iceland's volcanic activity is accompanied by other geological processes. Earthquakes are frequent, though usually mild, which has made natives rather blasé about them. Also common are volcanically heated regions of hot water and steam, most visible in the hot springs and geysers scattered across the island. In fact, the word "geyser" originated here, derived from *geysir,* and Reykjavík translates to "smoky bay." During my visit, the well-known Geysir, which erupts higher than the United States' Old Faithful, was reemerging from years of dormancy, to the delight of Icelanders everywhere.

The country first began to tap its geothermal energy for heating homes and other buildings (also called district heating) in the 1940s. Today, 90 percent of the country's buildings—and all of the capital's—are heated with geothermal water. Several towns in the countryside use geothermal heat to run greenhouses for horticulture, and geothermal steam is also widely harnessed for power generation. One tourist hotspot, the Blue Lagoon bathing resort, is supplied by the warm, silicate-rich excess water from the nearby Reykjanes geothermal power station. Yet it is estimated that only 1 percent of the country's geothermal energy potential has been utilized.

Falling water is another abundant energy source here. Although it was floating ice floes that inspired an early (but departing) settler to christen the island Iceland, the country's high latitude has exposed it to a series of ice ages. This icy legacy lingers today in the form of sizable glaciers, including Europe's largest, which have carved deep valleys with breathtaking waterfalls and powerful rivers. The first stream was harnessed for hydroelectricity in the 1900s. The country aggressively expanded its hydro capacity after declaring independence from Denmark in the 1940s, beginning an era of economic growth that elevated it from Third World status to one of the world's most wealthy nations today. Hydroelectricity currently provides 19 percent of Iceland's energy—and that share could be significantly increased, as the country has harnessed only 15 percent of potential resources (though many regions are unlikely to be tapped, due to their natural beauty, ecological fragility, and historical significance).

Iceland is unique among modern nations in having an electricity system that is already 99.9 percent reliant on indigenous renewable energy—geothermal and hydroelectric. The overall energy system, including transportation, is roughly 58 percent dependent on renewable sources. This, some experts believe, prepares the country well to make the transition from internal combustion engines to fuel cells, and from hydrocarbon to hydrogen energy. With its extensive renewable energy grid, Iceland has a headstart on the rest of the world, and is positioned to blaze the path to an economy free of fossil fuels.

PEAT AND PETROLEUM

When Vikings first permanently settled Iceland in the 9th century A.D., they used bushy birchwood and peat reservoirs to make fires for cooking and heating, and to fuel iron forges to craft weapons. But deforestation soon led to the end of wood supplies, and the cold climate would freeze the peat bogs, limiting their use as fuel.

Beyond its peat supplies, Iceland has virtually no indigenous fossil fuel resources. As the Industrial Revolution gathered momentum, the nation began to import coal and coke for heating purposes; coal would remain the primary heating source until the development of geothermal energy. In the late 1800s, as petroleum emerged as a fuel, Iceland turned to importing oil. Today, imported oil—about 850,000 tons per year—accounts for 38 percent of national energy use, 57 percent of this used to run its motor vehicles and the boats of its relatively large

THE HYDROGEN CYCLE

A blueprint for the post-fossil fuel energy economy

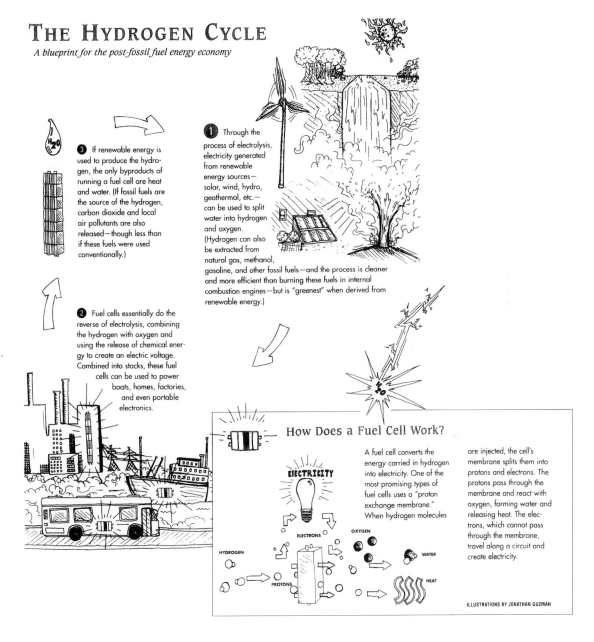

3 If renewable energy is used to produce the hydrogen, the only byproducts of running a fuel cell are heat and water. (If fossil fuels are the source of the hydrogen, carbon dioxide and local air pollutants are also released—though less than if these fuels were used conventionally.)

1 Through the process of electrolysis, electricity generated from renewable energy sources—solar, wind, hydro, geothermal, etc.—can be used to split water into hydrogen and oxygen. (Hydrogen can also be extracted from natural gas, methanol, gasoline, and other fossil fuels—and the process is cleaner and more efficient than burning these fuels in internal combustion engines—but is "greenest" when derived from renewable energy.)

2 Fuel cells essentially do the reverse of electrolysis, combining the hydrogen with oxygen and using the release of chemical energy to create an electric voltage. Combined into stacks, these fuel cells can be used to power boats, homes, factories, and even portable electronics.

How Does a Fuel Cell Work?

A fuel cell converts the energy carried in hydrogen into electricity. One of the most promising types of fuel cells uses a "proton exchange membrane." When hydrogen molecules are injected, the cell's membrane splits them into protons and electrons. The protons pass through the membrane and react with oxygen, forming water and releasing heat. The electrons, which cannot pass through the membrane, travel along a circuit and create electricity.

ELECTRICITY

ELECTRONS

HYDROGEN

OXYGEN

WATER

PROTONS

HEAT

ILLUSTRATIONS BY JONATHAN GUZMAN

fishing industry, the nation's leading source of exports. Dependence on oil imports costs the nation $150 million annually, and explains why transport and fishing each account for one-third of its carbon emissions.

The final third of Iceland's greenhouse emissions is found in other industries—primarily the production, or smelting, of metals like aluminum. The availability of low-cost electricity—at $.02 per kilowatt, it is the world's cheapest—has made Iceland a welcome haven for these energy-intensive industries. Metals production, along with transport and fishing, makes the island one of the world's top per-capita emitters of carbon dioxide, and offsets much of the greenhouse gas savings Iceland has achieved in space heating and electricity.

These features of Iceland's energy economy—a carbon-free power sector, costly dependence on oil for fishing and transportation, rising emissions from the metals industry—have placed the nation in a difficult situation with regard to complying with international climate change commitments. The 1997 Kyoto Protocol's guidelines for reducing greenhouse gas emissions in industrial nations are based on emission levels from the year 1990, which prevents Iceland from taking credit for its previously completed transition to greenhouse gas-free space heating and electricity generation. Although the government, arguing its special situation, negotiated a 10 percent reprieve between 1990 and 2010 under the Protocol, officials estimate that plans to build new aluminum smelters will cause it to exceed this target. Because of this so-called "Kyoto dilemma," Iceland is among only a few remaining industrial nations that have not signed the agreement.

In 1997, as the Protocol talks gathered momentum and the nation's dilemma was becoming apparent, a recently elected Parliamentarian named Hjalmar Árnason submitted a resolution to the Parliament, or Althing, demanding that the government begin to explore its energy alternatives. Árnason, a former elementary-school teacher who says he was "raised by an environmental extremist" father (he is not related to the scientist Bragi Árnason), soon found himself chairing a government committee on alternative energy, which was commissioned to submit a report. One of the first people he tapped for the committee was Professor Hydrogen.

SCIENCE MEETS POLITICS

Bragi Árnason began studying Iceland's geothermal resource "as a hobby," he tells me, while a graduate student pursuing doctoral research in the 1970s. His deep knowledge of the island's circulatory system of hot water flows enables him to explain, for example, why the water you shower with in Reykjavík probably last fell as rain back in A.D. 1000. As he came to grasp the size of the resource, he began to consider ways in which this untapped potential might be used. At the time, the climbing cost of oil imports was beginning to hit the fishing fleet, prompting discussion of alternative fuels—including hydrogen.

Iceland has been producing hydrogen since 1958, when it opened a state fertilizer plant on the outskirts of Reykjavík under the post-war Marshall Plan. The production process uses hydro-generated electricity to split water into hydrogen and oxygen molecules—a process called electrolysis (see diagram, page 113). The fertilizer plant uses about 13 megawatts of power annually to produce about 2,000 tons of liquid hydrogen, which is then used to make ammonia for the fertilizer industry. In 1980, Bragi Árnason and colleagues completed a lengthy study on the cost of electrolyzing much larger amounts of hydrogen, using not only hydroelectricity but geothermal steam as well—which can speed up what is a very high-temperature process. Their paper found that this approach would be cheaper than importing hydrogen or making it by conventional electrolysis, but it did not find a receptive audience as oil prices plummeted during the 1980s.

The early 1990s saw a reemergence of Icelandic interest in producing hydrogen, both for powering the fishing fleet and for export as a fuel to the European market. In a 1993 paper, Dr. Árnason argued that a transition in fuels from oil to hydrogen may be "a feasible future option for Iceland and a testing ground for changing fuel technology." He also contended that the country could benefit from using hydrogen sooner than other countries. Some of his reasons included Iceland's small population and high levels of technology; its abundance of hydropower and geothermal energy; and its absence of fossil fuel supplies. Another was the relatively simple infrastructural change involved in converting the fishing fleet from oil to hydrogen, by locating small production plants in major harbor areas and adapting the boats for liquid hydrogen.

Early on, the plan was to use liquid hydrogen to fuel the boats' existing internal combustion engines. But "then came the fuel cell revolution," as Dr. Árnason puts it. By the late 1990s, the fuel cell, an electrochemical device that combines hydrogen and oxygen to produce electricity and water, had achieved dramatic cost reductions over the previous two decades. The technology had become the focus of engineers aiming to make fuel cells a viable replacement not only for the internal combustion engine, but for batteries in portable electronics and for power plants as well. Demonstrations of fuel cell-powered buses in Vancouver and Chicago, and their growing use in hundreds of locations in the United States, Europe, and Japan, caught the attention of governments and major automobile manufacturers. The fuel cell was increasingly viewed as the "enabling technology" for a hydrogen economy.

One Icelander particularly taken with these developments was a young man named Jón Björn Skúlason, who while attending the University of British Columbia in

Vancouver became familiar with Ballard Power Systems, a leading fuel cell manufacturer headquartered just outside the city. Upon returning home, Skúlason encouraged the politician Hjalmar Árnason in his promotion of energy alternatives and hydrogen; his enthusiasm earned him a position on the expert committee. In 1998, the panel formally recommended that the nation consider converting fully to a hydrogen economy within 30 years.

By then, Hjalmar Árnason had already given the process a push. During a phone interview with a reporter from the *Economist,* he floated the year 2030 as a target date for the government's evolving hydrogen plans. The resulting article, published in August 1997, created a buzz abroad, and the parliamentarian received hundreds of phone calls from around the world. That fall, Iceland's prime minister released a statement announcing that the government was officially moving the country toward a hydrogen economy. The ministers of energy and industry, commerce, and environment signed on, as well as both sides of the two-party Althing. And Árnason obtained permission to start negotiating with interested members of industry.

A PIECE OF THE ACTION

Iceland has a tradition of "stock companies," or business cooperatives that evolved in the eighteenth century to help domestic farmers and fishers compete with the formidable Danish trading companies that at the time controlled fishing and goods manufacturing. The first of these, granted royal support in 1752, brought in weavers from Germany, farmers from Norway, and other overseas experts to teach the Icelanders the best methods of agriculture, boat-building, and the manufacture of woolen goods. Over the years, these long-lasting business associations helped the nation's enterprises survive and sometimes thrive.

The formation of the Icelandic Hydrogen and Fuel Cell Company (now Icelandic New Energy) can be seen as the latest example of this stock company tradition—but with a contemporary twist: German carmakers instead of weavers, Norwegian power companies rather than farmers. The first to contact Hjalmar Árnason after publication of the *Economist* article was DaimlerChrysler. Its roots traceable back to Otto Benz, designer of the first internal-combustion engine car, DaimlerChrysler now aspires to be the first maker of fuel cell-powered cars. The firm has entered into a $800 million partnership with Ballard Power Systems and Ford to produce fuel-cell cars, and plans to have the first

buses and cars on European roads in 2002 and 2004, respectively—making Iceland a potentially valuable training ground, especially for testing fuel cell vehicles in a cold climate.

The second company to touch base with the Iceland government was Royal Dutch Shell, the Netherlands-based energy company that, among those now in the oil business, has perhaps the most advanced post-petroleum plans. Birthplace of the "scenario planning" technique that prepared it for the oil shocks of the 1970s better than most businesses, Shell has posited an Iceland-like future for the rest of the world, with 50 percent of energy coming from renewable sources by 2050. The firm surprised its colleagues in mid-1998 by creating a formal Shell Hydrogen division, and then sending its representatives to the World Hydrogen Energy Conference in Buenos Aires.

The third group to establish communications with island officials was Norsk Hydro, a Norwegian energy and industry conglomerate. The company is involved in a trial run of a hydrogen fuel cell bus in Oslo, and has considerable experience in hydrogen production: it has its own fertilizer business, and Norsk Hydro electrolyzers run Iceland's hydrogen-producing fertilizer plant. Norsk Hydro is also involved in the politically sensitive issue of Iceland's planned aluminum smelters, having signed commitments with the national power company and the ministries of energy and industry and commerce to construct a new smelter on the island's east coast.

Negotiations among these companies and the Icelandic government culminated in February 1999 with the creation of the Icelandic Hydrogen and Fuel Cell Company. Shell, DaimlerChrysler, and Norsk Hydro each hold shares of the company. The majority partner, *Vistorka* (which means "eco-energy"), is a holding company owned by a diverse array of Icelandic institutions and enterprises: the New Business Venture Fund, the University of Iceland, the National Fertilizer Plant, the Reykjanes Geothermal Power Plant, the Icelandic Technological-Institute, and the Reykjavík Municipal Power Company. Also indirectly involved with the holding company is the Reykjavík City-Bus Company.

The stated purpose of the new joint venture is to "investigate the potential for replacing the use of fossil fuels in Iceland with hydrogen and creating the world's first hydrogen economy." On the day of its announcement, Iceland's environment minister stated: "The Government of Iceland welcomes the establishment of this company by these parties and considers that the choice of location for this project is an acknowledgement of Iceland's distinctive status and long-term potential." Like the *Economist* article, the announcement attracted industry attention. But for some companies, it was too late to climb on the bandwagon. Toyota officials reportedly attempted, to no avail, to take over the project by offering to foot its entire bill and supply all the needed engineers.

BUSES, CARS, AND BOATS

Bragi Árnason and a colleague, Thorsteinn Sigfússon, have outlined a gradual, five-phase scenario for the hydrogen transformation. In phase one (an estimated $8 million project that has received $1 million from the government), hydrogen fuel cells are to be demonstrated in Reykjavík's 100 municipal public transit buses. The current plan is to have three buses on the streets by 2002. The fertilizer plant will serve as the filling station for the buses, its hydrogen pressurized as a gas and stored on the roofs of the vehicles. Because enough hydrogen can be stored onboard to run a bus for 250 kilometers, the average daily distance traveled by a Reykjavík bus, there is no need for a complicated infrastructure for distributing the fuel.

In phase two, the entire city bus fleet—and possibly those in other parts of the island—will be replaced by fuel cell buses. The Reykjavík bus fleet program has a price tag estimated at $50 million, and this spring received $3.5 million from the European Community. Phase three involves the introduction of private fuel cell passenger cars—which requires a more complicated infrastructure. At present, storing pressurized hydrogen gas onboard a large number of smaller vehicles, with more geographically dispersed refueling requirements, is too expensive to be considered a realistic option. The first fuel cell cars are therefore expected to run not on hydrogen directly, but rather on liquid methanol—which contains bound hydrogen but must be reformed, or heated, onboard the vehicle to produce the hydrogen to power the fuel cell.

Methanol is also, at the moment, the preferred fuel for the final two phases: the testing of a fuel cell-powered fishing vessel, followed by the replacement of the entire boating fleet. These trawlers use electric motors that are in the range of one to two megawatts—larger than those for cars and buses, but close to the size of the fuel cells that are now starting to be commercialized for stationary use in homes and buildings. Several European vessel manufacturers have already expressed their interest in becoming involved in this phase, and Dr. Árnason would like to see a fuel cell boat demonstrated no later than 2006.

But using methanol as an intermediate step to hydrogen is not without its problems. Skúlason, who is now president of Icelandic New Energy, notes that Shell is concerned about the use of methanol, particularly its toxicity. And since methanol reforming releases carbon dioxide, the environmental benefit is much less than if a way can be found to store the direct hydrogen onboard, which in Iceland's case would mean complete elimination of greenhouse gas emissions. It's a difficult decision, notes Skúlason: "We must deal with the technologies we are given by the global companies."

Iceland will have to choose between two options: producing and distributing pure hydrogen and storing it onboard vehicles (the "direct hydrogen" option); or producing hydrogen onboard vehicles from other fuels—natural gas, methanol, ethanol, or gasoline—using a reformer (the "onboard reformer" option). In general, the automobile industry strongly favors the onboard option, using methanol and gasoline, because most existing service stations already handle these fuels. A third path, reforming natural gas at hydrogen refueling stations, is under consideration in countries like the United States, that already have an extensive natural gas network, but is not practical in Iceland.

The up-front costs of direct hydrogen will be high because such a change requires a new infrastructure for transporting hydrogen, handling it at fueling stations, and storing the fuel onboard as a compressed gas or liquid. According to DaimlerChrysler's Ferdinand Panik, retrofitting 30 percent of service stations in the U.S. states of New York, Massachusetts, and California for methanol distribution would cost about $400 million. Supplying hydrogen to these stations would cost about $1.4 billion.

But in terms of long-term societal benefits, direct hydrogen is the clear winner. Using hydrogen directly is more efficient, because of the extra weight of the processor and lower hydrogen content of the methanol or gasoline. It is also less complex than having a reformer onboard each vehicle—which adds $1,500 to the cost of a new car, takes time to warm up, and creates maintenance problems. As the vehicle population grows large enough to cover the capital costs of providing refueling facilities, the costs of direct hydrogen will become comparable to the onboard option. Once the infrastructure and vehicles are put in place, using hydrogen fuel will be more cost-effective than having cars with reformers—even excluding the environmental gains.

If Iceland, with its heavy renewable energy reliance, were to switch directly to hydrogen, the country would have *no* greenhouse emissions. And in fact, it is much easier to produce hydrogen than methanol from renewable energy through electrolysis. Thus, as renewables become more prominent around the world, a hydrogen infrastructure will emerge as the most practical option. In Iceland, rather than require that hydrogen first be used to create, and then be reformed from, methanol, the simplest approach would be to use geothermal power and hydropower, augmented by geothermal steam, to electrolyze water, creating pure hydrogen to drive cars and boats. But behind the seeming solidarity of the public/private venture, a fateful struggle may be emerging.

A FORK IN THE ROAD

In spite of the long-term economic and environmental advantages of the direct hydrogen approach, industry and government—both in Iceland and worldwide—have devoted substantially greater attention and financial support to the intermediate approach of using methanol and onboard reformers. Car companies are hesitant to mass-produce a car that cannot be easily refueled at many locations. Energy companies, similarly, are loathe to invest in pipelines and fuel stations for vehicles that have yet to hit the market. This is a classic case of what some engineers call the chicken-and-egg dilemma of creating a fueling infrastructure. But the potential public benefits—especially for addressing climate change—give governments around the world incentive to steer the private sector toward the optimal long-term solution of a hydrogen infrastructure, by supporting additional research into hydrogen storage and by collaborating with industry.

In Iceland's case, producing pure hydrogen through electrolysis by hydropower is at the moment three times as expensive as importing gasoline. But the fuel cells now being readied for the transportation market are three times as efficient as an internal combustion engine. In other words, running the island's transport and fishing sectors off pure hydrogen from hydropower is becoming economically competitive with operating conventional gasoline-run cars and diesel-run boats.

Since the methanol reformers these fuel cells will presumably use are still several years away from mass production, some scientists see the next few years as an important window of opportunity to prove the viability of direct hydrogen technology. But the history of technology is littered with examples of inferior technologies "locking out" rivals: witness VHS versus Beta in the videocassette recorder market. If methanol does gain market dominance,

and locks out the direct hydrogen approach, it may be decades before real hydrogen cars become widespread—a wrong turn that could take the Icelandic venture kilometers from its destination. By the time a full-blown methanol infrastructure were put in place, it would probably no longer be the preferred fuel—committing the country to a fleet of obsolete cars and causing the consortium to strand millions of *kronur* in financial assets.

Yet some outside developments are pointing in the direction of direct hydrogen. In California, where legislation requires that 10 percent of new cars sold in 2003 must produce "zero-emissions," a consortium called the California Fuel Cell Partnership is planning to test out 50 fuel cell vehicles and build two hydrogen fueling stations that will pump hydrogen gas into onboard fuel tanks. Hydrogen fueling stations have already been built in Sacramento (California's capital), Dearborn, Michigan (home to Ford headquarters), and the airport at Frankfurt, Germany—the last of which expects to eventually import hydrogen from Iceland. The prospect of Iceland becoming a major hydrogen exporter, perhaps the new energy era's "Kuwait of the North," surfaces several times during my interviews—and is no doubt a good selling point for the strategy to officials inclined to think more in narrow economic terms.

Skúlason assures me that there is a "very open discussion" underway within the consortium, and says "we have to take steps slowly because there might be a shift." He admits that he would prefer to see compressed hydrogen gas used, noting the advantages of having direct hydrogen fuel infrastructure and vehicles. Shell and Daimler Chrysler themselves seem to recognize the potential competitive advantage of putting up hydrogen filling stations and reformer-free cars right from the beginning, giving them a headstart in preparing for a world fueled by hydrogen. At a June 2000 conference in Washington, DC, Shell Hydrogen CEO Don Huberts asserted that direct hydrogen was the best fuel for fuel cells, and suggested that geothermal energy converted to hydrogen would be the main means for converting the Icelandic economy. Daimler Chrysler representatives have admitted that their methanol reformers are relatively expensive and large—they take up the entire back seat—and the company has recently rolled out "next generation" prototype cars that run on liquid and compressed hydrogen—prime candidates for the Iceland strategy.

"The transition is messy," the politician Hjalmar Árnason tells me. "We have one leg in the old world, and one in the new." It's an apt metaphor, given Iceland's geography.

But the question is whether the Icelandic venture will, in rather un-Viking fashion, cautiously creep ahead—sticking to the onboard methanol approach—or, brashly set *both* feet in the new world, voyaging straight to direct hydrogen. As a world leader in utilizing renewable energy sources, if Iceland does not take the "newest" path, governments and businesses elsewhere may extract the wrong conclusion from its experiment and give short shrift to the direct hydrogen option. Skúlason nails the conundrum: "How many times will we shift? Will it be cheaper for society to pay a little more now and not have to rebuild? This argument doesn't always work with government or the consumer."

Professor Árnason is quick to note that, whichever short-term infrastructural path the country takes, "the final destination is the same:" pure hydrogen, derived from renewable energy and used directly in fuel cells. But he acknowledges that there may be significant costs in taking the gradual approach. And he agrees that the assumption on which his scenario is based—that methanol is the most economical option—is "subject to revision." The cost and efficiency of fuel cells will continue to improve, and advances in carbon nanotubes, metal hydrides, and other storage technologies are making it more feasible to store hydrogen onboard. The high cost of electrolysis is likely to decline sharply with technical improvements, while other sources of hydrogen—tapping solar, wind, and tidal power, splitting water with direct sunlight, playing with the metabolism of photosynthetic algae—are on the horizon. And new climate policies or fluctuating fuel prices from volatile oil markets "would change the whole picture."

WHY ICELAND?

When he first met with his prospective joint venture partners, Bragi Árnason posed this query: "Why are you interested in coming to Iceland?" He asked the question because "we were quite surprised to learn about the strong interest of these companies in participating in a joint venture with little Iceland." Their answers shed light on some of the elements that may be useful for developing a hydrogen economy elsewhere in the world.

Without a doubt, the most critical element of getting the Iceland experiment underway has been the government's clearly stated commitment to transforming itself into a hydrogen economy within a set timeframe. A similar dynamic is at work in California, where the zero-emission mandate has forced energy and transport companies to join

forces with the public sector to seriously explore hydrogen. For Dr. Árnason, the lesson is clear: a strong public commitment can attract and encourage the participation of private sector leaders, resulting in partnerships that provide the financial and technical support needed to move toward environmental solutions. "You *must* have the politicians," he says.

In addition, companies have shown interest in the Iceland experiment because the results will be applicable around the world. While the country's hydrogen can be produced completely by renewable energy, its car and bus system and heavy reliance on petroleum—amplified by its island setting—are common characteristics of industrial nations, making the result somewhat adaptable. The island's head start in transitioning to renewable energy also makes it a good place to test out this larger shift.

Iceland may also have something more to tell us about the more general cultural building blocks that can enable the evolution of a hydrogen society. Icelanders treasure their hard-won independence, and the prospect of energy self-reliance is attractive. Hjalmar Árnason likes to emphasize his homeland's "free, open society," which he believes has maintained a political process more conducive to bold proposals and less subject to special-interest influence and partisan gridlock. He points, too, to the country's openness to new technology—to its willingness to take part in international scientific endeavors such as global research in human genetics. He hopes Iceland will become a training ground for hydrogen scientists from around the world, cooperating internationally to convert its NATO base to hydrogen. Skúlason cites a poll of Reykjavík citizens indicating that 60 percent of the citizens were familiar with and supportive of the hydrogen strategy—though some ask about the safety of the fuel (it is as safe as gasoline), pointing to the need for public education campaigns before people will be persuaded to buy fuel cell cars.

Another important cultural factor has been what Árni Finnsson, of the Icelandic Nature Conservation, describes as his nation's relatively recent but increasing "encounter with the globalization of environmental issues." This encounter originated with the emotional whaling disputes of the 1970s and 80s, and today includes debates about persistent organic pollutants and climate change. As Icelanders seek to become more a part of global society, so too do they seek legitimacy on global issues, forcing their government to sensitize itself to emerging cross-border debates—a process that has sometimes created Iceland's political equivalent of volcanic eruptions.

Finnsson points out that, thanks to the "Kyoto dilemma," Icelandic climate policy is not terribly progressive, consisting mainly of efforts to create loopholes that would allow additional greenhouse gas emissions from its new aluminum smelters. But there is little doubt that this dilemma has also unwittingly helped encourage the hydrogen strategy, by forcing the nation to explore deep changes in its energy system. In a land that, even as it becomes wired to the information age, routinely blocks new road projects due to age-old superstitions of upsetting elves and other "hidden people," it's a contradiction that somehow seems appropriate. A country that has stubbornly refused to sign the Kyoto Protocol provides the most compelling evidence to date that climate change concerns—and commitments—will increasingly drive the great hydrogen transformation.

But my favorite, if least provable, theory for "Why Iceland?" comes from the heroic ideals of its sagas. One of the recurring themes of these remarkable literary works is that a person's true value lies in renown after death, in becoming a force in the lives of later generations through one's deeds. Listening to Bragi Árnason, who is now 65, one cannot help but wonder whether this cultural concern for renown is playing a part in the saga now unfolding: how Iceland became the world's first hydrogen society, inspiring the rest of the globe to follow its lead. "Many people ask me how soon this will happen. I tell them, 'We are living at the beginning of the transition. You will see the end of it. And your children, they will live in this world.'"

QUESTIONS

1. Who first proposed the use of hydrogen as a viable energy source?

2. What percent of Iceland's energy needs is met by the renewable energy sources?

3. Describe the operation of a fuel cell.

4. Discuss various features of Iceland's "Hydrogen society" strategy.

5. Discuss potential impediments to the completion of the hydrogen experiment in Iceland.

14

Solar Energy

Ahmed S. Khan

Barbara A. Eichler

The sun was the major source of energy before the Industrial Age (Table 14.1). In today's electro-optics industrial age, it still remains a vast reservoir of energy which can fulfill today's and tomorrow's growing demands for energy. According to A.B. Lovins, "Solar power in general has several unique implications which do not arise from its obvious advantages. For example, it could help to redress the severe temperate and tropical zones; its diffuseness is a spur to decentralization and increased self-sufficiency of population; and as the least sophisticated major energy technology it could greatly reduce tensions resulting from the uneven distribution of fuels and from limited transfer of technology."

The sun is a continuous source of energy. The mass of the sun is 2.2×10^{27} tons. The sun converts its mass into energy at a rate of 4.2 million tons per second which is 0.02×10^{18} percent of its mass per second. At this rate the sun is expected to radiate energy for billions of years.

Solar radiation is earth's main income energy. Outside the earth's atmosphere the sun provides energy at a rate of 1,353 watts/m normal to sun. Considering the diametral plane of earth (1.27×10^{14} m) the solar input is 1.73×10^{17} watts. But only a small fraction of this reaches earth's surface. The majority of solar energy incident on earth is lost due to reflection by the atmosphere, scattering by clouds, water particles and dust particles (Figure 14.1). The average solar radiation reaching earth's surface is approximately 630 watts per meter. The energy emitted by the sun is spread over a broad band of electromagnetic spectrum. Most of the solar radiation lies in the range of 0.3 micrometer (μm) to 1.3 μm. The solar radiation does not come with constant intensity during the day. The average intensity varies with cloud cover, latitude, season, and time of the day. Solar energy could be used in a number of applications (Figure 14.2). In its intrinsic form the solar radiation can be used for heating. For other applications it needs to be converted into other forms of energy.

ADVANTAGES

1. Continuous source of energy.

2. Clean source of energy.

3. Safe source of energy.

Table 14.1
Use of Solar Energy

B.C.	Solar heat for distillation of liquids and drying agricultural products.
1913	Solar power electrical plants in Egypt.
1960	Use of solar water heaters in Florida.
1967	Use of silicon solar cells in Japan for isolated radio repeaters.
1968	Development of large-scale mirrored solar oven. Produces one megawatt of power per day.
1974	Breakthrough in mass production of photo-voltaic cells.
1980s	Afghan refugees in Pakistan use 100,000 solar cookers for preparing food.

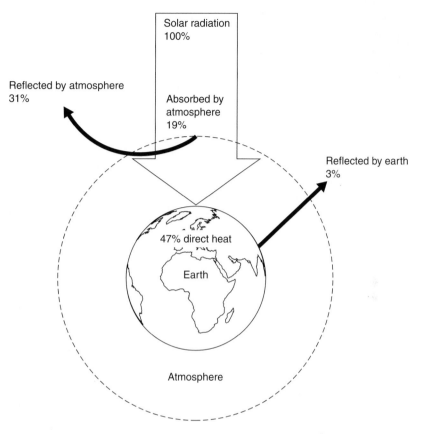

Figure 14.1
Solar Radiation Incident on Earth.

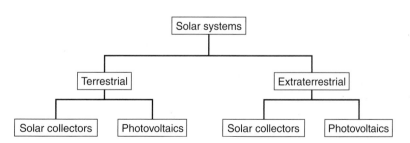

Figure 14.2
Classification of Solar Energy Systems.

Terrestrial Systems

Terrestrial systems consist of flat-plate or concentrating collectors which transfer solar heat to a carrying medium such as water. The impounded heat can be stored or converted immediately to work. Terrestrial collectors are subject to the intermittent nature of solar cycles and local climates.

Extraterrestrial Systems

The amount of solar power available for power generation in a geosynchronous orbit (35,800 Km from earth) is about 15 times that available on earth. The idea of an orbital solar power station was presented by P.E. Glaser in 1973. According to Glaser, a 5000 MW orbital power station employs concentrators to reflect sunlight on photoelectric solar cells assembled into two large arrays each 4.33 km × 5.2 km in dimension. The generated electrical power is converted into microwave power which can be transmitted towards earth in a focused beam with the help of a 1 km diameter antenna. At earth the microwave radiation is received by a 7.12 km diameter antenna. The received microwave power can be converted into DC or 60 Hz AC for power distribution. The power density of microwaves varies as a function of its frequency—the higher the frequency, the higher is the power density. The high frequency microwaves can cause damage to living tissue.

Scenario

Fort Worth, Texas: March 1, 20XX

TERASOLAR-I, the first orbital solar power generating station (Figure 14.3), has completed its twentieth year of service in space. During its twenty year service it has provided an average of 8,000 MW

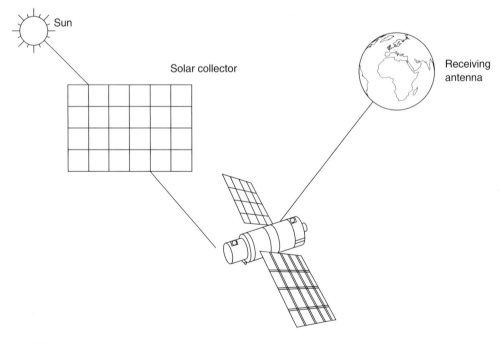

Figure 14.3
TERASOLAR-I, an Orbital Solar Power-Generating Station.

(45% of power consumption of Texas) of power to a receiving station. The station transmits power at 30 GHz. Originally the power station was designed to use a microwave frequency band of 300 MHz, but to decrease the size of the orbital transmitting antenna and earth's receiving antenna, and in order to cut the cost of the project, the use of a 30 GHz band was approved. A recent medical survey in Dallas reports that the number of cases of cancer has increased by 30%.

Response

Do you think that the radiation from the orbital solar power station is responsible for an increase in the number of cancer cases in the Fort Worth area? How could this have been avoided? Going back into the 20th century, develop an energy strategy that could have avoided this situation.

DISADVANTAGES

1. Average power is not steady and depends on cloud cover, season, latitude, and time of day.
2. Not available at night.
3. Needs to be converted into other forms (e.g., electrical) for useful applications.
4. Extraterrestrial solar energy will increase the heat burden of the biosphere.

REFERENCES

Dorf, R. (1981). *The Energy Factbook.* New York, McGraw-Hill.
Gabel, M. (1975). *Energy, Earth, and Everyone.* San Francisco, Straight Arrow Books.
Knoepfel, H. (1986). *Energy 2000: An Overview of the World's Energy Resources in the Decades to Come.* New York, Gordon and Breach Science Publishers.
World Resources 1988–89, A Report by World Resources Institute. New York, Basic Books.

Figure 14.4
An Array of Photovoltaic (PV) Solar Panels.

QUESTIONS

1. Define the following:
   a. Terrestrial systems
   b. Extraterrestrial systems
2. List some advantages (+) and disadvantages (−) of solar energy:

(+)	(−)

15

Wind Power

AHMED S. KHAN

BARBARA A. EICHLER

The atmosphere is a reservoir of solar radiation. The wind is continuously regenerated in the atmosphere as the solar radiation is converted into kinetic energy. The winds are local as well as regional. It is estimated that the average power available from shifting air masses all over the earth is 1.8×10^{15} watts. Wind power available at any location depends on its topographical features. The earth's surface offers resistance to wind, thereby decreasing its power density. In some locations, a wind power density of 500 W/m 10^2 is available, at a nominal height of 25 m from the ground.

The use of wind power dates back thousands of years. It was employed for grinding of grain and pumping of water. Table 15.1 lists the history of wind power.

Wind power can be used to turn vanes, blades, or propellers attached to a shaft. The revolving shaft spins the rotor of a generator, which produces electricity.

ADVANTAGES

1. Continuous source of energy.
2. Clean source of energy, no emissions into atmosphere.
3. Does not add to thermal burden of earth.

DISADVANTAGES

1. For most locations the wind power density is low.
2. The wind velocity must be greater than 7 mph to be usable in most cases.
3. Problem exists in variation in the power density and duration of wind.

Table 15.1
History of Wind Power

1000 B.C.	Wind power of sailing ships
1850	Use of windmills in America
1894	First use of wind power for electric generation by Arctic explorer Nasen
1929	Development of electric wind turbine, 20 meters in diameter, in Bourget, France
1931	Development of 100 kW wind turbine capacity in Yalta, USSR
1941	Development of 1,250 kW capacity wind turbine in Vermont, USA
1950	Development of 10-kW Hutter wind generator consisting of 200-foot blades atop 475-foot tower.
1951	Development of Thomas 6,500-kW generator, consisting of 200-foot blades atop 475-foot tower.
1954	In USSR, the number of wind power plants reached 29,500 with a total capacity of 1 billion kWh.
1957	Development of 200-kW fully automated unit with three 45-foot blades mounted on a 75-foot tower, in Denmark
1960	600 kW Gedser generator designed
1980s	Advances in wind power generators

4. Wind power may have some environmental effects, depending on the location and number of wind power plants (local climate, bird migration patterns, etc.).

REFERENCES

Dorf, R. (1981). *The Energy Factbook.* New York, McGraw-Hill.

Gabel, M. (1975). *Energy, Earth, and Everyone.* San Francisco, Straight Arrow Books.

Knoepfel, H. (1986). *Energy 2000: An Overview of the World's Energy Resources in the Decades to Come.* New York, Gordon and Breach Science Publishers.

World Resources 1988–89, A Report by World Resources Institute. New York, Basic Books.

QUESTIONS

1. How is wind power used to generate electricity?

2. List the advantages (+) and disadvantages (−) of wind power:

(+)	(−)

16

Wind Power: Small, But Growing Fast

CHRISTOPHER FLAVIN

Wind power is now the world's fastest growing energy source. Global wind power generating capacity rose to 4,900 megawatts at the end of 1995, up from 3,700 megawatts a year earlier (see Figure 16.1). Since 1990, total installed wind power capacity has risen by 150 percent, representing an annual growth rate of 20 percent.

By contrast, nuclear power is growing at a rate of less than 1 percent per year, while world coal combustion has not grown at all in the 1990s.

If the world's roughly 25,000 wind turbines were spinning simultaneously, they could light 122 million 40-watt light bulbs or power over a million suburban homes. In the windy north German state of Schleswig-Holstein, wind power already provides 8 percent of the electricity.

Although it now generates less than 1 percent of the *world's* electricity, the rapid growth and steady technological advance of wind power suggest that it could become an important energy source for many nations within the next decade. The computer industry has demonstrated the potentially powerful impact of double digit growth rates. The fact that personal computers provided less than 1 percent of world computing power in 1980 did not prevent them—a decade later—from dominating the industry, and changing the very nature of work.

Wind power is being propelled largely by its environmental advantages. Unlike coal-fired power plants, the leading source of electricity today, wind power produces no health-damaging air pollution or acid rain. Nor does it produce carbon dioxide—the leading greenhouse gas now destabilizing the world's atmosphere.

In many regions, wind power is now competitive with new fossil fuel-fired power plants. At an average wind speed of 6 meters per second (13 miles per hour) wind power now costs 5–7 cents per kilowatt-hour, similar or slightly lower than the range for new coal plants. As wind turbines are further improved, with lighter and more aerodynamic blades as well as better control systems, and as they are produced in greater quantity, costs could fall even further, making wind power one of the world's most economical electricity sources.

The modern wind power industry established its roots in Denmark and California in the early 1980s. Spurred by government research funds, generous tax incentives, and guaranteed access to the electricity grids, a sizable wind industry was created. However, development slowed dramatically at the end of the decade as government tax incentives were withdrawn and utilities became more resistant to higher-cost electricity.

Even as political support for wind power waned in the late 1980s, the technology continued to mature. Many of the small turbines installed in the early days were expensive and unreliable, but the lessons learned from those first generation turbines were soon translated into new and improved models. The turbines that entered the market in the early 1990s incorporated advanced synthetic materials, sophisticated electronic controls, and the latest in aerodynamic designs.

In an effort to make wind power more economical, most companies have built larger and larger turbines. In Germany, the average turbine installed in 1995 had a capac-

Reprinted with permission from Worldwatch Institute, Washington, DC., September/October 1996. pp.35-37.

Figure 16.1
World Wind Energy-Generating
Capacity, 1980–95.

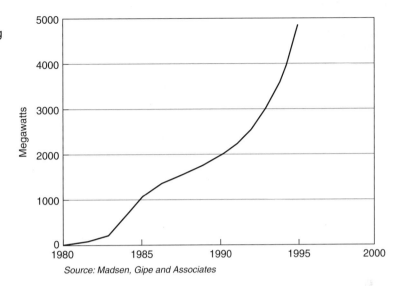

Source: Madsen, Gipe and Associates

ity of 480 kilowatts, up from 370 kilowatts in 1994 and 180 kilowatts in 1992. Several manufacturers will soon introduce machines that can generate between 1,000 and 1,500 kilowatts—with blade spans as great as 65 meters.

The 1,290 megawatts of wind generating capacity added in 1995 was almost double the capacity added a year earlier, and up sixfold from the 1990 figure (see Figure 16.2). In 1995, the country with the most new capacity was again Germany, which added 505 megawatts, the most any country has ever installed in a single year. India added 375 megawatts, followed by Denmark with 98 new megawatts, Netherlands with 95, and Spain with 58.

The European wind industry is now growing at an explosive pace: altogether, Europe had 2,500 megawatts of wind power capacity at the end of 1995, up nearly threefold from 860 megawatts in 1992 (see Figure 16.3). The United States still led the world with 1,650 megawatts of wind power capacity at the end of 1995, but Germany was closing in fast with 1,130 megawatts. Denmark ranked third with 610 megawatts, and India fourth at 580 megawatts.

Europe is now home to most of the world's leading wind power companies, which are introducing larger and more cost-effective models. Unlike the United States, where most development has consisted of large groups of 20 to 100 turbines, called "wind farms," Denmark and Germany have pursued a decentralized approach to wind power development. Most of their machines are installed one or two at a time, across the rural landscape. This has made them popular with local communities, which benefit from the additional income, public revenues, and jobs that result.

Europe's leadership also stems from the financial incentives and high purchase prices established for renewable energy in response to concern about the atmospheric pollution caused by fossil-fuel-fired power plants. In Germany, this approach has allowed determined investors and environmental advocates to beat back efforts by the electric utilities to reverse the 1991 "electricity in-feed law," which provides a generous price of about 11 cents

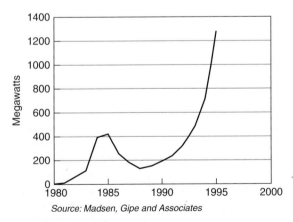

Source: Madsen, Gipe and Associates

Figure 16.2
Annual Additions to World Wind Energy-Generating
Capacity, 1980–95.

Figure 16.3
Wind-Generating Capacity, by
Region, 1980–95.

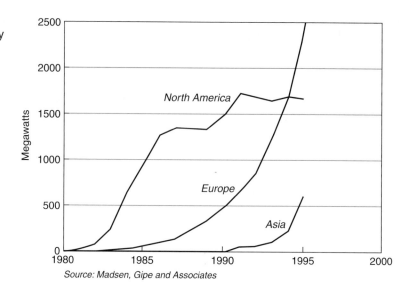

Source: Madsen, Gipe and Associates

per kilowatt-hour to electricity generators relying on solar, wind, and biomass energy. In a landmark vote in 1995, the Bundestag decided to uphold the law, though it remains under review by the courts.

Wind energy is also advancing rapidly in the Netherlands, Spain, and the United Kingdom. The U.K. has Europe's largest wind power potential, and hundreds of megawatts of projects are now being planned. European wind industry leaders are also hopeful that sizable wind power markets will soon emerge in Finland, Greece, Ireland, and Sweden, each of which has a large wind resource. Even France, the last bastion of the European nuclear industry, embarked on a sizable wind power development plan in 1995, aimed at adding 250 to 450 megawatts of wind power over the next decade.

Just as wind energy development is taking off in Europe, it has stalled in the United States, where the industry is buffeted by uncertainty about the future structure of the electricity industry. In fact, the country's total wind capacity has hardly increased since 1991. The country that led the world into wind power in the 1980s actually saw a net decline of 8 megawatts in its installed capacity in 1995. Some 50 megawatts were added—mainly in Texas—but 58 megawatts of old turbines were torn down in California. Kenetech, the leading U.S. wind power company, filed for bankruptcy in May 1996 after the combined effects of a slow market and mechanical problems with its new turbine led to large financial losses.

Prospects for developing nations are far brighter. Although most wind turbines are currently installed in industrial countries, much of the world's wind power potential is in the developing world. The leader so far is India, which is the first developing country with a real commercial market for wind power. India's roughly 3,000 wind turbines have virtually all been installed since the government opened the electricity grid to independent power producers and enacted tax incentives for renewable energy investments in the early 1990s. According to the government, 730 megawatts had been installed by April 1, 1996, which would make India the world's most active wind market in early 1996. However, uncertainty surrounding the Indian elections in May has slowed the pace of development since then.

Some of India's wind turbines are being imported, but others are manufactured in India, either by domestic companies or in joint ventures with foreign companies. Already, the Indian industry has more than 20 indigenous manufacturers and suppliers. In the windy southern state of Tamil Nadu, hundreds of jobs have been created as a result.

Many other developing countries, including Argentina, Brazil, China, Egypt, Mexico, and the Philippines are surveying their wind resources and installing small numbers of turbines on an experimental basis. Although none of these countries has yet encouraged or even permitted the development of a sustained, market-driven wind industry, some may be on the verge. China, for example, has just

36 megawatts installed but has plans to reach 1,000 megawatts by the year 2000.

In most developing countries, wind power development will be driven not by environmental concerns, as it is in industrial countries, but by a desperate need for electricity, which is in short supply throughout the Third World. In areas such as western China and northeast Brazil, wind power is the only indigenous source of electricity ready to be developed on a large scale.

The global wind energy potential is roughly five times current global electricity use—even excluding environmentally sensitive areas. In the United States, where detailed surveys have been conducted, it appears that wind turbines installed on 0.6 percent of the land area of the 48 contiguous states—mainly in the Great Plains—could meet one-fifth of current U.S. power needs—double the current contribution of hydropower. By comparison, the total cropland used to grow corn in the United States is nearly 3 percent of the country's land area. And unlike corn, wind power does not preclude the land from being used simultaneously for other purposes, including agriculture and grazing.

Other countries that have enough wind potential to supply most or all their electricity include Argentina, Canada, Chile, Russia, and the United Kingdom. China's wind energy potential is estimated by the government at 253,000 megawatts, which exceeds the country's current generating capacity from all sources by 40 percent. Much of that potential is located in Inner Mongolia, near some of the country's leading industrial centers.

India's potential is estimated at 80,000 megawatts, which equals the country's total current generating capacity. Europe could obtain between 7 and 26 percent of its power from the wind, depending on how much land is excluded for environmental reasons. Offshore potential in Europe's North and Baltic Seas is even greater.

Wind power cannot fully replace fossil fuels, but it has the potential to meet or exceed the 20 percent of world electricity provided by hydropower. Moreover, though wind power is more abundant in some areas than others, it is in fact one of the world's most widely distributed energy resources. More countries have wind power potential than have large resources of hydropower or coal.

Combined with other renewable energy sources such as solar and geothermal power, and by a new generation of gas-fired micro-power plants located in office and apartment buildings, wind power could help transform the world electricity system. These technologies could quickly replace coal and nuclear power—which together now supply two-thirds of the world's electricity—and allow a sharp reduction in world carbon emissions.

QUESTIONS

1. What percentage of the world's total electricity is produced by wind power?

2. Can wind power compete with new fossil-fuel-fired plants?

3. Compare the wind-generating capacities of Asia, Europe, and North America.

4. Why have wind energy projects stalled in the United States?

5. Discuss the factors that are responsible for the development of wind power projects in the Third World.

6. Will wind power someday replace fossil fuel power?

Hydroelectric Power

Ahmed S. Khan

Barbara A. Eichler

Hydroelectric power is the conversion of the gravitational pull of falling water of rivers and controlled release of water reservoirs through turbine generators. Hydroelectric power (Tables 17.1–17.3) provides a clean and efficient means of producing electric power. It supplied 21 percent of electricity worldwide in 1986, less than coal and oil but more than nuclear power. In 1984, the global yearly production of hydro energy amounted to about 1,700 billion kilowatt hours (kWh); an additional 550 was under construction. The estimated underdeveloped global hydroelectric potential is 5–50 trillion kW-hours per year. Southeast Asia, South America, and Africa contain respectively 16%, 20%, and 27% of the world's theoretical hydroelectric potential.

According to the World Bank, 31 developing countries doubled their hydropower capacity between 1980 and 1985, much of it with small-scale projects. Small-scale hydropower generation provided almost 10 billion watts worldwide by 1983. China is the world leader, with about 90,000 small hydropower stations supplying electricity to rural areas.

In the United States by early 1988, according to the Federal Energy Regulatory Commission (FERC), more than 2,000 hydro projects were operating. But according to the Environmental Protection Agency (EPA), there were 15,000 private hydro dams. So far, America has dammed about 17% of its 3.5 million miles of natural rivers, mostly during the last hundred years.

Throughout history there have been instances of dam failure and discharge of stored water, which have caused considerable loss of life and great damage to property. Advances in soil mechanics and structural engineering have revolutionized dam construction and hence increased

safety aspects. However, it is estimated that about 150,000 dams around the world present a potential hazard to life or property; there have been 200 failures since 1900. Table 17.4 lists major failures resulting in major loss of life.

ADVANTAGES

1. No air, thermal, chemical pollution in electricity generation.

2. Low production costs.

3. High efficiency (about 90%) converting from water to electrical energy.

4. Water reservoir can provide potential flood protection for downstream currents.

5. Groundwater reserves are increased by recharging from the dam's water reservoir.

Table 17.1

History of Hydro Power

B.C.	Use of water wheels.
1000	Development of water driven blast furnace.
1500	Use of water pumping works.
1882	First hydro-electric power station built at Fox River in Appleton, Wisconsin with a capacity of 25 kW.
1885	First large hydro-electric power station built at Niagara Falls, New York.
1936	Hoover Dam built with a capacity of 1345 MW.

Table 17.2
Power Generation Capacity of Major Dams of World

Name of Dam/Location	Capacity (MW)
Guri, Venezuela	10,000
Churchill Falls, Canada	5,225
Bratsk, Siberia, Russia	4,150
Grand Coulee Dam, Washington	4,200
Aswan Dam, Egypt	2,100
Terbala Dam, Pakistan	3,500
Hoover Dam, Nevada	1,345
Mangla, Pakistan	1,000
Three Gorges Dam, China expected completion date: 2009	18,200

6. Dam's water reservoir can store large volume of water for long periods of time; thus, downstream flow can be controlled for water quality and seasonal stream extreme conditions.

DISADVANTAGES

1. High construction cost.

2. Limited feasible sites for dam construction.

3. Electrical power production may be discontinued due to severe drought conditions.

Table 17.3
Estimated Hydro-Power Potential of World

Region	Potential (1000 MW)	Percent Developed
Africa	437	10%
Asia (excluding CIS republics)	684	5%
Europe (excluding CIS republics)	215	50%
Commonwealth of Independent States (CIS)	269	10%
North America	330	30%
South America	288	8%
Oceania	37	20%

Source: *The Energy Factbook* by Richard C. Dorf. New York: McGraw-Hill.

Table 17.4
Major Dam Failures

Year	Dam	Country
1626	San Ildefonso	Bolivia
1802	Puentes	Spain
1864	Bradfield	England
1889	Johnstown	U.S.
1890	Walnut Grove	U.S.
1895	Bouzey	France
1911	Austin	U.S.
1916	Bila Densa	Czechoslovakia
1917	Tigra	India
1923	Gienco	Italy
1928	St. Francis	U.S.
1935	Alla S. Zerbimo	Italy
1948	Fred Burr	U.S.
1959	Malpasset	France
1961	Kuala Lumpur	Malaya
1961	Babi Yar	Soviet Union
1963	Baldwin Hills	U.S.
1963	Vaiont	Italy
1972	Buffalo Creek	U.S.
1976	Teton	U.S.
1977	Kelly Barnes	U.S.
1979	Machhu II	India

The International Commission on Large Dams (ICOLD) was Formed in 1928 by 6 Countries with the Purpose of Developing and Exchanging Dam Design Experience, and it has Grown to 76 Member Countries. In 1982 ICOLD Established a Committee on Dam Safety to Define Common Safety Principles, Integrate Efforts, and Develop Guidelines, and in 1987 ICOLD Published "Dam Safety Guidelines."
Source: *McGraw-Hill Encyclopedia of Science & Technology*, 7th Edition, p. 20.

4. Dam construction causes loss of land suitable for agriculture.

5. Construction of dams impacts the ecological cycles of the rivers and surrounding landscape.

6. Silt accumulation and sedimentation changes flow and land drainage patterns.

7. Water stored in the dam's reservoir by impounding the river is low in oxygen; therefore, the water issued from the dam is low in oxygen and affects the species of the water stream.

8. Dam construction prevents up-stream migration of fish.

REFERENCES

Dorf, R. (1981). *The Energy Factbook*. New York, McGraw-Hill.

Gabel, M. (1975). *Energy, Earth, and Everyone.* San Francisco, Straight Arrow Books.

Knoepfel, H. (1986). *Energy 2000: an Overview of the World's Energy Resources in the Decades to Come.* New York, Gordon and Breach Science Publishers.

McGraw-Hill Encyclopedia of Science and Technology, 7th ed. New York, McGraw-Hill.

World Resources 1988–89, a Report by World Resources Institute. New York, Basic Books.

QUESTIONS

1. Define the following:
   a. KWh
   b. Hydroelectric
   c. EPA
2. List some advantages (+) and disadvantages (−) of hydroelectric power:

(+)	(−)

All the Wild Rivers

CURTIS RUNYAN

In 1966, Floyd Dominy, the commissioner of the U.S. Bureau of Reclamation, gave a speech lambasting environmentalists for their opposition to damming up the Grand Canyon national park. If the dams were not built, he told the audience, the Colorado River would be "useless to anyone." Dominy, head of the agency that led the charge in the United States' rush to dam up its rivers, concluded: "I've seen all the wild rivers I ever want to see."

Thirty years later, in 1998, Bruce Babbitt, the U.S. Secretary of the Interior, traveled across the country to several rivers on a "Sledgehammer tour"—not to break ground on new construction, but to tear four dams down. "America overshot the mark in our dam building frenzy," he said in a speech to the Ecological Society of America. "For most of this century, politicians have eagerly rushed in, amidst cheering crowds, to claim credit for the construction of 75,000 dams all across America. Think about that number. That means we have been building, on average, one large dam a day, every single day, since the Declaration of Independence. Many of these dams have become monuments, expected to last forever. You could say forever just got a lot shorter."

One of the world leaders in building new dams, the United States, is now leading the world in tearing them down. The country is now decommissioning more large dams than it builds each year, and has removed at least 465 of them, according to a study by American Rivers, Friends

of the Earth, and Trout Unlimited. France and other countries are following suit. "It's striking how, in just two or three decades, the U.S. has gone from building dams to not building dams to taking some of them down," wrote Marc Reisner, author of *Cadillac Desert,* in the Earth Day 2000 edition of *Time* magazine. "What we're just beginning to understand is how water development has, like nuclear energy, amounted to a Faustian bargain between civilization and the natural world."

Ecologically, rivers are under siege. They are being drained, diverted, polluted, and blocked at a rate that has degraded freshwater ecosystems worldwide. With more than half of the world's rivers stopped up by at least one large dam (over 15 meters high), dams have played a significant role in destabilizing riverine ecology. For example, at least one fifth of the world's freshwater fish are now endangered or extinct. In addition, reservoirs behind dams have flooded vast amounts of the world's most fertile agricultural and forest land. Reservoirs also trap the sediment loads of rivers, reduce the supply of nutrients flowing downstream, release water at cooler temperatures, and disrupt healthy river ecosystems.

The ill-effects of dams are not confined to river valleys. Half of the world's dams were built to irrigate the farmland that now provides about 12 to 16 percent of the human food supply. However, channeling water to irrigate basins without good drainage has led to extensive salinization and waterlogging of soils. Bad drainage and poorly planned irrigation—including groundwater pumping—have reduced or ended the productivity of nearly one-fifth of the world's irrigated land.

Reprinted with permission from Worldwatch magazine (www. worldwatch.org)Worldwatch Institute,Washington, D.C., March/April 1997. pp. 19-25

133

The impact of dam building on communities has also been substantial. An estimated 40 to 80 million people have been physically displaced by the construction of dams. They have been flooded out, forced to move. One of the world's most massive engineering projects, the Three Gorges Dam in China, if completed could force the relocation of nearly 2 million people. Most frequently, the people affected are not those who receive the irrigation, electricity, or other benefits provided by dams. In fact, those who are resettled have rarely ever seen their livelihoods restored. "The poor, other vulnerable groups, and future generations are likely to bear a disproportionate share of the social and environmental costs of large dam projects without gaining a commensurate share of the economic benefits," finds the World Commission on Dams (WCD), an independent, collaborative body consisting of dam construction industry representatives, anti-dam activists, and government officials, among others. The commission released its landmark report in November 2000, providing one of the first global surveys of dams with input from both supporters and critics.

The U.S. effort to consider dam removal or breaching is only part of a worldwide shift in thinking about dams. In almost every country in the world, the number of new dams being built is plummeting. Ninety-one large dams were built in the 1970s in Brazil, for instance. The number built dropped to 60 in the 1980s, and to 28 in the first six years of the 1990s (see Figure 18.1). Even where dams continue to

be built, public acceptance is waning, says Owen Lammers of the Glen Canyon Action Network, an ambitious U.S. activist group pushing to tear down the massive Glen Canyon Dam in Arizona. "The number of dams being constructed is going down," says Lammers, "while the number facing resistance and severe criticism is going up."

Evoking the almost religious fervor with which dams have been built in the past, Prime Minister Jewaharlal Nehru called the massive concrete and earthen structures being put up around his country "the temples of modern India." But after a half-century of being regarded as technological marvels, many of these structures are being reinspected and rejected as boondoggles.

Still, the number of dams and dam projects that have been stopped or removed is only a tiny fraction of those that have been built in the past half century. And projects that face strong opposition may also be getting strong support from urban residents, large-scale farmers, or other groups that stand to benefit from a dam's construction. The Sardar Sarovar Dam on India's Narmada River, for example, is at the center of the country's debate over how development should occur. The Narmada Bachao Andolan (NBA), the local movement opposed to the damming, has rallied international attention against the project, which includes plans to construct 3,200 dams on the river. But despite the opposition, in October 2000 the Indian Supreme Court lifted its four-year stay on the project.

Figure 18.1

Number of New Large Dams Constructed

Source: International Commission on Large Dams, *World Register of Dams, 1998*.

*1997–99 projected.

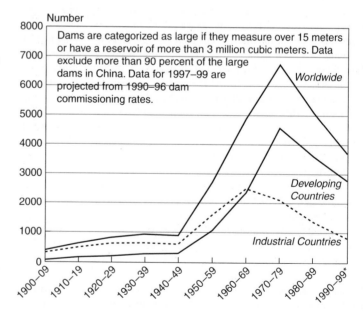

When the global rush to build dams reached its peak in the 1970s, on average two or three large dams were commissioned around the world every day. International lenders, governments, development agencies all felt they had found in dams a solution to many of the world's development dilemmas. Dams have played an important role in addressing hunger, drought, and lack of access to clean water and electricity. They generate 19 percent of the world's electricity supply, provide water for 30 to 40 percent of the world's irrigated land, and in some places help to reduce floods. But the benefits of controlling unruly waterways—building dams and creating reservoirs with the aim of halting floods, expanding irrigation, providing drinking water, and supplying hydroelectric energy—have always been assumed to overwhelmingly outweigh the costs, even though little was known about what these costs were. However, as researchers conduct more studies on the effects of dams and as more of the local people who are affected are consulted, the assumption that the benefits outweigh the costs has become less certain.

Now that more than 45,000 large dams (over 15 meters high) have been built around the world, a growing body of research indicates that their costs may be higher than many ever imagined. The World Commission on Dams report finds that "In too many cases an unacceptable and often unnecessary price has been paid [to secure the benefits of dams], especially in social and environmental terms, by people displaced, by communities downstream, by taxpayers, and by the natural environment." Irrigation schemes haven't supplied projected revenues, hydropower dams have not met electricity-generation projections, drinking water supplies have been costly and often unreliable, and reservoirs have lost their usefulness as they fill with sediment. Recent studies have shown that the organic debris washed into reservoirs releases large amounts of greenhouse gases, raising questions as to whether hydroelectric dams really do produce clean, renewable energy. "Considering the enormous capital invested in large dams, it is surprising that substantive evaluations of project performance are few in number, narrow in scope, and poorly integrated," finds the report.

Even the World Bank, the world's largest international funder of dam projects (the Bank has invested $75 billion in 538 dams), has begun to have second thoughts. "Our involvement in large dams has been decreasing and is focusing more on financing dam rehabilitation and safety and much less on financing new dams," said World Bank President James Wolfensohn in November 2000. To put the Bank's shift fully in perspective, however, it might be noted that while opposition to dams has been increasingly

effective, the most common reason for the dramatic drop in the growth of dams is simply that many countries are already at capacity—there are fewer and fewer safe or unprotected places left to build the structures.

The jury is still out on how the World Bank, which helped instigate the dam commission, will respond to the findings of the WDC report. Wolfensohn recently told an audience of Indian business reporters that "It is unfortunate that the World Bank could not understand the depth of the water crisis in Gujarat and had to pull out of the Narmada project," which is fiercely opposed by the NBA.

"We note and appreciate that the World Commission on Dams report vindicates many concerns raised by NGO campaigns," announced an international coalition of more than 100 nongovernmental anti-dam activist groups in November 2000. In many ways, the World Commission on Dams report provides an up-front review of adverse impacts that most dam projects are never subjected to. The activists contend that if the planning process proposed by the WCD had been followed in the past, many dams would never have been built. The report concludes that dam projects should require the consent of affected communities, participatory decision-making, examination of alternatives to dams, requirements to "sustain aquatic ecosystems," and mechanisms to ensure proper reimbursement to affected communities. The coalition of activists has called for suspension of all large dam projects until countries follow the report's recommendations for equitable, accountable, and participatory decision-making.

The debate over dams has come a long way since Dominy's call 30 years ago to silence all the rivers. And while the thinking about dams has expanded since then, so has the number of dams that choke the world's rivers. It is time to take the lessons learned from constructing more than 45,000 large dams around the world, and to incorporate them into our thinking about future planning for our rivers. *For more information see: The World Commission on Dams, www.dams.org*

QUESTIONS

1. What are the ill effects of dams on the ecology?

2. Using Figure 18.1, which illustrates the number of new large dams constructed during 1900–1999, interpret the chart for developing and developed countries.

3. Why is there a worldwide decline in the number of new large dam construction?

19

The Electric Car Arrives—Again

S ETH D UNN

It was two bicycle mechanics from Massachusetts, Charles and Frank Duryea, who rolled out the first commercially manufactured automobiles—13 of their Duryea Motor Wagons—in Detroit in June of 1896. Dubbed "horseless carriages," they ran on a noisy new invention known as the internal-combustion engine—and a pungent fuel called gasoline. Soon, cars were spreading across the countryside— their costs falling rapidly, thanks to Henry Ford's assembly lines—and the world was moving toward a heavy dependence on oil.

One casualty of the internal-combustion engine's triumph was the electric car, which had become quite popular in the 1890s. Proclaimed as quieter, cleaner, and simpler than the engine-driven car, the electric vehicle was widely expected to dominate the automotive market of the twentieth century. Instead, it quietly disappeared as automobile companies chose to pour billions of dollars into developing, and incrementally improving, the internal combustion engine. The electric auto, it seemed, was destined for the scrap heap of technological wrong turns.

It may have taken a century, but suddenly the electric car has returned from the dead. Its comeback has been fueled in large part by the engine-induced smog now filling urban areas like Athens, Bangkok, and Los Angeles— just as the manure piling up in the streets of America's cities a hundred years ago prodded the search for alternatives to the horsedrawn carriage. After years of false starts

and heated debate, electric cars are on the road again: already an estimated 7,500 "engineless carriages" are now in use worldwide.

The most telling sign of life for this comeback invention came last December with the appearance of General Motors' long-awaited electric sports car, the EV1, in Saturn showrooms in southern California and Arizona. With its high-profile launch, the world's biggest automaker joined a rapidly growing list of companies around the world marketing, or set to market, electric cars—among them Honda, Mercedes, Peugeot, and Renault. These giants will do battle with a quickly growing army of some 250 entrepreneurial startups, each with visions of becoming the next Henry Ford.

While the electric cars on the market so far are expensive and can only travel limited distances, they are taking carmakers in new directions. With the push of government mandates soon to be overtaken by the pull of market opportunities, the drive to produce the most practical, economical electric car is quickly becoming a competitive auto race. As Michael Gage, President of CALSTART, a California electric vehicle consortium, puts it, "We are entering the tornado."

THE WHIRRING NINETIES

This time it was a maker of flying machines, Paul MacCready, who got the creaky wheels of car innovation turning. MacCready, inventor of the first human-powered aircraft to fly across the English Channel—a bicycle-like device that earned him "engineer of the century" plaudits from his contemporaries—had designed a solar-powered

Reprinted with permission from Worldwatch Institute, Washington, D.C., January/February 2001. pp. 31-33 car, the Sunrayer, for GM. In 1987, the Sunrayer won the first Solar Challenge race, crossing Australia on the equivalent of five gallons of gasoline. GM then asked MacCready's firm, AeroVironment, to build a concept electric car for the company.

Three years later, the resulting prototype—called the Impact—was greeted with such unexpected plaudits at a Los Angeles auto show that then-President Jack Smith brashly vowed to begin mass-producing the car immediately. In doing so, he gave this second automotive revolution a much-needed push. The actual jumpstart came later that year, when the California Air Resources Board—at the time facing worsening air pollution in Los Angeles and other cities, and greatly encouraged by GM's vow—passed the toughest auto emissions standards in the world. Most notable was the industry-shaking requirement that 2 percent of cars sold in the state by the seven major carmakers in 1998 be "zero-emission," with the share rising to 10 percent by 2003.

Auto-industry lobbyists immediately turned on their own creation and—joined by the oil industry—began a bitter right to roll back the electric car mandate. At times, it seemed some of these companies had devoted more money to badmouthing zero-emission cars than to designing them. This certainly appeared to be the case for Chrysler, whose chairman—in the midst of scaling back its program—went so far as to declare, "There is absolutely no economic basis for electric vehicles in the world."

Eventually, the lobbying paid off, and California legislators lifted the 1998 mandate. But the big automakers are still required to make the more stringent 10 percent target in 2003, which means that some 800,000 zero-emissions cars should be on California's roads by 2010. This would be a giant leap from the approximately 2,300 electric cars in use in the entire country today.

ROADBLOCKS

Whether these targets will be met depends on whether prospective consumers can be helped around the immediate barriers of cost and range. At today's low-volume production levels, electric cars are more expensive to buy or lease than conventional cars with internal-combustion engines. The problem is not the electric motor, a highly evolved technology used in everything from tiny dentist drills to huge freight locomotives. In fact, today's electric motors are already between four and five times as efficient as internal combustion engines.

The biggest roadblock for electric cars is, and always has been, storing the electricity needed to run them. The EV1—the product of the California mandate and a $345 million investment by GM—carries 1,175 pounds of lead-acid batteries (the same kind used to start conventional cars), but has a range of just 70 to 90 miles between recharges (the electric equivalent of refueling, recharging takes three hours, using a 220-volt inductive "paddle" at home or in public charging stations—of which California already has more than 400). In part because of the cost of the battery, the EV1s now on the market in California and Arizona lease for as much as $34,000 over 3 years.

But the energy–weight ratio of lead-acid batteries has been cut by 60 percent over the last decade, and further gains are likely down the road. Virtually all of the major carmakers are working hard to lower the cost of more advanced batteries with greater energy density—including nickel–metal-hydride models that could double the EV1's range to between 150 and 200 miles. Other alternatives in the works include nickel–cadmium and lithium-mion batteries, and even flywheels—mechanical batteries consisting of a rapidly spinning disk made of synthetic materials.

These new batteries are still too expensive for wide commercial use, but experts at California's Air Resources Board believe that they should be on the market soon. The next advance is expected to come from Honda, whose Formula One race car engineers are now fully devoting their work to electric cars, and whose solar car has displaced GM's as the Solar Challenge champion. Honda plans to launch its EV Plus in California this May. Billed as the first family-oriented electric car, the compact four-seater will be equipped with nickel-metal-hydride batteries which give it a range of 125 miles—but at substantial cost: the batteries alone go for an eyepopping $40,000. Whether people will be willing to pay $500 a month (though this does include insurance and roadside service) to lease a car that is virtually indistinguishable from a standard economy car is uncertain; Honda expects to lease just 300 of the cars over the next three years.

Ironically, the limitations of today's batteries have made these first electric cars far more advanced than they otherwise might have been. Forced to stretch the range of bulky batteries, designers threw away the book on conventional automotive design and construction. In the search for a commercially viable electric car, automakers—for the first time in decades—turned their engineers loose on truly revolutionary concepts.

Author Michael Schnayerson, who was given inside access to GM's program, notes in his book *The Car That Could* how engineers struggled for eight years to fuse unconventionality and practicality, garnering 23 electronics and aerospace patents and a slew of engineering achievements. The EV1's aluminum car frame is the world's lightest; its teardrop-shaped body has aerodynamics equal to a modern fighter plane; and its brakes can regenerate, recharging the battery. With a super-efficient electric motor and low-resistance tires, it accelerates from zero to sixty miles per hour in less than nine seconds—faster than most conventional cars.

With these technological wonders, the EV1 has impressed most of those who have test-driven it. But the far greater challenge ahead for electric cars will be to make their way from the engineering track to the suburban garage. Although many buyers may be lured by the "zero-emission" label, performance, convenience, and cost are the benchmarks by which the cars must ultimately be judged.

ALTERNATE ROUTES

Automakers are taking a variety of approaches to "niche marketing" their first-generation electric cars. GM, for example, appears to be aiming at young, wealthy environmentalists—perhaps Hollywood stars and executives looking for a fast, sporty, pollution-free car that can speed through Beverly Hills. This is, after all, a group already accustomed to spending $30,000 to $40,000 for a Mercedes or Lexus. (Including tax incentives, the monthly lease rate for the EV1 falls to between $480 and $640 per month—less than the figure for luxury cars like the Cadillac DeVille.)

So far, the strategy seems to be working. Thanks to a major ad campaign, demand for the EV1 has been stronger than GM anticipated, with about 50 cars leased out on the very first day. GM's electric car expert, Robert Purcell, believes the EV1 will assume a "second car" role for commuting (the average U.S. commute is less than 35 miles) and for short trips. Purcell notes similarities to the microwave oven, which was unpopular at first but which eventually caught on as a second oven. While production plans have been kept under wraps, Detroit's labor press estimates that at least 80 EVs were made last year.

European automakers appear to be targeting a different set: green-oriented urban dwellers. The City Bee, for example, is a lightweight, fully recyclable two-seater with limited range, developed by the Norwegian consortium PIVCO. Scheduled for assembly in Europe and California later this year, it will cost around $10,000 at a production volume of 10,000, and is intended only for use in the inner city, and for quick rental at rail stations.

A multitude of other small companies, meanwhile, are preparing their own models for production. And the "urban car" niche has now begun to attract bigger players as well, resulting in some intriguing partnerships. German auto giant Mercedes-Benz and Swiss watchmaker Swatch have teamed up to develop a "Smart" car: a two-seater that can be built in under five hours, using plastic parts with interchangeable color schemes like those of the Swatch watches. The Smart car has proven crashworthy even though it is less than 10 feet long. Cast as a modern version of the "runabouts" seen at the turn of the century, and aimed at a younger audience, it can use electric as well as other drive systems. Some 200,000 Smart cars, a sizable percentage of which could be electric, are scheduled to roll out in Europe in March 1998 at a price of between $10,000 and $13,000.

Other European automakers are taking a more conservative approach to marketing electric cars, at much lower cost, by converting conventional cars into electrics. One of the leaders is Peugeot, which has put out a car using nickel-cadmium batteries, with a range of 50 miles. Having successfully tested 500 of its cars in the city of La Rochelle, the Peugeot-Citroen group believes there will be around 100,000 electric cars in Europe by 2000 and plans to grab a quarter of the market; it produced more than 4,000 in 1996.

Renault, meanwhile, sold 215 of its electric conversions in the first six months of last year. It will make 1,000 cars this year, and expects to continue increasing output in 1998. On a smaller scale, Fiat and Volkswagen are selling conversions, which they are reportedly making at a rate of about one a day, in Swiss cities. These conversions have a more limited range than cars designed to be electric from the start, but they allow the automakers to enter the market and gain experience without a huge up-front investment.

Much as L.A.'s smog sped the move to electric cars, air pollution and congestion in Europe have prompted some carmakers to rethink the car's conventional ownership and role in transportation. In France, Peugeot and Citroen are involved with the design of a transit system that will allow Parisian commuters to rent electric cars over short distances under a credit card-like system. Renault plans, as part of a consortium of electric utilities, carmakers, and government agencies, to operate 50 such "multiuser" cars in one of the city's high-tech suburbs. Similar efforts are underway in Switzerland. Swiss carmaker Horlacher will soon offer a lightweight "instant taxi," while Fiat has begun to rent electric cars in Geneva.

Whatever the route to consumers, proponents note that, despite their current high price and limited range, electric cars already have a number of advantages over today's cars. Their relative noiselessness, practicality, and simplicity appeal to many drivers. Electric cars cost less to refuel and service, and have many fewer parts to break down. Their owners are likely to spend less time on maintenance, and if they recharge at home, will rarely have to go to the service station. These time savings have real value in today's busy world. On a lifecycle basis, then, the cost gap between cars that pollute and those that don't is not all that great, even now; with another decade of battery development and mass production, it could be closed entirely.

Air regulators in California and New England argue further that if the avoided costs of urban air pollution, acid rain, and global climate change were included, electric cars would look like a steal. Of course, a fair comparison must include the emissions from the power plants that are used to charge the batteries. Fortunately, emissions from stationary power sources are easier to control than those from vehicles.

More importantly, running cars on electricity opens up a host of new fuel options not based on oil—including renewable resources such as wind power and solar energy. Already, some of the municipal governments promoting electric cars are erecting solar cells on the roofs of their parking garages to recharge them. The California city of Sacramento, for example, through the involvement of automakers, utilities, and local authorities, has installed more than 70 charging stations.

TOOL KITS

As the Sacramento example suggests, governments can play an important role in accelerating the transition to electric cars. Several of them, in fact, offer incentives that can lessen the electric car's initial cost. In the United States, federal and state tax credits and local rebates are available in many areas: customers in Los Angeles, for example, can use a 10 percent federal tax credit and a $5,000 rebate from the South Coast Air Quality Management District.

A 5,000-franc subsidy, meanwhile, is available to those buying electric cars in France. Paris and several other French cities add on tax credits, as do Switzerland, Austria, and Denmark. Switzerland already has about 2,000 electric cars in operation, and aims to use tax incentives to help make 8 percent of its cars electric by 2010. Germany, with more than 2,400 electric cars, gives its customers federal tax exemptions and state-level subsidies. Electric cars are exempt from sales taxes (which are typically high in Europe) in Italy, Norway, Sweden, and the United King-

dom. In Japan, which has the ambitious if distant goal of producing 200,000 electric vehicles by 2000, buyers may see costs cut in half by federal tax credits, incentives, and depreciation allowances—and even further by municipal governments.

Spurred by these and other less-noticed but important policies and programs—installing recharging stations, setting up demonstration projects, funding battery research and development—an electric car industry appears to be taking shape. Consortiums like CALSTART—a network of 200 government agencies, environmental groups, and aerospace, defense, and electric power companies—are helping these groups work horizontally (in contrast to today's "vertical" auto industry) to start programs and attract funding. Electric-vehicle associations in North America, Europe, and Asia report growing memberships, and are holding well-attended conferences and exhibitions each year to share the latest surveys, technologies, and production plans.

Surprisingly, the production hub for cost-competitive electric cars may turn out not to be in today's auto industry powers, the United States and Japan, but in the developing countries. With the world's most polluted urban air, some of these countries are just starting to develop automobile industries and make the associated investment in oil refineries, service stations, and the like—and therefore have less vested interest in the internal-combustion engine. Their industrialists recognize that if they are already making computers and televisions, electric cars should be within reach.

This notion has already taken hold in Asia, where prospective electric-car manufacturers have held extensive discussions with potential manufacturing partners. Thailand, which offers tax exemptions to electric car makers, is producing electric versions of its three-wheeled "tuk-tuk" taxi. Korean car makers Daewoo and Hyundai are working to produce lightweight electric cars for sale by the end of 1997 and 1998, respectively. China, which hopes to have several carmaking plants by 2000, plans to link domestic and foreign makers. One fledgling Chinese car maker has cut a deal with Peugeot-Citroen to produce a small electric model.

As their production picks up, electric cars' prices will fall noticeably. Comparing their price history with that of the traditional automobile, Daniel Sperling of the University of California at Davis estimates full-scale production could reduce the cost of electric cars to well below half the current level. And new technological breakthroughs will not be needed for costs to plummet, according to Tufts

University's Global Development and Environment Institute. The Institute projects price declines analogous to those that have occurred in personal computers, for example, with the costs of an electric car approaching—and with government support, falling below—those of conventional cars. And some analysts believe electric cars will be competing with gasoline-powered cars, *without* subsidies, within a decade. Table 19.1 lists major commercial or near-commercial cars.

BACK TO THE FUTURE?

Once the electric car catches up to today's cars by lowering upfront costs and extending range, its continued success will depend on the ability of manufacturers to lure consumers with inexpensive, battery-powered versions of commuter, family, or sports cars. Gage of CALSTART foresees "an improved driving experience" analogous to the shift from long-playing records and cassettes to the compact disc: the former worked, but the latter is better. Carmakers may also, however, find themselves dusting off hundred-year-old advertising pitches. As *Scientific American* observed in 1896, "The electric automobile . . . has the great advantage of being silent, free from odor, simple in construction, capable of ready control, and having a considerable range of speed."

Perhaps that praise was premature, but precisely a century later, *Scientific American* has returned to the topic—and suggests that the tables have turned. Writes Sperling of UC-Davis in a recent issue: ". . . it seems certain that electric-drive technology will supplant internal-combustion engines—perhaps not quickly, uniformly, nor entirely—but inevitably. The question is when, in what form and how to manage the transition." Those words, ironically, recall a very similar statement anticipating the adoption of the internal combustion engine a century ago. After watching the Duryea brothers race their wagons one afternoon in 1895, a young inventor named Thomas Edison boldly announced, "it is only a question of time when the carriages and trucks in every large city will be run with motors."

While technological dreams do not always come true, the electric car now seems to have more than a fighting chance. The major limitations of today's models are within sight of being overcome: indeed, a host of companies are betting billions of dollars on their ability to make that happen. And they cannot but be pleasantly surprised by the initial response to the EV1: according to Saturn dealers, the waiting list of several hundred hopeful lease holders continues to grow. As any car executive will tell you, the Model T took off when its costs were cut in half and cheap gasoline became an available fuel: now, they seem to be

Table 19.1

Major Commercial or Near-Commercial Electric Cars

Car and Model	Battery	Range (miles)	Location	Date
"GROUND-UP" ELECTRIC CARS				
General Motors EV1	lead–acid	70–90	US	1996
American Honda EV	nickel–metal hydride	125	US	1997
			Japan	1997
Solectria Sunrise	various	120	US	1998
SMALL ELECTRIC CARS				
PIVCO City Bee	nickel–cadmium	60–70	Europe	1997
			US	1997
Mercedes/Swatch "Smart"	various	various	Europe	1998
MODIFIED CONVENTIONAL CARS				
Peugeot 106	nickel–cadmium	50	Europe	1994
Citroen AX	nickel–cadmium	50	Europe	1994
Renault Clio	nickel–cadmium	50	Europe	1996
Fiat Panda Elettra	lead–acid	36	Europe	1996
Volkswagen CitySTROMER	lead–acid	30–54	Europe	1996
Solectria Force	various	60	US	1994

musing, might the electric car do likewise once its prices fall and batteries improve dramatically?

But larger forces, as much as the battery, will determine how far the electric car ultimately goes. On these counts, it seems to have a good deal riding in its favor. Its biggest constituency has always been women, who have far more purchasing power today. And its advantages, curious virtues in 1897, have become serious necessities by now, as the burdens of the gasoline culture grow heavier. Urban air pollution, congestion, dependence on oil imports, and global warming: these mounting societal concerns are recharging the electric car.

Still, if it is to avoid the fate of its predecessors, the EV1 and its companions must now handle the rocky road test of the market by selling, leasing, or renting. Making inroads on today's car population will not happen overnight, yet GM's model is already gracing the pages of the *The New York Times'* Automobile Section, alongside the conventional competition. This, perhaps, is the real story of the electric car's reemergence, lost amid the press conferences and television ads: that, true to the invention's own nature, its entry into the mainstream may resemble less a loud revving than a quiet hum.

REFERENCES

CALSTART. (1996, October). *Electric Vehicles: An Industry Prospectus.* Burbank, CA, CALSTART, Inc.

Flavin, C., and N, Lenssen. (1994). *Power Surge: Guide to the Coming Energy Revolution.* New York, W.W. Norton & Co.

MacKenzie, J. (1994). *The Keys to the Car.* Washington, DC, World Resources Institute.

Natural Resources Defense Council. (1996). *Green Auto Racing.* Washington, DC, NRDC.

Schiffer, M.B. (1994). *Taking Charge: The Electric Automobile in America.* Washington, DC, Smithsonian Institution Press.

Shnayerson, M. (1996). *The Car That Could: The Inside Story of GM's Revolutionary Electric Vehicle.* New York, Random House.

Sperling, D. (1995). *Future Drive: Electric Cars and Sustainable Transportation.* Washington, DC, Island Press.

QUESTIONS

1. Does the electric car reduce pollution? What are some of the pro and con arguments?

2. Discuss the roadblocks that are in the way of mass production of electric cars.

3. How do you think we can safely dispose of a high volume of used batteries that are a waste product of electric car technology? Is any ecologically safe approach being considered? If so, which?

4. Compare the marketing strategies of U.S. and European electric-car manufacturers.

5. Discuss the role of the oil industry in the development and success of the electric car.

ONE STEP FORWARD, A HUNDRED STEPS BACK?

Reduction in carbon emissions by World Bank-funded projects measure in the thousands of tons per year . . . but the emissions generated by bank-funded projects measure in the millions of tons.

Selected Fossil Fuel Power Plants Financed in Part by the World Bank in the 1990s

Location	Plant Description	Estimated Cost ($ billion)	Annual Carbon Emissions (million tons)
Tuoketo, China	3,600 MW, coal fired	4	7.19
Dolana Odra, Poland	1,600 MW, coal fired	0.3	3.196
Paiton, Indonesia	1,230 MW, coal fired	1.8	2.457
Pangasaman, Philippines	1,200 MW, oil fired	1.4	2.397
Hub River, Pakistan	1,469 MW, oil fired	2.4	2.345*

*Mr. Z. A. Khan, Divisional Manager Chemistry & Environment, National Power International, Disputes These Emission Numbers, for Details Visit International Power Web site: www.internationalpowerplc.com

Hub River Power Plant, Pakistan.
Photo Courtesy of Ahmed S. Khan.

Selected Global Environment Facility Energy Projects, Some Cofinanced by the World Bank

Location	Plant Description	Estimated Cost ($ billion)	Annual Carbon Emissions (million tons)
Leyte, Luzon, Philippines	440 MW, Geothermal	1,300	872
Countrywide, Indonesia	Installation of 200,000 PV systems	118	120
Nine cities, China	Energy-efficiency upgrades of industrial boilers	101	41–68
Guadalajara and Monterrey, Mexico	Dissemination of high-efficiency lighting (1.7 million fluorescent lamps)	23	32
Tejona, Costa Rica	20 MW	31	16

Source: *Banking Against Warming,* Christopher Flavin, Worldwatch, November/December 1997, pp. 32–33

CO$_2$ EMISSIONS

The combined 1995 population of Africa, Asia, Oceania, and Central and South America, which that year had a total of 200 million motor vehicles: 4.40 billion.

The 1995 population of the United States, which also had a total of 200 million motor vehicles: 0.27 billion.

––––––––––

The amount of carbon dioxide that a car getting 27.5 miles per gallon emits over 100,000 miles: 31,752 kilograms.

The amount of carbon dioxide that a human walking that same distance would produce: 59 kilograms.

––––––––––

Source: *Matters of Scale*. Reprinted with Permission from Worldwatch Institute, Washington, D.C. November/December 1997, p. 39.

Conclusion

During the past 50 years, the global economy has increased fivefold, world population has doubled, and world energy use has tripled. Will these trends continue in the 21st century? Increased production and use of fossil fuel could have severe local and regional impacts. Locally, air pollution takes a toll on human health. Acid precipitation and other forms of air pollution can degrade downwind habitats, especially in lakes, streams, and forests. On a global level, the increased burning of fossil fuels will result in increased emissions of greenhouse gases, which in turn could lead to global warming and other adverse climate changes.

As we enter a new millennium, population pressures and the deterioration of the environment present a challenge to humankind: how to satisfy the ever-growing appetite of the energy-hungry genie—how to develop and use energy sources that are friendly to the environment. We face a dilemma: Energy technologies often enhance material well-being across the planet, but the continuation of current trends could lead to a degraded planet, yielding an uncertain existence for future generations.

According to some estimates, by the year 2010, world population will total 7 billion, and the gross world product will have doubled. It is predicted that, if energy consumption styles do not change, world energy consumption will increase by 50 to 60 percent. Carbon dioxide emissions will also increase by 50 to 60 percent. Therefore, as we plan for the future, we must use energy sources that enable us to sustain our environment. Appropriate energy policies at the individual, national, and international levels, along with efficient energy technologies and conservation efforts, could play an important role in achieving a balance between economic growth and a sustainable environment.

The next part explores the impact of increased energy consumption on the ecology of our planet.

INTERNET EXERCISE

1. Use any of the Internet search engines (e.g., Alta Vista, Yahoo, Infoseek, etc.) to re-search the following:
   a. When did the oil tanker *Exxon Valdez* (Figure P3.8) run aground on Bligh Reef in Prince William Sound in Alaska?
   b. How much oil was spilled into the Sound?
   c. What was the major cause of the accident?
   d. How did the construction of the *Exxon Valdez* contribute to the accident?
   e. What are the short-term and long-term effects of the oil spill in Prince William Sound?
   f. List the 10 largest oil spills to date.
   g. Compare the Prince William Sound oil spill with the Persian Gulf War oil spill in terms of size and short-term and long-term effects.
   h. What actions should be taken by multinational oil companies, governments, and international organizations to prevent future oil spills?

2. Use any of the Internet search engines (e.g., Alta Vista, Yahoo, Infoseek, etc.) to re-search the following questions:
   a. Where is Chernobyl located?
   b. What happened on April 26, 1986, at Chernobyl?
   c. What was the cause of the accident at Chernobyl?
   d. What are the short-term and long-term effects of the nuclear disaster at Chernobyl?

Figure P3.8
The *Exxon Valdez* Floats Serenely in the Midst of an Oil Spill in the Waters of Prince William Sound, Alaska, Flanked by Another Tanker.
Courtesy of U.S. Coast Guard.

3. Use any of the Internet search engines (e.g., Alta Vista, Yahoo, Infoseek, etc.) to research the following:
    a. Define the following terms:
        (i)   Fossil fuel
        (ii) Acid rain
        (iii) Global warming
        (iv) Greenhouse effect
        (v)   Radioactive waste
        (vi) GNP
        (vii) Per capita oil consumption
    b. Determine the electricity consumption per capita for the following developed and developing countries:

Developed Country	Electricity Consumption per Capita (kWh)	Developing Country	Electricity Consumption per Capita (kWh)
United States		Bangladesh	
France		China	
Japan		Ethiopia	
Australia		Ghana	
Norway		India	
Sweden		Indonesia	
Switzerland		Kenya	
Italy		Malaysia	
Singapore		Pakistan	
United Kingdom		Venezuela	
Germany		Zimbabawe	

    c. What alternative energy sources look promising for the energy demands of the 21st century?
    d. What percentage of the world's total electricity is produced by hydroelectric power?
    e. Compare the features of large dams with those of small dams.

f. Complete the following table:

Country	Hydroelectric Power-Generating Capacity (MW)
United States	
Mexico	
Canada	
Egypt	
China	
Brazil	
Pakistan	
India	
Russia	
Japan	
Argentina	
Venezuela	
Italy	

g. What percentage of the world's total electricity is produced by solar power?
h. Will solar power someday replace fossil fuel as a source of energy? Why? Why not? Explain your answer.

i. Complete the following table:

Country	Solar Power-Generating Capacity (MW)
United States	
Mexico	
Canada	
Nigeria	
China	
Brazil	
Pakistan	
India	
South Africa	
Japan	
Malaysia	
Indonesia	
Australia	

j. Determine the wind power-generating capacity for the following developed and developing countries:

Developed Country	Wind Power-Generating Capacity (MW)	Developing Country	Wind Power-Generating Capacity (MW)
United States		Brazil	
France		China	
Japan		Nigeria	
Australia		Ghana	
Norway		India	
Sweden		Indonesia	
Switzerland		Jordan	
Italy		Malaysia	
Singapore		Pakistan	
United Kingdom		Viet Nam	
Germany		Zimbabwe	

k. Visit NASA's Multimedia Gallery at http://www.nasa.gov. The site contains a large collection of photos of the earth taken from space by various space shuttle missions and depicting the impact of different types of human interaction. Answer the following questions:

    (i) List the impact of the construction of dams on ecology and soil erosion in the developing and the developed countries.

   (ii) List those bans in the United States where the impact of construction on ecology and soil erosion is most evident.

Trends of the 21st Century

1. Do you feel that these trends are important for the 21st century? Indicate the degree of importance that these issues will have for the 21st century.

Trend	Very Important	Important	No Opinion	Not Important	Nonsignificant
Fossil fuel consumption					
Family values					
Pollution					
Humanity					
Population					
Availability of resources					
State of environment					
Stress					
Moral and ethical values					
Materialism					

2.   Chart the following trends into the 21st century and explain your opinion.

Figure P3.9
Trends of the 21st Century.

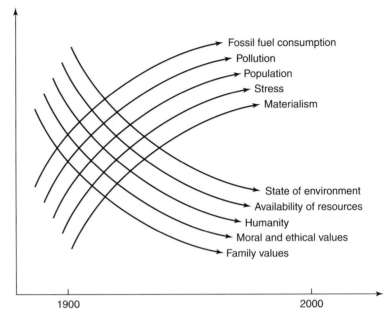

A Bridge to Your 21st Century Understanding

Suppose you were the Chernobyl plant director in 1986. What decisions and actions would you have taken to prevent the disaster? Complete the following flowchart by listing your decisions and actions in chronological order.

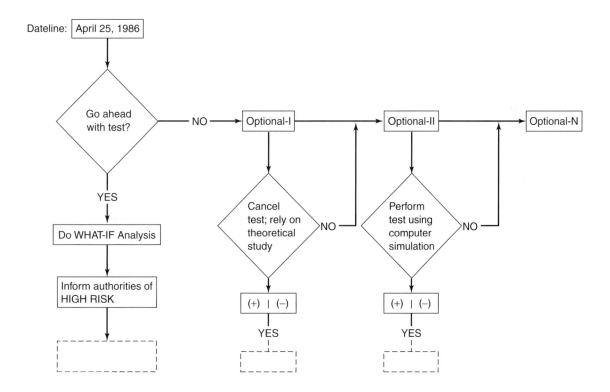

Beyond the Bridge: Explore Future Options . . .

1. The gigantic increase in the use of fossil fuels in the 20th century has wreaked havoc on the earth's atmosphere. What kind of risks does the increased fossil fuel consumption impose on society at personal, national, and international levels? What efforts must be made to minimize these risks? Complete the following table:

	Personal Level	*National Level*	*International Level*
Risks			
Efforts			

Figure P3.10
Impact of Increased Fossil Fuel
Consumption.

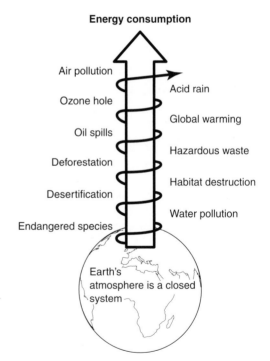

2. What types of options and policies must be adopted at personal, national, and international levels to change 20th-century's styles of consumption of fossil fuels?

	Personal Level	National Level	International Level
Options/Policies			

3. What types of renewable and alternative energy technologies will promote economic growth for a sustainable environment in the 21st century?

4. Can the impact of technology in the 20th century be reversed? Yes or No? Explain your answer.

USEFUL WEB SITES

URL	Site Description
http://www.er.doe.gov	U.S. Department of Energy
http://www.nrel.gov/business/international	Alternative Energy Sources
http://www.solar-energy.vegan.org	What's Solar Energy?
http://www.fsec.ucf.edu	Florida Solar Energy Center
http://www.nei.org	Nuclear Energy Institution
http://www.eagle2.online.discovery.com/dco/doc/10 12/world/science/chernobyl/weblinks.html	Chernobyl Web links
http://www.ponderosa-pine.uoregon.edu/Bi220/Blesofsky/menu.html	Genetic Effects of Chernobyl Incident
http://www.worldwatch.org	Worldwatch Institute
http://www.worldbank.org	World Bank
http://www.undp.org	United Nations Development Program
http://www.lib.umich.edu/libhome/Documents.center/ stats.html	Statistical Resources on the Web
http://www.sandia.gov/Renewable_Energy	Renewable Energy Sources
http://www.hooverdam.com/Gallery	Hoover Dam Photo Gallery
http://www.eren.doe.gov/wind	Wind Power

PART

IV

Ecology

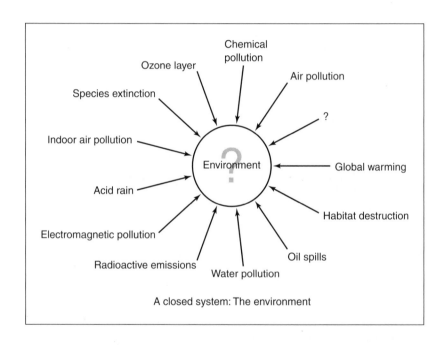

A closed system: The environment

OBJECTIVES

Part IV will help you to:

- Recognize the historical background and major and minor issues of ecology
- Discover the various problems, urgencies and implications of specific ecological issues
- Distinguish between industrialized and un-industrialized nation's ecological issues and viewpoints
- Analyze and use tables and graphs to promote specific and additional ecological perspectives
- Understand the implications of increasing technological and industrial wastes
- Understand the impact of rapidly increasing human resource use on the environment
- Determine ethical considerations with environmental issues
- Realize the strength, ethics, and potential of planned environmental approaches
- Appreciate the importance and contributions of individual ecological involvement and action

INTRODUCTION

The term "environment" simply means the world that is all around us. This definition, then, includes no less than all of our natural world—our ecosystem. The study of ecology examines the mutual relationship between organisms and this natural world, the environment, and therefore analyzes the changes and effects on our whole environment due to organisms and their tools. Of course, one chapter or part of a book can hardly deal with such an all-embracing subject, but it can give an overview of some of the major issues of our ecosystem as man, along with the powerful and unpredictable use of technology, changes its balance, supportive systems, and even beauty. This part's purpose, then, is to enhance awareness, thought, understanding, responsibility, strategies, and involvement for a future environment that is desirable, inhabitable, and sustainable. As this part of the text is presented, historical, social, and economic views will be discussed along with the issues and articles in order to give a more complete understanding of the issues from differing perspectives.

Looking into the environment of the past, present, and future. Photo courtesy of Ahmed S. Khan.

Historical Background and Major Issues

The environmental movement formally began in the United States on Earth Day 1970, when the American public began to awaken to the ecological destruction surrounding it. At that time, pollution controls did not exist for cars; people and cities dumped untreated sewage into the nation's rivers, some of which were so filled with chemical waste that they actually caught fire; and industrial cities were clouded with pungent, acid smoke. Since then, many of these problems have diminished and have been effectively addressed. The Environmental Protection Agency (EPA), established in 1970, monitors air quality around the country, and toxic emissions from smokestacks, factories, and incinerators have been sharply reduced. Mandatory pollution-control standards on automobiles have led to a drop in lead emissions, and recycling as a way of reducing solid waste has made a significant impact on the health of the environment.

As, however, our use of materials continues to grow on a worldwide basis, along with an exponentially increasing population growth and level of need, new environmental strategies cannot help but involve an ecological cost of use and cost of the pollution of our supportive natural resources. Based on the findings of a scientific advisory group, the EPA has ranked world environment issues in the following manner (Wright, 1995).

High risk: Habitat destruction, global warming, ozone-layer depletion, species extinction, loss of biological diversity

Medium risk: Herbicides and pesticides, surface water pollution, airborne toxic substances

Low risk: Oil spills, radioactive materials, groundwater pollution

Human health risk: Indoor air pollution, outdoor air pollution, exposure of drinking water to chemicals

For an overview of and a brief introduction to the high-risk issues, a small discussion of each of the major issues in that category follows.

Historical and major issues from Wright, J.W. (1995). *The Universal Almanac 1996* Kansas City, Andrews and McMeel, pp. 604–618.

Important Decisions for a Sustainable Environment. Photo courtesy of B. Eichler

When the environment is finally forced to file for bankruptcy because its resource base has been polluted, degraded, dissipated, and irretrievably compromised, the economy will go down with it.

TIMOTHY WIRTH, U.S. DEPARTMENT OF STATE

Habitat Destruction

Habitat destruction includes the destruction of wildlands that harbor the biological diversity of earth. Estimates of the total number of plant and animal species are between 10 million and 80 million, of which only about 1.4 million species have been identified. The biodiversity's mass extinction at unprecedented rates has already begun to cause the loss of 50,000 invertebrate species per year (nearly 140 each day) in the rain forests. By the year 2050, it is believed that 25% of the earth's species will become extinct if rain forest destruction is continued. Other habitats, such as islands, wetlands, and freshwater lakes, are also losing great proportions of their life-forms due to environmental loss and degradation. Rain forests are home to at least half of the planet's species and have been reduced by nearly half of their original area (from 14% of the earth's land to only 6% presently), losing an area about the size of the state of Washington each year due to deforestation. During the past 30 years, about one-third of the world's rain forests have disappeared. Tropical rain forests are lush habitats filled with more diversity of life than any other habitat. A typical patch of rain forest covering four square miles contains 750 species of trees, 750 species of other plants, 125 species of mammals, 400 species of birds, 100 species of reptiles, and 60 species of amphibians. The rich plant life supports food chains that include many of the world's most spectacular animals: brightly colored frogs, eagles, monkeys, tigers, sloths, army ants, and so on. Many rain forest species cannot survive in any other type of environment and are, like all species, biologically unique. The destruction of species that accompanies deforestation becomes a large-scale loss of species to the earth and mankind. For example, of the 3,000 plant species that have anticancer properties, 70% are found in tropical rain forests. Rain forests also have an impact on world climate, affecting the chemistry and ecology of the ocean, consuming huge quantities of CO_2, producing a large percent of the world's oxygen, and slowing global warming. The Amazon River basin, one of the largest rain forests, holds 66% of the Earth's freshwater and produces 20% of the earth's oxygen. (Refer to Tables P4.1 and P4.2 for comparisons of deforestation by country.)

Table P4.1
Deforestation in the Megadiversity Countries, Eighties

Country	Share of World's Land Area	Share of World's Flowering Plant Species[1]	Annual Deforestation Rate[2]	
	(percent)		(square kilometers)	(percent)
Brazil	6.3	22	13,820[3]	0.4
Colombia	0.8	18	6,000	1.3
China	7.0	11	n.a.	n.a.
Mexico	1.4	10	7,000	1.5
Australia	5.7	9	n.a.	n.a.
Indonesia	1.4	8	10,000	0.9
Peru	1.0	8	2,700	0.4
Malaysia	0.2	6	3,100	1.5
Ecuador	0.2	6	3,400	2.4
India	2.2	6	10,000	2.7
Zaire	1.7	4	4,000	0.4
Madagascar	0.4	4	1,500	1.5

[1]Based on Total of 250,000 Known Species; Because of Overlap Between Countries, Figures Cannot be Added. [2]Closed Forests Only. [3]1990 Figure.

Sources: Jeffrey A. McNeely et al., *Conserving the World's Biological Diversity* (Gland, Switzerland, and Washington, D.C.: International Union for Conservation of Nature and Natural Resources (IUCN) et al., 1989); World Resources Institute, *World Resources 1990–91* (New York: Oxford University Press, 1990); Ricardo Bonalume, "Amazonia: Deforestation Rate Is Falling," *Nature*, April 4, 1991; Walter V. Reid, "How Many Species Will There Be?" in J. Sayer and T. Whitmore, eds., *Tropical Deforestation and the Extinction of Species* (Gland, Switzerland: IUCN, Forthcoming).

From *State of the World 1992:* A Worldwatch Institute Report on Progress Toward a Sustainable Society by Lester Brown, et. al, eds. Copyright © 1992 by the Worldwatch Institute. Reprinted by Permission of W.W. Norton & Company, Inc.

Table P4.2
Tropical Forest Area and Rate of Deforestation for 87 Countries, 1981–90 (in thousand hectares)

Region/Subregion	Number of Countries Studied	Total Land Area	Forest Area 1980	Forest Area 1990	Area Deforested Annually 1981–90
LATIN AMERICA	**32**	**1,675,700**	**923,000**	**839,000**	**8,300**
Central America and Mexico	7	245,300	77,000	63,500	1,400
Caribbean subregion	18	69,500	48,800	47,100	200
Tropical South America	7	1,360,800	797,100	729,300	6,800
ASIA	**15**	**896,600**	**310,800**	**274,900**	**3,600**
South Asia	6	445,600	70,600	66,200	400
Continental Southeast Asia	5	192,900	83,200	69,700	1,300
Insular Southeast Asia	4	258,100	157,000	138,900	1,800
AFRICA	**40**	**2,243,400**	**650,300**	**600,100**	**5,000**
West Sahelian Africa	8	528,000	41,900	38,000	400
East Sahelian Africa	6	489,600	92,300	85,300	700
West Africa	8	203,200	55,200	43,400	1,200
Central Africa	7	406,400	230,100	215,400	1,500
Tropical Southern Africa	10	557,900	217,700	206,300	1,100
Insular Africa	1	58,200	13,200	11,700	200
Total	**87**	**4,815,700**	**1,884,100**	**1,714,800**	**16,900**

Note: Figures are Preliminary Estimates.
Source: World Resources Institute, *World Resources 1992–93* (1992).

Wetlands are also important environmental ecosystems that provide habitats for many species of animals and plants; a third of the endangered or threatened species in the United States live in or are dependent on them. Wetlands are the nurseries for many fish, all amphibians, and many birds and mammals. Wetlands also filter, purify, and hold water, even eliminating pesticides and other toxins by speeding up degradation by microbes. They store water, slow down its flow, and buffer floods and erosion. Coastal wetlands are spawning grounds for between 60% and 90% of U.S. commercial fisheries, and hunting depends entirely upon wetland habitats. Wetlands are among the most productive natural ecosystems on earth in terms of total biological mass per unit of area. Since colonial times, half of the wetlands in the United States have been destroyed, with only about 105 million acres remaining (excluding Alaska). (Refer to Table P4.3 for more detailed information on the loss of state wetland acreage.)

Species Extinction

Biological diversity faces a rate of species extinction greater than any since the mass extinctions of the dinosaurs 65 million years ago. As mentioned previously, a minimum of 50,000 invertebrate species alone are rendered extinct per year, which comes to about 140 extinctions each day. The world's 10 to 80 million species are threatened by deforestation, loss of

Table P4.3
States with Most Wetland Acreage, 1780s–1980s

State	Wetlands in 1780s Acres	Wetlands in 1780s Percent of Area	Wetlands in 1980s Acres	Wetlands in 1980s Percent of Area	Percent Lost
Alaska	170,200,000	45.3%	170,000,000	45.3%	−0.1%
Florida	20,325,013	54.2	11,038,300	29.5	−46
Louisiana	16,194,500	52.1	8,784,200	28.3	−46
Minnesota	15,070,000	28.0	8,700,000	16.2	−42
Texas	15,999,700	9.4	7,612,412	4.4	−52
North Carolina	11,089,500	33.0	5,689,500	16.9	−49
Michigan	11,200,000	30.1	5,583,400	15.0	−50
Wisconsin	9,800,000	27.3	5,331,392	14.8	−46
Georgia	6,843,200	18.2	5,298,200	14.1	−23
Maine	6,460,000	30.4	5,199,200	24.5	−20

Note: Ranked by Wetlands Acreage in the 1980s.
Source: U.S. Dept. of Interior, Fish and Wildlife Service, *Wetlands in the United States, 1780s to 1980s* (1990).

wetlands, and urban sprawl, as well as shifting climate and vegetation zones. Another cause for animal and plant loss is the change to monocultures in agriculture, which is the production of only one strain of crop for food that then becomes vulnerable to disease and other threats. Although it may be weakened when it is renewed, the U.S. Endangered Species Act of 1966 is the chief protector of species in the United States, and the Convention on International Trade in Endangered Species of Wild Fauna and Flora (CITES), signed in 1975 by 122 nations, continues to be one of the main forces in the preservation of species internationally. To abate the trend toward species extinction, governments around the world have set aside a total of about 16.4 million square miles of protected lands in about 3,500 parks and preserves. In addition, many countries try to identify species threatened with extinction so that these species may be carefully monitored and protected.

The following are some of the selected endangered species of the world: cheetah, African chimpanzee, Dugong (sea cow), gibbon, gorilla, jaguar, leopard (three species), monkey (many species), orangutan, giant panda, rhinoceros, tiger, whale (eight species), wolf, birds (over 1,000 of 9,672 species), reptiles (many species of alligators, crocodiles, iguanas, and sea turtles), amphibians (all frogs and toads in the United States), fish (many species), plants (about 21%, or 4,200 of 20,000 species; as many as 750 plant species in the United States could become extinct by the first years of the 21st century). (Refer to Tables P4.4, P4.5, and P4.6 for a more detailed breakdown of data on endangered species.)

Global Warming

Many experts today think that the earth has begun to warm significantly over the past century—by about 1°F—and could further warm 3° to 5°F over the next 50 to 60 years. If

Table P4.4
Endangered and Threatened Species, 2001

Group	Endangered U.S.	Endangered Foreign	Threatened U.S.	Threatened Foreign	Total Listed
Mammals	63	251	9	17	340
Birds	78	175	14	6	273
Reptiles	14	64	22	15	115
Amphibians	10	8	8	1	27
Fishes	70	11	44	0	125
Snails	20	1	11	0	32
Clams	61	2	8	0	71
Crustaceans	18	0	3	0	21
Insects	33	4	9	0	46
Arachnids	12	0	0	0	12
Total animals	**379**	**516**	**128**	**39**	**1,062**
Flowering plants	565	1	141	0	707
Conifers	2	0	1	2	5
Ferns and others	26	0	2	0	28
Total plants	**593**	**1**	**144**	**2**	**740**
Total species	**972**	**517**	**272**	**41**	**1,802**

Note: Separate Populations of a Species, Listed Both as Endangered and Threatened, are with Few
Exeptions, Tallied Only Once, for the Endangered Population Only. (For Details, See Source.) Figures
are as of May 31, 2001.
Source: U.S. Fish and Wildlife Service, *Box Score of U.S. List of Endangered and Threatened Species.*

Table P4.5
Endangered and Threatened Species in the U.S., 1980–2001

Year	Endangered Animals	Endangered Plants	Endangered Total	Threatened Animals	Threatened Plants	Threatened Total	Total Species Listed
1980	174	50	224	48	9	57	281
1985	207	93	300	59	25	84	384
1990	263	179	442	93	61	154	596
1995	324	432	756	113	93	206	962
1996	324	513	837	115	101	216	1,053
1997	343	553	896	121	115	236	1,132
1998	357	567	924	135	135	270	1,194
1999	358	581	939	126	140	266	1,205
2000	368	593	961	129	142	271	1,232
2001[1]	379	593	972	128	144	272	1,244

[1]As of May 31, 2001. Source: U.S. Fish and Wildlife Service, U.S. Species Listed Per Calendar Year,
1980–Present.

Table P4.6
Observed Declines in Selected Animal Species, Early Nineties

Species Type	Observation
Amphibians[1]	Worldwide decline observed in recent years. Wetland drainage and invading species have extinguished nearly half New Zealand's unique frog fauna. Biologists cite European demand for frogs' legs as a cause of the rapid nationwide decline of India's two most common bullfrogs.
Birds	Three fourths of the world's bird species are declining in population or threatened with extinction.
Fish	One third of North America's freshwater fish stocks are rare, threatened, or endangered; one third of U.S. coastal fish have declined in population since 1975. Introduction of the Nile perch has helped drive half the 400 species of Lake Victoria, Africa's largest lake, to or near extinction.
Invertebrates	On the order of 100 species lost to deforestation each day. Western Germany reports one fourth of its 40,000 known invertebrates to be threatened. Roughly half the freshwater snails of the southeastern United States are extinct or nearly so.
Mammals	Almost half of Australia's surviving mammals are threatened with extinction. France, western Germany, the Netherlands, and Portugal all report more than 40 percent of their mammals as threatened.
Carnivores	Virtually all species of wild cats and most bears are declining seriously in numbers.
Primates[2]	More than two thirds of the world's 150 species are threatened with extinction.
Reptiles	Of the world's 270 turtle species, 42 percent are rare or threatened with extinction.

[1]Class that Includes Frogs, Toads, and Salamanders. [2]Order that Includes Monkeys, Lemurs, and Humans.
Source: Worldwatch Institute, Based on Sources Documented in Endnote 13.
From *State of the World 1992:* A Worldwatch Institute Report on Progress Toward a Sustainable Society by Lester R. Brown, et al, eds. Copyright © 1992 by the Worldwatch Institute. Reprinted by Permission of W.W. Norton & Company, Inc.

this estimate is true, the effects would be enormous, considering that the earth's temperature has only risen 9 degrees since the last ice age 12,000 years ago. Such an increase in warming would cause polar ice caps to melt, thus causing a rise in the sea level worldwide, destroying coastal areas, islands, water supplies, forests, and agriculture in many parts of the world. Global warming is caused from gases in the atmosphere that prevent sunlight from being reflected away from the earth. Usually, sunlight that reaches the surface of the earth is partly absorbed and partly reflected. The absorbed light heats the surface and is later released as infrared radiation. Gases (called "greenhouse gases") that are not released collect the heat and keep it in the atmosphere. The earth's atmosphere is only 0.03% CO_2, but together with other gases, can trap 30% of the reflected heat and maintain the earth's average temperature at about 59°F.

Since the 1800s, an unprecedented amount of the four primary greenhouse gases (carbon dioxide, chlorofluorocarbons (CFCs), methane, and nitrous oxide) have been released into the atmosphere. The worst of these gases are CFCs, which, in addition to contributing to the greenhouse effect, are also destroying the stratospheric ozone, which shields the earth's surface from ultraviolet rays. The United States, Russia, Japan, and various countries in Europe have agreed to a 100% phaseout of CFCs by the year 2000. Concentrations of carbon dioxide have increased by 25% within the last century. As the greenhouse gases increase in concentration, they will produce a domino effect that could cause environmental chaos in weather patterns, warming patterns, and sea level. Many scientists speculate that global warming is occurring; however, agreement with regard to global warming is not definite, and the issue remains very contested (Wright, 1995).

The Ozone Layer

The chemical ozone (O_3) exists as smog at ground level, where we have an abundance of it, causing pollution, and in the stratosphere, where we do not have enough of it, causing a very serious weakening of the ozone shield, which protects us from the sun's ultraviolet rays and other types of radiation. A significant reduction of the ozone layer would lead to an increase in skin cancers and cataracts, the loss of small ocean algae and bacteria, and the breakdown of other natural systems. Chlorofluorocarbons (CFCs), chemicals manufactured since the 1930s for refrigeration, insulation, and packaging, add chlorine to the atmosphere and absorb the ozone in the stratosphere. CFCs are very stable and have lifetimes of 75 to 110 years. They drift into the upper atmosphere, where their chlorine components are released. The freed chlorine atoms find ozone molecules and react, creating chlorine monoxide. In a subsequent reaction, the chlorine monoxide releases its oxygen atom to form molecular oxygen, and the chlorine atom is freed again to repeat the process. With the long lifetime of chlorine, each chlorine atom can destroy about 100,000 molecules of ozone before the chain reaction ends. Ice crystals in the Antarctic and Arctic accelerate the process, producing "ozone holes" near both poles. Studies indicate that stratospheric ozone has declined at least 4% to 8% over the northern hemisphere in the past decade. To combat this process, major nations of the world have agreed to eliminate CFC use and production by 2000. However, NASA indicates that adverse effects on the ozone layer from CFCs will continue until at least 2020.

There are many environmental issues that need to be discussed. The issues, articles, and case studies of this part of the book begin to show our environmental needs, causes, perspectives, problems, and successes. The articles and case studies not only indicate the issues themselves, but also societal and economic responses, since these responses are where effects and actions are rooted. The three chapters, entitled "The Grim Payback of Greed," "Young at Risk," and "Fisheries: Exploiting the Ocean—What Will Be Left?" provide a global analysis of the consequences of increasing intense resource use by man in the environment. The last five chapters, entitled "Buried Displeasure: The Love Canal," "What Will Happen to the Endangered Species Act?", "Air Poisons around the World," "Earth Day: 25 years," and "Rain Forests May Offer New Miracle Drugs" are readings in the field of ecology that deal with more specialized issues in specific locations that have focused outcomes and lessons to humankind and our future decisions. The articles and case studies were selected for the central issues and reactions they represent, but the variety of issues and case studies available is, of course, as endless as the sea of everyday occurrences and events. The articles and case studies selected were chosen because they wrestle with

Understanding the Many Types of Value of our Ecology. Photo Courtesy of Ahmed S. Khan.

central issues and responses that enhance awareness and decision making for our future world. These studies also consistently underscore the vital necessity to acknowledge the increasing value and importance of the world around us—our life-sustaining world, the environment.

An Economic View of Ecology

In acknowledging the value of our ecosystem, 13 researchers, in a May 1997 report in the journal *Nature,* tried to establish the economic value of the annual worth of earth's natural goods and services—the annual worth of our ecosystem. This value is estimated at $33 trillion and includes earth's basic services and products, from beaches and forest lumber to oceans' regulation of carbon dioxide. In comparison to the total ecosystem's worth, the world's annual gross national products alone total about $18 trillion.

The actual worth of our ecosystem is probably much larger than $33 trillion, since the calculations were conservative, not all-inclusive, and came from the conversion of a vari-

Reprinted with permission from *Science News,* the weekly news magazine of science, copyright © 1996, 1997 by Science Service.
Excerpted from Mlot, C. (1997, May 17). A price tag on the planet's ecosystems. *Science News.* Vol. 151, p. 303.

ety of ecosystem values from more than 100 studies. For example, the relationship of shrimp-harvest yields to the local wetlands was calculated to the average value of a hectare (about 2.5 acres) of wetland, which was then applied to the international size of that habitat. Recreational value came from people's reports of willingness to pay for access to lakes, reefs, and other recreational areas. The estimates omit ecosystem services in desert and tundra areas, since these areas have not been studied in terms of ecosystem values. The study also omits services in urban areas like green spaces in New York's Central Park.

The most valuable ecosystems per hectare turned out to be estuaries and wetlands, since they have been studied the most. Other ecosystems will probably increase in value as they also become more studied and understood. This field of ecological economics is a brand new field, only about a decade old. This study is just a starting point, according to the researchers, and they acknowledge many limitations to it. However, even as a starting point, the power of the analysis is the acknowledgment of the per-hectare value of the various habitats, the assignment of value to processes and areas taken for granted, and the way in which the assignment of value will influence local decisions and conservation when it comes to planning.

REFERENCE

Wright, J.W. (ed). 1966. *The Universal Almanac.* Kansas City, MO: Andrews and McNeel.

QUESTIONS

1. Is the term "ecosystem" the same as the term "environment"? What are some different connotations of the two terms?

2. How have attitudes toward the environment changed from the 1970s?

3. Why do you think the EPA's world environmental issues are prioritized as they are? Why are some of the issues labeled as high risk, others as medium risk, and yet others as low risk? Explain.

4. Why is the loss of species an issue? Do we have to have so many species?

5. Why is the loss of rain forests an issue? Give three significant reasons.

6. Give three reasons why wetlands are important.

7. What do you think we can do to prevent habitat destruction? Creatively think of three workable ideas.

8. What can the industrial world do to help abate the extinction of species?

9. What causes global warming? What are the four gases mostly responsible for heat absorption of the atmosphere?

10. Explain how CFCs destroy ozone. What can be done to combat CFC use? Give some specific and practical ideas.

11. Why is it a good idea to place economic value on earth's natural habitat and goods and services? Explain.

12. Refer to the title page of this chapter. Why is the environment a "closed system"? Explain.

ECOLOGICAL STATISTICS

Global Facts

- Over 40% of all tropical forests have been destroyed and another acre is lost each second.

- Each year, humankind adds six to eight billion tons of carbon to the atmosphere by burning fossil fuels and destroying forests, pumping up the concentration of greenhouse gases responsible for global warming—an effect that could raise temperatures by three to ten degrees by the year 2050.

- While the U.S. makes up only 5% of the world's population, we produce 72% of all hazardous waste and consume 33% of the world's paper.

- Worldwide, thousands of pounds of plutonium are being produced, used and stored under conditions of inadequate security. Using current technology, only two pounds of plutonium is required to make a nuclear device.

- The annual catch in 13 of the world's 15 major fishing zones has declined and in four of those— three in the Atlantic and one in the Pacific oceans—the catch has shrunk by a startling 30%.

Land Facts

- Taxpayers will lose over one billion dollars over the next decade as the Forest Service spends more money on building logging roads and preparing commercial timber sales than it makes on selling the timber.

- In 1992, taxpayers subsidized the clearcutting of our Alaskan rain forest with an estimated $40 million.

- Mining companies are allowed to buy our public lands for less than five dollars an acre—and they pay no royalties on the gold and other minerals they extract. This taxpayer giveaway, combined with the cost of massive environmental damage and cleanup, amounts to a billion dollars every year.

- Grazing has led to soil erosion, watershed destruction and ruin of wildlife habitat on millions of acres of our public lands. Taxpayers subsidized grazing fees with $1.8 billion during the years 1985–1992.

Air Facts

- As many as 70,000 people nationwide may die prematurely from heart and lung disease aggravated by particulate air pollution.

- More than 100 million Americans live in urban areas where the air is officially classified by the EPA as unsafe to breathe.

- In many urban areas, children are steadily exposed to high levels of pollutants, increasing the risk of chronic lung disease, cell damage and respiratory illnesses.

- Dioxin and other persistent pollutants that are released into the air accumulate in our waterways, wildlife, food supply and human bloodstreams. These poisons may cause cancer and reproductive disorders in human beings and other animal species.

"Ecology Facts" from *25 Year Report 1970–1995*. (1995). New York, New York, Natural Resources Defense Council.

QUESTIONS ABOUT THE TABLES

1. Compare the information contained in Table P4.1 and Table P4.2. What common conclusion can be drawn from this material? What are some of the differences between the various pieces of information presented?

2. Summarize in a short written paragraph the overall conclusions in Table P4.1 and in Table P4.2. Compare the results.

3. Summarize the information contained in Table P4.3. What conclusions do you draw, and what action might you take from these conclusions?

ECOLOGICAL STATISTICS, *continued*

Water Facts

- Millions of pounds of toxic chemicals, like lead, mercury and pesticides, pour into our waterways each year contaminating wildlife, seafood and drinking water.

- One-half of our nation's lakes and one-third of our rivers are too polluted to be completely safe for swimming or fishing.

- Raw sewage, poison runoff and other pollution have caused 8,000 beach closures or advisories over the past five years.

- We are losing once pristine national treasures—like the Everglades, Lake Superior, and the Columbia River System—to toxic pollution, chemical spills, development, and diversion of freshwater flows.

- All but one species of the magnificent ocean-going salmon in the Pacific Northwest face a growing risk of extinction throughout most of their range, due to habitat degradation and overfishing.

Energy Facts

- The United States is responsible for almost 25% of the world's total energy consumption. We use one million gallons of oil every two minutes.

- Energy currently wasted by U.S. cars, homes and appliances equals more than twice the known energy reserves in Alaska and the U.S. Outer Continental Shelf.

- We could cut our nation's energy consumption in half by the year 2030 simply by using energy more efficiently and by using more renewable energy sources. In the process, we would promote economic growth by saving consumers $2.3 trillion and by producing one million new jobs.

- When just 1% of America's 140 million car owners tune up their cars, we eliminate nearly a billion pounds of carbon dioxide—the key cause of global warming—from entering the atmosphere.

Health and Habitat Facts

- In 1991, 2.2 billion pounds of pesticides were used in the U.S.—eight pounds for every man, woman and child.

- A 1991 NRDC study found that pesticide use can be reduced in nine major U.S. crops by 25 to 80%.

- Americans are exposed to 70,000 chemicals, some 90% of which have never been subjected to adequate testing to determine their impact on our health.

- In the early 1990s, 116 million Americans drank water from systems that violated the Safe Drinking Water Act.

- As of 1994, 1.7 million American children, ages one to five, suffered from lead poisoning.

- Of the trash that we Americans throw away every day, 30% by weight is packaging alone. In 1993, we threw away 14 billion pounds of plastic packaging.

4. Read Tables P4.4, P4.5, and P4.6. Compare Table P4.4 to Table P4.6. Explain the differences in the information and how the information is presented. Does Table P4.5 confirm the findings in Table P4.4, or does it confuse the data? Explain.

5. Summarize in a short written paragraph the findings and conclusions from each of Tables P4.4, P4.5, and P4.6.

Ecology Vocabulary Exercise

As you proceed through this part of the book and your reading and research of the issues, keep a listing of ecological terms and organizations that you would like to know more about. Use various resources to define and learn more about them, including other texts, articles, encyclopedias, and the Internet (refer to the Internet exercise at the end of Part IV).

The Grim Payback of Greed

ALAN DURNING

A man is rich in proportion to the things he can afford to let alone.
HENRY DAVID THOREAU

Early in the age of affluence that followed World War II, an American retailing analyst named Victor Lebow proclaimed that an enormously productive economy "demands that we make consumption our way of life . . . . We need things consumed, burned up, worn out, replaced and discarded at an ever increasing rate."

Americans have responded to Mr. Lebow's call, and much of the world has followed. The average person today is four-and-a-half times richer than were his great-grandparents at the turn of the century. That new global wealth is not evenly spread among the Earth's people, however. One billion live in unprecedented luxury; one billion live in destitution.

Overconsumption by the world's fortunate is an environmental problem unmatched in severity by anything but perhaps population growth. Their surging exploitation of resources threatens to exhaust or inalterably disfigure forests, soils, water, air and climate.

Of course the opposite of overcompensation—poverty—is no solution either to environmental or human problems. Dispossessed peasants slash-and-burn their way into the rain forests of Latin America, and hungry nomads turn their herds out onto fragile African rangeland, reducing it to desert.

If environmental destruction results when people have either too little or too much, we are left to wonder how much is enough. What level of consumption can the Earth

Reprinted with permission from Worldwatch Institute. Durning, A. (May–June 1991). "The Grim Payback of Greed." *International Wildlife.* Vol. 21, No. 3, pp. 36–39.

support? When does having more cease to add appreciably to human satisfaction?

The consuming society: Skyrocketing consumption is the hallmark of our era. The trend is visible in statistics for almost any per capita indicator. Worldwide, since midcentury the intake of copper, energy, meat, steel and wood has approximately doubled; car ownership and cement consumption have quadrupled; plastic use has quintupled; aluminum consumption has grown sevenfold; and air travel has multiplied 32 times.

Moneyed regions account for the largest waves of consumption since 1950. In the United States, the world's premier consuming society, on average people today own twice as many cars, drive two-and-a-half times as far, use 21 times as much plastic and travel 25 times as far by air as did their parents in 1950. Air conditioning has spread from 15 percent of households in 1960 to 64 percent in 1987, and color televisions from 1 to 93 percent. Microwave ovens and video cassette recorders found their way into almost two-thirds of American homes during the eighties alone.

The eighties were a period of marked extravagance in the United States; not since the Roaring Twenties had conspicuous consumption been so lauded. Between 1978 and 1987, sales of Jaguar automobiles increased eightfold, and the average age of first-time fur-coat buyers fell from 50 to 26. The select club of American millionaires more than doubled its membership from 600,000 to 1.5 million over the decade, while the number of American billionaires reached 58 by 1990.

Japan and Western Europe have displayed parallel trends. Per person, the Japanese of today consume more than four times as much aluminum, almost five times as much energy and 25 times as much steel as people in Japan did in 1950. They also own four times as many cars and eat nearly twice as much meat. In 1972 one million Japanese traveled abroad; in 1990, the number was expected to top ten million. As in the United States, the eighties were a particularly consumerist decade in Japan, with sales of BMW automobiles rising tenfold.

Like the Japanese, West Europeans' consumption levels are only one notch below Americans'. Taken together, France, West Germany and the United Kingdom almost doubled their per capita use of steel, more than doubled their intake of cement and aluminum and tripled their paper consumption since mid-century. Just in the first half of the eighties, per capita consumption of frozen prepared meals—with their excessive packaging—rose more than 30 percent in every West European country except Finland; in Switzerland, the jump was 180 percent.

The cost of wealth: Long before all the world's people could achieve the American dream, however, the planet would be laid waste. Those in the wealthiest fifth of humanity are responsible for the lion's share of the damage humans have caused to common global resources. They have built more than 99 percent of the world's nuclear warheads. Their appetite for wood is a driving force behind destruction of the tropical rain forest and the resulting extinction of countless species. Over the past century, their economies have pumped out two-thirds of the greenhouse gases that threaten the Earth's climate, and each year their energy use releases perhaps three-fourths of the sulfur and nitrogen oxides that cause acid rain. Their industries generate most of the world's hazardous chemical wastes, and their air conditioners, aerosol sprays and factories release almost 90 percent of the chlorofluorocarbons that destroy the Earth's protective ozone layer. Clearly, even 1 billion profligate consumers are too much for the Earth.

Beyond the environmental costs of acquisitiveness, some perplexing findings of social scientists throw doubt on the wisdom of high consumption as a personal and national goal: Rich societies have had little success in turning consumption into fulfillment. A landmark study in 1974, for instance, revealed that Nigerians, Filipinos, Panamanians, Yugoslavians, Japanese, Israelis and West Germans all ranked themselves near the middle of a happiness scale. Confounding any attempt to correlate affluence and happiness, poor Cubans and rich Americans were both found to be considerably happier than the norm, and citizens of India and the Dominican Republic, less so. As Oxford psychologist Michael Argyle writes, "There is very little difference in the levels of reported happiness found in rich and very poor countries."

As measured in constant dollars, the world's people have consumed as many goods and services since 1950 as all previous generations put together. Since 1940, Americans alone have used up as large a share of the Earth's mineral resources as did everyone before them combined. If the effectiveness of that consumption in providing personal fulfillment is questionable, perhaps environmental concerns can help us redefine our goals.

In search of sufficiency: Some guidance on what the Earth can sustain emerges from an examination of current consumption patterns around the world. For three of the most ecologically important types of consumption—transportation, diet and use of raw materials—the world's people are distributed unevenly over a vast range. Those at the bottom fall below the "too little" line, while those at the top, in what could be called the cars-meat-and-disposable class, consume too much.

About 1 billion people do most of their traveling, aside from the occasional donkey or bus ride, on foot, many of them never going more than 100 kilometers from their birthplaces. Unable to get to jobs easily, attend school or bring their complaints before government offices, they are severely hindered by the lack of transportation options.

The massive middle class of the world, numbering some 3 billion, travels by bus and bicycle. Kilometer for kilometer, bikes are cheaper than any other vehicles, costing less than $100 new in most of the Third World and requiring no fuel.

The world's automobile class is relatively small: Only 8 percent of humans, about 400 million people, own cars. Their cars are directly responsible for an estimated 13 percent of carbon dioxide emissions from fossil fuels worldwide, along with air pollution, acid rain and a quarter-million traffic fatalities a year.

The global food consumption ladder has three rungs, as well. At the bottom, the world's 630 million poorest people are unable to provide themselves with a healthful diet, according to World Bank estimates.

On the next rung, the 3.4 billion grain-eaters of the world's middle class get enough calories and plenty of plant-based protein, giving them the most-healthful basic diet of the world's people. They typically receive less than 20 percent of their calories from fat, a level low enough to protect them from the consequences of excessive dietary fat.

The top of the ladder is populated by the meat-eaters, those who obtain close to 40 percent of their calories from fat. These 1.25 billion people eat three times as much fat per person as the remaining 4 billion, mostly because they eat so much red meat. The meat class pays the price of its diet in high death rates from the so-called diseases of affluence—heart disease, stroke and certain types of cancer.

The Earth also pays for the high-fat diet. Indirectly, the meat-eating quarter of humanity consumes nearly 40 percent of the world's grain—grain that fattens the livestock they eat. Meat production is behind a substantial share of the environmental strains induced by the present global agricultural system, from soil erosion to overpumping of underground water.

In raw material consumption, the same pattern emerges. About 1 billion rural people subsist on biomass collected from the immediate environment. Most of what they use each day—about a half-kilogram of grain, 1 kilogram of fuelwood, and fodder for their animals—could be self-replenishing renewable resources. Unfortunately, because these people are often pushed by landlessness and population growth into fragile unproductive ecosystems, their minimal needs are not always met.

These materially destitute billion are part of a larger group that lacks many of the benefits provided by modest use of nonrenewable resources—particularly durable things like radios, refrigerators, water pipes, high-quality tools, and carts with lightweight wheels and ball bearings. More than 2 billion people live in countries where per capita consumption of steel, the most basic modern material, falls below 50 kilograms a year. In those same countries, per capita energy use—a fairly good indirect indicator of overall use of materials—is lower than 20 gigajoules per year (compared to 280 gigajoules in the United States).

Roughly 1.5 billion people live in the middle class of materials' use. Providing each of them with durable goods every year uses between 50 and 150 kilograms of steel and 20–50 gigajoules of energy.

At the top of the heap is the throwaway class, which uses raw materials extravagantly. A typical resident of the industrialized fourth of the world uses 15 times as much paper, 10 times as much steel and 12 times as much fuel as a Third World resident. The extreme case is the United States, where the average person consumes most of his own weight in basic materials each day—18 kilograms of petroleum and coal, 13 kilograms of other minerals, 12 kilograms of farm products and 9 kilograms of forest products, or about 115 pounds total.

In the throwaway economy, packaging becomes an end in itself, disposables proliferate and durability suffers. Four percent of consumer expenditures on goods in the United States goes for packaging—$225 a year. Likewise, the Japanese use 30 million disposable single-roll cameras each year, and the British dump 2.5 billion diapers. Americans toss away 180 million razors annually, enough paper and plastic plates and cups to feed the world a picnic six times a year and enough aluminum cans to make 6,000 DC-10 airplanes.

In transportation, diet and use of raw materials, as consumption rises on the economic scale, so does waste—both of resources and of health. Yet despite arguments in favor of modest consumption, few people who can afford high consumption levels opt to live simply. What prompts us, then, to consume so much?

The cultivation of needs: "The avarice of mankind is insatiable," wrote Aristotle 23 centuries ago, describing the way that as each of our desires is satisfied, a new one seems to appear in its place. That observation, on which all of economic theory is based, provides the most obvious answer to the question of why people never seem satisfied with what they have. If our wants are insatiable, there is simply no such thing as enough.

Much confirms this view of human nature. The Roman philosopher Lucretius wrote a century before Christ: "We have lost our taste for acorns. So [too] we have abandoned those couches littered with herbage and heaped with leaves. So the wearing of wild beasts' skins has gone out of fashion . . . . Skins yesterday, purple and gold today—such are the baubles that embitter human life with resentment."

Nearly 2,000 years later, Russian novelist Leo Tolstoy echoed Lucretius: "Seek among men from beggar to millionaire, one who is contented with his lot, and you will not find one such in a thousand . . . . Today we must buy an overcoat and galoshes, tomorrow, a watch and a chain: the next day we must install ourselves in an apartment with a sofa and a bronze lamp; then we must have carpets and velvet gowns: then a house, horses and carriages, paintings and decorations."

What distinguishes modern consuming habits from those of interest to Lucretius and Tolstoy, some would say, is simply that we are much richer than our ancestors and consequently have more ruinous effects on nature. There is no doubt a great deal of truth in that view, but there is also reason to believe that certain forces in the modern world encourage people to act on their consumption desires as rarely before.

The search for social status in massive and anonymous societies, omnipresent advertising messages, a shopping culture that edges out nonconsuming alternatives, government biases favoring consumption and the spread of the commercial market into most aspects of private life—all these things nurture the acquisitive desires that everyone has. Can we, as individuals and as citizens, act to confront these forces?

A culture of permanence: When Moses came down from Mount Sinai, he could count the rules of ethical behavior on the fingers of his two hands. In the complex global economy of the late twentieth century, in which the simple act of turning on a light sends greenhouse gases up into the atmosphere, the rules for ecologically sustainable living run into the hundreds.

The basic value of a sustainable society, though, the ecological equivalent of the Golden Rule, is simple: Each generation should meet its needs without jeopardizing the prospects of future generations to meet their own needs.

What is lacking is the thorough practical knowledge—at each level of society—of what living by that principle means. For individuals, the decision to live a life of sufficiency—to find their own answer to the question "How much is enough?"—is to begin a highly personal process. The goal is to put consumption in its proper place among the many sources of fulfillment and to find ways of living within the means of the Earth. One great inspiration in this quest is the body of human wisdom passed down over the ages.

Materialism was denounced by all the sages, from Buddha to Muhammad. These religious founders, observed historian Arnold Toynbee, "all said with one voice that if we made material wealth our paramount aim, this would lead to disaster." The Christian Bible echoes most of human wisdom when it asks, "What shall it profit a man if he shall gain the whole world and lose his soul?"

For those people experimenting with voluntary simplicity, the goal is not ascetic self-denial. What they are after is personal fulfillment: they just do not think consuming more is likely to provide it.

Still, shifting emphasis from material to nonmaterial satisfaction is hardly easy: It means trying both to curb personal appetites and to resist the tide of external forces encouraging consumption.

Many people find simpler living offers rewards all its own. They say life can become more deliberate as well as spontaneous, and even gain an unadorned elegance. Others describe the way simpler technologies add unexpected qualities to life.

Realistically, however, voluntary simplicity is unlikely to gain ground rapidly against the onslaught of consumerist values. The call for a simpler life has been peren-

nial through the history of North America, from the Puritans of Massachusetts Bay to the back-to-the-landers of the 1970s. None of these movements ever gained more than a slim minority of adherents.

It would be naive to believe that entire populations will suddenly experience a moral awakening, renouncing greed, envy and avarice. What can be hoped for is a gradual weakening of the consumerist ethos of affluent societies. The challenge before humanity is to bring environmental matters under cultural controls, and the goal of creating a sustainable culture—a culture of permanence—is a task that will occupy several generations.

The ultimate fulfillment: In many ways, we might be happier with less. Maybe Henry David Thoreau had it right when he scribbled in his notebook beside Walden Pond. "A man is rich in proportion to the things he can afford to let alone."

For the luckiest among us, a human lifetime on Earth encompasses perhaps a hundred trips around the Sun. The sense of fulfillment received on that journey—regardless of a person's religious faith—has to do with the timeless virtues of discipline, hope, allegiance to principle and character. Consumption itself has little part in the playful camaraderie that inspires the young, the bonds of love and friendship that nourish adults, the golden memories that sustain the elderly. The very things that make life worth living, that give depth and bounty to human existence, are infinitely sustainable.

QUESTIONS

1. How much richer are Americans than at the turn of the century?

2. What is another major cause (other than consumption) of severe environmental problems?

3. What two groups are most destructive to the environment? Give two examples from each of the groups.

4. Name three statistics of the industrialized countries' (1/5 world population) destruction of global resources.

5. Give some evidence that refutes the idea of consumerism as an avenue for happiness.

6. Which world class of people do the least damage to the environment? Give three examples.

7. Why do the industrialized nations keep accelerating their consumerism?

8. What can be done to curb this exponential consumerism?

(Questions continued on p. 172)

Table 20.1

Generation and Recovery of Selected Materials in U.S. Municipal Solid Waste, 1960–93 (millions of tons)

Item and Material	1960	1965	1970	1975	1980	1985	1990	1993
Gross waste generated[1]								
Paper and paperboard	29.9	38.0	44.2	43.0	54.7	61.5	73.3	77.8
Glass	6.7	8.7	12.7	13.5	15.0	13.2	13.2	13.7
Metals								
Ferrous	9.9	10.1	12.6	12.3	11.6	10.9	12.3	12.9
Aluminum	0.4	0.5	0.8	1.1	1.8	2.3	2.7	3.0
Other nonferrous	0.2	0.5	0.7	0.9	1.1	1.0	1.2	1.2
Plastics	0.4	1.4	3.1	4.5	7.8	11.6	16.2	19.3
Rubber and leather	2.0	2.6	3.2	3.9	4.3	3.8	4.6	6.2
Textiles	1.7	1.9	2.0	2.2	2.6	2.8	5.6	6.1
Wood	3.0	3.5	4.0	4.4	6.7	8.2	12.3	13.7
Other	0.1	0.3	0.8	1.7	2.9	3.4	3.2	3.3
Materials recovered[2]								
Paper and paperboard	5.4	5.7	7.4	8.2	11.9	13.1	20.9	26.5
Glass	0.1	0.1	0.2	0.4	0.8	1.0	2.6	3.0
Metals								
Ferrous	0.1	0.1	0.1	0.2	0.4	0.4	1.9	3.4
Aluminum	0.0	0.0	0.0	0.1	0.3	0.6	1.0	1.1
Other nonferrous	0.0	0.3	0.3	0.4	0.5	0.5	0.8	0.8
Plastics	0.0	0.0	0.0	0.0	0.0	0.1	0.4	0.7
Rubber and leather	0.3	0.3	0.3	0.2	0.1	0.2	0.2	0.4
Textiles	0.0	0.0	0.0	0.0	0.0	0.0	0.2	0.7
Wood	0.0	0.0	0.0	0.0	0.0	0.0	0.4	1.3
Other[3]	0.0	0.3	0.3	0.4	0.5	0.5	0.8	0.7
Percent of gross discards recovered								
Paper and paperboard	18.1%	15.0%	16.7%	19.1%	21.8%	21.3%	28.6%	34.0%
Glass	1.5	1.1	1.6	3.0	5.3	7.6	19.9	22.0
Metals								
Ferrous	1.0	1.0	0.8	1.6	3.4	3.7	15.4	26.1
Aluminum	0.0	0.0	0.0	9.1	16.7	26.1	38.1	35.4
Other nonferrous	0.0	60.0	42.9	44.4	45.5	50.0	67.7	62.9
Plastics	0.0	0.0	0.0	0.0	0.0	0.9	2.2	3.5
Rubber and leather	15.0	11.5	9.4	5.1	2.3	5.3	4.4	5.9
Textiles	0.0	0.0	0.0	0.0	0.0	0.0	4.3	11.7
Wood	0.0	0.0	0.0	0.0	0.0	0.0	3.2	9.6
Other	0.0	100.0	37.5	23.5	17.2	14.7	23.8	22.1

Note: 1. Generation Before Materials Recovery or Combustion. Does not Include Construction or Demolition Debris or Industrial Process Wastes. 2. Recovery of Postconsumer Wastes for Recycling and Composting; Does not Include Converting Fabrication Scrap. 3. Recovery of Electrolytes in Batteries; Probably not Recycled.
Source: Environmental Protection Agency, *Characterization of Municipal Solid Waste in the United States: 1994 Update* (1995).

9. Explain briefly the major points in the "Payback of Greed." What did you learn from this article?

10. Define the consuming society. Do you think that this will continue to be a problem in the next few decades? Why? Why not?

11. How has wealth affected ecological issues? Does materialism create ecological concerns? How?

12. What is the ultimate fulfillment?

13. How can the payback of greed be stopped?

Table 20.2
Municipal Solid Waste Generated, Recovered, Combusted, and Discarded, 1960–93

Category	1960	1965	1970	1975	1980	1985	1990	1993
Millions of tons								
Generation	87.8	103.4	121.9	128.1	151.5	164.4	195.7	206.9
Materials recovery	5.9	6.8	8.6	9.9	14.5	16.4	33.4	45.0
Recovery for recycling	5.9	6.8	8.6	9.9	14.5	16.4	29.2	N.A.
Recovery for composting	0.0	0.0	0.0	0.0	0.0	0.0	4.2	N.A.
Discards after recovery[1]	81.9	96.6	113.3	118.2	137.0	148.1	162.3	N.A.
Combustion	27.0	27.0	25.5	18.5	13.7	11.7	31.9	32.9
With energy recovery	0.0	0.2	0.4	0.7	2.7	7.6	29.7	N.A.
Without energy recovery	27.0	26.8	25.1	17.8	11.0	4.1	2.2	N.A.
Discards to landfill or other[2]	54.9	69.6	88.2	99.7	123.3	136.4	130.4	129.0
Percent of total generation								
Generation	100.0%	100.0%	100.0%	100.0%	100.0%	100.0%	100.0%	100.0%
Materials recovery	6.7	6.6	7.1	7.7	9.6	9.9	17.0	21.7
Recovery for recycling	6.7	6.6	7.1	7.7	9.6	9.9	14.9	N.A.
Recovery for composting	0.0	0.0	0.0	0.0	0.0	0.0	2.1	N.A.
Discards after recovery[1]	93.3	93.4	92.9	92.3	90.4	90.1	82.9	N.A.
Combustion	30.8	26.1	20.6	14.4	9.1	7.1	16.3	15.9
With energy recovery	0.0	0.2	0.3	0.5	1.8	4.6	15.2	N.A.
Without energy recovery	30.8	25.9	20.3	13.9	7.3	2.5	1.1	N.A.
Discards to landfill or other[2]	62.5	67.3	72.4	77.8	81.4	82.9	66.6	62.4

[1]Does not Include Residues from Recycling or Composting Processes. 2. Does not Include Residues from Recycling, Composting, or Combustion Processes.
Source: Environmental Protection Agency, *Characterization of Municipal Solid Waste in the United States: 1994 Update* (1995).

Table 20.3
Worldwide Growth in Selected Human Activities and Products 1970–1990

	1970	*1990*
Human population	3.6 billion	5.3 billion
Registered automobiles	250 million	560 million
Kilometers driven/year (OECD countries only)		
by passenger cars	2584 billion	4489 billion
by trucks	666 billion	1536 billion
Oil consumption/year	17 billion barrels	24 billion barrels
Natural gas consumption/year	31 trillion cubic feet	70 trillion cubic feet
Coal consumption/year	2.3 billion tons	5.2 billion tons
Electric generating capacity	1.1 billion kilowatts	2.6 billion kilowatts
Electricity generation/year by nuclear power plants	79 terawatt-hours	1884 terawatt-hours
Soft drink consumption/year (U.S. only)	150 million barrels	364 million barrels
Beer consumption/year (U.S. only)	125 million barrels	187 million barrels
Aluminum used/year for beer and soft drink containers (U.S. only)	72,700 tonnes	1,251,900 tonnes
Municipal waste generated/year (OECD countries only)	302 million tonnes	420 million tonnes

Reprinted from *Beyond the Limits* Copyright © 1992 by Meadows, Meadows and Randers. With Permission from Chelsea Green Publishing Co, White River Junction, Vermont. 1-800-639-4099 www.chelseagreen.com

QUESTIONS ON THE TABLES

1. Read and analyze Tables 20.1 and 20.2. Compare in both tables the growth of waste generated historically.

2. Compare in both tables the growth and percentages of materials recovered historically.

3. Draw conclusions from the previous two exercises. What further actions would you recommend to continue this process of material recovery and recycling efforts?

4. What are your observations about discards to landfills? What conclusions would you draw and what action would you recommend with the knowledge that a large percentage of current landfills are closing with few new ones opening?

5. Compare Tables 20.1, 20.2, 20.3 to the information in "The Grim Payback of Greed." Does the data in the tables correspond to that in the article? Explain.

Young at Risk

Children get 12 percent of their lifetime exposure to dioxin in their first year of life. On a daily basis, the infant is getting about 50 times the exposure an adult gets during what may be a critical developmental stage.

EPA TOXICOLOGIST LINDA BIRNBAUM

During the eight years after an industrial dioxin (a group of chlorine-based chemicals from wastes of papermaking incineration of chlorinated plastics and other processes) pollutant explosion in Seveso, Italy, an unusual scarcity of male babies being born was noticed—twice as many girls were born as were boys—differing from the usual ratios of baby boys slightly outnumbering girls. In addition, excess cancers turned up among Seveso's adults. Clinical pathologist Polo Mocarelli theorized that the dioxin interfered with hormonal balances in developing embryos, either making normal male growth impossible or killing males. Such an effect of dioxin affecting sex ratios is well known in wildlife. For example, crossed bills in double-crested cormorants with the presence of dioxin in the Great Lakes region during the 1980s occurred almost always in females; scientists speculated that the males died before they hatched with this deformity (Monks, 1997, p. 18).

Dioxins are a chemical byproduct of many common industrial processes; most (84%) is released as air pollution from waste incineration. Ironically, 53% of dioxin release comes from medical-waste incineration, due to the high level of plastic garbage accumulated by the medicine industry. To add to this negative picture, the EPA says that the dioxins stored in the environment are 15 to 36 times that of known annual emissions, and the amount found on the ground from unidentified sources in the United States is two to five times that from identified sources (Mazza, 1996, p. 2).

Dioxins have already been established in animal studies as one of the most potent carcinogens; they have also been linked to human illnesses that affect almost every major body system, including diabetes, bronchitis, irregular heartbeat, and nervous-system and thyroid disorders. One of the most comprehensive studies of dioxins by the Environmental Protection Agency in 1994 found that dioxins are more prevalent and dangerous in the population than previously reported. According to the EPA study, the average American has accumulated dioxins amounting to nine parts per trillion in his or her body. Studies have shown that dioxins begin to slow the action of the immune system at about seven parts per trillion, which leads to the conclusion that the dioxin insult to the body is already above the level that has been shown to cause harm (Mazza, 1996, p. 1).

A growing body of scientific literature and observation strongly suggests that the young of most animals are far more susceptible to toxins (such as dioxins, PCBs, and other chemicals) than adults. "Children and animal young eat and breathe more for their body weights than adults do, so they get bigger proportional doses of whatever is out there," explains Herbert Needleman of the University of Pittsburgh, who pioneered studies linking lowered intelligence with early childhood exposures to lead. Significantly, in 1993, the National Academy of Sciences concluded that infants and children are not sufficiently protected by pesticide regulations, since the risks have been calculated for adults. These various toxins can not only cause cancer, but also affect the young's immune systems, brains, and reproductive organs. "It's important for us to realize that if we're seeing abnormalities in wildlife, similar mechanisms may exist in humans. We are just another species in the ecosystem; if other species are harmed, we may be too," says University of Florida zoologist Louis Guillette (Monks, 1997, p. 20).

DEFORMED NEWBORNS

In Minamata, Japan, in the 1950s, before people understood the effects of industrial pollution, mercury discharges from a chemical plant poisoned the seafood that habitated the surrounding area and those who ate it. By the 1960s, evidence of the harm of the mercury began to mount from poisonings for which animals died and people became sick. Fishermen noticed that seabirds were dying; feral cats that ate scavenged fish became stiff legged; cerebral palsy and mental retardation significantly increased in children; and adults were frequently ill. At the time of the Minamata poisonings, science held that the womb was a protected environment capable of filtering out harmful substances. However, in Japan, many women who ate the contaminated fish did not become ill themselves, but they gave birth to children with severe mental retardation and physical deformities. That incident changed scientists' thinking to hypothesize that in actuality, the womb was not protective from toxins, but instead the fetus was sharing the mother's toxic load and in a way was actually protecting the mother, since the fetus was absorbing some of the mercury, thus reducing the mother's exposure to it. The fetus received at least the same doses as did its mother—and the fetus was far more susceptible to toxic pollutants.

Not only are the young exposed to toxic chemicals in the womb, but mothers also unload toxins in their milk to the young. The milk from many species, ranging from beluga whales to dairy cows, have measurable concentrations of chemicals, including dioxins, PCBs, and various pesticides. Children, in their first year of life, get 12 percent of their lifetime exposure to dioxin, and on a daily basis, during their critical developmental months, infants get about 50 times the exposure of an adult. Milk is still recommended for its antibodies, protection, and nourishment, and its benefits clearly outweigh its potential risks, but exposure of the young to toxins is increasing and is of growing concern (Monks, 1997, p. 20).

Another study, which took place from 1987 to 1992 by the EPA, indicated strong evidence that lake trout embryos exposed to dioxin could develop a lethal syndrome called "blue sac." The yolk sac of young healthy trout is a rich golden color. During the first month or so, when the fry (baby fish) rely exclusively on the yolk for nutrients, they become vulnerable to blue sac syndrome, where fluid leaks out of the blood vessels and into the yolk sac, turning it milky and slightly blue. Many conditions can cause blue sac, but this study confirmed that certain dioxins, like chemicals found in the Great Lakes at very small concentrations

of just 60 parts per trillion (ppt), will cause 50% of lake trout fry to develop the blue sac disorder—making this fish the most vulnerable known. (This finding is in comparison to the same mortality rate of rainbow trout from the same chemicals at 400 ppt.) (Raloff, 1997, pp. 306–7.)

DAMAGED IMMUNITY

Along the coast of Florida, bottlenose dolphins' firstborn calves die between the ages of three and six. In four generations of dolphins in the past 25 years, only one firstborn is known to have survived. Although the cause of the deaths is not certain, high levels of toxins exist in the fat of marine mammals. Research suggests that mother dolphins unload as much as 80% of their accumulation of pollutants into each of their calves, most likely through nursing. The firstborn calf receives the highest dose through the mother's accumulated toxins. The chemicals found in Florida dolphins' blubber are some of the most deadly and long-lived contaminants of the industrial age, including dioxins and PCBs (although banned, PCBs are still found in insulation of electrical systems). The toxins are so persistent and widely distributed, moving into the food chain from the soil and water, that people and other animals continue to be exposed to them worldwide. "What we are seeing now is the impact of damage that was done over the last few decades," says biologist Randall Wells of the Chicago Zoological Society.

The evidence is that these and other toxic chemicals that are still being manufactured can interfere with the immune system. In 1987, about 700 bottlenose dolphins, half of the migrant Atlantic population, died and washed up along the Atlantic coastline of the United States from New Jersey to Florida. Analysis determined that they were killed by infectious disease and that their bodies contained high levels of PCBs, DDT, and other compounds that suppress the immune system, evidence that scientists think explains the dolphins' susceptibility to disease. If these chemicals are damaging immunity in adult dolphins, they may be doing even more harm to juveniles, because mammalian immune systems aren't fully functional until months or years after birth, according to immunologist Garet Lahvis of the University of Maryland School of Medicine (Monks, 1997, p. 22).

Regarding the Inuits, an Eskimo people in the Canadian Arctic, researchers from Quebec are analyzing the relationship of unusually high rates of infectious disease among the Inuit children and exposure to toxic chemicals. Even though no polluting industries operate near the region, contaminants enter the ecosystem from high-altitude winds and migrant wildlife. The contaminants accumulate in greater density with every link up the food chain, as PCBs, pesticides, and other organochlorines progress from plant and fish to seal, whales, polar bears, and humans. Inuit babies in their first year of life have rates of infectious disease that are 20 times greater than those of babies in southern Quebec, and Inuit women have rates of PCBs in their breast milk that are 7 times greater than those of women from the urban, industrialized south of Quebec. With the Inuits, acute ear infections are common, causing hearing loss for nearly one in four among the Inuit children, and the usual childhood immunizations do not work very well. In another study of the Inuits, conducted in 1993, it was found that babies nursed by mothers with the highest contaminant levels in their milk were afflicted with more acute ear infections than were bottle-fed Inuit babies. The babies with the highest exposures to contaminants also produced few of the helper T cells that play an important role in eliminating bacteria and other harmful invaders. Even though there may be other factors causing these problems, the data suggest that contaminants remain a significant factor. In the Netherlands, researchers have concluded that even infants with mild exposures to contaminants may experience weakened immunity; a correlation was discovered between PCB–dioxin exposure and suppressed levels of disease-fighting white blood cells that would cause immune-system changes (although not extreme) that could persist throughout life or provoke autoimmune diseases.

In nonmammalian species, biologist Keith Grasman of Wright State University has measured immune suppression that is mediated by T cells in young Caspian terns and herring gulls from contaminated colonies around the Great Lakes between 1992 and 1994. Grasman states that the same PCB and organochlorine pollutants that are found in these birds have also been measured in seals, dolphins, humans, and other species with similar T-cell immune problems. Many chicks with suppressed immune systems die before they leave the breeding grounds. Some contaminated terns do grow and migrate south, but most never return to breed (Monks, 1997, p. 23–24).

LOWERED INTELLIGENCE

Since the metabolism of young animals is faster than that of adults, and because young animals do not excrete contaminants or store them away in fat in the same ways that adults do, babies and young get continuous exposure to toxins at the time that all of their organs, including their brains, are still developing. In an adult, a blood–brain barrier guards the brain from potentially harmful chemicals in the body, but in a child, that barrier does not become fully developed until six months after birth.

The developing brains of the young of various wildlife species are also far more sensitive to toxic contaminants than are the brains of the adults of the species. As an example, in the late 1980s, great blue heron hatchlings from dioxin-contaminated colonies in Canada developed gross asymmetries and other abnormal changes in brain structures. The susceptibility of human young to such toxic effects was evidenced from an accidental PCB poisoning in Taiwan in 1979. In the Taichung province of Taiwan, more than 2,000 people were exposed to PCB-contaminated cooking oil, resulting in "Yu-Cheng," or oil disease. In the first three years after the accident, many newborns died, and others developed blotch patches of dark skin and fingernail and toenail deformities. As the children grew, they were mentally slower than other kids their age and displayed hyperactive and other behavioral problems. These developmental delays and IQ deficits have not gone away as the children have aged. The exposed mothers continued to deliver babies with problems as late as 1985, even though the accident occurred in 1979. As Walter Rogan of the National Institute of Environmental Health Sciences (NIEHS), who studied the case, explains it, a large portion of the PCBs that these women consumed ended up being stored in their fat, a process that happens with many toxic chemicals. During pregnancy, women mobilize a lot of body fat, and the contaminants in the fat are also passed to their children—even years later, as in the case of Yu-Cheng. The same mechanism also applies to low-level toxic exposures from food, air, and water, for which even small amounts of pollutants—accumulated by women throughout their lives—can have lasting consequences for a child exposed to the pollutants in the womb.

A Michigan study found persistent intellectual deficits in children exposed before birth to much lower doses of PCBs than those in the case of Yu-Cheng. In 1981, two Wayne State University psychologists, Sandra and Joseph Jacobson, measured PCB levels in mothers and newborn infants. Since consumption of fatty fish from contaminated water is a major source of PCB exposure, the Jacobsons selected mothers for their study who had eaten Lake Michigan salmon or lake trout during the years before their children were born. The infants with the highest exposures grew more slowly than other babies and at 4 years old had

Dioxin's Effect on Fish Raises Questions of Effects on Higher Animals

Pollutants are considered dioxinlike if they connect to the Ah receptor in cells (a protein that reacts to these pollutants and turns genes on or off). The receptor was identified in fish in 1988 after being recognized in mammals in 1986. Since then, researchers have been examining other lower species to see how far down the evolutionary ladder this receptor exists and the accompanying vulnerability to dioxin with it. Scientists at Woods Hole Oceanographic Institution have been examining various types of animals for this Ah receptor and have so far found sharks to be the most primitive animal with the Ah receptor. (Sea lampreys have something that resembles the receptor, but this substance does not appear to bind to dioxinlike compounds.)

Scientists have also been using lower animals as a useful model of the common effects of dioxins on all animals. So far, it has been found from a joint study between Cornell University and the University of Wisconsin-Madison that dioxinlike chemicals target the cardiovascular system of lower animals as they do in mammals. Using zebra fish, which have transparent embryos, scientists found that the pollutants slow blood flow feeding the head and gills and also slow the heart's rate. Richard Peterson, of the University of Wisconsin-Madison, says that there appears to be a pruning of these blood vessels, which may account for the head malformations that often accompany blue sac syndrome. Dioxinlike compounds also appear to weaken blood vessels once they form, which may explain why blood vessels become leaky in blue sac.

The dioxin chemicals trigger early death in blood vessel cells caused by oxidant damage. This effect may trace to the ability of dioxins to turn on genes that increase the production of detoxifying enzymes which start a process that releases oxidative compounds which do not respond to normal controls. The overproduction of oxidants can damage the vessels. Because there is no reason to suspect that this effect occurs only in fish, scientists are also searching for it in birds, reptiles, and mammals—including humans.

Reprinted with Permission from *Science News,* the Weekly Newsmagazine of Science, Copyright © 1996, 1997 by Science Service.
Excerpted from Raloff, J. (1997, May 17). Those Old Dioxin Blues. *Science News,* Vol. 151, p. 307.

poorer short-term memory. By the time they were 11 years old, the 30 most exposed children had average IQs six points lower than those of the least exposed group. Twenty-three percent of the high-exposure kids were two years behind in reading, while 10% of the least exposed group were two years behind. The Jacobsons also found that fish-free diets did not guarantee lower PCB levels, since there were some very highly exposed children from mothers who did not eat the fish. The exposure might have come from other fatty foods, such as butter, cheese, beef or pork, but there is no way of knowing the source, since exposure to toxins is a societal problem. As Jacobson said, "We are all walking around with PCBs in us." (Monks, 1997, p. 24)

SEXUAL IMPAIRMENT

Sexual development in the growing fetus may be as sensitive to toxic effects as the brain is. When certain chemicals bind to hormone receptors, they can interfere with the work of natural hormones in the development of male or female organs, resulting in any number of reproductive disorders. These chemicals, known as endocrine disrupters, include PCBs, dioxins, and many pesticides. The growing body of evidence suggests reason for concern about the effects of endocrine-disrupting chemicals found in the environment. Among the Yu-Cheng children of Taiwan, the boys with high PCB exposures had smaller than average penises. University of Florida biologists found the same phenomenon in alligators born in a lake poisoned by pesticides. In the highly polluted St. Lawrence River, biologists found a male beluga whale with a fully developed set of female organs in addition to the whale's male apparatus. This male carried a very high load of endocrine-disrupting contaminants in its blubber.

In South Florida, 13 of the 19 male panthers that still survive have undescended testicles. Because such males produce abnormal sperm and have low sperm counts, biologists are worried about the potential for saving the endangered animals from extinction. It is suspected that environmental endocrine-disrupting chemicals may be contributing to these cats' sexual abnormalities. The panthers are exposed to heavy doses of pesticides and toxic metals such as methyl mercury from their diet of raccoons, which ingest the pollutants in fish. It is possible that most of the problems of the Florida panther could be attributed to pesticides. If this assertion is true, then the introduction of female Texas cougars to improve the panthers' genetic diversity may not accomplish much. If all of the panthers' habitat is contaminated, the animal may not be able to be saved.

According to a 1996 study by U.S. and European scientists, data from several countries show substantial increases since the 1950s in the number of baby boys born with undescended testicles and other sexual abnormalities. One London study found that 5.2% of low-birthweight boys born in the 1980s had undescended testicles, as compared with 1.74% from the 1950s. Testicular cancers nearly doubled among older teenagers in the United States between 1973 and 1992.

Many of the health effects documented in young wildlife from toxic contamination may not apply to human children, but wildlife can give indicators as to where to look for problems and answers. Deformed frogs in Minnesota may yield clues about the reasons for high rates of birth defects among the region's farm children. Links are being established everywhere between environmental toxic contaminants and the health of young wildlife and children. The warnings found in wildlife may help us to do something about environmental toxicities before we permanently contaminate and deform the animal world of which we are a representative part. (Monks, 1997, p. 25)

REFERENCES

Mazza, P. (1996, February 21). "Love Canal is everywhere: The pervasive threat of dioxin." Available: http://www.tnews.com/test/dioxin.html.

Monks, V. (1997, June–July). Children at risk. *National Wildlife,* Vol. 35, No. 4, pp. 18–27.

Raloff, J. (1997, May 17). Those old dioxin blues. *Science News.* Vol. 151, pp. 306–307.

QUESTIONS

1. Why are the young of most animals more susceptible to toxins than adults?

2. Why do you think infants and children are not sufficiently protected by pesticide regulations? How do you think we can use pesticide regulations to better protect infants and children? Explain some specific ideas.

3. What does the case study in Minamata, Japan, teach us about the effects of toxic pollution on fetuses? How are infants doubly exposed to pollutants?

4. What does the study regarding blue sac syndrome in lake trout embryos teach us about dioxins?

5. Give three examples of damaged immunity caused by dioxin and PCB contamination.

6. Give two examples of lowered intelligence caused by toxic pollution. Why are the young particularly susceptible to toxins affecting intelligence? Do you think that the examples in the article can support the claim that such toxins affect intelligence in humans? Explain.

7. Give two examples of sexual impairment caused by toxic pollution. Do you think that the examples in the article can support the claim that such toxins affect sexual impairment in animals and humans? Explain.

8. What is the reason that animals should be so closely watched for their reactions to chemical toxins? Why is the issue of the young such a special case?

9. How can evidence of the effects of chemical toxins, especially on the young, be made more apparent to determine clearer relationships and, hence, approaches to prevention? Outline appropriate studies that could be carried out to this end.

10. How can such chemical toxic pollution be prevented? What worldwide plan would realistically prevent such pollution and casual dumping?

QUESTIONS ABOUT THE TABLES

1. Examine Tables 21.1–21.3. Draw numeric and locale conclusions. What do you conclude from this data on hazardous waste sites?

Table 21.1
The 50 Worst Superfund Waste Sites, 1994

State	Site	City/County	Score (Rank)	When Listed
Arkansas	Vertac, Inc.	Jacksonville	65.46 (32)	Oct. 1981
California	McCormick & Baxter Creosoting Co.	Stockton	74.86 (5)	Feb. 1992
California	Stoker Company	Imperial	65.51 (30)	July 1991
California	Riverbank Army Ammunition Plant	Riverbank	63.94 (36)	June 1988
Colorado	Rocky Flats Plant (USDOE)	Golden	64.32 (35)	Oct. 1984
Delaware	Tybouts Corner Landfill[1]	New Castle County	73.67 (6)	Oct. 1981
Delaware	Army Creek Landfill	New Castle County	69.92 (18)	Oct. 1981
Florida	Stauffer Chemical Co.	Tarpon Springs	70.71 (12)	Feb. 1992
Florida	Stauffer Chemical Co.	Tampa	63.62 (39)	Feb. 1992
Hawaii	Pearl Harbor Naval Complex	Pearl Harbor	70.82 (11)	July 1991
Idaho	Triumph Mine Tailings Piles	Triumph	90.33 (1)	May 1993
Indiana	U.S. Smelter & Lead Refinery Inc.	East Chicago	70.71 (12)	Feb. 1992
Iowa	Mason City Coal Gasification Plant	Mason City	69.33 (20)	Jan. 1994
Maine	Portsmouth Naval Shipyard	Kittery	67.71 (24)	June 1993
Massachusetts	Industri-Plex	Woburn	72.42 (9)	Oct. 1981
Massachusetts	Nyanza Chemical Waste Dump	Ashland	69.22 (21)	Oct. 1981
Massachusetts	Baird & McGuire	Holbrook	66.35 (27)	Dec. 1982
Michigan	Berlin & Farro	Swartz Creek	66.74 (26)	July 1982
Michigan	Liquid Disposal, Inc.	Utica	63.28 (42)	July 1982
Minnesota	FMC Corp. (Fridley Plant)	Fridley	65.50 (31)	July 1982
Missouri	Big River Tailings/St. Joe Minerals	Desloge	84.91 (3)	Feb. 1992
Montana	Silver Bow Creek/Butte Area	Silver Bow/Deer Lodge	63.76 (37)	Dec. 1982
Montana	East Helena Site	East Helena	61.65 (49)	Sept. 1983
New Hampshire	Somersworth Sanitary Landfill	Somersworth	65.56 (29)	Dec. 1982
New Hampshire	Keefe Environmental Services	Epping	65.19 (34)	Oct. 1981
New Hampshire	Sylvester[1]	Nashua	63.28 (42)	Oct. 1981
New Jersey	Lipari Landfill	Pitman	75.60 (4)	Oct. 1981
New Jersey	Helen Kramer Landfill	Mantua Township	72.66 (8)	July 1982
New Jersey	Price Landfill[1]	Pleasantville	71.60 (10)	Oct. 1981
New Jersey	CPS/Madison Industries	Old Bridge Township	69.73 (19)	Dec. 1982
New Jersey	GEMS Landfill	Gloucester Township	68.53 (23)	July 1982
New Jersey	Lone Pine Landfill	Freehold Township	66.33 (28)	Oct. 1981
New York	Pollution Abatement Services[1]	Oswego	70.80 (12)	Oct. 1981
North Carolina	General Electric Co.—Shepherd Farm	East Flat Rock	70.71 (12)	Feb. 1992
Ohio	Arcanum Iron & Metal	Darke County	62.26 (47)	Dec. 1982
Pennsylvania	Bruin Lagoon	Bruin Borough	73.11 (7)	Oct. 1981
Pennsylvania	East Tenth Street	Marcus Hook	67.68 (25)	Jan. 1994
Pennsylvania	Tysons Dump	Upper Merion Township	63.10 (44)	Sept. 1983
Pennsylvania	McAdoo Associates[1]	McAdoo Borough	63.03 (45)	Oct. 1981
South Dakota	Whitewood Creek[1]	Whitewood	63.76 (37)	Oct. 1981
Texas	French, Ltd.	Crosby	63.33 (40)	Oct. 1981
Texas	Motco, Inc.[1]	La Marque	62.66 (46)	Oct. 1981
Texas	Sikes Disposal Pits	Crosby	61.62 (50)	Oct. 1981
Utah	Murray Smelter	Murray City	86.60 (2)	Jan. 1994
Utah	Kennecott (South Zone)	Copperton	70.71 (12)	Jan. 1994
Utah	Wasatch Chemical Co. (Lot 6)	Salt Lake City	63.31 (41)	Jan. 1987
Utah	Petrochem Recycling/Ekotek Plant	Salt Lake City	62.18 (48)	July 1991
Washington	Pacific Sound Resources	Seattle	70.71 (12)	May 1993
Washington	Hanford 200-Area (USDOE)	Benton County	69.05 (22)	June 1988
Washington	Hanford 300-Area (USDOE)	Benton County	65.23 (33)	June 1988

[1]Site is Also Its State's Top Priority.
Source: Environmental Protection Agency, *National Priorities List* (February 1994).

Table 21.2
Superfund Hazardous Waste Sites, Selected Years, 1981–95

Year	Number of Sites	Year	Number of Sites
1981	115[1]	1989	981
1982	418[1]	1990	1,187
1983	406	1992	1,183
1984	538	1993	1,202
1986	703	1994	1,287
1987	802	1995	1,296

[1]Proposed Sites Only. Final Sites Not Calculated Until Release of First National Priorities List in 1983.
Source: Environmental Protection Agency, *National Priorities List. Supplementary Materials* (February 1995).

Table 21.3
States with the Most Hazardous Waste Sites, 2001

State	Nonfederal	Federal	Total
New Jersey	103	8	111
California	72	24	96
Pennsylvania	89	6	95
New York	84	4	88
Michigan	67	0	67
Florida	46	6	52
Washington	33	14	47
Wisconsin	39	0	39
Illinois	35	4	39
Texas	34	4	38

Source: Environmental Protection Agency, *National Priorities List, Supplementary Materials*, (June, 2001).

Table 21.4
Common Hazardous Chemicals Requiring National Response Center Notification (partial list)

Substance	Remarks	Amount to be Reported if Released
Acetic acid	Vinegar is generally 2% acetic acid, but acetic acid is used in many manufacturing processes. Toxic as vapor at 10 parts per million in air.	5.000 lbs. (2,270 kg)
Acetone	Toxic chemical (1,000 parts per million in air) used in large amounts as solvent for resins and fats.	5,000 lbs. (2,270 kg)
Aluminum sulfate	Used sometimes in dyeing or in foam fire extinguishers	1,000 lbs. (454 kg)
Ammonia	Use as fertilizer does not need to be reported; however it is toxic and extremely hazardous; emergency planning required if 500 lbs. possessed.	100 lbs. (45.4 kg)
Benzene	Used in drugs, dyes, explosives, plastics, detergents, and paint remover; can cause cancer; toxic.	10 lbs. (4.54 kg)
Chlorine	Widely used to disinfect water, the gas is toxic at concentration of one part per million in air extremely hazardous; requires emergency planning if 100 lbs. is possessed.	10 lbs. (4.54 kg)
Cumene	Additive for high-octane fuels; toxic to skin at 50 parts per million in air.	5,000 lbs. (2,270 kg)
Cyclohexane	Petroleum derivative.	1,000 lbs. (454 kg)
Ethylbenzene	Toxic at 100 parts per million in air.	1,000 lbs. (454 kg)
Ethylene dichloride	Additive to gasoline that combines with lead to make "Ethyl" gasoline; also used in making plastics.	100 lbs. (45.4 kg)
Ethylene oxide	Widely used in making plastics; toxic at 50 parts per million in air; extremely hazardous; requires emergency planning if 1,000 lbs. possessed.	1–10 lbs. (0.454 kg–4.54 kg)

Continued

Table 21.4

Common Hazardous Chemicals Requiring National Response Center Notification (partial list), *continued*

Substance	Remarks	Amount to be Reported if Released
Formaldehyde	Used in wood substitutes and plastics; toxic and may cause cancer; extremely hazardous; requires emergency planning if 500 lbs. possessed.	100 lbs. (45.4 kg)
Hydrochloric acid	Used in petroleum, manufacturing, and metals industries; toxic; as the gas hydrogen chloride, it is extremely hazardous and requires emergency planning if 500 lbs. possessed.	5,000 lbs. (2,270 kg)
Methanol	Commonly called wood alcohol; used as antifreeze, solvent, and starting material for other compounds; toxic.	5,000 lbs. (2,270 kg)
Nitric acid	Used in preparing fertilizers and explosives; toxic and extremely hazardous; requires emergency planning if 1,000 lbs. possessed.	1,000 lbs. (454 kg)
Phenol	Used in making plastics; vapor is toxic to skin at 5 parts per million in air; extremely hazardous; requires emergency planning if 500 pounds possessed and further planning if 10,000 pounds possessed.	1,000 lbs. (454 kg)
Phosphoric acid	Used as flavoring agent, in pharmaceuticals, and in manufacturing fertilizers; toxic.	5,000 lbs. (2,270 kg)
Sodium hydroxide	Commonly known as lye or as caustic soda; toxic.	1,000 lbs. (454 kg)
Styrene	Used in manufacture of styrene plastics and artificial rubber; toxic.	1,000 lbs. (454 kg)
Sulfuric acid	Most common chemical used in U.S; toxic and extremely hazardous; emergency planning required if 1,000 lbs. possessed.	1,000 lbs. (454 kg)
Toluene	Used in making explosives, drugs, and dyes; toxic.	1,000 lbs. (454 kg)
Vinyl chloride	Used to make plastics and aerosols; causes cancer; toxic.	1 lb. (0.454 kg)
Xylene	Used to make other compounds; toxic.	1,000 lbs. (454 kg)

Source: Environmental Protection Agency, *Title 3 List of Lists* (Jan. 1990).

2. Read the list of chemicals in Table 21.4. How often do you use any of these chemicals? How do you dispose of them? Were you aware of their classification as hazardous before examining Table 21.4?

3. Are you aware of the health hazards of the chemicals listed in Table 21.4? Please list the health hazards of which you are aware for any of the chemicals mentioned. Share this knowledge in discussion.

4. Discuss approaches to chemical dumping and chemical disposal.

5. Research superfund hazardous waste site dumps. What progress is being made? How much money is being spent?

6. How should we dispose of hazardous wastes? Discuss some plausible methods.

Fisheries: Exploiting the Ocean— What Will Be Left?

Developing countries hire private companies to conduct surveillance and enforcement of new fisheries laws. These companies are zealous: One has proposed to watch over fisheries from a blimp, which could descend to launch a patrol boat.

MICHAEL PARFIT

Technology has applied its increasing power to the fish of the ocean . . . and there is now trouble at sea. "There are too many fishermen and not enough fish" (Parfit, p. 9). Fifty years of rapidly improving fishing technology has created an immensely powerful industrial fleet—37,000 freezer trawlers that catch and process a ton or more of fish an hour, manned by about a million people worldwide. This fleet contrasts with small-boat fishermen, who probably number about 12 million, but who catch only about half the world's fish.

The problem is that fish stocks are being damaged by pollution, by destruction of wetlands that serve as nurseries and provide food, by the waste of unprofitable fish (called "by-catch"), and, most of all, by overfishing. These practices have caused the collapse of some fish stocks and the fishing of many important groups of fish beyond sustainable capacity. At this point, the annual catch from the sea has peaked at about 78 million metric tons and seems somewhat stable, but the cost of this fishing intensity will be felt with future shortages of certain types of fish and in future sustainable yields. Although technology has helped quadruple the world's catch of seafood since 1950, a nearly empty basket is typical for what a lone fisherman has to show for hours of work—a complaint heard around the world. Morocco and most small coastal or island countries are extremely worried about overharvesting and want to reduce fish quotas taken in their waters by foreign fleets. One Moroccan official asserts, "People using traditional techniques will not survive. The waters are being emptied by industrial fishing." One marine scientist says, "We've come to our reckoning, the next ten years are going to be very painful, full of upheaval for everyone connected to the sea" (Parfit, 1995, pp. 2–11).

According to the U.N. Food and Agriculture Organization (FAO), almost two thirds of the world's 200 commercially important distinct fish populations are either exploited or fished to the edge. The FAO states that there is an urgent need for the development of effective measures to reduce and control fishing capacity and effort. The visual images of overfishing are very apparent when

- trawlers line up abreast to sweep the life from the sea with nets so large that each could drag up a dozen jumbo jets;
- vessels called "long-liners" trail out thousands of baited hooks on lines stretching 80 miles across the ocean; and
- Japan's squid fleet is so enormous that its lights to draw squid to its nets in the north Pacific can be seen by orbiting astronauts.

Within two decades, giant vessels have combined the latest technology, including satellite navigation, spotter planes, fish-scanning sonar, and lightweight nets, to capture the "limitless" bounty of the seas into finite resources. Big trawling vessels can pull up at least 20,000 pounds of fish in 20 minutes (Hanley, 1997, pp. 1–2).

The problem isn't easily resolvable, and it could grow into a catastrophe. Fishing is a 70-billion-dollar-a-year industry with strong roots in national pride, profits, and age-old traditions of freedom. As governments struggle to solve the problems at sea, they inevitably create laws that challenge traditional freedoms. Throughout the world seas, fishing vessels are attacked in competitive territorial battles and ownership of waters. Such examples as the following are common:

- In Patagonia, an Argentinian gunboat chases and fires on a vessel from Taiwan; the crew is rescued, but the trawler sinks.
- In the North Atlantic, the Stern trawler REX is arrested west of Scotland for trespassing in British waters. REX is officially multinational to evade fishing laws and is owned by Icelanders, registered in Cyprus, and crewed by fishermen from the Faroe Islands.
- In the South Atlantic, a patrol boat from the Falkland Islands chases a Taiwanese squid boat 4,364 nautical miles from home waters, all the way past South Africa. The boat gets away.

There has been an increasing aggressiveness of the law on the free oceans, which began after World War II, such that by the 1970s, most nations pushed their territorial control from 12 to 200 nautical miles offshore in order to grab valuable fishing grounds, thereby pushing the boats of other nations (the "distant-water fleets") far out to sea. Now, 200 nautical miles is not enough. Many fish roam from national to international waters, where they are taken by intense fishing methods outside of any nation's control. Therefore, nations whose roaming salmon, cod, or pollock are caught before they get home fight with those nations that intercept them. Furthermore, more and more countries are taking the fight to sea. In Canada, the Great Banks of Newfoundland were in danger of fishing collapse, necessitating that Canada shut down its own fishery there, putting about 40,000 people out of work. However, distant-water trawlers from Spain, Portugal, and other nations continued to fish just outside the 200-mile limit, making Canada very angry and distrustful. Typical strong offensive responses from Canada or other nations in this situation included strategies of using spy planes to record any suspicious ship's activities, shooting at poachers, implementing blockades, employing patrol boats, and making arrests.

Such fishing-rights arguments have made the oceans a sort of oceanic Wild West, with one state acting as the self-appointed lawmaker, as well as sheriff and judge. In actuality, except for the past few decades, most of the sea was indeed like the Old West: free, wild, unregulated, and a place of opportunity for any brave enough to venture forth.

But declining populations of fish and feuding fishermen, along with the power of technology, are proving that even the limits of the grandest piece of the planet—our seas—have been reached (Parfit, 1995, pp. 10–20).

HIGH-TECH FISHING VESSELS AND METHODS

The number of fishing vessels has doubled since 1970 to more than a million. Not only has the number doubled, but also the types of fishing vessels used have radically changed to types of superefficient floating ship factories where all fish from the ship's enormous fishing path is processed onboard and where all products are used—nothing is wasted. A typical new factory ship is 376 feet long and both operates as a mothership for fleets of smaller vessels and catches its own fish, processing and packaging fish at sea as it catches them. The typical "technological kit" of such ships includes a variety of gigantic nets (some large enough to swallow the Statue of Liberty), highly powerful lights that act as bait, literally miles of fishing lines that dangle lures or baited hooks, sonar arrays and computers to locate and track the fish, and sensors on the net. The fish are not so much strained from the sea by the nets, but are instead herded by the net's cables and winch-like leading edges. A typical factory ship is capable of processing more than 600 metric tons of pollock a day into surimi, the protein paste used in imitation seafood products. All processing occurs onboard, including the tasks of processing into various products, quick freezing, and packaging. To keep morale high among the 125 crew members of such ships, living decks on the ships are separated from work decks and typically can include well-decorated cafeterias, gymnasiums, bathrooms with Japanese soaking tubs, and television in most of the crew's cabins (Parfit, 1995, pp. 16–17).

THE DIMINISHING CATCH

Nearly 40% of the world's oceans have been locked up by territorial claims and exclusive fishing zones. The 200-nautical-mile coastal boundaries have not stopped the overexploitation of species. Rich nations buy into poor countries' waters. High-tech fleets grab fish migrating outside protected coastal zones. And the fish just keep getting scarcer and smaller. After peaking in 1989, and despite a temporary rise in the Pacific ocean catch, the

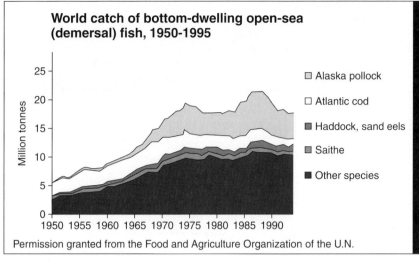

World catch of bottom-dwelling open-sea (demersal) fish, 1950-1995

Legend:
- Alaska pollock
- Atlantic cod
- Haddock, sand eels
- Saithe
- Other species

The global trend in demersal fish landings shows the impact of overfishing that has resulted in reduced catch of several major species, including the Atlantic cod, haddock, and Alaska pollock. When these species are excluded, data reveal that the demersal catch increased steadily up to the early 1970s when it levelled off and has remained stable since then.

Permission granted from the Food and Agriculture Organization of the U.N.

Figure 22.1
World Catch of Bottom-dwelling Open-sea (Demersal) Fish, 1950–1995.
Source: Food and Agriculture Organizations of the United Nations.

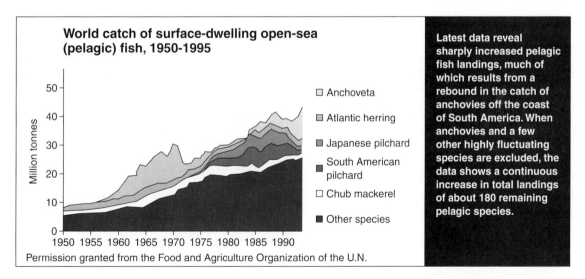

World catch of surface-dwelling open-sea (pelagic) fish, 1950-1995

Legend:
- Anchoveta
- Atlantic herring
- Japanese pilchard
- South American pilchard
- Chub mackerel
- Other species

Latest data reveal sharply increased pelagic fish landings, much of which results from a rebound in the catch of anchovies off the coast of South America. When anchovies and a few other highly fluctuating species are excluded, the data shows a continuous increase in total landings of about 180 remaining pelagic species.

Permission granted from the Food and Agriculture Organization of the U.N.

Figure 22.2
World Catch of Surface-dwelling Open-Sea (Pelagic) Fish, 1950–1995.
Source: Food and Agriculture Organizations of the United Nations.

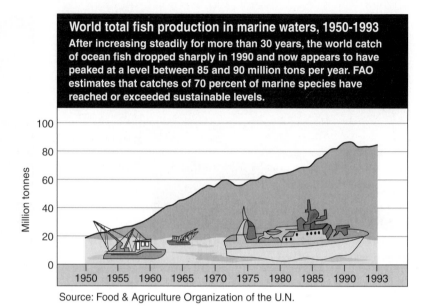

World total fish production in marine waters, 1950-1993
After increasing steadily for more than 30 years, the world catch of ocean fish dropped sharply in 1990 and now appears to have peaked at a level between 85 and 90 million tons per year. FAO estimates that catches of 70 percent of marine species have reached or exceeded sustainable levels.

Source: Food & Agriculture Organization of the U.N.

Figure 22.3
Total Fish Production, 1950–1993.
Source: Food and Agriculture Organizations of the United Nations.

tonnage of sea-food harvested worldwide has reached a foreboding plateau. Meanwhile, demand continues to grow, with an annual per-capita consumption of 66.6 kilograms (147 pounds). Japan has the world's biggest appetite; China is gaining rapidly in bulk catches. (See the accompanying graphs.) Some of the species being caught include the following:

Salmon: Hydropower, pollution, and logging have diminished salmon-spawning streams worldwide. The North Pacific waters remain the healthiest waters for salmon.

Shrimp: About one quarter of all shrimp consumed are raised in man-made ponds. Viruses recently devastated farmed stocks in China, Ecuador, and Texas.

Herring, sardines, anchovies: These small open-ocean fish are used mainly for industrial purposes, such as the production of fish meal and fertilizer (although they remain a delicacy in many countries). Herring has been heavily fished in both the

Atlantic and Pacific. Anchovies are caught mainly off the Pacific coast of South America, where populations fluctuate dramatically, depending on the El Niño phenomenon.

Cod, pollock, haddock: These fish are the mainstays of human consumption and have been heavily exploited. The cod fishery in the northwest Atlantic recently collapsed, and in Newfoundland, it completely disappeared. Pollock is still plentiful, and much of it is processed into fish sticks and fast food. Cod, haddock, and flounder stocks have collapsed in New England.

Jacks, sauries, capelins: Used both for eating and industrial purposes, these fish are prey for larger, more favored fish such as cod or tuna.

Redfish, seabreams, roughies. When more easily accessible fish stocks run out, fishermen turn to alternative species. Redfish, a restaurant basic, have been fished commercially since the 1930s. Orange roughy fishing began only in the 1970s. The danger is that little is known about the biology of these

THE SEAS . . . NOW AND THEN . . .

When my father began diving in the Mediterranean Sea in the early forties, the water was clean. Great beds of sea-grass and algae thrived there, along with dense schools of fish and rich invertebrate fauna. Gorgonians were abundant, and so were huge groupers and spiny lobsters. It was the rich sea-floor community the world was seeing in the early Cousteau films.

Since those days, the Mediterranean coast has become densely populated. Industries, hotels, and homes line the coast. Sewage and other wastes stream into the sea. Sadly, the same waters where the first Aqua-lung divers discovered the sea's beauty and diversity are today biologically impoverished. And this scene—where urban development meets the water—is spreading rapidly around the world today.

JEAN-MICHEL COUSTEAU (1990)

The very survival of the human species depends upon the maintenance of an ocean clean and alive, spreading all around the world. The ocean is our planet's life belt.—Marine Explorer Jacques-Yves Cousteau (1980)

---

alternative food fishes. Some scientists think that orange roughies may live longer than humans. If this is true, then orange roughies will need ample time to replenish from intense fishing.

Tuna, billfish: These fish are the glamour fish of the industry, where they command luxury prices. A two-bite portion of sushi made from prime meat from the bluefin tuna can cost $75 in Japan. Large fish are hunted with harpoons and spotter planes (Parfit, 1995, pp. 12–13).

Our seas have changed, and the unlimited bounty of the sea is endangered. This doesn't mean that the sea is ruined, however. It is more like a forest than a mine in that it will keep producing as long as we do not plunder it without restraint, and that is just what we do not know how to do yet. The Dust Bowl did not kill American agriculture; it just changed it into a big industry—highly regulated and tidy. Fishing may go the same route. Fish farming has produced the only productive growth in recent years and will continue to grow, as will the regulation of the sea itself. We will still have fish, but not the fish or traditions or industries that we have known.

SOME FACTS

- The world fishing fleet has doubled in size over the past two decades and now includes 37,000 "industrial" vessels of more than 100 tons. U.S. and other government subsidies encouraged this growth.

- The ocean catch has exploded from 18 million tons in 1950 to 90 millions tons in recent years. Fish stocks began to decline in the 1990s, and production could no longer increase at the same rates. If the same fishing practices continue, the average marine catch could decline by 10 million tons a year.

- Impact on seafood prices has been moderate due to aquaculture of new fish, shrimp, and scallop farms in China, Thailand, and elsewhere.

- Fishing jobs directly or indirectly employ about 200 million people in the world. There are an estimated 3 million fishing vessels made up of mostly small vessels in developing countries. The 3 million vessels include the supertrawlers of 37,000 ships of more than 100 tons. These ships caught 90.7 million tons of marine and inland fish in 1995. Aquaculture produced 21.3 million tons in 1995 (Hanley, 1997, pp. 1–2).

THE MAGNUSON FISHERY CONSERVATION AND MANAGEMENT ACT

The enactment of the Magnuson Fishery Conservation and Management Act (MFCMA) in 1976 extended fishery jurisdiction to 200 nautical miles offshore and established the current federal fishery management structure. This act, however, did little to preserve fisheries—its intended objective. By banishing foreign competition from U.S. coastal waters and offering generous subsidies to U.S. fishers, it has encouraged thousands of new U.S. boats and, in most cases, imposed no restrictions on the amount of fish an individual fisher could take. As a result of this "open access" policy, U.S. fisheries have become packed with competing boats in an intense, dangerous race for any fish left in the water. In 1992, eight lives were lost in a frenzied race for halibut in which the entire season was compressed into a 48-hour time period. MFCMA also intensifies problems by letting fishery managers use economic and social factors to modify scientists' estimates of levels for sustainable catches. Then, because of monetary and social pressure, the managers reset the allowable catch to be high.

The Environmental Defense Fund would like Congress to reauthorize a Magnuson Act that defines and prohibits overfishing and requires the allowable catch to be ecologically sustainable rather than based on short-term economic demands. One suggested scheme would assign each fisher a tradable share of the total allowable catch. Such shares, called Individual Transferable Quotas (ITQs), could be used or sold, giving fishers an equitable base and vested interest in the fishery's future. Through market forces, ITQs would reduce fishing by ending the race for fish, since as fish populations again increase, ITQs would increase in value, giving fishers a strong incentive to help fish populations recover (EDF is advocating reforms to reduce overfishing, 1994, pp. 1–2).

The MFCMA was reauthorized and signed by President Clinton on October 11, 1996. The reauthorization focuses more on the issues of habitat degradation, overfishing, bycatch (discards and other fish), and funding, and for many people, it did not strengthen sustainable yield approaches enough. Provisions for ITQs remain controversial. The strength, effects, and direction of the MFCMA reauthorization will be a central ecological issue that affects us all and that we should watch carefully (Buck, 1996).

REFERENCES

Buck, E. H. (1996, December 4) 95036: Magnuson Fishery Conservation and Management Act reauthorized. *Congressional Research Service Issue Brief*, pp. 1–14. Available: http://www.cnie.org/nle/mar-3.html.

EDF is advocating reforms to reduce overfishing. (1994, January). *EDF Letter.* Vol. XXV, No. 1, pp. 1–2. Available: http://www.edf.org/pubs/EDF-Letter/1994/Jan/c_overfish.html.

Hanley, C. J. (1997, July 21). Fishermen endanger "limitless" seas". *Chicago Tribune* pp. 1–2.

New bill seeks to reverse crashes of fish populations. (1994, July). *EDF Letter.* Vol. XXV, No. 4, pp. 1,3.

Parfit, M. (1995, November). Diminishing returns, exploiting the ocean's bounty. *National Geographic.* Vol. 188, No. 5, pp. 2–37.

QUESTIONS

1. What has caused the depletion of worldwide fish stocks?

2. In what way has the increase of technology contributed to the depletion of the world's fish stocks?

3. Explain why nations are using warlike tactics to protect their national fishing waters?

4. Why are the oceans of the past compared to the Wild West? What has changed?

5. What is the role of floating fish factories in the depletion of fish stocks? What do you think can be a sustainable approach for the future with the existence and use of these fish factory trawlers?

6. Devise a plan for future world fisheries that accounts for sustainable yields, the use of high technology, and national fishing rights. Contrast that plan with current practices.

7. Explain the cause and effect of the 200-mile offshore coastal waters fishing jurisdiction.

8. What is the future of the small independent fisherman? Explain.

9. What was the result of the Magnuson Fishery Conservation and Management Act in 1976 on world fisheries? Explain both the positive and negative results.

10. How could the Magnuson Act be rewritten to deal with some of the problems of world fisheries? Research the new reauthorization of the 1996 Magnuson Act. What are its strengths and weaknesses?

11. Explain the concept of Individual Transferable Quotas (ITQs).

12. Consult Tables 22.1 and 22.2 for fishing data patterns for discussion.

13. Research open sea and fishery areas. In what ocean is the largest catch found?

14. Refer to Figure 22.1, 22.2, and 22.3. Which fish is the most caught dermesal fish in the world catch? Which fish is the most caught pelagic fish in the world catch?

15. Refer to Figure 22.3. Why do you think that total fish production has peaked? Which species do you think are fished the most? Which species of fish do you feel we still have abundant supplies of?

Table 22.1
Dying Seas

Sea Threat	Baltic	Bering	Black	Caspian	Mediterranean	South China	Yellow
Over-fishing	H	H	H	H	H	H	H
Eutrophication	M	N	M	L	L	M	M
Dams	L	N	H	H	M-H	M	M
Organo-Chlorines	H	M	M	L	M	M	M
Heavy Metals	M	L	H	M	M-H	H	H
Oil Drilling	L	L	H	H	M	H	H
Population (millions)	80	1.4	165	45	360	517	250
Area (Thousands of sq. km)	370	420	2,292	371	2,500	3,685	404

Key: H = High, M = Medium, L = Low, N = Negligible
Source: Dying Seas, Worldwatch, January/February 1995. Reprinted with Permission from Worldwatch Institute.

Table 22.2
World Commercial Catch of Fish, Crustaceans, and Mollusks,[1] by Country, 1988–93
(in thousands of metric tons; live weight)

Country	1988	1989	1990	1991	1992	1993
China	10,358	11,220	12,095	13,135	15,007	17,568
Peru	6,641	6,854	6,875	6,949	6,871	8,451
Japan	11,966	11,173	10,354	9,301	8,502	8,128
Chile	5,209	6,454	5,195	6,003	6,502	6,038
United States[2,3]	5,956	5,775	5,868	5,486	5,588	5,939
Russia	—	—	—	7,047	5,611	4,461
India	3,125	3,640	3,794	4,044	4,232	4,324
Indonesia	2,795	2,948	3,044	3,252	3,442	3,638
Thailand	2,642	2,700	2,786	2,968	3,240	3,348
South Korea	2,731	2,841	2,843	2,521	2,696	2,649

(1) Does not Include Marine Mammals and Aquatic Plants. (2) Includes Weight of Clam, Oyster, Scallop, and other Mollusk Sshells. (3) Statistics on Quantities Caught by Recreational Anglers in the U.S. are Excluded.
Source: U.S. Dept. of Commerce, Natl. Oceanic and Atmospheric Admin., Natl. Marine Fisheries Service.

QUESTIONS ABOUT THE TABLES

1. According to Table 22.1, where are the biggest threats to the seas? How does this result compare with the threats mentioned in the previous case study?

2. What does Table 22.1 tell you about the health of the seas? Select one of the threats and outline a plan to help abate some of the damage.

3. Examine Table 22.2. What countries have increased or decreased their yields during the five-year period of 1988–1993? Explain these behaviors.

A Bridge to Your 21st Century Understanding

Complete and discuss the following flowchart.

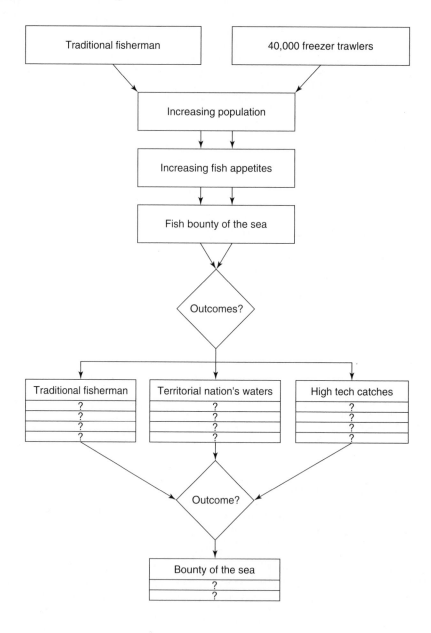

Buried Displeasure:
The Love Canal

Love Canal was the dream community of William T. Love. He imagined a village nestled in the rolling hills and orchards near the Niagara River in upstate New York. The success of the venture, begun back in 1894, depended on the construction of a canal that would connect the two branches of the Niagara River. The canal would tap the power of the rapids just before Niagara Falls and produce hydroelectric power, which would attract business and industry to the area.

But Love's dream was not to be. Economic difficulties caused many of his backers to pull out of the scheme. The only trace of Love's plan was a three thousand foot long, sixty foot wide canal. In 1927, the area was annexed to the city of Niagara Falls. Around 1946, the city was approached by the Hooker Chemical Company, which was looking for a place to dump chemical wastes. Love Canal seemed like the perfect spot, sparsely populated, and the thick clay-like soil provided perfect protection against any possible leakage. Hooker began dumping chemical wastes into the canal that year, and for every year until 1952. At least 150 chemicals were placed in Love Canal, including Dioxin, the most toxic man-made chemical known. There was little public awareness at the time about the dangers of these chemicals, and their connection with nervous system disorders, kidney problems, respiratory distress, deafness and birth defects.

The potential for a technological disaster at Love Canal mounted during the 1950s. The post-war Baby Boom had created a pressing need for housing and schools. Officials in Niagara Falls began looking at Love Canal as a place to expand. In 1953, the Niagara Falls Board of Education announced it intended to acquire Love Canal for the purpose of building an elementary school in the area. Hooker, aware the land could be taken from them via eminent domain, sold the land to the city for $1, but made city officials sign a deed acknowledging that the area had been used as a chemical dump site, and absolving Hooker of any deaths or loss or damage to property once construction began. The 99th Street School was built on the central part of the canal, and the city sold the southern part to developers. As the bulldozers began tearing away the top soil, the rain and snow gradually began to seep into the drums and containers holding the chemical wastes, and in time a chemical mixture, leachate, began to flow out of Love Canal. The leachate which began to appear in the late 1950s produced skin burns on a few children who were playing in the area. But that was nothing compared to the complaints that began surfacing in the 1970s as the danger of Love Canal became apparent. Heavy rains during that decade forced the buried chemicals to the surface, and with them came increased reports of miscarriages, birth defects, liver abnormalities and cancer. In 1978 over 200 families were forced from their homes after toxic fumes were detected in their basements. Two years later, the Federal Government and the state of New York, reacting to an Environmental Protection Agency report, moved an additional 800 families out of the Love Canal area.

Love Canal remained a near ghost town throughout the 1980's as the E.P.A., Hooker Chemical, the School Board and City of Niagara Falls as well as numerous insurance companies thrashed out the legal and possibly criminal costs of this disaster. In 1988, Love Canal residents

191

received a $20 million settlement, in addition to the $30-$40,000 each family received for their homes. Individual claims were in the $2-$4,000 range. By 1990, 1,000 cases were still pending.

ONE PERSON MAKES A DIFFERENCE

Lois Gibbs was a housewife living in Love Canal in 1978. She discovered an epidemic of miscarriages, birth defects, nervous-system problems, and respiratory disorders across the neighborhood and also that the neighborhood had been built next to a huge toxic waste dump. She became a neighborhood organizer and ultimately became responsible, as much as anyone, for bringing this toxic disaster into the public forum and causing responsible parties to pay for victim relocation. She continued to become one of the most prominent grassroots environmental activists in the United States, founded the Citizens Clearinghouse on Harzardous Wastes, and has written extensively on toxic hazards, the prevention of new ones, and approaches for communities with existing toxic dumps. Ms. Gibbs has also written on the effects of dioxin as a potent carcinogen, as well as a cause of illnesses affecting almost every major system of the body, such as diabetes, chronic bronchitis, irregular heartbeat, nervous disorders, thyroid and immune system disorders, and a variety of reproductive system disorders, including birth defects, miscarriages, and lowered fertility. Gibbs recommends a strong dose of democracy as a cure for toxic environmental problems. She concludes, "The job is too big for some national organization or remote coalition to achieve on our behalf . . . Our country's power is vested in the people, and the people must act" (Mazza, 1996).

SOME INFORMATION TO CONSIDER

- Manufacturing wastes are being created at an accelerating pace of 6% per year, which means that the total waste production is doubling every 12 years.
- In the 12 years since the problems with Love Canal were found, American industry has pumped out a total amount of toxic waste equal to all of the toxic waste created prior to Love Canal (from the years 1880–1978).

- The Niagara River has the greatest concentration of toxic dumps anywhere on the North-American continent, with 65 huge chemical dumps along the banks of the river. Love Canal, at 20,000 tons of toxins, is not the largest of these dumps, since the Hyde Park dump contains 80,000 tons of toxins, the "S" site contains 70,000 tons, and the 102 Street Site contains 80,000 tons; all of these sites are within a few hundred yards of the river (Montague, 1990).

When Lois Gibbs founded the Citizens Clearinghouse for Hazardous Waste (CCHW) in 1981, its main focus was to help community groups suffering from the effects of toxic dump sites like Love Canal, but it has expanded its programs to address a broad range of environmental issues, including toxic waste, solid waste, air pollution, incinerators, medical waste, radioactive waste, pesticides, sewage, and industrial pollution. The CCHW is a now a 16-year-old nonprofit environmental organization and remains the only national organization started and led by grassroots organizers. As of 1996, CCHW has worked with over 80 community-based groups nationwide, and Lois Gibbs remains Executive Director of CCHW (Gibbs, 1996).

REFERENCES

Excerpted from Ferell, O.C. and John Fraedrich. *Business Ethics.* New York: Houghton Mifflin, 1991. Additional information provided by researchers Denise Schultz, Paula Ingerson and Joanne Hammond.

"Lois Gibbs' biography page." (1996). Available: http://www.medaccess.com/newsletter/n10415/gibbiog.htm.

Mazza, P. (1996, February). "Love Canal is everywhere: The pervasive threat of dioxin." pp. 1–3. Available: http://www.tnews.com/text/dioxin.html.

Montague, P. (1990, June). The Niagara River—part I: How industry survived Love Canal. *Rachel's Hazardous Waste News #186.* Environmental Research Foundation, Annapolis. pp. 1–3. Available: http://xp0.rtknet.org/E3540T132.

What Will Happen to the Endangered Species Act?

The motivation and purpose of the Endangered Species Act is to provide some regulations for the mutual coexistence of humans and all other species within the demands of each population. Increased grazing, logging, mining, and many other environmentally dependent businesses in the United States threaten the existence of many animal and plant species. Efforts to protect these "human dependents of the world" are the result of the landmark Endangered Species Act (ESA) legislation, which gives endangered wildlife an advantage over human business and profits. The conservation efforts of the ESA present a serious (and at times highly emotional) conflict in many affected areas to economic operations and result in unemployment and loss of private property rights. Opponents believe that the ESA ignores the economic considerations and interests of the specific location, while supporters believe that the ESA ensures and preserves the species and a variable gene pool. The debate then becomes the following: Can the Endangered Species Act accommodate increasing human demands and decreasing biodiversity?

In 1973, President Richard Nixon signed the Endangered Species Act into law; the ESA's main purpose is to maintain a list of endangered or threatened species.

Based on the assumptions that each life-form may prove invaluable in ways we cannot yet measure and that each is entitled to exist, the act gave the federal government sweeping powers to prevent extinction. No commitment of this magnitude to other life-forms had ever been made before. This act provides guidance to two agencies: the Fish and Wildlife Service and the National Marine Fisheries Service, whose main responsibilities are to stop further endangerment of species. The services then strive to protect wildlife and remove them from the list when they have gained recovery. The decision is to be based solely on scientific data rather than on economic and political aspects. In 1973, the U.S. list contained 109 names; in 1995, the total included more than 900 (1,400 including foreign species), with 3,700 officially recognized candidates waiting to be listed (Chadwick, 1995, pp. 7–9).

Conjecture regarding possible removal of the ESA brings much reaction from both critics and supporters, businessmen and environmentalists. Of course, businessmen who depend on the land for revenue would not mind seeing the removal of the ESA altogether, and other opponents of the act feel that the ESA is poorly written and simply stands in the way of economically beneficial

---

Endangered Species: A species that could become extinct in the future.
Threatened Species: A species that could become endangered soon.
Recovery Point: The time at which the measures provided by the Endangered Species Act are no longer necessary.

Fast Disappearing Wetlands.
Photo courtesy of B. Eichler

activities. On the other side, supporters stand on the foundation that one cannot put a price on the value of a species. There are, however, many additional problems aside from economic issues with the way that the ESA is written and managed.

The ESA categories of threatened, endangered, and secure are not well defined and allow for different views of interpretation. Since the ESA gives the government the ability to prioritize the future of a species above any human disturbance that may cause a decline of that species, the vague interpretations allow groups to twist ESA categories to their own needs or to define data differently. The resulting ambiguity yields conflicting reports of population statuses, inaccurate information, and ineffective—and even destructive—action taken on a species population. The ESA lacks population guidelines for each species so that it can be accurately categorized and protected. Without that clear frame of reference, the protection of a species becomes very difficult and arbitrary while human impacts are steadily increasing.

Another problem is that the ESA is used as a last-chance approach. After a species qualifies for the ESA due to decreased numbers, the situation is already extremely dire. In these cases, the act is dealing with the last population of a species, and action has to be taken immediately to curb extinction. Such desperation forces the government to make drastic strategies to save the species, and so a more balanced approach involving the needs of local human populations and economics are not considered. Such last-minute recovery plans magnify tensions and are not nearly as effective when the act has to salvage a species that is nearly lost. If the act could practice planned "preventive medicine" instead of

"reactive medicine," resources would be more strategically used, productivity would be increased, and both humans and animal species would experience decreased tension.

An additional problem is that the ESA does not adequately take into account the amount of habitat required for a species' numbers to increase. The ESA does not make provisions for species that need larger tracts of land or for the construction of habitat bridges, which link smaller tracts of land together to allow cross-movement. Instead, the ESA protects islands of habitat surrounded by urban influence, allowing very little chance of survival for nomadic species such as the fox, wolf, grizzly bear, lynx, cougar, and many others. The larger nomadic predators need very large tracts of land to search for food. The ESA does make good attempts at preserving national parks as large areas of habitat, but national parks are not usually biodiverse ecosystems that meet the food needs of these species and are appealing more for their aesthetic beauty. Then, typically, when the large predators roam from their small habitats in search of food, they are shot when entering an agricultural or populated area. Conflict then arises because the ESA prohibits shooting an endangered species, but humans feel betrayed by the government for simply defending their land, lives, and resources (The Endangered Species Act, 1997, p. 1).

Other problems exist in processing and interpreting the list of species on the ESA. Some of the removals from the list exaggerate the rate of success of the ESA. In 1993, the Burneau Host Spring Snail was removed from the list after a federal judge ruled that it was listed erroneously; this species was one of 8 others removed out of a total of 21 species because of listing errors. Another example is that of the Rydberg milvetch (a member of the pea family), which existed only in Southwestern Utah in 1905 and is believed to be extinct. However, in 1980, taxonomists decided that a dozen populations of a close relative (the plataua) should be counted as Rydberg milvetches, and the species was removed from the list even though the plant was not the same one.

The species that were delisted because their status had improved also need correct interpretation. The Arctic peregrin falcon was struck from the list in October 1995, but its improvement probably had more to do with the reduction of pesticides than changes in hunting laws and the protection of habitat. Also, the smallness of recovery numbers may be a basic reflection of the dangers faced by the species at the time of the listing rather than the ineffectiveness of recovery tactics. The belief is that, allowing for data errors, extinctions, and other extraneous factors, success stories under the ESA can be misinterpreted and the effectiveness of the act can not be realistically appraised (Is Endangered Species Act in Danger, 1995).

According to most sources, the ESA does fundamentally work, but the act would prove more effective if changes were made. The ESA has many success stories: the palau dove, palau flycatcher, palau owl, Atlantic brown pelican, and the gray whales. The increase in the numbers of red wolves within the continental United States is due mostly to anti-shooting laws as set forth by the ESA. The ESA, in conjunction with the banning of DDT, has increased many populations of birds, including the bald eagle. The ESA has also funded the recovery of some severely endangered species, such as the California condor. According to research numbers, the ESA does help to slow rates of extinction; however, it also has been suggested that many species go extinct while waiting for ESA recognition (Endangered Species Act, 1997, p. 2).

When first written in 1973, the ESA appeared to be the answer for increased environmental concern and pressure for diminishing species. There remain some fundamental problems, however, that are now prompting views that species-recovery planning and implementation processes are not working very well, and these problems also agitate involved parties pitting business interests against environmentalists and federal agencies.

Some of the ESA classic case studies include the Northern Spotted Owl, the Snail Darter, The Red-Cockaded Woodpecker and the Gnatcatcher.

Northern Spotted Owl
Habitat: **Northwest Old Growth Forests**
Status: **Threatened**

The government declared 8 million acres off-limits to chainsaws, saving the environment as well as the owl, but taking away thousands of jobs and millions of dollars in timber sales in a depressed area. Controversy surrounded the importance of economics versus endangered species. The conclusion was that new regulations for national forest use were created, emphasizing conservation, not consumption. Management of both federal and state forest timberlands has shifted from lumber production to environmental protection. The city of Sweet Home has new industries moving into the region, with the emergence of strong growth opportunities (Mitchell, 1997).

Snail Darter
Habitat: **Little Tennessee River**
Status: **Endangered**

Work in the middle of the construction of the Tellico Dam in the Little Tennessee River was halted due to concern over the future of the snail darter (a tiny fish). Criticisms of the ESA began, due to the costs of salvaging a species versus completing the half-finished multimillion-dollar dam. Supporters of the ESA argued that the snail darter existed only in the Little Tennessee River, but other populations have since been found. The conclusion was that the fish were transplanted and the dam was completed.

Red-Cockaded Woodpecker
Habitat: **Southeast Old Pine Forests**
Status: **Endangered**

The woodpecker inhabits the old growth pine forest ecosystem and passes its nest from one generation to the next. The loss of 88 million acres of pine forest due to development threatens the woodpecker as well as many other potentially endangered species. The conclusion of this case is as yet unresolved.

Gnatcatcher
Habitat: **Sage Shrub Ecosystem of California**
Status: **Endangered**

The gnatcatcher blocked the future construction of a development in southern California. Supporters said that the planned site of the development was the only remaining location for the species, while opponents said that preservation attempts were prohibiting the construction of a development with high property values. To break the impasse, Secretary of the Interior Bruce Babbit changed the classification of the species to "threatened" rather than "endangered" to allow continuing "controlled development" to the area. In return, the developing companies had to set aside a certain percentage of sage shrub habitat to ensure the recovery and future of the gnatcatcher. A 1992 report by the Fish and Wildlife Service gives the following data:

Species	Number of in 1990	Number of in 1992
Improving	57	69
Stable	181	201
Declining	219	232
Extinct	11	14
Unknown	113	195

This report indicates an increase of 11 to 14 extinct species and 219 to 232 species declining, showing approximately the same number of species improving as declining. This data does not represent the ESA well, since with most species in decline due to the pressure of human encroachment and the loss of habitat, the ESA as written in 1973 could be interpreted as an inadequate protection of species (Endangered Species Act, 1997, p. 2).

POLITICAL DEVELOPMENTS

Secretary of the Interior Bruce Babbit has proposed revisions to the Endangered Species Act that include a spirit of compromise to replace the old era of confrontation. The core of his proposal of 1997 is to appease both sides, such that environmentalists have more land set aside for preservation and property owners can use portions of their land without enduring excessive bureaucratic and legal entanglements. His proposal incorporates 10 policy changes, representing the most sweeping changes in the law's 24-year history, and reacts to many criticisms of the ESA. The criticisms recently have been so vocal that some legislators have wanted to repeal the entire act, and various classic controversies such as those of the spotted owl and snail darter fueled that approach. However, since then there have been administrative reforms that have been field tested, have been proven to work, and will be less subject to shifting political whims.

Spending by States on Endangered Species

37% Birds
33% Mammals
13% Fishes
8% Plants
5% Invertebrates
4% Reptiles/Amphibians

Source: 1996 Endangered Species Survey, International Association of Fish and Wildlife Agencies

Some of the policies that Babbit wants to incorporate into the law include the following:

- allowing property owners with five acres or less to be exempted from some restrictions;

- encouraging local and state governments and Indian tribes to play a larger role in the protection of wildlife with private land management;

- focusing more on helping threatened species boost their numbers so that they do not end up on the endangered species list;

- enacting a "no surprises" policy where property owners can agree to set aside lands that shelter endangered wildlife or donate money to a conservation project. In exchange, the U.S. Fish and Wildlife Service will give the property owners authorization to destroy the habitat on other portions of their land that they plan to develop. In order to obtain the authorization, owners have to devise thorough, comprehensive plans. Once the deal is struck, the wildlife plan would be unchangeable. If another endangered species not covered by the plan shows up later, the United States would pay the cost of protecting it.

Since 1994, when Babbit first introduced the no-surprises policy, 200 habitat recovery plans have been established and another 200 are forthcoming. More than 18 million acres of developable lands have been set aside for wildlife under the plans. However, some scientists say that the policy is bad science, since it is impossible to make a plan that will consider long-term dynamic needs for species that will occur over the next 50 to 100 years. Also, critics say that allowing land owners to write their own environmental plans represents a conflict of interest that can lead to abuse. Babbit's approaches, at the very least, conserve the law while defusing ownership and economic tension (Haynes, 1997).

REFERENCES

Is Endangered Species Act in danger? (1995, March 3). *Science,* p. 267.

The Endangered Species Act (1997). Available: http://gladstone. uoregon.edu/-cait/mainbody.htm.

Chadwick, D. H. (1995, March). Dead or alive: The Endangered Species Act. *National Geographic,* pp. 2–41.

Haynes, V.D. (1997, July 27). U.S. wants to amend endangered species law. *Chicago Tribune,* Section 1, p. 4.

Mitchell, J. G. (1997, March). In the line of fire: Our national forests. *Scientific American,* Vol. 191, No. 3, pp. 57–58.

FURTHER READING

Graham, F., Jr. (1994, July/August). Winged victory. *Audubon,* pp. 36–49.

Horton, T. (1995, March/April). The Endangered Species Act: Too tough, too weak, or too late? *Audubon,* p. 68.

Kohm, K. (1991). *Balance on the Brink of Extinction.* Washington, D.C.: Island Press.

Korn, P. (1992, March 30). The case for preservation. *The Nation,* p. 414.

Mann, C. C., and Plummer, M. L. (1995, January/February). California vs. Gnatcatcher. *Audubon,* pp. 40–48.

National Research Council. (1993). *Science and the Endangered Species Act.* National Research Publishing Company.

Rauber, P. (1996, January/February). "An End to Evolution." *Sierra,* p. 28.

QUESTIONS

1. Do you think that the Endangered Species Act is needed? Do you think that this legislation was too strong or too weak in 1973? If you were to change the law now, how would you change it?

2. Where do you stand in your political view on the environment? Could you be regarded as a critic or supporter of the ESA? Explain in detail your reasons. Should the ESA be eliminated? If so, what should take its place? If not, why not?

3. Name some significant successes of the ESA. Support your list of successes with the role and contribution of the ESA for these successes.

4. Name some of the problems of the ESA, and give some suggestions to correct those problems.

5. Research some of the classic ESA cases (e.g., bald eagle, gnatcatcher, northern spotted owl, California condor, milvetch), and present arguments for and against ESA involvement with and protection of these species.

6. Do you feel that the ESA works? Give arguments for and against your conclusions.

7. Do you feel that Babbit's proposal strengthens or weakens the ESA's power and effectiveness? Explain your answer. Do you like Babbit's revisions? Explain.

8. If you were to change Babbit's proposal, how would you change it?

9. What is meant by a "no surprises" policy? Do you feel that this is a necessary component to Babbit's proposal? Why or why not?

10. Research the species that the ESA lists. Does your research support or refute the utilization of the ESA?

11. What is your overall reaction to the issue of ESA? Give your views regarding the environmental, ethical, economical, and political aspects of this issue.

A Bridge to Your 21st Century Understanding

Complete and discuss the following flowchart.

Air Poisons Around the World

There are studies showing that on days when there are more particulates in the air, more people die. You can't get more basic that that. . . . This country has a law called the Clean Air Net, which says that Americans should be able to breathe without harm.

JOHN H. ADAMS, EXECUTIVE DIRECTOR, NATURAL RESOURCES DEFENSE COUNCIL

Air pollution in its many forms—sulfur dioxide, ozone, fine particles, carbon monoxide and nitrogen oxide—is causing a deathly fog around the world. "From Hong Kong to Mexico City, the toll taken in terms of human death and disease caused or aggravated by air pollution almost certainly is measured in the millions," says Dr. Alfred Munzer, past president of the American Lung Association and respiratory specialist.

In Japan, various respiratory illnesses suffered by thousands of citizens living in heavily polluted areas of Japan were deemed indisputably to be caused by sulfur dioxide. The next year the Japanese government set up a tracking program and medical reimbursements for certified victims of air pollution ranging from bronchitis to asthma. Even though the program monitored only sulfur dioxide and was limited to just a few industrial areas, it still certified more than 90,000 air pollution victims before it was stopped in 1988 because of pressure from polluters. Since no other nation has routinely collected this kind of data on air pollution and human health, this number serves as an indicator of a world-wide unprecedented problem.

Sulfur dioxide along with fine particles which results in newly formed acidic compounds claim an unbelievable number of victims in all parts of the world. In 1992 when weather conditions trapped sulfur dioxide over the Mae Moh region of Thailand, 4000 residents required medical treatment, cattle died with blistered hides and crops withered. In Poland, medical tracking of army inductees has revealed four times more asthma and three times more bronchitis in areas that were polluted by sulfur dioxide. In Krakow, men were tracked who lived in the city's most polluted areas for 13 years; the study found losses of lung function similar to smoking.

Ozone, the dominant chemical in smog, is an invisible toxic gas that is the result of reactions of unburnt gasoline with other pollutants. Ozone scars tissue, burns eyes, promotes coughing, wheezing and rapid, painful breathing. Within minutes of entering the lungs, ozone burns through the cell walls. The immune system tries to defend the lungs, but they are stunned by the ozone. Cellular fluid then seeps into the lungs and breathing becomes rapid and painful. In the nose and airways, ozone destroys ciliated cells which the body replaces with thick-walled abnormal squamous cells. With time the lungs stiffen and the ability to breathe drops. Children raised in ozone-polluted areas have unusually small lungs and adults lose up to 75 percent of their lung capacity. The massive cell death that ozone causes triggers pre-cancerous physiological responses.

Mexico City's air which has some of the world's worst ozone pollution rises in the unhealthful ozone levels about 98% of the time. Healthy men newly exposed to Mexico City's air developed pre-cancerous cell alterations in nasal and airway passages. Mexico City residents generally are in the second of three stages of cancer where the third is the production of cancer.

In Los Angeles, one in four fatal accident victims aged from 14 to 25 had severe lung lesions of the sort caused by ozone which is a destructive, irreversible disease in young people. Los Angeles residents exposed to ozone had double the risk of cancer compared to residents of cleaner cities.

Particulates are a catch-all term for everything from road dust to soot to mixtures of pollutants—solids as well as liquids, microscopic and larger grains that vary in the environment. Scientists believe that fine particles which are small enough to lodge deep into the lung are the most dangerous. Fine particles result from the burning of coal, oil and gasoline as well as from the atmospheric change of oxide of sulfur and nitrogen into sulfates and nitrates.

Bangkok police work in gritty air pollution thick with air particulates from motorcycles and thick clouds of burnt diesel fuels from buses and trucks. The effects on these officers include abnormal lung function tests and spots on lungs as well as general congestion. Roughly one out of every nine Bangkok residents has a respiratory ailment caused by air pollution.

The evidence that particulates kill is "absolutely complete" according to University of British Columbia's Dr. David Bates, a foremost air pollution and health expert. Studies from around the world and within the U.S. confirm that as the particulate pollution rises, so do sickness and death.

Carbon monoxide is invisible and oxides of nitrogen almost so—therefore, it is impossible to tell when the air contains dangerous levels of both. These gases are produced when gasoline and other fuels are burned incompletely. Roughly 90 percent of urban CO occurs from motor vehicle tailpipes especially in cities like London. Carbon monoxide, when breathed, starves the body of oxygen which then causes dizziness and unconsciousness.

Oxides of nitrogen form the reddish-brown layer that can be seen from an airplane one mile up. Nitrous oxides cause oxygen and nitrogen in the air to combine which then further cause air and ozone pollution. Like ozone, oxide of nitrogen destroys organic matter such as human tissues and makes organisms more susceptible to bacterial infections and to lung cancer.

Some of the worst air pollution is in the "Black Triangle," where the Czech Republic meets Poland and the former East Germany. More sulfur per square meter falls there than on any other place in Europe. In Poland, high lead levels in soil from the air are unprecedented elsewhere in the Western world. According to experts, the "Black Triangle" is among the most polluted locations in the world. Respiratory disease in children up to 14 years is epidemic. In this triangle, black smoke clings to everything. The World Health Organization has concluded that approximately 15 percent of infant mortality and 50 percent of postneonatal respiratory mortality in the Czech

Republic may be connected to air pollution. According to studies, children living in cities with air pollution are as much as 11 months behind in bone growth than those breathing cleaner air. In Poland, lead-laden air pollution is so severe that it causes lead poisoning and intelligence loss where some experts have estimated that 10 to 15 percent of the nation's citizens have been affected.

Governments who are forced to choose between protecting the public and shielding industry from regulation too often choose to sacrifice their men, women and children. The people in Mexico, Japan, Thailand, England, U.S., Poland, the Czech Republic and the rest of the world require a basic quality of life of breathing clean air. Governments must prioritize their efforts to the health of their citizens and utilizing technology for their peoples as well as their economics—otherwise their economic decisions become short-lived like their suffering people. Technology has the capability to provide for both economic and human needs with a priority on the value of all life and the treasured air on which all depends.

U.S. HISTORICAL AND SOCIETAL PROGRESS REGARDING AIR POLLUTION

In the United States in 1973, after a lawsuit initiated by the Natural Resources Defense Council, the Environmental Protection Agency (EPA) created a five-year program to gradually reduce the lead content in gasoline. At that time, tetrethyl lead (TEL) was routinely added to lower grade gasoline to effect more efficient burning and prevent gasoline "knock" caused by uneven combustion. For decades, the nation had allowed the combustion of leaded gas to emit millions of tons of lead into the air, where it was breathed in or deposited in soil and dust. Lead, however, is toxic, even in the smallest amounts. Children with elevated blood–lead levels can suffer lowered IQs, slower neural transmission, hearing loss, and disruption of the formation of hemoglobin red blood cells, and acute lead poisoning causes more massive physiological damage. By the mid 1970s, physiological links were established between lead content in gasoline and health problems in children.

By 1985, after many political battles and opposition by special-interest groups, the EPA proposed stricter regulations with the goal of eliminating gasoline lead entirely by

U.S. Public Health Service.

the mid-1990s. Two factors drove this development. One was new health data showing that blood–lead levels previously accepted as safe had adverse effects, and the other was data by Joel Schwartz (formerly of the EPA, now a professor at the Harvard School of Public Health) that illustrated that it would cost industry about $575 million to meet the newest lead standards by 1986—but that sum would be small in comparison to the $1.8 billion saved in 1986 alone from reduced needs for medical care, lower pollution emissions from catalytic converters, greater fuel efficiency, and less vehicle maintenance. As of 1994, the Center for Disease Control (CDC) estimated that still 1 million children under six years of age in the United States had blood–lead levels exceeding recommended thresholds, but the number was significantly reduced from previous data.

Aside from lead, the two worst air pollution components are ozone and fine airborne particles called "particulates." There were several major air pollution disasters during the early part of this century—in Meuse Valley, Belgium, in 1930; in Donora, Pennsylvania, in 1948; and in London in 1950—that demonstrated that high concentrations of air pollution could kill large numbers of people. By the late 1980s, there was accumulating evidence that the particulate matter (tiny suspended particles

of soot, soil, mineral, or metal) in this pollution was closely associated with death and illness (Skelton, 1997, pp. 27–29).

FINE-PARTICULATE REGULATIONS NEEDED

The people who are hurt worst are those with the most vulnerable lungs—children, the elderly, and asthmatics. For people with heart or lung disease, fine particulates can cause an earlier death by a year or more. Over the past 10 years, there have been hundreds of studies indicating that the damage caused by these two pollutants is much more serious than previously known and that on days when more particulates are in the air, more people die. Fine particulates are currently the leading air pollution health threat in the country. Researchers believe that about 60,000 Americans may die annually as a result of particulate pollution—a larger number than for any other form of pollution. Currently, however, less than one third of all funds to reduce air pollution are directed toward removing fine particulates (Adams, 1997, p. 2).

Airborne particles have different sources and are of different sizes. Currently, the smallest particulates regulated by the federal government have a diameter of 10 microns or less (referred to as PM-10). The largest of these particulates are dust and dirt, but the smallest of these particulates (PM-2.5) cause the greatest health threat. These very fine particles evade the body's clearance mechanism and penetrate deeply into the lung's most sensitive areas. Additionally, these particulates are mostly byproducts of combustion (from coal-fired power plants, industrial boilers, highway vehicles, and other pieces of machinery) that in themselves form chemicals that are health hazards. In 1971, the EPA set a general limit of particulate standard concentrations at PM-10 or above. But by the end of the decade, it was becoming clear that it was the smallest particulates that were the worst health risks. In 1987, the EPA set a tougher standard limit for particles at PM-10, the first-ever standard limit on fine, inhalable particles. Evidence continued to mount that the lives of thousands of Americans were being shortened by a year or two, on average, from exposure to particulates at levels below the PM-10 standard (Skelton, 1997, p. 29).

Then, additional studies became available that tracked the health of more than 8,000 people from six cities for 14 to 16 years. These studies concluded that fine particles increased the risk of premature death for residents of the most polluted city by 26%. Another study tracked 500,000 people in 151 cities with the results that people living in the most polluted areas with fine particles had a 17% greater risk of mortality. In 1996, a study of 239 cities found that about 64,000 people may die prematurely from heart or lung disease annually due to particulate pollution. In 1996, the EPA finally proposed a tougher particulate standard that included the first-ever standard for the finest, most dangerous particulates, PM-2.5. It is estimated that these new standards and the new law would annually prevent 9,000 hospital admissions, 250,000 cases of aggravated asthma, and 20,000 premature deaths every year. But there is no knowing whether the PM-2.5 standard will survive the final version of the EPA law. There is much industrial pressure for this new standard not to survive. This issue is better put in the following manner: As a standard for the entire and future industrial world, the United States and the world require such protection for rights to and needs for clean air (Skelton, 1997, p. 30).

In 1987, U.S. law required certain industries to disclose their emissions of 320 toxic chemicals. Since the disclosure began, selected industries have reduced those chemical emissions by 31%; however, companies still admitted to 1992 emissions of 3.8 billion pounds of toxic chemicals.

Public disclosure has been quite effective in getting industry to clean up its approach, but there is still a long way to go (It's the ecosystem, stupid, 1994). A study in 1997 indicated that industries' quantitative breakdown of emissions was not very accurate in terms of analysis of location, composition, and emission rates and that more accurate methods have to be developed to rely more on actual measurements rather than calculations on paper. In 1997, some of the largest industrial polluters, including oil refineries, coal-power plants, steel companies, and car and truck exhaust pipes, are attempting to ease these standards (Raloff, 1997).

REFERENCES

Excerpted from

Adams, J. H. (1997, Spring). Past time for clean air. *The Amicus Journal,* p. 2. It's the ecosystem, stupid. (1994, February/March). 26th environmental quality index. *National Wildlife,* p. 40.

Moore, Curtis A. (1995, September/October.) Poisons in the air. *International Wildlife.* Vol 25, No. 5, pp. 38–45.

Raloff, J. (1997, June 28). Industries tally air pollution poorly. *Science News,* Vol. 151, p. 396.

Skelton, R. (1997, Summer.) Clearing the air. *The Amicus Journal,* pp. 27–30.

QUESTIONS

1. What compounds mentioned in this article cause air pollution? How are each of these compounds created?

2. What can be done to stop the creation of these air pollution compounds?

3. Is such air pollution a significant health risk in your estimation, or is it the necessary price and inconvenience that is a part of the price of technology? What priority would you place on curbing these air pollution compounds?

4. What are some approaches that the industrialized nations can take to curb such air pollution? How can these efforts be made effective?

5. How can the industrialized nations aid the Third-World nations in this effort to clean up the air?

6. What were some of the deciding factors that drove the passage of stricter regulations of lead in gasoline? Explain how economics and society–health arguments won this standard. Do you think that this method is a

winning strategy for the passage of proposals rather than the facts of the environmental effects alone? What are the implications of this situation?

7. Why are fine particulates so dangerous?

8. Do you feel that there should be a PM-2.5 standard? How can such a standard be enacted? (You may wish to refer to Question #6 for some ideas.)

9. Has industrial disclosure of toxic emissions helped to reduce industrial chemical emissions? Why do you think this has happened? What are the problems with self-disclosure as far as pollution control? What ap-

proaches would you further recommend for reducing levels of emissions?

10. Refer to Tables 25.1, 25.2 and 25.3. Observe the historical improvement of air toxins and air pollutants. Comment on these trends in each of the tables.

11. Refer to Table 25.4: The Common Air Pollutants. Discuss how each pollutant could realistically be minimized in industry, transportation, and domestic uses. How does your plan compare with what is being done today? How can we improve governmental and societal efforts to reduce pollutants?

Table 25.1
Toxics Release Inventory, 1993–94

Reported industrial releases of toxic chemicals into the environment by major manufacturing facilities (excluding power plants and mining facilities) decreased 8.6% from the 1993 figure and 44.1% from the figure for 1988, the baseline year.

Pollutant Releases	1993 mil lb	1994 mil lb	Top Industries, Total Releases	1993 mil lb	1994 mil lb
Air releases	1,672	1,556	Chemicals	1,316	851
Underground injection	576	349	Primary metals	329	313
Land releases	289	289	Paper	216	246
Water releases	271	66	Transportation equipment	136	122
Total	**2,808**	**2,260**	Plastics	127	119
Pollutant transfers			**Carcinogens, air/water/ land releases**		
To recycling	3,252	2,456	Dichloromethane	64	63
To energy recovery	487	464	Styrene	33	40
To treatment	328	319	Chloroform	14	11
To disposal/other	325	298	Formaldehyde	12	12
To publicly owned treatment works	314	255	Tetrachloroethylene	12	10
			Benzene	11	10
Total	**4,706**	**3,792**			

Source: Environmental Protection Agency.

Table 25.2
Emissions of Principal Pollutants 1985–1994

(in thousand short tons)

Source	1985	1986	1987	1988	1989	1990	1991	1992	1993	1994
Carbon monoxide[1]	114,690	109,199	108,012	115,849	103,144	100,650	97,376	94,043	94,133	98,017
Lead	20.1	7.3	6.9	6.5	6.0	5.7	5.3	4.9	4.9	5.0
Nitrogen oxides[2]	22,860	22,348	22,403	23,618	23,222	23,038	22,672	22,847	23,276	23,615
Volatile organic compounds[2]	25,799	24,991	24,777	25,720	23,934	23,600	22,876	22,422	25,575	23,174
Particulate matter	3,220	3,092	2,964	3,067	3,036	2,704	2,674	2,725	2,666	2,688
Sulfur oxides	23,230	22,442	22,204	22,647	22,785	22,433	22,068	21,836	21,517	21,118
Total	**189,819**	**184,079**	**180,366**	**190,907**	**176,127**	**172,430**	**167,671**	**163,877**	**167,171**	**168,617**

[1] The Observed Increase in Carbon Monoxide Emissions Between 1993 and 1994 is Atributed to 2 Sources: Transportation Emissions (up 2%) and Wildfire Emissions (up 160%). [2] Ozone, a Major Air Pollutant and the Primary Constituent of Smog, is not Emitted Directly to the Air but Is Formed by Sunlight Acting on Emissions of Nitrogen Oxides and Volatile Organic Compounds.
Source: U.S. Environmental Protection Agency, Office of Air Quality Planning and Standards

Table 25.3
U.S. Lead Emission Estimates, 1985–94

(in short tons)

Source	1985	1986	1987	1988	1989	1990	1991	1992	1993	1994
Fuel combustion	515	516	510	511	505	500	495	491	491	493
Industrial processes	3,402	2,972	3,004	3,090	3,161	3,278	3,081	2,771	2,866	2,868
Transportation	16,207	3,808	3,343	2,911	2,368	1,888	1,704	1,637	1,580	1,596
Natural sources	0	0	0	0	0	0	0	0	0	0
Miscellaneous	0	0	0	0	0	0	0	0	0	0
Total[1]	20,124	7,296	6,857	6,513	6,034	5,666	5,279	4,899	4,938	4,956

[1] Totals May Not Add Because of Rounding.
Source: U.S. Environmental Protection Agency, Office of Air Quality Planning and Standards.

Table 25.4
The Common Air Pollutants (Criteria Air Pollutants)

Name	Source	Health Effects	Environmental Effects	Property Damage
Ozone (ground-level ozone is the principal component of smog)	Chemical reaction of pollutants; VOCs and NOx	Breathing problems, reduced lung function, asthma, irritated eyes, stuffy nose, reduced resistance to colds and other infections, may speed up aging of lung tissue	Ozone can damage plants and trees; smog can cause reduced visibility	Damages rubber, fabrics, etc.
VOCs[4] (volatile organic compounds); smog-formers	VOCs are released from burning fuel (oil, gas, etc.), solvents, paints, and other products used at work or at home. Cars are an important source of VOCs. VOCs include chemicals such as benzene, toluene, methylene chloride.	In addition to ozone (smog) effects, many VOCs can cause serious health problems such as cancer and other effects	In addition to ozone (smog) effects, some VOCs such as formaldehyde and ethylene may harm plants	
Nitrogen Dioxide (one of the NOx); smog-forming chemical	Burning of gasoline, natural gas, coal, oil etc. Cars are an important source of NO^2.	Lung damage, illnesses of breathing passages and lungs (respiratory system)	Nitrogen dioxide is an ingredient of acid rain (acid aerosols), which can damage trees and lakes. Acid aerosols can reduce visibility.	Acid aerosols can eat away stone used on buildings, statues, monuments, etc.
Carbon Monoxide (CO)	Burning of gasoline, wood, natural gas, coal, oil, etc.	Reduces ability of blood to bring oxygen to body cells and tissues. CO may be particularly hazardous to people who have heart or circulatory (blood vessel) problems and people who have damaged lungs or breathing passages		

continued

205

Table 25.4, *Continued*

The Common Air Pollutants (Criteria Air Pollutants)

Name	Source	Health Effects	Environmental Effects	Property Damage
Particulate Matter (PM-10); (dust, smoke, soot)	Burning of wood, diesel and other fuels; industrial plants; agriculture (plowing, burning off fields); unpaved roads	Nose and throat irritation, lung damage, bronchitis, early death	Particulates are the main source of haze that reduces visibility	Ashes, soots, smokes and dusts can dirty and discolor structures and other property, including clothes and furniture
Sulfur Dioxide	Burning of coal and oil, especially high-sulfur coal from the Eastern United States; industrial processes (paper, metals)	Breathing problems, may cause permanent damage to lungs	SO_2 is an ingredient in acid rain (acid aerosols), which can damage trees and lakes. Acid aerosols can also reduce visibility.	Acid aerosols can eat away stone used in buildings, statues, monuments, etc.
Lead	Leaded gasoline (being phased out), paint (houses, cars), smelters (metal refineries); manufacture of lead storage batteries	Brain and other nervous system damage; children are at special risk. Some lead-containing chemicals cause cancer in animals. Lead causes digestive and other health problems.	Lead can harm wildlife.	

[4]All VOCs contain carbon (C), the basic chemical element found in living beings. Carbon-containing chemicals are called organic. Volatile chemicals escape into the air easily. Many VOCs, such as the chemicals listed in the table, are also hazardous air pollutants, which can cause very serious illnesses. EPA does not list VOCs as criteria air pollutants, but they are included in this list of pollutants because efforts to control smog target VOCs for reduction.

Source: U.S. Environmental Protection Agency.

A Bridge to Your 21st Century Understanding

Complete and discuss the following flowchart.

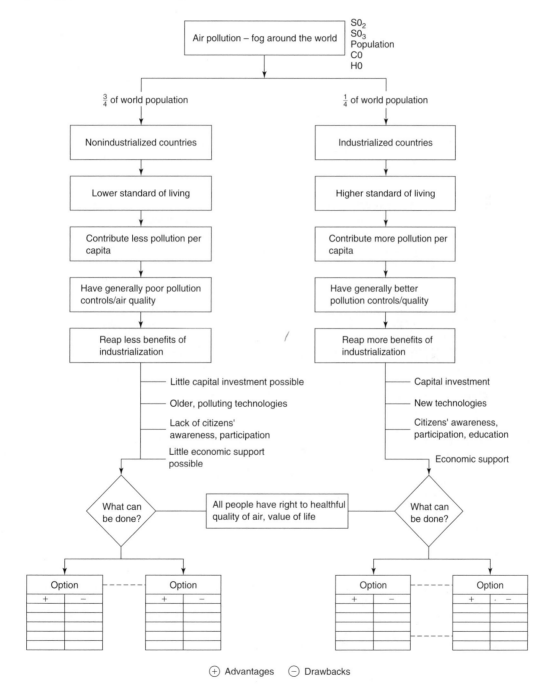

SO_2
SO_3
Population
CO
HO

Air pollution – fog around the world

$\frac{3}{4}$ of world population — **$\frac{1}{4}$ of world population**

Nonindustrialized countries	Industrialized countries
Lower standard of living	Higher standard of living
Contribute less pollution per capita	Contribute more pollution per capita
Have generally poor pollution controls/air quality	Have generally better pollution controls/quality
Reap less benefits of industrialization	Reap more benefits of industrialization

- Little capital investment possible
- Older, polluting technologies
- Lack of citizens' awareness, participation
- Little economic support possible

- Capital investment
- New technologies
- Citizens' awareness, participation, education
- Economic support

What can be done? — All people have right to healthful quality of air, value of life — **What can be done?**

Option	Option	Option	Option
+	−	+	−	+	−	+	. −

⊕ Advantages   ⊖ Drawbacks

26

Earth Day: 25 Years

FRANK GRAHAM

Rejoicing at grass roots, Mary Lyn Ray epitomizes a generation of activists who answered the call of Earth Day, April 22, 1970, when millions rallied to the cause of the environment. Ray, for one, is rousing her neighbors to keep their corner of New Hampshire green.

It was to have been a teach-in. The National Environmental Teach-In, the organizers called it, hoping to capture the spirit if not the politics of those earlier "sit-ins" of the fractious 1960s. Instead, what emerged on April 22, 1970, was Earth Day, the greatest nationwide street demonstration the United States had witnessed since the tumultuous close of World War II.

Twenty million Americans turned out that day to hear politicians and philosophers pledge allegiance to the planet by deploring its polluted condition. Congress stood in recess, its lawmakers out on the ecological stump. Senator Edward Kennedy, ex-Harvard, stumped at Yale; Senator Barry Goldwater at Adelphi University in Garden City, New York. Margaret Mead, the anthropologist, predicted that the energy required to roll back environmental degradation might well match that of the Industrial Revolution. John V. Lindsay, the mayor of New York City, proclaimed it a day to remember: "This is the first time," he said, "I've walked down Fifth Avenue without getting booed. . . ."

A few speakers feared that the enthusiasm of that day might soon fade away. They needn't have worried. There have been Earth Days each April 22 ever since. A silver

anniversary for this sometimes uncertain notion that we the people are a part of the earth, not apart from it, and that (to borrow from the mountain philosopher John Muir) "when we try to pick out anything by itself, we find it hitched to everything else in the universe."

The original event, 1970, had been the brainchild of Senator Gaylord Nelson of Wisconsin, a longtime battler for clean water and a man regarded by many conservationists as one of the few voices of conscience on Capitol Hill. At the time, Nelson was troubled that the environment remained, as he called it, a "non-issue in the politics of our country" even after all the ecological alarms and disasters of the 1960s. The parching of the Everglades. Blackened beaches from the oil-rig blowouts off Santa Barbara. The perils of pesticides. The smog in Los Angeles. The sewage afloat in Chesapeake Bay.

But these *were* issues to more folks than Gaylord Nelson suspected. And as inquiries from the town began to swamp the senator's staff, Nelson was obliged to hire Denis Hayes, a 25-year-old Harvard Law School student, to take over the operation. With Hayes aboard, the planned festivities evolved into Earth Day, and Environment with an uppercase E was off and running, though not without its fair share of detractors.

From the left came insinuations that Earth Day was a scam to deflect the nation's attention from the war in Vietnam. On the right it was said that the event was a communist plot—the April 22 date that Nelson had selected to avoid a conflict with college exams and spring vacations happened to coincide with the 100th anniversary of Lenin's birth.

Graham, F. (April, 1995). "Earth Day: 25 Years." *National Geographic.* Vol. 187, No. 4, pp 123–138.

Photo courtesy of
Ahmed S. Khan.

The events of that day were as varied as the expectations of the participants. There were trail hikes and talkathons, prayer vigils and publicity stunts. In Washington, D.C., demonstrators poured oil on a sidewalk in front of the Department of the Interior to protest its policies on offshore drilling. At Boston's Logan International Airport, 200 people posed as pallbearers for the United States' planned but never-to-be-built supersonic jetliner.

The gimmicks grabbed the headlines, but for most of the people who turned out for the ubiquitous street fairs and pep rallies, Earth Day One was about as confrontational as a bouquet of tulips. But it wasn't a lark either, or a "springtime skipalong," as one of the national newsweeklies suggested. By year's end, even some skeptics began to concede that the fallout from Earth Day would be writ on the face of America for years to come. At first the most noticeable changes appeared in the law books. Three months before Earth Day, President Richard Nixon had signed the National Environmental Policy Act, requiring review and analysis of public projects. An amended— and tougher—Clean Air Act cleared Congress that year, to be followed by a Clean Water Act, a Marine Mammal Protection Act, an Endangered Species Act, a Safe Drinking Water Act, and a Toxic Substances Control Act, among scores of new federal and state statutes.

And inevitably, as the environmental community scored one victory after another on the regulatory and legislative fronts, a few Americans would come to resent what Earth Day had wrought, perceiving the regulations as infringements on their property rights and decrying the laws as pernicious attacks on liberty and the pursuit of happiness.

But the most surprising legacy of that landmark occasion 25 years ago is neither the backlash nor the passage of statutes, but rather the emerging prominence of grassroots advocates acting as unaffiliated individuals or in small groups to educate a community, save a special place, or squash a project harmful to the commonwealth.

The message on April 22, 1970, was that the planet's plight demands an urgent response from each of us. For a quarter century that message has endured to influence the way thousands of Americans now live their lives. Here are brief profiles of seven of them.

MARY LYN RAY

The heart of Mary Lyn Ray's world is a cluster of green-and-gold valleys ringed by the blue-tinted mountains of central New Hampshire. It is a good piece of country for a woman like Ray, a writer of children's books and a land saver with a strong sense of place.

Unlike most over-40 activists in the environmental movement, Ray sat out Earth Day the first time around. "It just sort of passed me by," she recalls. "I remember thinking, 'How encouraging that so many people care.'"

By 1987, however, Ray had something of her own to care about. More than 150 acres of open fields and woodlands adjoining her home in South Danbury was at risk of being developed into small house lots. And if that happened, it would mark the beginning of the end for the values and traditions vested in the land. She decided to acquire the property herself to protect it.

With help from the nonprofit Society for the Protection of New Hampshire Forests, Ray contrived to purchase the land by using underlying gravel deposits as collateral for a bank loan. When the deposits proved too shallow to meet payments on the loan, Ray sold off her collection of art and antiques and learned to live on the edge of bankruptcy. And, convinced that the land should be preserved beyond her lifetime, she arranged for a conservation easement that precluded development.

That was only for starters. Having limited the use of her own land, Ray felt she could ask her neighbors to think about placing easements on *their* property. By late last year her persistence had helped bring permanent protection to more than 5,000 acres of forest and farmland on Ragged Mountain and the surrounding hills.

"When Mary Lyn sets her sights on something," says Paul Bofinger, president of the Forest Society, "you can see the fire in her eyes. She's bet the farm more than once to make good things happen. We're awful lucky to have her on our side."

JUDITH HANCOCK

Right now Judith Hancock is afloat in Okefenokee National Wildlife Refuge. But you might find her, too, at the wheel of a pickup truck on a backcountry road in Osceola National Forest, near Lake City, Florida, checking out potential threats to wildlife and its habitat. "I hope this one's got a bottom," she says as she accelerates the truck across a mudhole filled by recent rains. An earthy plume oozes over the windshield.

Then, pounding the steering wheel with an open palm, Hancock inquires of a rattled passenger, "Still with me?"

The intensity of Hancock's approach to mudholes mirrors her passion for her unremunerated work. In the Osceola she contended with the U.S. Forest Service over timber plans she considered unfriendly to the endangered red-cockaded woodpecker and the wood stork.

The Earth Day idea was a decade old when Hancock first came to the environmental wars, and the first battle she fought was right here in the national forest, where a consortium of companies had applied for permits to mine phosphate. Hancock was familiar enough with wetland

ecology to know that the mining and processing of phosphate rock along the headwaters of rivers can have a devastating effect on them, introducing sediment, excess nutrients, and toxic chemicals to the water. So she joined a citizens campaign to block the miners. The citizens prevailed. Based partly on information supplied by the group, Congress enacted a bill in 1984 prohibiting phosphate mining in the national forest.

Now Judy Hancock leaves the cab of her truck to plunge down a trail through wet woods along the Suwannee River. She pauses to admire large golden silk spiders and a canebrake rattlesnake, and before the day's done, she's picked up enough chiggers to keep a hound scratching for a week.

FATHER PETER KREITLER

Every day is Earth Day for Father Peter Kreitler, a 52-year-old Episcopal priest in Los Angeles, California. Tall and outgoing, with a tidy white beard and the confident, resonant voice of a preacher, Kreitler serves as a kind of diocesan minister to the environment, touring the grittier parts of the city in a relentless war against visual pollution. He knows there are far greater problems to solve in the vast sprawl of Los Angeles, but at the same time he believes that solutions have to start somewhere, even if they're small.

"When people lose respect for themselves and their surroundings," he says, "everything else breaks down. Gangs mark out their territories with graffiti, and litter piles up in the streets. We want to turn those things around." Kreitler founded Earth Service, Inc., which fights blight by helping neighborhood groups organize to paint over wall graffiti and haul trash out of vacant lots. Much of the group's effort is aimed at involving young people in recycling and the planting of trees.

El Padre, as he is identified by the letters on his license plate, was ordained in 1970, just five days after the first Earth Day. He did not participate in the event himself, but he remembers reading about it and being impressed by the number of people who expressed their concern. Later that year he began addressing environmental issues regularly from the pulpit, to the extent that some parishioners would soon be joking about "Father Peter's Save the Whales sermons."

So far Kreitler has been frustrated by the reactions to his environmental ministry from some of his colleagues of the cloth. A couple of years ago he called on several pastors to persuade them to correct such sources of wasted energy as incandescent bulbs. "I didn't get much of a response," Kreitler laments. "When the churches say, 'Let there be light,' they really mean it."

ISAAC EASTVOLD

Isaac Eastvold made his pact with the petroglyphs before Earth Day was conceived. His vow was to keep them from harm's way. The petroglyphs are those drawings carved in stone centuries ago by the ancestors of the Pueblo Indians. Though the art is scattered widely throughout the cliffs and canyons of the American Southwest, one of the greatest remaining petroglyph concentrations lies along the West Mesa escarpment on the outskirts of Albuquerque, New Mexico—right in the path of intense development.

Eastvold, bearded and often stirred to lyricism, has devoted at least half his 54 years to the study and protection of Indian rock art. Profoundly moved by the spiritual images, he first roamed the California desert, photographing petroglyphs for the Bureau of Land Management. Then, in 1985, he and his wife, Sharon, moved to Albuquerque and turned their attention to the rock-art treasures that lay just across the Rio Grande.

"When we got here," Eastvold recalls, "the petroglyphs were under terrific pressure. Vandals were using them for target practice. Big boulders with petroglyphs on them were being carted into the city for residential landscaping. In some areas development was being permitted right up to the rocks."

Eastvold proposed that the petroglyphs be protected in a national monument. Then, assuming the presidency of a new group called Friends of the Albuquerque Petroglyphs, he spearheaded an effort to enlist the support of New Mexico's congressional delegation and forge a consensus among competing city and state interests.

In 1990 his dream came true: Congress established 7,200-acre Petroglyph National Monument under the joint management of the city, the state, and the National Park Service. But the fragile consensus evaporated when the city of Albuquerque proposed building two huge roads through the fledgling monument. Eastvold and Pueblo leaders, who consider the park a major religious site, supported alternative routes recommended by the Park Service, but the city has recently granted subdivision permits blocking those routes.

"One developer threatened to run me out of town and ruin my reputation," Eastvold said. "When I pointed out that this is a place of worship for Native Americans, he laughed in my face and scorned their beliefs as a 'Stone Age religion.' He couldn't understand that the land around the petroglyphs is as vital to their meaning as the great cathedrals are to the paintings they hold."

On occasion Eastvold walks in the monument with Phillip Lauriano, a spiritual leader of the nearby Sandia Pueblo. Lauriano recalls first visiting the petroglyphs with his grandfather nearly 70 years ago. He explains to Eastvold that prayers here "go beyond the great divide to a reservoir of strength and power." The men walk together through a sea of wildflowers—evening primroses, mallows, sunflowers. Hover flies, gemlike and volatile, float in front of the blossoms. Wind rustles the bunchgrass.

"Wind in the grass," says Eastvold, "is the voice of the desert."

MARGARET MILLER

Margaret Miller is telling her husband, Paul, that it's time to go over to Dillon's Super Store and check on the bins. "Mrs. Recycling," they call her in Wichita, Kansas. "I got into recycling," Margaret explains, "because I thought it would be most enlightening for the person who wouldn't ordinarily become involved in environmental issues. It's visible. It gets people thinking."

Miller first came to the attention of Kansans when she took on one of the state's electric utilities by opposing its big Wolf Creek nuclear power plant (she lost) and then organized a consumers coalition seeking representation before the state board that sets utility rates (she won).

In 1989 Miller and two friends asked the Wichita City Council to help them set up a recycling program. "They laughed at us," Miller recalls. "They said people here would never go for it. So we established our own collection in a few parking lots. At the end of 18 months it was taking a fleet of trucks to handle all the waste." Now both the city and the state have pitched in, and local governments throughout Kansas coordinate the collection, reduction, and recycling of waste.

A key to Miller's success is her willingness to cooperate with the business community. "Margaret Miller pioneered recycling in Kansas," says Eric Evenson, manager of a new Weyerhauser Company plant in Wichita. The plant sorts and bales recyclable office papers and other "secondary" fibers for reprocessing. Evenson also credits Miller with

To People Who Make a Difference

Planet Earth from Space. Photo Courtesy of NASA Headquarters.

The registry of people who make environmental differences live in every town, county, and province of the world. What is happening notably since the 1970s is that people are becoming aware and concerned; they understand that they can make a difference and change the spread of ecological malaise. They are thinking change and enacting a system. The people mentioned previously are only a few of the people that have altered ways of doing things. Magazines, newspapers, and publications of local organizations all herald individual ecological achievement, but there is still much achievement that goes unacknowledged. The central idea here is that ecological change stems from individual conscience that has held to a commitment to improve some system or aspect that affects our earth. Ecological leadership and change begin with such individuals that care and extend energy and commitment behind their caring. From their actions, there are positive changes, new systems, and new research. This is where the hope for our future lies—with a strong, positive, ecologically minded, innovative dynamic. This box is dedicated to the many individuals who have made a difference and who will make *such* a difference.

educating the public on how to separate wastes—"and that's a job we could never have done by ourselves."

But the job never ends. Over in the parking lot at Dillon's Super Store, Miller spots a newcomer to the system. "Look at that woman," she says in a low voice. "She's going to put those colored bottles in with the clear ones."

Miller approaches. The miscreant is about to get educated.

PERCY AND ELLA HERON

Percy and Ella Heron's community garden sits at the corner of Inwood Street and Lakewood Avenue in the South Jamaica section of Queens, not far from the great airports of New York City. The garden is enclosed by a high chain-link fence. The gate is padlocked when the garden is not in use. The neighborhood of single-family houses on narrow lots has seen better days, though hardly anyone but the Herons has lived here long enough to remember them.

Percy and Ella Heron moved in here, two houses from the corner, in 1949. Together they and their neighbors tend the community garden, nurturing not only plants in the earth but also ideas in the minds of a younger generation. "We teach the neighborhood children about the soil," says Ella. "We teach about water and where it comes from and why it has to be clean."

Once there had been a house at the corner. Then the house burned, and the city condemned it and tore what remained to the ground. Vacant, the lot attracted garbage and litter until the Herons stepped in, appealed to the city's Department of General Services, and, with its blessing, slowly turned the place into a community garden, one of some 700 now flourishing in New York under the auspices of the department's Operation Green Thumb.

In season, the garden sparkles with roses, azaleas, hydrangeas, daisies, asters, tulips, and daffodils. There are planter boxes with peanuts and okra and tomatoes. Children come to the garden after school to help the Herons with their chores. This is how the ideas get planted.

Now it is time for Earth Day again. It will be, as always, a very special day at the community garden. Last October Ella and Percy Heron and some of their young friends got down on their hands and knees and planted tulip bulbs around the edge of a bed of roses. And what Ella told the kids was, "Don't you forget to come back here on April 22. These tulips might be fixing to bloom for that day."

EXERCISE

Research from publications, your local newspapers, and local organizations people who have enacted environment-friendly systems or approaches for ecological improvement. If possible, interview these individuals and give a presentation on their struggles and accomplishments.

QUESTIONS

1. What effect did each of these people have on the environment?

2. Why do you feel that all seven of these people are important in helping our planet? Which one do you think is most important?

A Bridge to Your 21st Century Understanding

Choose two people from this article that are the most interesting to you. Create a flowchart on each one.

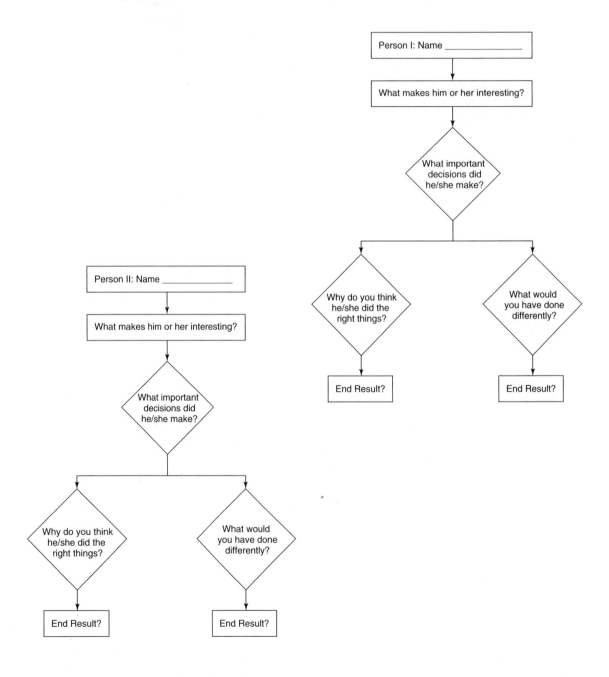

Person I: Name _____

What makes him or her interesting?

What important decisions did he/she make?

Why do you think he/she did the right things?

What would you have done differently?

End Result?

End Result?

Person II: Name _____

What makes him or her interesting?

What important decisions did he/she make?

Why do you think he/she did the right things?

What would you have done differently?

End Result?

End Result?

Rain Forests May Offer New Miracle Drugs

In the remote rain forests of Suriname, families have traditionally used local plants to treat common illnesses such as fever and stomachache, and more exotic ailments like leishmaniasis, an Amazon parasite which produces skin lesions. Folklore knowledge not only includes which plants are specific to certain diseases, but also proscribes the methods of preparation and application of the health restoring tinctures.

Synthetic production of drugs has been the primary focus of western pharmaceutical companies since the 1940s. Producing drugs in the laboratory allows these companies to safeguard their investment in research and development through intellectual property legislation enacted via patents and licenses.

In recent times, international drug companies such as Bristol-Myers are refocusing their interest on the natural laboratory that lies beneath the canopy of the rain forest, which is able to offer an albeit undocumented but nevertheless tried and perfected methodology towards drug refinement and employment. Recent discoveries include tree extracts from Indonesia which show promise in fighting HIV infection; a Brazilian shrub which is used to treat diabetes; and fungi and bacteria from the Philippines which show promise in treating a variety of ailments.

The process of searching for new natural medicines is called bioprospecting. Recent technological advances in screening samples of plants, animals and other natural substances for curative ingredients, has made it cost-effective to collect and analyze the large number of specimens needed to hone in viable medicines. In addition, the natural proclivity of diseases such as malaria to build up resistance to drugs has accelerated the need for alternative forms of treatment.

Bioprospecting accounts for 10% of the research budget in most large U.S. drug companies, which can amount to a significant investment for firms such as Merck & Co., which have annual R & D budgets of up to $1 billion. In addition, government agencies such as the National Institutes for Health, the National Science Foundation, and the U.S. Agency for International Development are also turning to bioprospecting. These three agencies will spend $2.3 million searching the rain forests of Costa Rica, Peru, Suriname, and Cameroon, and the deserts of Mexico, Chile, and Argentina.

Bioprospecting successes include the landmark Taxol, a treatment for advanced ovarian cancer found in the bark and needles of the Pacific Yew, and Vinblastine, a drug used for treating cancers including Hodgkin's disease and leukemia found in the rosy periwinkle, a Madagascan plant. Two other natural substances, Michellamine B from a Cameroon vine, and Prostratin, from a Samoan Rubber tree, are showing promise in the fight against AIDS.

Plants are not the only source of potential life saving drugs. On the French Guinea border, the Wayana people eat the heads of soldier ants to obtain a chemical which they believe will ward off malaria.

The search for curative compounds does not only take place in the natural environment. Medicinal plant researchers scour local markets seeking folk cures which may lead to the next wonder drug. In addition, these researchers rush to record potential remedies which may be lost to the passing of generations. Traditional cures are often complex combinations of herbs and animal extracts taken together to produce the required effect. Plants which reduce the side effects of curative substances are often taken with their medicinal counterparts. Researchers fear the knowledge of which combinations are effective may be lost as cultures are assimilated into the mainstream, and as the natural environment is destroyed. For example, at least 90% of the plants in the vanishing Brazilian Amazon have never been chemically analyzed.

Another problem is the time it takes to bring a discovered drug to market, often 10 to 15 years. Lab-developed drugs are often cheaper to refine and document for approval; a recent report indicated that Aspirin, derived from Willow tree bark, may have never been approved under the modern approval process due to uncertainties about the specific mechanism which makes it effective.

Researchers are also struggling with the dilemma of how to return part of the profits to the people and countries who provided the cure. In Suriname, bioprospectors have set up a Forest People's Fund, which received an initial $50,000 royalty from Bristol-Myers Squibb, and will receive between 1% and 5% of any royalties from marketed drugs. This program is designed to encourage local conservation of threatened ecosystems with financial incentives.

Thus while bioprospecting is showing increasing promise as a source of medicines and disease fighting drugs, researchers are fighting environmental and cultural destruction in the race to discover the next wonder drug. And while the search for traditional cures accelerates, no such speed-up in the process of bringing these cures to market seems imminent.

RAIN FOREST DIVERSITY AND DESTRUCTION

As stated in the introduction, the rain forests of the world need protection, since they significantly affect world climate, consuming large quantities of CO_2, contributing a large percentage of the world's oxygen, and slowing global warming. Also, the rain forest guards over most of the planet's biodiversity (an umbrella term for the variety of ecosystems, species, and genes present). Scientists theorize that tens of millions of species exist; however, they have described only a small proportion, only between 1.4 and 1.5 million of them. Half of these species that have been identified live in tropical forests, but scientists esti-

mate that 90% of all existing species live in the rain forest and would be identified if only the scientists had time to study them. The destruction of rain forests is rampant at about the pace of 100 acres a minute (approximately the size of the state of Washington each year). Rain forests now cover about only 5% of the earth's land surface, yet they contain at least half of its plant and animal species. During the past 30 years, about one third of the world's rain forests have disappeared, and this rate is increasing (Jukofsky, 1993, p. 20).

An example of this biological richness and abundance is in a study which found that a single hectare (about 2.5 acres) of rain forest in Peru contained 300 tree species— almost half the number of tree species native to all of North America. In another study, scientists found more than 1,300 butterfly species and 600 bird species within one five-square-kilometer patch of Peruvian rain forest. (The whole United States harbors 400 butterfly species and about 700 bird species.) In the same Peruvian jungle, Harvard entomologist Edward O. Wilson determined 43 ant species in a single tree, about the same number as exists in all of the British Isles (Rice, 1997, p. 48).

ECONOMIC VALUE

One reason that such plant and animal life diversity is vital is that it is essential for creating food, medicines, and raw materials. Wild plants have the genetic resources to breed pest- and disease-resistant crops. An estimated 120 clinically useful prescription drugs originate from 95 species of plants, of which 39 grow in tropical forests. Botanists estimate that from 35,000 to 70,000 plant species that are mostly from tropical forests provide traditional remedies in countries throughout the world. Without the rain forest, these plant species and the vast array of existing and potential medicines derived from them would become forever lost (Rice, 1997, p. 48).

Dr. Robert Balick of the New York Botanical Garden and Dr. Robert Mendelsohn of Yale University have studied the potential economic value that might result from medicines derived from the flora of the rain forest. Effective new drugs are valued at an average of $94 million each to drug companies. They concluded that at least 328 drugs await discovery with a projected value of some $147 billion (Medical Herpetology, 1997).

About half of the Earth's 250,000 flowering plants exist in tropical forests, but less than 1% of them have been thoroughly tested for medicinal uses. Nearly half of all drugs pre-

Table 27.1
The Rain Forest Pharmacy: Partial List

Drug	Derived from	Medical Uses
Atropine	Belladonna	Eases asthma attacks
Cocaine	Coca	Local anesthetics
D-Tubocurarine	Chondodendron tomentosum	Skeletal muscle relaxant
Diosgenin	Mexican Yam	Oral contraceptive
Papain	Papaya	Eases chronic diarrhea
Picrotoxin	Levant berry seeds	Eases schizophrenic seizures
Pilocarpine	Pilocarpus	Glaucoma
Quinine	Cinchona	Malaria
Reserpine	Snakeroot	Tranquilizers and treatment of hypertension
Vinblastine	Rosy periwinkle	Helps Hodgkin's disease
Vincristine	Rosy periwinkle	Helps acute leukemia

Source: Jukofsky, 1993 p. 26.

scribed in the United States have originated from plant life, with 47 medications currently on the market that are derived from the tropical forest, including codeine, quinine, and curare (Drugs and Money Abound in Rain Forests, 1995).

It is hoped that the economic value of studies of the rain forest will emphasize the importance of research and accelerate the conservation of invaluable world resources. However, efforts to slow destruction of the world's rain forests, home to these innumerable species, have met with limited success. Entire plant and animal populations and species are disappearing fast. Factors that cause this loss, aside from deforestation, including acid rain, other types of pollution, and increased UV-B exposure, are being investigated. It is greatly hoped that the discovery of pharmaceutical and other products derived from the rain forest can result in increased support and attention for biodiversity conservation and funds for continuing research (Medical Herpetology, 1997).

REFERENCES

"Drugs and Money Abound in Rain Forests." *ION Science,* July 1995. Available: http://w.w.winjersey.com/media/IonSci/glance/news795/raindru g.html.

Goering, L. (1995, September 12). Rain forests may offer new miracle drugs. *Chicago Tribune,* section 1, 16.

Jukofsky, D., and Wille, C. (April/May 1993). "They're Our Rain Forests Too." *National Wildlife,* Vol. 31, No. 3, pp. 18–37.

Medical Herpetology. 1997. Available: http://www.worlcorp.com/biodiversity/newsletter/two/herp.htm.

Rice, R. E., Gullison, R. E., and Reid, J. W. (1997, April). "Can Sustainable Management Save Tropical Forests?" *Scientific American.* Vol. 276, No. 4, pp. 44–49.

QUESTIONS

1. Should local inhabitants be rewarded for drugs discovered in their region? Can ownership of a renewable living resource (such as a plant) be treated in the same way as a non-renewable resource (such as mineral rights)?

2. The article mentions that it can take 10 to 15 years to gain approval and bring a drug to market. Is there a justification for performing drug trials on patients before full testing has been completed, and the drug proven safe and effective? If a patient has a terminal illness, what guidelines would you suggest in administering experimental medicines?

3. Researchers are concerned that potential cures will be lost to the destruction of the environment and the assimilation of folklore into popular culture. While there is much focus on saving the environment, the preservation of cultural diversity attracts little attention. What responsibility do the western nations have in safeguarding native cultures in the countries which they influence? Conversely, what right do they have in denying modern "civilization" to less technological societies?

4. The goal of the medical community, including drug companies, is to keep people alive for the longest possible time, regardless of their usefulness to society

as a whole. It is also true that the earth has limited resources, and natural law dictates that exponential population growth is not sustainable. This dictates that at some point (or points) mankind will be beset by natural and uncontrollable events such as epidemics which will cull the world's population back to sustainable levels. Extend this logic to produce an argument supporting euthanasia, or other solutions to world over-population. What are the ramifications of such arguments?

5. Project how long the rain forests have to exist given the current rate of destruction. What are some approaches you can think of to stop this rate of destruction?

6. What specifically are the ramifications of our increasing loss of rain forests? From this answer, derive solutions for abating this loss.

7. How effective do you think estimations of the economic value of potential drugs will be to save rain forest diversity? How would you use this strategy of placing an economic value on resources of rain forests to support rain forest conservation?

8. Research a more extensive list of rain forest medicines.

9. The article states that one reason that plant and animal life diversity is vital is for creating food, medicines, and raw materials. What are some other reasons?

A Bridge to Your 21st Century Understanding

Complete and discuss the following flowchart.

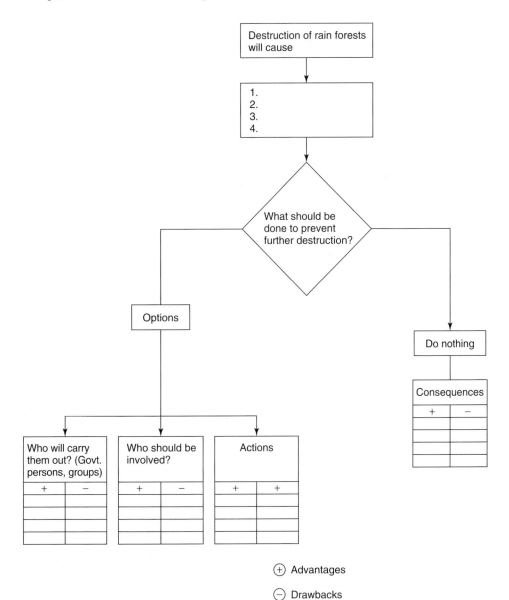

⊕ Advantages

⊖ Drawbacks

A Bridge to Your 21st Century Understanding

Complete and discuss the following flowchart.

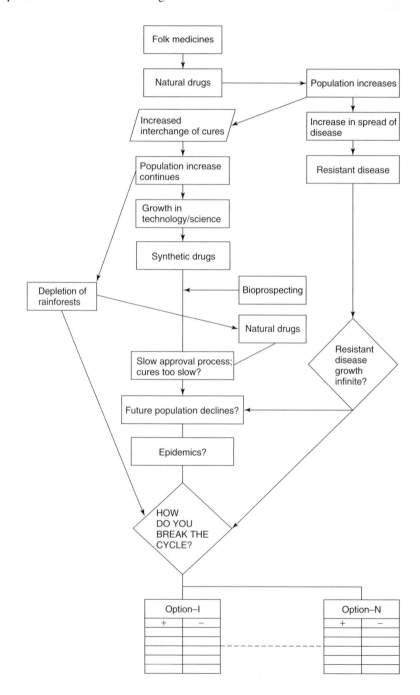

Case Study: The Great Molasses Flood

"Old timers in Boston say that to this very day you can smell a faint whiff of molasses in the old buildings on Commercial Street. But is that only an old timer's tall tale?"

MARJORIE STOVER, AUTHOR

It might be unusual to smell molasses on Commercial Street in Boston today, but on January 15, 1919, there was no doubt that the smell of molasses was in the air. On that mild winter day, "without warning, a massive tidal wave of sweet, sticky death gushed from a fractured steel tank, leaving twenty-one dead, over one hundred and fifty injured and many buildings crushed under nearly 12,000 tons of thick, brown, sugary molasses." (Schlager, 1994).

The problem started when a giant steel-sided storage tank that was 50 feet high and 90 feet in diameter was built by the Purity Distilling Company to contain rum, but instead was being used to store tons of molasses (Cavadias, 1998, p. 1; Stover, 1985, p. 34). The original order for the tank was issued without an engineer's approval, and the only requirement used in making the order was that the tank be able to hold liquids that were 50% heavier than water (Schlager, 1994, p. 246). There were further problems with the tank as well:

> The steel thicknesses for the rings were 5 to 10 percent less than illustrated on the original permit plans. . . . The only way that the tank was tested was by running six inches of water into it. (Schlager, 1996, p. 247)

These problems with the tank caused it to fail. Apparently, employees had noticed that the tank had been leaking molasses, but chose not to report it. However, on January 15, 1919, at 12:40 PM, everyone in the area of 529 Commercial Street in the North End of Boston found out the hard way that the molasses tank had problems. First, they heard sounds similar to the firing of machine guns as

the rivets holding together the metal sheets of the tank's body popped out of their holes (Mercan, 1998, p. 2). Then, they heard a roaring sound as the tank exploded and separated warm molasses poured into the streets. (The molasses was warm because the company had placed steam heating pipes inside of the tank.) Some say that the river of molasses flowed at 35 miles per hour and was at times 15 feet high and 160 feet wide (Cavadias, 1998, p. 1; Mercan, 1998, p. 2; Schlager, 1996, p. 247).

Twenty-one people were killed due to the molasses flow, either by drowning or by being hit by debris. Over 150 people were injured. It was difficult for the rescue workers and their trucks and equipment to get through to help the injured because the gooey liquid stood three feet deep in some places (Stover, 1985, p. 36). A 2.5-ton section of the tank had been thrown onto a playground 182 feet away (Schlager, 1996, p. 247). Another part of the tank broke and tore off a part of the Atlantic Avenue Elevated Railway, causing the railway to collapse. It seems that the pieces of tank were thrown far distances because the fermenting molasses caused pent-up gases to form within the tank, causing the tank to burst apart when the pressure got too high (Mercan, 1998, p. 2). Some survivors had to have their "molasses-stiffened clothing cut off of them, and trapped horses had to be shot" (Cavadias, 1998, p. 1).

The flood of goo engulfed the area and destroyed most everything in its path. Buildings fell off their foundations, basements were filled with molasses, molasses "statues" of people were found, and the tank was left laying in pieces on the ground, as a painful reminder. It was six or seven days before all of the bodies could be found and months before Commercial Street could return to normal (Cavadia, 1998, p. 1; Mercan, 1998, p. 2; Stover, 1985, p. 35; Schlager, 1996, p. 248).

When the tank was examined, two fractures were found in the steel plates; it was these fractures that allowed the tank to be torn apart. Also, it did not help that the tank had been made of materials that were thinner than those called for on the original permit drawings. The tank was a technological failure because it was improperly designed and constructed and because the calculations for the tank's construction were never checked by an experienced engineer. The tank disaster had occurred because of structural weakness and human error. (Schlager, 1996, p. 249).

REFERENCES

January 15, 1919: A tidal wave of molasses inundates Boston. (1998, August). Available: http://cavadia.com/tidalwave.html.

The great Boston molasses flood. (1998, August). Available: http://www.mercan.com/~kimchee/molass1.nun.

Schlager, N. (1996). *Accidents and Failures of the Twentieth Century.* London, Gale Research, Inc.

Stover, M. *Patrick and the Great Molasses Explosion.* (1985). Minneapolis, Dillon Press, Inc.

QUESTIONS

1. If you had been working at the Purity Distilling Company in 1919, what would you have done to prevent this accident from happening?

2. What is the relationship between this case and social responsibility? Why do you think that the employees ignored the leaking tank seams? Why wouldn't anyone question the tank's use when the tank was installed?

A Bridge to Your 21st Century Understanding

Complete and discuss the following flowchart.

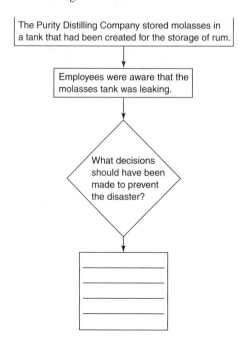

The Purity Distilling Company stored molasses in a tank that had been created for the storage of rum.

Employees were aware that the molasses tank was leaking.

What decisions should have been made to prevent the disaster?

Case Study: The Toxic Cloud Over Bhopal

In the early hours of December 3, 1984 a toxic cloud of methyl isocyanate (MIC) gas from Union Carbide's plant in Bhopal, India, escaped and enveloped the hundreds of shanties and huts surrounding the plant. The deadly cloud drifted in the cool night air through streets in surrounding neighborhoods killing and badly injuring thousands of people. By dawn inhabitants of Bhopal had become victims of the worst industrial accident.

A report drawn up eight months later, by two international trade union confederations, suggested that it was a combination of management mistakes, badly designed equipment and poor maintenance that caused the accident. The gas was formed when a plant employee added water to a methyl isocyanate storage tank. The water caused a reaction that built up heat and pressure in the tank, and quickly converted the chemical into lethal gas that escaped into the atmosphere. The sensors and alarms that could have detected and indicated the increase in pressure and heat in the MIC tank were inoperative. The plant's flare tower that could have been used to burn off the escaping gas was inoperative too. The warning sirens were used with one hour delay. There was no evacuation plan prepared beforehand for such an emergency. The civil authorities and the population of Bhopal were also not informed about the degree of toxicity and the movement of the MIC gas cloud.

Approximately twenty-seven tons of MIC vapor, and fourteen tons of reaction products, were released in the atmosphere over a period of ninety minutes and affected an area of fifteen square miles.

The MIC gas leak was not the first accident at the plant; in the preceding four years there had been six accidents, some of which involved MIC, including one fatality in 1981 which led to an official investigation. The findings of that investigation were not acted upon until after the accident in 1984.

Union Carbide developed its plant in 1969 at Bhopal, initially as a mixing and packaging plant for pesticides imported from the United States. Later, in 1980 it was expanded to manufacture the pesticides Sevin and Temik. One chemical used in large quantities in the production process was methyl isocyanate (MIC), a highly reactive, volatile and toxic compound. The process that reacted MIC with another compound was considered the leading technology for producing pesticides Sevin and Temik. The development of the plant was part of an Indian government effort to achieve industrial self-sufficiency. The plant was located on the outskirts of Bhopal on land leased to Union Carbide India Limited (UCIL) by the Indian state government of Madhya Pradesh, at an annual rent of about Rupees 500 ($40).

In 1984 Union Carbide reported sales of $9.5 billion, reflecting its position as one of the largest industrial companies in the United States and the world. The international operations were responsible for approximately 30 percent of total sales that year. India was one of three dozen countries where Union Carbide had business interests. In 1984 Union Carbide India Limited (UCIL) celebrated its 50th anniversary. It operated 14 plants with a workforce of 9,000 people. UCIL had sales of about $200 million annually. In 1984 the entire work force at Bhopal plant was Indian. In keeping with the government's interest in promoting self-sufficiency and local control, the last American employed at the plant had left two years before, making the plant an all Indian operation.

The Bhopal Catastrophe

Chemicals Leaked	Approximately twenty-seven tons of Methyl Isocyanate (MIC) vapor and fourteen tons of reaction products
Threshold limit value (TLV) of MIC established by United States Occupational Safety and Health Act (OSHA).	0.02 parts per million over eight hour period.
Exposure area	Fifteen square miles
Deaths	2,352–10,000
People Disabled	17,000–20,000
People Exposed	200,000
People Evacuated	70,000

The legal dimensions of the Bhopal disaster began when Indian prosecutors brought charges of criminal negligence against the Indian and American management of Union Carbide in the Indian courts. In 1985 the government of India, on behalf of the victims of Bhopal, filed a civil suit against Union Carbide in the Federal District Court in New York City. The U.S. court decided that India was the proper site for any Bhopal action and sent the litigation there for disposition. The Indian government filed suit in India for $3.3 billion. After several years of negotiations, Union Carbide agreed to a settlement of $470 million with the government of Ragiv Ghandi. The dropping of all criminal charges against Union Carbide executives was part of the settlement.

After the elections in India, the government of new prime minister V.P. Singh repudiated the Indian Supreme Court and rejected the $470 million settlement as "totally inadequate". His government announced its intention to return to the original $3.3 billion claim and to pursue criminal charges against Union Carbide executives. Following a lengthy review by the Indian Supreme Court, the original settlement was upheld and the criminal proceedings were reopened. Union Carbide has paid, as ordered, $190 million in interim compensation which the government of India has to distribute among survivors pending the final outcome of the settlement. Presently the Indian government is providing survivors 200 Rupees a month (approximately U.S. $10). These payments are to last as long as the case remains unsettled in the Indian courts. It is expected that the litigation and debate over the criminal prosecution for negligence and financial compensation will not end soon.

No matter what kind of settlement is reached in courts, the survivors face a nightmare future of uncertainty. Thousands are now suffering from a host of ailments including acute respiratory distress, vomiting blood, conjunctivitis, damaged eyesight, pain in the abdomen and injuries to liver, brain, heart, kidneys and immune system. In addition to these problems, women are also suffering from gynecological complications such as menstrual and reproductive disorders. An epidemiological study of the yearly rate of spontaneous abortions and infant deaths in the years after the accident found the rate in Bhopal to be three to four times the regional rate.

Today, at personal, local, national and international levels, the Bhopal disaster remains a reality. Like Union Carbide's plant in Bhopal, a number of chemical plants belonging to multi-national corporations are operating in the Third World countries. The lack of environmental and occupational safety regulations in the developing countries, coupled with multi-national companies' desire to increase earnings per share at any cost, sets the stage for reoccurrence of the Bhopal catastrophe.

REFERENCES

Goldsmith, E., and Hildyard, N. (1998). *The Earth Report—The Guide to Global Ecological Issues,* Los Angeles, Price Stern Sloan, Inc.

Gottschalk A.J. (1993). *Crisis Response,* Detroit, Visible Ink Press.

Kurzman, D. (1987). *A Killing Wind,* New York, McGraw-Hill Company.

Schlager, N. (1994). *When Technology Fails,* Detroit, Gale Research.

QUESTIONS

1. What lessons should the chemical industry have learned from the Bhopal disaster?

2. What could be done at personal, local, national, and international levels to promote safety in the chemical plants?

3. In the Third World, more than half a million people are poisoned every year by pesticides, but these same pesticides also save millions from starvation. Should the use of pesticides be banned worldwide?

4. What steps and measures could Union Carbide have taken to avert the disaster at Bhopal?

5. How do you feel about the $470 million settlement? Is it an appropriate amount for 200,000 survivors for the rest of their lives?

A Bridge to Your 21st Century Understanding

Complete and discuss the following flowchart.

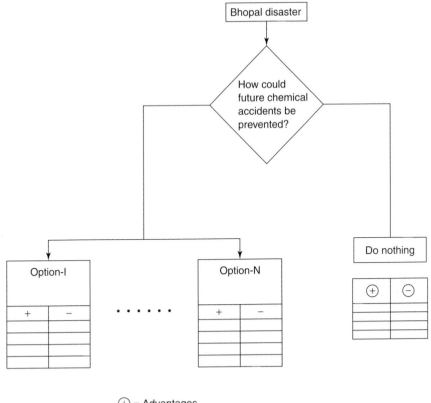

⊕ = Advantages

⊖ = Drawbacks

EARTH DAY 2000: A 30-YEAR REPORT CARD

On the first Earth Day in 1970, experts warned that the planet's natural systems were being dangerously destabilized by human industry. Here is how we have fared on some key fronts since then:

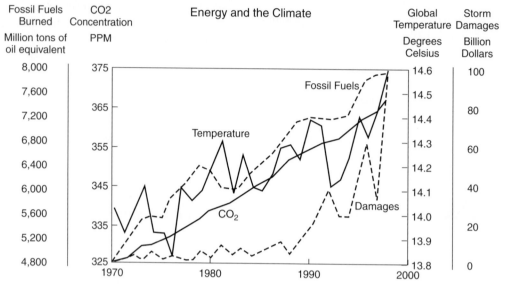

As our growing population increased its burning of coal and oil to produce power, the carbon locked in millions of years worth of ancient plant growth was released into the air, laying a heat-retaining blanket of carbon dioxide over the planet. Earth's temperature increased significantly. Climate scientists had predicted that this increase would disrupt weather. And indeed, annual damages from weather disasters have increased over 40-fold.

Solution: *A faster shift to nonpolluting, renewable solar, wind, and hydrogen energy systems.*

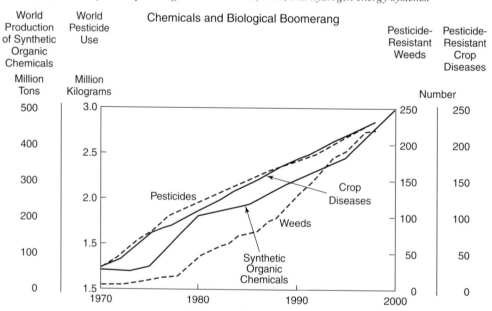

Our consumption of chemicals has exploded, with about three new synthetic chemicals introduced each day. Almost nothing is known about the long-term health and environmental effects of new synthetics, so we have been ambushed again and again by belated discoveries. One of the most ominous chronic effects: as pesticide use has increased, so has the evolution of pesticide-resistant pests.

Solution: *A large-scale shift to organic farming; a shift away from excessive consumption of synthetic chemical products; and application of the precautionary principle to the chemical industry.*

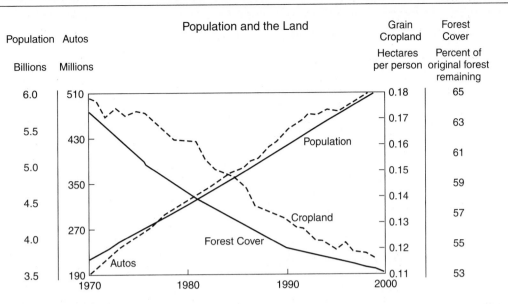

Population and the Land

Population has increased by as much in the past 30 years as it did in the 100,000 years prior to the mid-20th century. And as the number of people has grown, the amount of land used by each person—either directly or through economic demand—has also expanded. As a result of this double expansion, incursions of human activity into agricultural and forested land have accelerated.

Solution: *Stabilize population, especially by improving the economic and social status of women; design cities in ways that reduce distances traveled between home, work, shopping, and school; and in urban transit systems, shift emphasis from cars to public transportation, bicycling, and walking.*

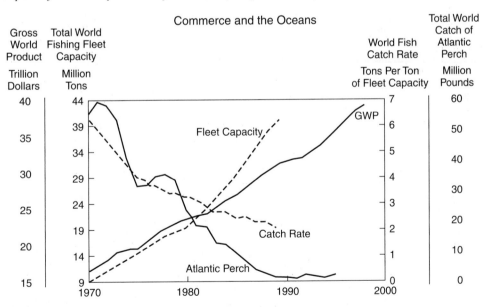

Commerce and the Oceans

The global economy has more than doubled in the past 30 years, putting pressure on most countries to increase export income. Many have tried to increase revenues by selling more ocean fish—for which there is growing demand, since the increase in crop yields no longer keeps pace with population growth. Result: overfishing is decimating one stock after another, and the catch is getting thinner and thinner.

Solution: *Stabilize population growth; stop subsidizing fishing fleets; and end the practice of feeding ocean-caught fish to farmed fish (it takes five pounds of ocean catch to produce one pound of farmed fish), which is still a very profitable and common practice.*

7 Moments that Helped Define the Trends of the Past 30 years . . .

The Car: *Mannheim, Germany, 1885*

Karl Friedrich Benz takes the world's first gasoline-driven automobile out for a test drive and reaches a speed of 9 miles per hour. It's not yet faster than a horse, but the global infatuation with motorized speed is about to begin. Though petroleum has been around for decades, used mainly for lighting lamps, the advent of the internal combustion engine causes a surge in demand, and the fossil-fuel age begins.

The Gusher: *Masjid-I-Salaman, Persia, May 26, 1908*

Drillers strike oil, and the rights are quickly acquired by the British government. The new enterprise, British Petroleum, turns out to be sitting atop the largest oil reservoir in the world, and thus is established a Western dominance of oil that will prevail throughout the 20th century. That dominance will be strengthened by the establishment of the U.S.-controlled Arab-American Oil Company (Aramco) in 1933 and the Iranian coup in 1953. The resulting flow of cheap oil allows the fossil-fuel economy to dominate global industrialization.

The Golden Arch: *Oak Park, Illinois, Late 1950s*

McDonald's decides to open franchises all over the world. In order to establish uniform standards of production for its French fries, the company requires suppliers in each country it enters to grow its global standard potato—the Idaho russet. Other varieties, often better adapted to local conditions of soil, rainfall, temperature, and growing seasons, are displaced. The French fries policy becomes a model for the "mono-culturization" of agriculture on a global scale. It is an approach that eventually increases food supply for the expanding human population but also opens the way to increased erosion, soil depletion, dependence on fertilizers and pesticides, nitrogen pollution of rivers and bays, and the decline of genetic diversity in the world's major food crops.

The TV: *Western Europe, 1952*

The first international standard for transmission of TV images (in lines per frame and frames per second) is established, opening the way to mass-audience broadcasts. Appetites for consumption are stimulated first in the industrial countries where TVs catch on quickly, then in the developing world where subtitled or dubbed American or European shows serve as implicit but vivid advertisements for first-world over-consumption.

The Highway: *Washington, DC, 1956*

The U.S. Congress passes the Interstate Highway Act, authorizing construction of a national network of high-speed roads across the United States. The American penchant for traveling long distances, even in routine trips between home, work, shopping, and recreation, is greatly facilitated. Suburbanization is accelerated, natural areas are paved over, and pollution increases as major cities build beltways and open the way to "edge" cities. High mobility becomes a model for other countries, which develop their own highway systems—causing massive increases in deforestation, oil spills, air pollution, and carbon dioxide emissions.

The Backlash: *India, mid-1970s*

The Indian government, faced with surging population, adopts a policy of enforced birth control. Many men and women undergo compulsory sterilization. The policy triggers a great backlash, and the birthrate climbs instead of declining. Demographers project that by 2010, India will have passed China as the most populous country on Earth.

The Flood: *Yangtze River Basin, China, 1998*

Chinese developers clear thousands of hectares of forest to make space for the country's burgeoning population—thus setting the stage for one of the largest disasters in history. Stripping tree cover reduces the watershed's capacity to slow the flow of surface water. Global warming increases evaporation—and thus increases rainfall. When the monsoon of 1998 comes, the heightened volume and velocity of the runoff—and unprecedented numbers of people living in the water's path—drive over 100 million people from their homes. The following year, Hurricane Mitch inundates Honduras and Belize, where similar deforestation has taken place. The disruptive impacts of climate change appear to be well under way.

And 7 Moments (Past and Future) that could be Keys to the *Next* 30 Years . . .

Civil Society: *Uttarakhand, India, 1958*

A popular movement arises to protest government mismanagement of Himalayan forest, and the operations of large timber companies engaged in what is widely regarded as a form of looting. Led mainly by women, the Chipko movement asserts the traditional rights of villagers to manage their local forests rather than submit to management by a distant bureaucracy. The Chipko movement raises the profile of nongovernmental environmental movements in India, as thousands of women stand in the way of tree-cutters. In the ensuing years, grass-roots groups proliferate, and become more numerous in India than in any other country. By the 1990s they have become a "third force" in human organization worldwide—a "civil society" that may soon be strong enough to begin to counterbalance unresponsive government and industry.

Precautionary Principle: *New York, 1962*

Rachel Carson publishes a book, *Silent Spring,* calling attention to the rising burden of chemical pollutants on the environment. As the burden continues to worsen in the following decades, it provokes discussion of a new Precautionary Principle—the principle that the burden of proof of safety should be on those who wish to introduce a new chemical, not on those who claim to have been injured by it. In the 1990s, the principle will be invoked by members of the Intergovernmental Panel on Climate Change, a network of the world's leading climate scientists, in their argument that "uncertainty" in climate science should not be a reason to avoid preventive action on climate change.

Earth Summit: *Stockholm, Sweden, 1972*

The United Nations Conference on Human Development becomes the first global effort to place the protection of the biosphere on the official agenda of international policy and law. It will be followed by the UN Conference on Human Settlements (HABITAT) in 1976, the first World Climate Conference in 1979, and the UN Conference on Environment and Development (Earth Summit) in 1992—leading to what has become an essentially continuous process of international discussion on issues that concern transnational threats to human security.

Micropower: *Sri Lanka, About 1990*

In 100 villages, solar panels are installed on rooftops to provide low-cost electricity to homes that are not on the electric grid. Similar installations are being made, around the same time, in the Domican Republic, Zimbabwe, and other developing countries. They form the first scatterings of a movement toward the use of decentralized electric power systems, based on nonpolluting solar or wind power, that will eventually revolutionize the energy industry worldwide.

GMO-Free Food: *Western Europe, 1998*

European protesters compel transnational biotech companies to halt the rush to use genetically modified organisms (GMOs) in agriculture. Monsanto's bullish advertising campaign is scrapped; major food producers and retailers change their food-processing formulas; Monsanto halts its program to force farmers to buy terminator seed.

The Climatic Wake-Up Call: *Somewhere on Earth, Soon*

An extreme weather event strikes a major population center head-on, with cataclysmic results. The event may be a gigantic hurricane or storm surge striking a coastal city, or it may be an inland flood inundating a heavily populated river basin. This time the disaster achieves a perceptual critical mass in the global public—an undeniable recognition that the greatest threats to human security are not those of military invasion but of environmental degradation. As a result, large-scale campaigns are undertaken to gird for—and stabilize—the future impacts of climate change.

Bioregionalism: *U.S. and Canadian Pacific, Early 21st Century*

Along the northern Pacific coast, there is yet another clash between native peoples and the companies logging the region's remaining old growth rainforest. But after decades of controversy over the management of coastal forests and waters, the native activists discover they have a constituency much broader than anything their predecessors enjoyed. From Oregon through British Columbia, they have awakened a latent bioregional awareness—a widely-shared view that the region is unique, both ecologically and culturally. This awareness begins to reshape local politics, to make it better reflect the long-term interests of the region itself. As the region thrives, people elsewhere come to believe—and act on—the principle that environmental progress often comes easier when natural regions are given precedence over political ones.

Earth Day 2000: A 30-Year Report Card. WorldWatch, March/April 2000. The WorldWatch Institute, http://www.worldwatch.org.

What Would (Can) You Do?

Review the four graphs and their solutions on: Energy and the Climate, Chemicals and the Biological Boomerang, Population and the Land, and Commerce and the Oceans. Read the next two pages of "7 moments (and individuals) of the past and the future" that shaped and could shape the future. Investigate the following topics according to "what can or would you do." Relate your findings to the graphs and how they could affect trends in the future. Provide examples of these topics as specific as possible as to how you can contribute to the solutions and make a possible difference. Also create your own topics. Also relate your actions and answers to ethical personal and technological guidelines.

1. Investigate home and work, industry energy efficiency possibilities and patterns
2. Investigate home and work, industry recycling possibilities and patterns
3. Discover initiatives for more energy efficiency in the home and work, industrial environments
4. Discover initiatives for more recycling industries
5. Discuss type of car used and type and amount of fuel it uses, individual as well as local groups and demographics. What are the implications?
6. Research transportation and passenger patterns, both individual and local
7. Investigate creativity and development of more products from waste technologies
8. Investigate how to become involved with government policies
9. Investigate resources, reading, awareness, involvement of local energy policies
10. Research local chemical industrial practices and local waste contaminants
11. Investigate local fertilization, pesticide and other community chemical applications and practices
12. Research local and regional garbage practices - collection, dumping, standards for garbage fills, incineration, etc.
13. Initiate conversations with local farmers, growers, industry and company owners about their chemical concerns and practices
14. Contact your local census bureau and discover local demographics of population and various breakouts of growth of socio-economic levels, ethnic groups, types of jobs, etc.
15. Contact area or county map bureaus and track the type of growth of your community within the last decade. Develop a future plan of projected growth considering ecological and environmental priorities and including special needed or saved "green areas."
16. Contact local or area fish markets. Discover area fish consumption including approximate consumption (home and restaurant), types of fish, origins of catch, and amount and use of waste.
17. Develop a plan for more ecologically and environmentally sustainable approach to fish consumption and use.
18. Create an effective approach to being ecological and ethically aware and involved.

Conclusion

As stated in the beginning of Part IV, the purpose of this part is to enhance awareness, responsibility, strategies, and involvement for a sustainable, desirable environment. This part has overviewed many major issues that cry out for more results and better solutions. Many of those solutions have begun to be implemented and have already made an important difference—but the issues of this part (and others) remain central and critical to our future and our future quality of life. Your thoughtful awareness and involvement can and will make an important contribution. Therefore, this text is designed so that you may become aware of the issues and the decisions you can make or that can be made to design the future. Your involvement and awareness become the fulcrum for the solution. The flowcharts illustrate this philosophy.

The issues presented are not meant to present a negative picture of ecology, but instead are intended to bring out three different layers of central issues that need to be addressed. The three layers of the overview, the global case study, and the specific case study could be likened to the earth itself with (1) the overview being similar to the atmosphere of ecological understanding; (2) the global case study being comparable to the surface of the earth, with all the interactions of activity that take place on the surface of the globe; and (3) the specific case study being similar to the local activity that then affects global issues and understanding. The three layers of the issues presented then included the following aspects:

1. The overview structural layer in the introduction discussed the history and overview of the high-risk environmental issues of habitat destruction, species extinction, global warming, and depletion of the ozone layer, with an approach with regard to economic value.

2. The text then proceeded to the major global issues of consumerism and waste, long-term effects of pollutants on our animal and human young, and superfunds—what to do with our toxic wastes—and the exploitation of and lack of planning for our world fishing industries.

3. The specific case studies brought to your attention the situations regarding Love Canal, the Endangered Species Act, air pollution, Earth Day, people who have made a difference in environmental causes, rain forest pharmaceuticals, Great Molasses Flood, and Bhopal's toxic cloud.

It is hoped that from these issues and experiences, you have a better pulse of some of the major ecological issues and are oriented to participate in their solutions.

Our "Ecological" Social–Economic View

Associating fiscal value with the earth's natural goods and services will begin to raise perceptions of ecological value to its real importance and value. The estimation of our ecosystem at $33 trillion seems conservative, but this estimate is a start in bringing to peoples' minds the replaceable and working value of the ecosystem. Pollution controls can be viewed from dollars saved in annual health costs. Raw material costs should include costs of earth materials lost and recovery techniques to the environment. Costs of fish should be raised with monetary penalties when quotas are exceeded. Rain forest conservation approaches can be partially funded through the market of its products and pharmaceuticals. And lastly, threatened and endangered species can receive the funding that is necessary if the extinction of species becomes a heavy debt and monetary liability.

Not all ecological problems can be solved with a credit and debit ledger, but placing true monetary value on an interactive world ecosystem on which we all depend may help unindustrialized countries as well as industrialized countries treasure and account for their resources and level the playing field. Many Third-World countries retain valuable resources without receiving compensation for their use or conservation (e.g., the animals of Africa and Asia, the rainforests of Central America, etc.). It makes good sense to count treasure as treasure.

Bridge to the Future

Many of these issues discussed in this part of the book have been the recipients of some successful efforts at enacting solutions. The Rio de Janeiro Earth Summit of 1992, attended by more than 100 heads of state, addressed many issues with serious intent to make progress on issues of environment and poverty, but the solutions that came out of this summit are not necessarily working. In the part on the future, Part VIII, we will discuss some successes, failures, and future initiatives for ecological progress in the 21st Century.

INTERNET EXERCISE

1. Use any of the Internet search engines (e.g., Alta Vista, Yahoo, Infoseek, etc.) to research the following topics:
   a. Updates on rain forest destruction and some programs to conserve rain forests.
   b. Disappearance and malformation of frogs on the earth.
   c. The latest research on global warming—is it or is it not happening?
   d. The latest views on the disappearance of the ozone layer and the laws that try to protect the ozone layer and their effects.
   e. Some of the facts on improvements in conservation due to recycling.
   f. Water pollution and the water we drink.
   g. More information on Superfund.
   h. The latest information on the Magnuson Fishery Conservation and Management Act.
   i. Commercial and recreational fishing catches in your local area for the last decade.
   j. Some of the projects that the Citizens Clearinghouse for Hazardous Waste have undertaken.
   k. New revisions and implications of the Endangered Species Act.

    l. Air pollution in your area.

    m. Water pollution in your area.

    n. The latest laws on air pollution particulates.

    o. A list of people whom you admire who have made a difference ecologically.

    p. New medicines and products derived from flora or fauna in the rain forest.

    q. Garbage production in the last ten years (locally, statewide, or nationally).

    r. Ecological laws passed in the last two years (local, state, or federal).

2. Use any of the Internet search engines (e.g., Alta Vista, Yahoo, Infoseek, etc.) to research the following topics:

    a. Ecosystem, ecosphere, and sythesphere.

    b. Environmental Protection Agency.

    c. Biological diversity.

    d. Chlorofluorocarbons and ozone.

    e. Dioxins.

    f. PCBs.

    g. Pesticides.

    h. Hazardous wastes.

    i. U.N. Food and Agriculture Organization.

    j. Individual Transferable Quotas.

    k. Endangered Species Act and threatened species.

    l. U.S. Fish and Wildlife Service.

    m. Air particulates.

    n. Ethnobotany.

    o. The words you have listed (see the vocabulary exercise in the introduction) from your research and reading of ecological issues.

USEFUL WEB SITES

URL	Site Description
http://www.edf.org	Environmental Defense Fund
http://www.rachel.org	Environmental Research Foundation
http://www.audubon.org	National Audubon Society
http://ens.lycos.com	Lycos Environmental News
http://www.nationalgeographic.com	National Geographic Society
http://www.nrdc.org	National Resources Defense Council
http://www.nwf.org	National Wildlife Federation
http://www.100topenvironmentsites.com	100 Top Environment Sites
http://www.tnc.org	The Nature Conservancy
http://www.sciam.com	Scientific American
http://www.sierraclub.org	Sierra Club
http://www.worldwildlife.org	World Wildlife Fund
http://www.worldwatch.org	The WorldWatch Institute
http://www.nasa.gov	NASA
http://www.fao.org	Food and Agriculture Organization of the United Nations

PART

V

Population

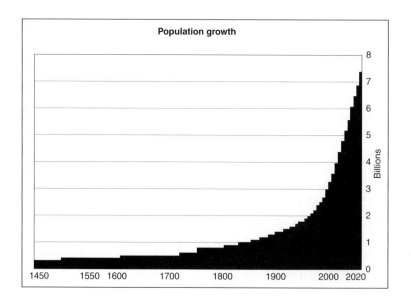

Population growth

Billions

1450 1550 1600 1700 1800 1900 2000 2020

234

OBJECTIVES

Part V will help you to:

- Recognize world human population growth projections and their long- and short-range implications
- Perceive the urgency of human growth expansion and their accompanying economic, human rights, and ecological global and local implications
- Discern population growth differences and its causes among the industrialized and unindustrialized nations, along with global and local issues and implications
- Analyze and use the tables and graphs to develop and support specific information on population perspectives
- Realize world birth control issues from varying cultural, ethical, and national perspectives
- Understand the roles of women's rights and education in conjunction to fertility rates
- Realize the positive and negative impacts of various local and global strategies regarding birth control
- Appreciate the importance and influence of effective and appropriate education, human rights, and birth control strategies

INTRODUCTION

"Humanity is approaching a crisis point with respect to the interlocking issues of population, environment, and development."
STATEMENT FROM 60 SCIENCE ACADEMIES LED BY THE U.S. NATIONAL ACADEMY OF SCIENCES AND BRITAIN'S ROYAL SOCIETY, OCTOBER 1993.

"The decisions that the international community takes over the next several years, whether leading to action or inaction, will have profound implications for the quality of life for all people, including generations not yet born, and perhaps for the planet itself."
STATEMENT MADE AT THE INTERNATIONAL CONFERENCE ON POPULATION AND DEVELOPMENT, CAIRO, 1994.

Understanding World Population
Issues: A Street in Calcutta, India.
Photo Courtesy of United Nations/J.P.
LaFonte.

As you read and reflect upon these pages, you will find many issues—individual, local, national, global, religious, societal, economic, and ecological—that all closely interrelate. Graphs in this part of the book show that we reached our first billion people in world history around 1839; our second billion 90 years later, around 1930; and our third billion 30 years later, in 1960. Currently, we are adding a billion people to the world's population approximately every 15 years or less, which could easily escalate our population count to the U.N. high projection of 11.9 billion by the year 2050. For sustenance, the United Nations recommends an ambitious plan to stabilize population at low to medium projections of 7.9 to 9.8 billion people by 2050. In order for this stabilization to happen, it is necessary to address the following issues that are confronting us:

growth projections

ecological implications

food production

limitation of the earth's natural resources for food production

the issues raised by the 1994 Cairo Conference

population control issues

numbers carrying capacity where country becomes stabilized

specific case studies (such as those of Pakistan, India, Ethiopia, and Mexico)

the interdependent nature of population, poverty, and the local environment

This part on population and the issues that it addresses have many implications, the purpose of which is to help you to be informed and to make informed decisions on the individual, local, national, economic, ecological, and other fronts. There is much to say and more to do, but it all starts with being informed and playing a part in determining a future plan; this is a vital assignment, since we add to the world population approximately

3	people	each second
117	people	each minute
10,600	people	each hour
254,000	people	each day
1,800,000	people	each week
7,700,000	people	each month
93,000,000	people	each year.

All of these numbers translate to a situation in which nine babies are born and three people die every two seconds, resulting in a net increase of three people each second. By the year 2000, annual population growth will increase to 94 million people, and by 2020 it will rise to 98 million. In 2020, 98% of the increase will be from births in developing countries (U.S. Bureau of the Census).

For real-time population numbers, the following Internet addresses access the current second-by-second population clocks and projection numbers to give you a further idea of our exponential patterns of population growth (also refer to the "Useful Web Sites" section at the end of this part):

National population	http://www.census.gov/cgi-bin/popclock
World population	http://sunsite.unc.edu/lunarbin/worldpop
	http://www.census.gov/cgi-bin/ipc/popclockw

Like the other parts of this book, this part presents factual, historical, social, and economic weavings along with the issues and articles to provide a cross-analysis of the issues from differing perspectives. The tables and figures that follow the introductory text give a necessary empirical base to realize the issues numerically before the descriptive discussions of the cases and articles are presented. This format builds an understanding of the issues from both data-based and cross-descriptive views.

MALTHUS VS. DEMOGRAPHIC TRANSITION THEORY

The rapid growth of Europe's population because of more abundant food supplies and better living conditions caused Thomas Malthus (1766–1834), an English economist, to predict uncontrolled population growth and, hence, future doom. In 1798, he wrote the book *An Essay on the Principle of Population,* in which he predicted what became known as the Malthus theorem. His principal argument was that while population grows geometrically (from 2 people to 4 to 8 to 16, etc.), food supply increases arithmetically (from 1 unit to 2 to 3 to 4, etc.), concluding that if births proceed unchecked, the population will outgrow its food supply (Henslin, 1996, pp. 352–353). However, a different and somewhat opposite argument called a three-stage demographic transition theory also exists in which the first stage, Stage I, has high birth rates and high death rates while population remains stable; the second stage, Stage II, experiences high birth rates and low death rates while population surges; and the third stage, Stage III, has low birth rates and low death rates while population stabilizes and the economic quality of life improves. The theory is that all nations proceed through these stages in economic and societal growth before reaching stabilization (Brown, 1987). Consider the following historical world population patterns below.

Historical World Population Trends

Approximate Year	Estimated Population	
1 million B.C.	125	thousand
8000 B.C.	5.3	million
4000 B.C.	86.5	million
1 A.D.	200	million
1000 A.D.	275	million
1650 A.D.	545	million
1750 A.D.	728	million
1839 A.D.	1	billion
1850 A.D.	1.17	billion
1900 A.D.	1.55	billion
1930 A.D.	2.0	billion
1950 A.D.	2.5	billion
1960 A.D.	3.0	billion
1970 A.D.	3.6	billion
1976 A.D.	4.0	billion
1982 A.D.	4.5	billion
1995 A.D.	5.72	billion
PROJECTED		
1998 A.D.	6.0	billion
2025 A.D.	8.5	billion
2050 A.D.	10	billion

Sources: Pytlik, 1985, p. 148; Wright, 1995, p. 356

QUESTIONS

1. What are the slowest periods of population gain? What are the most rapid?

2. Using these population figures and those leading to current and future growth, what do you feel are projections for world food supply, economic growth, political stability, and quality of life?

3. Apply your conclusions to the Malthus theorem and the demographic transition theory. Do your conclusions change when you consider these viewpoints? What validity do you think these arguments hold?

4. What will happen after 2050 in terms of population growth? What does the demographic theory project will happen? What does the Malthus theory project will happen with these exponential projections? What theory do you have regarding the future of population growth?

5. What do you project the population numbers to be in the year 3000 A.D.? Give a planned and supported explanation of your projection.

POPULATION VOCABULARY EXERCISE

As you proceed through this part and your reading and research of various issues associated with population topics, keep a listing of terms and organizations dealing with population that you would like to know more about. Use various resources to define and learn more about them, including other texts, articles, encyclopedias, and the Internet (refer to the Internet exercise at the end of this part).

UN World Projection Plan
Permission Granted by the United
Nations Population Fund.

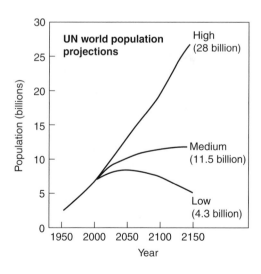

WORLD POPULATION CONTINUES TO SOAR

Concern for the exponentially increasing world population is becoming more widespread. One of the most important questions in the world today is whether human beings are wise enough to see what is coming. Today's global population cannot be sustained at the current lifestyle level of the United States, Europe, or Japan. Over the past couple of centuries, the following statements were issued about population growth:

- Around 200 years ago (1798), Englishman Thomas Malthus stated, "Population, when unchecked, increases in a geometrical ratio. Subsistence only increases in an arithmetical ratio."

- About a century later, Charles Darwin said, "Man tends to increase at a greater rate than his means of subsistence; consequently he is occasionally subjected to a severe struggle for existence."

- Nearly 100 years after that (1968), Paul Ehrlich spoke of population calamity in *The Population Bomb.*

Other Comments on the Issue of Continued Increasing Growth and the Resulting Decreasing World Stability

- Joel E. Cohen, in "How Many People Can Earth Hold?", states, "According to every plausible calculation that's ever been done, Earth could not feed even the 695 billion people that the U.N. projected for 2150 if present fertility rates were to continue" (Cohen, p. 100). He quotes Princeton demographer Ansely Coale's conclusions: "Every demographer knows that we cannot continue a positive rate of increase indefinitely. The inexorable arithmetic of compound interest leads us to absurd conditions within a calculable period of time. Logically we must, and in fact we will, have a rate of growth very close to zero in the long run" (Cohen, p. 100).

- Lester R. Brown and Hal Kane, in their book *Full House,* explain the following: "The demands of the 90 million people added each year for grain and seafood are being satisfied by reducing consumption among those already here. This is a new situation, one that puts population policy in a new light . . . [This population expansion rate] can devastate local life-support systems" (Brown and Kane, pp. 49–50).

Debate on this issue by nations and their people heatedly continues as the world's population continues to leap geometrically from 2.5 billion in 1950 to 5.6 billion in the 1990s. It is projected that there will be 6 billion people by 2000 and perhaps 9 billion by 2025 or 2030. Overpopulation is one of the chief obstacles to universal well-being. A 1994 Cornell University study concluded that the world can only support 2 billion people at the current standard of living now experienced by the industrialized nations of the world. In 1992, 1,600 prominent world scientists, including half of the living science Nobel laureates, presented a declaration to the world's leaders stating that the continuation of destructive human activities "may so alter the living world that it will be unable to sustain life in the manner that we know" (Brown and Kane, p. 30).

The current global growth pattern adds about the equivalent of Mexico's population (92,202,200 people) to the world every year. The U.S. population climbed to 261.7 million in 1994. The annual year-end estimate shows a gain of 2.5 million people in 1994, for a total of 261,653,497 people. The growth includes a natural increase of 1.7 million people—4 million births minus 2.5 million deaths—plus 733,000 additional people due to immigration and the return of 84,000 Americans from abroad.

REFERENCES

(1) Brown, L.R., and Kane, H. (1994). *Full house: Reassessing the Earth's Population Carrying Capacity.* New York: W.W. Norton & Company.
(2) Cohen, J.E. (1992, November). How many people can earth hold? *Discover.* Vol. 13, pp. 114–120.

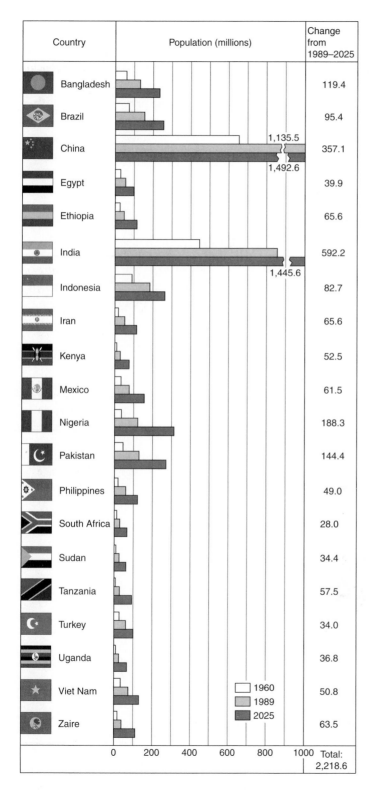

Country	Population (millions)	Change from 1989–2025
Bangladesh		119.4
Brazil		95.4
China	1,135.5 / 1,492.6	357.1
Egypt		39.9
Ethiopia		65.6
India	1,445.6	592.2
Indonesia		82.7
Iran		65.6
Kenya		52.5
Mexico		61.5
Nigeria		188.3
Pakistan		144.4
Philippines		49.0
South Africa		28.0
Sudan		34.4
Tanzania		57.5
Turkey		34.0
Uganda		36.8
Viet Nam		50.8
Zaire		63.5
	0 200 400 600 800 1000	Total: 2,218.6

Legend: 1960, 1989, 2025

Figure P5.1
High-growth Countries.
Seventy Percent of the Projected
Increase in World Population by the
Year 2025 will Occur in these 20 Less
Developed Countries. The Data are
from the United Nations Department of
International Economic and Social
Affairs
Reprinted with Permission from Ian Worpole.

QUESTIONS

1. Using an almanac or the Internet, find the current population figures for the world and for the United States. Chart the new percentage-growth figures. What are your conclusions regarding the various statements quoted in this box?

2. Do you agree with the statements quoted in this box? Why or why not? Build your position on the figures of population growth.

3. Consult world population growth charts. Where are the highest percentages of population growth occurring? What are some of your conclusions about food resources in these areas?

Table P5.1
Annual Rates of Birth, Death and Growth of the Human Population Since 1950

5-Year Period Beginning in Year	Annual Growth Rate (Percent)	Annual Birth Rate (Percent)	Annual Death Rate (Percent)
1950	1.79	3.7	2.0
1955	1.86	3.6	1.7
1960	1.99	3.5	1.5
1965	2.06	3.4	1.3
1970	1.96	3.2	1.2
1975	1.73	2.8	1.1
1980	1.74	2.8	1.0
1985	1.74	2.7	0.98

Source: United Nations Population Fund.

Table P5.2

Percentage Distribution of the World's Population, by Major Area, 1950–2050

Area	1950	1990	2050
Developed countries	29.9%	20.5%	12.4%
Europe	15.6	9.4	4.9
North America	6.6	5.2	3.3
Oceania	0.5	0.5	0.4
USSR[1]	7.2	5.4	3.8
Developing countries	70.1	79.5	87.6
Africa	8.8	12.1	22.6
Latin America	6.6	8.5	9.2
China	22.1	21.5	15.2
India	14.2	16.1	17.0
Other Asia	18.4	21.2	23.7

[1]Republics of Former Soviet Union.
Source: United Nations Department of International Economic and Social Affairs, *Long-range World Population Projections, 1950–2150* (1992).

Table P5.3

Estimated and Projected Population of Major World Areas, 1950–2050 (in millions)

Area	1950	1990	2000	2025	2050
World	**2,518**	**5,292**	**6,261**	**8,504**	**10,019**
Developed	752	1,089	1,143	1,237	1,233
Europe	393	498	510	515	486
Northern America	166	276	295	332	326
Oceania	13	26	30	38	41
USSR[1]	180	289	308	352	380
Developing	1,766	4,203	5,118	7,267	8,786
Africa	222	642	867	1,597	2,265
Latin America	166	448	538	757	922
China	555	1,139	1,299	1,513	1,521
India	358	853	1,042	1,442	1,699
Other Asia	465	1,121	1,372	1,958	2,379

[1]Republics of former Soviet Union.
Source: United Nations Department of International Economic and Social Affairs, *Long-Range World Population Projections, 1950–2150* (1992).

Table P5.4

The World's Largest Urban Areas, Ranked by 1990 Estimated Population (in millions)

The following chart is arranged according to urban areas that had populations of more than three million in 1990. An urban area is a central city, or several cities, and the surrounding urbanized areas, also called a metropolitan area.

Urban Area	Country	1950	1970	1990	2000[1]
1. Mexico City	Mexico	3.1	9.4	20.2	25.6
2. Tokyo	Japan	6.7	14.9	18.1	19.0
3. São Paulo	Brazil	2.4	8.1	17.4	22.1
4. New York	United States	12.3	16.2	16.2	16.8
5. Shanghai	China	5.3	11.2	13.4	17.0
6. Los Angeles	United States	4.0	8.4	11.9	13.9
7. Calcutta	India	4.4	6.9	11.8	15.7
8. Buenos Aires	Argentina	5.0	8.4	11.5	12.9
9. Bombay	India	2.9	5.8	11.2	15.4
10. Seoul	South Korea	1.0	5.3	11.0	12.7
11. Beijing	China	3.9	8.1	10.8	14.0
12. Rio de Janeiro	Brazil	2.9	7.0	10.7	12.5
13. Tianjin	China	2.4	5.2	9.4	12.7
14. Jakarta	Indonesia	2.0	3.9	9.3	13.7
15. Cairo	Egypt	2.4	5.3	9.0	11.8
16. Moscow	Russia	4.8	7.1	8.8	9.0
17. Delhi	India	1.4	3.5	8.8	13.2
18. Metro Manila	Philippines	1.5	3.5	8.5	11.8
19. Osaka	Japan	3.8	7.6	8.5	8.6
20. Paris	France	5.4	8.3	8.5	8.6
21. Karachi	Pakistan	1.0	3.1	7.7	11.7
22. Lagos	Nigeria	0.3	2.0	7.7	12.9
23. London	United Kingdom	8.7	8.6	7.4	13.9
24. Bangkok	Thailand	1.4	3.1	7.2	10.3
25. Chicago	United States	4.9	6.7	7.0	7.3
26. Teheran	Iran	1.0	3.3	6.8	8.5
27. Istanbul	Turkey	1.1	2.8	6.7	9.5
28. Dhaka	Bangladesh	0.4	1.5	6.6	12.2
29. Lima	Peru	1.0	2.9	6.2	8.2
30. Madras	India	1.4	3.0	5.7	7.8
31. Hong Kong	Hong Kong	1.7	3.4	5.4	6.1
32. Milan	Italy	3.6	5.5	5.3	5.4
33. Madrid	Spain	1.6	3.4	5.2	5.9
34. St. Petersburg	Russia	2.6	4.0	5.1	5.4
35. Bangalore	India	0.8	1.6	5.0	8.2

Table P5.4
The World's Largest Urban Areas, Ranked by 1990 Estimated Population (in millions)

Urban Area	Country	1950	1970	1990	2000[1]
36. Bogotá	Colombia	0.6	2.4	4.9	6.4
37. Shenyang	China	2.1	3.5	4.8	6.3
38. Philadelphia	United States	2.9	4.0	4.3	4.5
39. Caracas	Venezuela	0.7	2.0	4.1	5.2
40. Baghdad	Iraq	0.6	2.0	4.0	5.1
41. Lahore	Pakistan	0.8	2.0	4.1	6.0
42. Wuhan	China	1.2	2.7	3.9	5.3
43. Alexandria	Egypt	1.0	2.0	3.7	5.1
44. Detroit	United States	2.8	4.0	3.7	3.7
45. Guangzhou	China	1.3	3.0	3.7	4.8
46. San Francisco	United States	2.0	3.0	3.7	4.1
47. Ahmedabad	India	0.9	1.7	3.6	5.3
48. Belo Horizonte	Brazil	0.4	1.6	3.6	4.7
49. Naples	Italy	2.8	3.6	3.6	3.6
50. Hyderabad	India	1.1	1.7	3.5	5.0
51. Kinshasa	Zaire	0.2	1.4	3.5	5.5
52. Toronto	Canada	1.0	2.8	3.5	3.9
53. Athens	Greece	1.8	2.5	3.4	3.8
54. Barcelona	Spain	1.6	2.7	3.4	3.7
55. Dallas	United States	0.9	2.0	3.4	4.4
56. Katowice	Poland	1.7	2.8	3.4	3.7
57. Sydney	Australia	1.7	2.7	3.4	3.7
58. Yangon	Myanmar	0.7	1.4	3.3	4.7
59. Casablanca	Morocco	0.7	1.5	3.2	4.6
60. Guadalajara	Mexico	0.4	1.5	3.2	4.1
61. Ho Chi Minh City	Vietnam	0.9	2.0	3.2	4.1
62. Chongqing	China	1.7	2.3	3.1	4.2
63. Porto Alegre	Brazil	0.4	1.5	3.1	3.9
64. Rome	Italy	1.6	2.9	3.1	3.1
65. Algiers	Algeria	0.4	1.3	3.0	4.5
66. Chengou	China	0.7	1.8	3.0	4.1
67. Harbin	China	1.0	2.1	3.0	3.9
68. Houston	United States	0.7	1.7	3.0	3.6
69. Monterrey	Mexico	0.4	1.2	3.0	3.9
70. Montreal	Canada	1.3	2.4	3.0	3.1
71. Taipei	China	0.6	1.8	3.0	4.2

[1]Projected figures.
Source: United Nations Department of International Economic and Social Affairs, *World Urbanization Projections 1990* (1991).

Table P5.5
World Births, Deaths, and Population Growth, 1994

Characteristic	World	Developed	Developing
Population	5,642,151,000	1,240,354,000	4,401,797,000
Births	139,324,000	16,944,000	122,380,000
Deaths	52,514,000	11,715,000	40,799,000
Natural increase	86,810,000	5,229,000	81,582,000
Births per 1,000 population	25	14	28
Deaths per 1,000 population	9	9	9
Growth rate (percent)	1.5%	0.4%	1.9%

Source: U.S. Bureau of the Census. *World Population Profile 1994* (1994).

Table P5.6
Nations With Highest and Lowest Fertility Rates, 1995 Estimates

	HIGHEST FERTILITY RATES			LOWEST FERTILITY RATES			
Country	Fertility Rate per Woman	Country	Fertility Rate per Woman	Country	Fertility Rate per Woman	Country	Fertility Rate per Woman
Yemen	7.4	Ethiopia	6.8	Hong Kong	1.2	Slovenia	1.5
Niger	7.3	Guinea	6.8	Spain	1.2	Austria	1.6
Ivory Coast	7.1	Somalia	6.8	Italy	1.3	Belgium	1.6
Uganda	7.1	Afghanistan	6.6	Germany	1.3	Estonia	1.6
Angola	6.9	Liberia	6.6	Greece	1.5	Latvia	1.6
Benin	6.9	Burundi	6.5	Bulgaria	1.5	Netherlands	1.6
Mai	6.9	Zaire	6.5	Japan	1.5	Portugal	1.6
Malawi	6.9	Laos	6.4	Romania	1.5	Switzerland	1.6
Oman	6.9			Russia	1.5	Ukraine	1.6

Source: United Nations Fund for Population Activities (UNFPA). *The State of World Population 1995* (1995).

Table P5.7

Population Indicators by Region and Nation

| Region/Country | Population Estimate ('000s) | | Birth Rate Per 1,000 | Death Rate Per 1,000 | Life Expec- tancy | Percent Urban | Fertility Rate Per Woman |
	1995	2025	1990–95	1990–95	1990–95	1995	1995
World total	**5,716,400**	**8,294,300**	**26**	**9**	**65**	**45%**	**3.0**
More developed regions	1,166,600	1,238,400	14	10	75	75	1.7
Less developed regions	4,549,800	7,055,900	29	9	62	38	3.4
Least developed countries	575,400	1,162,300				22	5.6
AFRICA	**728,100**	**1,495,800**	**43**	**14**	**53**	**34**	**5.6**
Eastern Africa	**227,100**	**494,600**	**48**	**16**	**49**	**22**	**6.2**
Burundi	6,400	13,500	46	17	48	8	6.5
Eritrea	3,500	7,000				17	5.6
Ethiopia	55,100	126,900	49	18	47	13	6.8
Kenya	28,300	63,400	44	10	59	28	6.0
Madagascar	14,800	34,400	45	13	55	27	5.9
Malawi	11,100	22,300	54	21	44	14	6.9
Mauritius	1,100	1,500	18	7	70	41	2.3
Mozambique	16,000	35,100	45	18	47	34	6.3
Rwanda	8,000	15,800	52	18	46	6	6.3
Somalia	9,300	21,300	50	19	47	26	6.8
Tanzania	29,700	62,900	48	15	51	24	5.7
Uganda	21,300	48,100	51	21	42	13	7.1
Zambia	9,500	19,100	46	18	44	43	5.7
Zimbabwe	11,300	19,600	41	11	56	32	4.8
Middle Africa	**82,300**	**189,100**	**46**	**15**	**51**	**33**	**6.2**
Angola	11,100	26,600	51	19	46	32	6.9
Cameroon	13,200	29,200	41	12	56	45	5.5
Central African Republic	3,300	6,400	44	18	47	39	5.5
Chad	6,400	12,900	44	18	48	21	5.7
Congo	2,600	5,700	45	15	52	59	6.0
Gabon	1,300	2,700	43	16	54	50	5.5
Zaire	43,900	104,600	47	15	52	29	6.5
Northern Africa	**160,600**	**268,600**	**34**	**9**	**61**	**46**	**4.0**
Algeria	27,900	45,500	34	7	66	56	3.6
Egypt	62,900	97,300	31	9	62	45	3.7
Libya	5,400	12,900	42	8	63	86	6.2
Morocco	27,000	40,700	32	8	63	48	3.4
Sudan	28,100	58,400	42	14	52	25	5.6
Tunisia	8,900	13,300	27	6	68	57	3.0
Southern Africa	**47,400**	**82,800**	**32**	**9**	**63**	**48**	**4.1**
Botswana	1,500	3,000	38	9	61	28	4.7
Lesotho	2,100	4,200	34	10	61	23	5.0
Namibia	1,500	3,000	43	11	59	37	5.0
South Africa	41,500	71,000	31	9	63	51	4.0

Table P5.7
Population Indicators by Region and Nation, *continued*

Region/Country	Population Estimate ('000s)		Birth Rate Per 1,000 1990–95	Death Rate Per 1,000 1990–95	Life Expec- tancy 1990–95	Percent Urban 1995	Fertility Rate Per Woman 1995
	1995	*2025*	*1990–95*	*1990–95*	*1990–95*	*1995*	*1995*
Western Africa	**210,700**	**460,600**	**46**	**15**	**51**	**37%**	**6.3**
Benin	5,400	12,300	49	18	46	31	6.9
Burkina Faso	10,300	21,700	47	18	48	27	6.3
Ghana	17,500	38,000	42	12	56	36	5.8
Guinea	6,700	15,100	51	20	45	30	6.8
Guinea-Bissau	1,100	2,000	43	21	44	22	5.6
Ivory Coast	14,300	36,800	50	15	52	44	7.1
Liberia	3,000	7,200	47	14	55	45	6.6
Mali	10,800	24,600	51	19	46	27	6.9
Mauritania	2,300	4,400	46	18	48	54	5.2
Niger	9,200	22,400	51	19	47	17	7.3
Nigeria	111,700	238,400	45	14	53	39	6.2
Senegal	8,300	16,900	43	16	49	42	5.8
Sierra Leone	4,500	8,700	48	22	43	36	6.3
Togo	4,100	9,400	45	13	55	31	6.3
LATIN AMERICA	**482,000**	**709,800**	**26**	**7**	**68**	**74**	**3.0**
Caribbean	**35,800**	**49,600**	**24**	**8**	**69**	**62**	**2.7**
Cuba	11,100	12,700	17	7	76	76	1.8
Dominican Republic	7,800	11,200	28	6	68	65	2.9
Haiti	7,200	13,100	35	12	57	32	4.7
Jamaica	2,400	3,300	22	6	74	54	2.2
Puerto Rico	3,700	4,600	18	7	75	73	2.1
Trinidad and Tobago	1,300	1,800	23	6	71	72	2.3
Central America	**126,400**	**197,500**	**30**	**6**	**69**	**68**	**3.3**
Belize	200	400	—	—	—	47	3.9
Costa Rica	3,400	5,600	26	4	76	50	3.1
El Salvador	5,800	9,700	33	7	66	45	3.8
Guatemala	10,600	21,700	39	8	65	42	5.1
Honduras	5,700	10,700	37	7	66	44	4.6
Mexico	93,700	136,600	28	5	70	75	3.0
Nicaragua	4,400	9,100	40	7	67	63	4.8
Panama	2,600	3,800	25	5	73	53	2.8
South America	**319,800**	**462,700**	**24**	**7**	**67**	**78**	**2.9**
Argentina	34,600	46,100	20	9	71	88	2.7
Bolivia	7,400	13,100	34	9	61	61	4.6
Brazil	161,800	230,300	23	7	66	78	2.8
Chile	14,300	19,800	23	6	72	84	2.5
Colombia	35,100	49,400	24	6	69	73	2.6
Ecuador	11,500	17,800	30	7	67	58	3.3

Table P5.7
Population Indicators by Region and Nation

Region/Country	Population Estimate ('000s)		Birth Rate Per 1,000	Death Rate Per 1,000	Life Expec-tancy	Percent Urban	Fertility Rate Per Woman
	1995	2025	1990–95	1990–95	1990–95	1995	1995
Paraguay	5,000	9,000	33	6	67	53%	4.1
Peru	23,800	36,700	8	65	71	72	3.3
Uruguay	3,200	3,700	17	10	72	90	2.3
Venezuela	21,800	34,800	26	5	70	93	3.1
NORTHERN AMERICA	**292,800**	**369,600**	**16**	**9**	**76**	**76**	**2.0**
Canada	29,500	38,300	14	8	77	77	1.9
United States	263,300	331,200	16	9	76	76	2.1
ASIA	**3,458,000**	**4,960,000**	**26**	**8**	**65**	**35**	**3.0**
Eastern Asia	**1,424,200**	**1,745,800**	**20**	**7**	**72**	**37**	**1.9**
China	1,221,500	1,526,100	21	7	71	30	2.0
Hong Kong	5,900	5,900	13	6	78	95	1.2
Japan	125,100	121,600	11	7	79	78	1.5
Korea, North	23,900	33,400	24	5	71	61	2.3
Korea, South	45,000	54,400	16	6	71	81	1.8
Mongolia	2,400	3,800	34	8	64	61	3.4
South-eastern Asia	**484,300**	**713,400**	**28**	**8**	**63**	**34**	**3.2**
Cambodia	10,300	19,700	39	14	51	21	5.1
Indonesia	197,600	275,600	27	8	63	35	2.8
Laos	4,900	9,700	45	15	51	22	6.4
Malaysia	20,100	31,600	29	5	71	54	3.4
Myanmar	46,500	75,600	33	11	58	26	4.0
Philippines	67,600	104,500	30	7	65	54	3.8
Singapore	2,800	3,400	16	6	74	100	1.7
Thailand	58,800	73,600	21	6	69	20	2.1
Vietnam	74,500	118,200	29	9	64	21	3.7
South Central Asia	**1,381,200**	**2,196,300**	**32**	**10**	**59**	**29**	**3.9**
Afghanistan	20,100	45,300	53	22	43	20	6.6
Bangladesh	120,400	196,100	38	14	53	18	4.1
Bhutan	1,600	3,100	40	17	48	6	5.7
India	935,700	1,392,100	29	10	60	27	3.6
Iran	67,300	123,500	40	7	67	59	4.8
Nepal	21,900	40,700	37	13	54	14	5.2
Pakistan	140,500	284,800	41	10	59	35	5.9
Sri Lanka	18,400	25,000	21	6	72	22	2.4
Western Asia	**168,400**	**304,600**	**34**	**7**	**66**	**66**	**4.3**
Iraq	20,400	42,700	39	7	66	75	5.5
Israel	5,600	7,800	21	7	77	91	2.8
Jordan	5,400	12,000	40	5	68	72	5.4
Kuwait	1,500	2,800	28	2	75	97	3.0

Table P5.7

Population Indicators by Region and Nation, *continued*

Region/Country	Population Estimate ('000s)		Birth Rate Per 1,000 1990–95	Death Rate Per 1,000 1990–95	Life Expec- tancy 1990–95	Percent Urban 1995	Fertility Rate Per Woman 1995
	1995	*2025*					
Lebanon	3,000	4,400	27	7	69	87%	2.9
Oman	2,200	6,100	40	5	70	13	6.9
Saudi Arabia	17,900	42,700	36	5	69	80	6.2
Syria	14,700	33,500	42	6	67	52	5.6
Turkey	61,900	90,900	28	7	67	69	3.2
United Arab Emirates	1,900	3,000	21	4	71	84	4.1
Yemen	14,500	33,700	48	14	53	34	7.4
EUROPE	**727,000**	**718,200**	**13**	**11**	**75**	**74**	**1.6**
Eastern Europe	**308,700**	**299,400**	**14**	**11**	**71**	**70**	**1.6**
Bulgaria	8,800	7,800	13	12	72	71	1.5
Czech Republic	10,300	10,600	—	—	—	65	1.8
Hungary	10,100	9,400	12	14	70	65	1.7
Poland	38,400	41,500	14	10	72	65	1.9
Romania	22,800	21,700	16	11	70	55	1.5
Slovakia	5,400	6,000	—	—	—	59	1.9
Northern Europe	**93,500**	**98,600**	**14**	**11**	**76**	**84**	**1.9**
Denmark	5,200	5,100	12	12	76	85	1.7
Estonia	1,500	1,400	14	12	71	73	1.6
Finland	5,100	5,400	13	10	76	63	1.9
Ireland	3,600	3,900	14	9	75	58	2.1
Latvia	2,600	2,300	14	12	71	73	1.6
Lithuania	3,700	3,800	15	10	73	72	1.8
Norway	4,300	4,700	15	11	77	73	2.0
Sweden	8,800	9,800	14	11	78	83	2.1
United Kingdom	58,300	61,500	14	11	76	90	1.8
Southern Europe	**143,900**	**139,300**	**11**	**10**	**76**	**65**	**1.4**
Albania	3,400	4,700	23	5	73	37	2.8
Bosnia and Herzegovina	3,500	4,500	—	—	—	49	1.6
Croatia	4,500	4,200	—	—	—	64	1.7
Greece	10,500	9,900	10	10	78	65	1.4
Italy	57,200	52,300	10	10	77	67	1.3
Macedonia	2,200	2,600	—	—	—	60	2.0
Portugal	9,800	9,700	12	10	75	36	1.6
Slovenia	1,900	1,800	—	—	—	64	1.5
Spain	39,600	37,600	11	9	78	76	1.2
Yugoslavia	10,800	11,500	—	—	—	57	2.0
Western Europe	**180,800**	**180,900**	**12**	**11**	**76**	**81**	**1.5**
Austria	8,000	8,300	12	11	76	56	1.6
Belgium	10,100	10,400	12	11	76	97	1.6

Table P5.7
Population Indicators by Region and Nation, *continued*

Region/Country	Population Estimate ('000s)		Birth Rate Per 1,000 1990–95	Death Rate Per 1,000 1990–95	Life Expectancy 1990–95	Percent Urban 1995	Fertility Rate Per Woman 1995
	1995	2025					
France	58,000	61,200	13	10	77	73%	1.7
Germany	81,600	76,400	11	11	76	87	1.3
Netherlands	15,500	16,300	14	9	77	89	1.6
Switzerland	7,200	7,800	13	10	78	61	1.6
OCEANIA	**28,500**	**41,000**	**19**	**8**	**73**	**70**	**2.5**
Australia-New Zealand	**21,700**	**29,000**	**15**	**8**	**77**	**85**	**1.9**
Australia	18,100	24,700	15	8	77	85	1.9
Melanesia	5,800	10,100	32	9	59	21	4.5
New Zealand	3,600	4,400	17	8	76	86	2.1
Papua New Guinea	4,300	7,500	33	11	56	16	4.8
Countries of the former USSR with economies in transition							
Armenia	3,600	4,700	23	6	71	69	2.5
Azerbaijan	7,600	10,100	27	7	70	56	2.4
Belarus	10,100	9,900	16	10	72	71	1.7
Georgia	5,500	6,100	18	9	72	59	2.1
Kazakhstan	17,100	21,700	24	8	69	60	2.4
Kyrgyzstan	4,700	7,100	31	7	68	39	3.5
Moldova	4,400	5,100	21	10	68	52	2.1
Russian Federation	147,000	138,500	16	11	70	76	1.5
Tajikistan	6,100	11,800	40	7	70	32	4.7
Turkmenistan	4,100	6,700	36	8	65	45	3.8
Ukraine	51,400	48,700	14	12	71	70	1.6
Uzbekistan	22,800	37,700	—	—	—	41	3.7

Note: Totals May Not Add Because of Rounding Numbers. Data for Small Countries or Areas, Generally Those With Population of 200,000 or Less in 1990, Are Not Given in This Table.
Source: United Nations Populations Fund (UNFPA), *The State of World Population 1995* (1995) and *The State of World Population 1994* (1994).

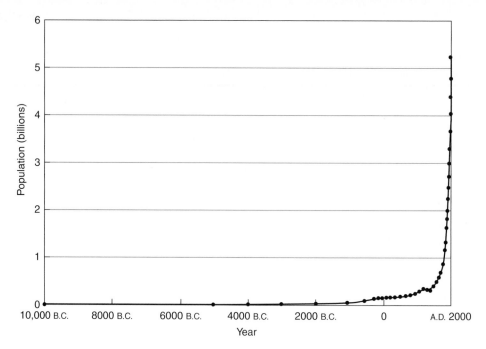

Figure P5.2

Estimated Human Population from The Last Ice Age to The Present.

From *How Many People Can the Earth Support?* by Joel E. Cohen. Copyright © 1995 by Joel E. Cohen. Reprinted by permission of W.W. Norton & Company, Inc.

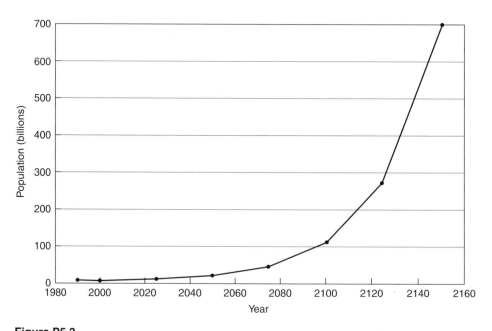

Figure P5.3

United Nations' Projection of World Population, Assuming Fertility Remains Constant at Its 1990 Levels in Different Regions.

Source: Original Figure Drawn According to Data of United Nations (1992a). Permission Granted by the United Nations Population Fund.

Figure P5.4
Where in the World Do They Live?

According to the U.S. Bureau of the Census, 75 of each 100 persons in the world today live in only 22 countries. The other 25 live in any of the remaining 184 countries. Note that 37 people out of every 100 live in China and India.

Of every 100 people in the world in 1991:

21 live in China (mainland)
16 live in India
 5 live in the former Soviet Union
 5 live in the United States
 4 live in Indonesia
 3 live in Brazil
 2 live in Bangladesh
 2 live in Japan
 2 live in Mexico
 2 live in Nigeria
 2 live in Pakistan
 1 lives in Egypt
 1 lives in Ethiopia
 1 lives in France
 1 lives in Germany
 1 lives in Iran
 1 lives in Italy
 1 lives in the Philippines
 1 lives in Thailand
 1 lives in Turkey
 1 lives in the United Kingdom
 1 lives in Vietnam

Source: U.S. Bureau of the Census. *World Population Profile 1991* (1991). United Nations Population Fund.

Figure P5.5
Another World Perspective View

If We Shrank the Earth's Population to 100 People. . .

The following is a non-biased perspective of the world's population. The author is unknown.

If we could shrink the earth's population to a village of precisely 100 people, with all the existing human ratios remaining the same, it would look something like the following:

57 Asians
21 Europeans
14 from the Western Hemisphere, both north and south
8 Africans

52 would be female
48 would be male

70 would be non-white
30 would be white

70 would be non Christian
30 would be Christian

89 would be heterosexual
11 would be homosexual

6 people would possess 59% of the entire world's wealth and all 6 would be from the United States

80 would live in substandard housing

70 would be unable to read

50 would suffer from malnutrition

1 would be near death; 1 would be near birth

1 (yes, only 1) would have a college education

1 would own a computer

"When one considers our world from such a compressed perspective, the need for both acceptance, understanding and education becomes glaringly apparent."

QUESTIONS ABOUT THE TABLES

1. Using the first quote in the introduction, from the U.S. National Academy of Sciences, further explain what is meant by a *crisis point*. Explain this term in factual, empirical terms from the information contained in the introduction.

2. In the history of world population trends, where are the population acceleration points?

3. Table P5.1 seems to indicate that the annual growth rate, birth rates, and death rates are all slowing. Explain why population continues to rise sharply, as indicated in Figure P5.2.

4. Figure P5.3 indicates a U.N. World Projection Plan. Using this figure and other sources, explain what you think is a realistic projection and how it can be attained.

5. Figure P5.1 contains a projection of population growth for various countries. Compare this data with the information in Tables P5.2, P5.3, P5.6, and P5.7 and Figure P5.4. Write a summary of your findings.

6. The box titled "World Population Continues to Soar" states that the world can support only 2 billion people at a universal standard of living now enjoyed by industrialized nations. Interpret this statement from world projections and selected countries' population growth patterns.

7. Discuss the patterns of growth in Tables P5.2 and P5.3.

8. Examine Table P5.4. Relate this information to your findings from Question 7. What countries and areas have the most populated cities? What patterns do you find?

9. Relate Tables P5.5 and P5.6. How are births, deaths, and population growth related to fertility rates? Generalize your findings.

10. Table P5.7 gives important country-by-country profiles. Select 10 countries in various regions and compare their demographic data. What patterns do you find?

11. Consulting Table P5.7, locate the 10 countries with the highest life expectancy and the 10 countries with the lowest life expectancy. What significant patterns do you find?

12. Summarize your findings from Questions 5–11 into well-written paragraphs that strongly support your conclusions.

30

The Great Leap Forward

JARED DIAMOND

World population figures around 1 A.D have been estimated at about 200 million people. One million years prior to 1 A.D., population figures of early humankind of Homo sapien or Homo erectus have been estimated at about 125 thousand. In order for us to understand and appreciate the very long path of this development and growth of humankind, it is helpful to discuss early hominoid history and its development, expansion, explosion along with a discussion of the relationship of tools, concepts and technology to that expansion. The Great Leap Forward gives some perspectives on the early struggles of humankind and exponential growth toward modern creativity, expansion, and technology.

One can hardly blame nineteenth-century creationists for insisting that humans were separately created by God. After all, between us and other animal species lies the seemingly unbridgeable gulf of language, art, religion, writing, and complex machines. Small wonder, then, that to many people Darwin's theory of our evolution from apes appeared absurd.

Since Darwin's time, of course, fossilized bones of hundreds of creatures intermediate between apes and modern humans have been discovered. It is no longer possible for a reasonable person to deny that what once seemed absurd actually happened—somehow. Yet the discoveries of many missing links have only made the problem more fascinating without fully solving it. When and how did we acquire our uniquely human characteristics?

We know that our lineage arose in Africa, diverging from that of chimpanzees and gorillas sometime between 6 million and 10 million years ago. For most of the time since then we have been little more than glorified baboons. As recently as 35,000 years ago western Europe was still occupied by Neanderthals, primitive beings for whom art and progress scarcely existed. Then there was an abrupt change. Anatomically modern people appeared in Europe, and suddenly so did sculpture, musical instruments, lamps,

trade, and innovation. Within a few thousand years the Neanderthals were gone.

Insofar as there was any single moment when we could be said to have become human, it was at the time of this Great Leap Forward 35,000 years ago. Only a few more dozen millennia—a trivial fraction of our 6-to-10 million-year history—were needed for us to domesticate animals, develop agriculture and metallurgy, and invent writing. It was then but a short further step to those monuments of civilization that distinguish us from all other animals—monuments such as the *Mona Lisa* and the Ninth Symphony, the Eiffel Tower and Sputnik, Dachau's ovens and the bombing of Dresden.

What happened at that magic moment in evolution? What made it possible, and why was it so sudden? What held back the Neanderthals, and what was their fate? Did Neanderthals and modern peoples ever meet, and if so, how did they behave toward each other? We still share 98 percent of our genes with chimps; which genes among the other 2 percent had such enormous consequences?

Understanding the Great Leap Forward isn't easy; neither is writing about it. The immediate evidence comes from technical details of preserved bones and stone tools. Archeologists' reports are full of such terms as "transverse occipital torus," "receding zygomatic arches," and "Chatelperronian backed knives." What we really want to understand—the way of life and the humanity of our vari-

---

Jared Diamond/© 1989. Reprinted with permission of *Discover Magazine.*

256

ous ancestors—isn't directly preserved but only inferred from those technical details. Much of the evidence is missing, and archeologists often disagree over the meaning of the evidence that has survived.

I'll emphasize those inferences rather than the technical details, and I'll speculate about the answers to those questions I just listed above. But you can form your own opinions, and they may differ from mine. This is a puzzle whose solution is still unknown.

To set the stage quickly, recall that life originated on Earth several billion years ago, the dinosaurs became extinct around 65 million years ago, and, as I mentioned, our ancestors diverged from the ancestors of chimps and gorillas between 6 and 10 million years ago. They then remained confined to Africa for millions of years.

Initially, our ancestors would have been classified as merely another species of ape, but a sequence of three changes launched them in the direction of modern humans. The first of these changes occurred by around 4 million years ago: the structure of fossilized limb bones shows that our ancestors, in contrast to gorillas and chimps, were habitually walking upright. The upright posture freed our forelimbs to do other things, among which toolmaking would eventually prove to be the most important.

The second change occurred around 3 million years ago, when our lineage split in two. As background, remember that members of two animal species living in the same area must fill different ecological roles and do not normally interbreed. For example, coyotes and wolves are obviously closely related and, until wolves were exterminated in most of the United States, lived in many of the same areas. However, wolves are larger, they usually hunt big mammals like deer and moose, and they often live in sizable packs, whereas coyotes are smaller, mainly hunt small mammals like rabbits and mice, and normally live in pairs or small groups.

Now, all modern humans unquestionably belong to the same species. Ecological differences among us are entirely a product of childhood education: it is not the case that some of us are born big and habitually hunt deer while others are born small, gather berries, and don't marry the deer hunters. And every human population living today has interbred with every other human population with which it has had extensive contact.

Three million years ago, however, there were hominid species as distinct as wolves and coyotes. On one branch of the family tree was a man-ape with a heavily built skull and very big cheek teeth, who probably ate coarse plant food; he has come to be known as *Australopithecus robus-*

tus (the "robust southern ape"). On the other branch was a man-ape with a more lightly built skull and smaller teeth, who most likely had an omnivorous diet; he is known as *Australopithecus africanus* (the "southern ape of Africa"). Our lineage may have experienced such a radical division at least once more, at the time of the Great Leap Forward. But the description of that event will have to wait.

There is considerable disagreement over just what occurred in the next million years, but the argument I find most persuasive is that *A. africanus* evolved into the larger-brained form we call *Homo habilis* ("man the handyman").

Complicating the issue is that fossil bones often attributed to *H. habilis* differ so much in skull size and tooth size that they may actually imply another fork in our lineage yielding two distinct *habilis*-like species: *H. habilis* himself and a mysterious "Third Man." Thus, by 2 million years ago there were at least two and possibly three protohuman species.

The third and last of the big changes that began to make our ancestors more human and less apelike was the regular use of stone tools. By around 2.5 million years ago very crude stone tools appear in large numbers in areas of East Africa occupied by the protohumans. Since there were two or three protohuman species, who made the tools? Probably the light-skulled species, since both it and the tools persisted and evolved. (There is, however, the intriguing possibility that at least some of our robust relatives also made tools, as recent anatomical analyses of hand bones from the Swartkrans cave in South Africa suggest. See "The Gripping Story of Paranthropus," by Pat Shipman, in last month's issue.)

With only one human species surviving today but two or three a few million years ago, it's clear that one or two species must have become extinct. Who was our ancestor, which species ended up instead as a discard in the trash heap of evolution, and when did this shakedown occur?

The winner was the light-skulled *H. habilis,* who went on to increase in brain size and body size. By around 1.7 million years ago the differences were sufficient that anthropologists give our lineage the new name *Homo erectus* ("the man who walks upright"—*H. erectus* fossils were discovered before all the earlier ones, so anthropologists didn't realize that *H. erectus* wasn't the first protohuman to walk upright). The robust man-ape disappeared somewhat after 1.2 million years ago, and the Third Man (if he ever existed) must have disappeared by then also.

As for why *H. erectus* survived and *A. robustus* didn't, we can only speculate. A plausible guess is that the robust man-ape could no longer compete: *H. erectus* ate both meat and plant food, and his larger brain may have made him more efficient at getting the food on which *A. robustus* depended. It's also possible that *H. erectus* gave his robust brother a direct push into oblivion by killing him for meat.

The shakedown left *H. erectus* as the sole protohuman player on the African stage, a stage to which our closest living relatives (the chimp and gorilla) are still confined. But around one million years ago *H. erectus* began to expand his horizons. His stone tools and bones show that he reached the Near East, then the Far East (where he is represented by the famous fossils known as Peking man and Java man) and Europe. He continued to evolve in our direction by an increase in brain size and in skull roundness. By around 500,000 years ago some of our ancestors looked sufficiently like us, and sufficiently different from earlier *H. erectus,* to be classified as our own species, *Homo sapiens* (the "wise man"), although they still had thicker skulls and brow ridges than we do today.

Was our meteoric ascent to *sapiens* status half a million years ago the brilliant climax of Earth history, when art and sophisticated technology finally burst upon our previously dull planet? Not at all: the appearance of *H. sapiens* was a non-event. The Great Leap Forward, as proclaimed by cave paintings, houses, and bows and arrows, still lay hundreds of thousands of years in the future. Stone tools continued to be the crude ones that *H. erectus* had been making for nearly a million years. The extra brain size of those early *H. sapiens* had no dramatic effect on their way of life. That whole long tenure of *H. erectus* and early *H. sapiens* outside Africa was a period of infinitesimally slow cultural change.

So what was life like during the 1.5 million years that spanned the emergence of *H. erectus* and *H. sapiens?* The only surviving tools from this period are stone implements that can, charitably, be described as very crude. Early stone tools do vary in size and shape, and archeologists have used those differences to give the tools different names, such as hand-ax, chopper, and cleaver. But these names conceal the fact that none of these early tools had a sufficiently consistent or distinctive shape to suggest any specific function. Wear marks on the tools show that they were variously used to cut meat, bone, hides, wood, and nonwoody parts of plants. But any size or shape tool seems to have been used to cut any of these things, and the categories imposed by archeologists may be little more than arbitrary divisions of a continuum of stone forms.

Negative evidence is also significant. All the early stone tools may have been held directly in the hand; they show no signs of being mounted on other materials for increased leverage, as we mount steel ax blades on wooden handles. There were no bone tools, no ropes to make nets, and no fishhooks.

What food did our early ancestors get with those crude tools, and how did they get it? To address this question, anthropology textbooks usually insert a long chapter entitled something like "Man the Hunter." The point they make is that baboons, chimps, and some other primates prey on small vertebrates only occasionally, but recently surviving Stone Age people (like Bushmen) did a lot of big-game hunting. There's no doubt that our early ancestors also ate some meat. The question is, how much meat? Did big-game hunting skills improve gradually over the past 1.5 million years, or was it only since the Great Leap Forward—a mere 35,000 years ago—that they made a large contribution to our diet?

Anthropologists routinely reply that we've long been successful big-game hunters, but in fact there is no good evidence of hunting skills until around 100,000 years ago, and it's clear that even then humans were still very ineffective hunters. So it's reasonable to assume that earlier hunters were even more ineffective.

Yet the mystique of Man the Hunter is now so rooted in us that it's hard to abandon our belief in its long-standing importance. Supposedly, big-game hunting was what induced protohuman males to cooperate with one another, develop language and big brains, join into bands, and share food. Even women were supposedly molded by big-game hunting: they suppressed the external signs of monthly ovulation that are so conspicuous in chimps, so as not to drive men into a frenzy of sexual competition and thereby spoil men's cooperation at hunting.

But studies of modern hunter gatherers, with far more effective weapons than those of early *H. sapiens,* show that most of a family's calories come from plant food gathered by women. Men catch rats and other small game never mentioned in their heroic campfire stories. Occasionally they get a large animal, which does indeed contribute significantly to protein intake. But it's only in the Arctic, where little plant food is available, that big-game hunting becomes the dominant food source. And humans didn't reach the Arctic until around 30,000 years ago.

So I would guess that big-game hunting contributed little to our food intake until after we had evolved fully modern anatomy and behavior. I doubt the usual view that hunting was the driving force behind our uniquely human brain and societies. For most of our history we were not mighty hunters but rather sophisticated baboons.

To return to our history: *H. sapiens,* you'll recall, took center stage around half a million years ago in Africa, the Near East, the Far East, and Europe. By 100,000 years ago humans had settled into at least three distinct populations occupying different parts of the Old World. These were the last truly primitive people. Let's consider among them those whose anatomy is best known, those who have become a metaphor for brutishness: the Neanderthals.

Where and when did they live? Their name comes from Germany's Neander Valley, where one of the first skeletons was discovered (in German, *thal*—nowadays spelled *tal*—means "valley"). Their geographic range extended from western Europe, through southern European Russia and the Near East, to Uzbekistan in Central Asia, near the border of Afghanistan. As to the time of their origin, that's a matter of definition, since some old skulls have characteristics anticipating later "full-blown" Neanderthals. The earliest full-blown examples date from around 130,000 years ago, and most specimens postdate 74,000 years ago. While their start is thus arbitrary, their end is abrupt: the last Neanderthals died around 32,000 years ago.

During the time that Neanderthals flourished, Europe and Asia were in the grip of the last ice age. Hence Neanderthals must have been a cold-adapted people—but only within limits. They got no farther north than southern Britain, northern Germany, Kiev, and the Caspian Sea.

Neanderthals' head anatomy was so distinctive that, even if a Neanderthal dressed in a business suit or a designer dress were to walk down the street today, all you *H. sapiens* would be staring in shock. Imagine converting a modern face to soft clay, gripping the middle of the face from the bridge of the nose to the jaws, pulling the whole mid-face forward, and letting it harden again. You'll then have some idea of a Neanderthal's appearance. Their eyebrows rested on prominently bulging bony ridges, and their nose and jaws and teeth protruded far forward. Their eyes lay in deep sockets, sunk behind the protruding nose and brow ridges. Their foreheads were low and sloping, unlike our high vertical modern foreheads, and their lower jaws sloped back without a chin. Yet despite these startlingly primitive features, Neanderthals' brain size was nearly 10 percent greater than ours! (This does not mean they were smarter than us; they obviously weren't. Perhaps their larger brains simply weren't "wired" as well.) A dentist who examined a Neanderthal's teeth would have been in for a further shock. In adult Neanderthals front teeth were worn down on the outer surface, in a way found in no modern people. Evidently this peculiar wear pattern resulted from their using their teeth as tools, but what

exactly did they do? As one possibility, they may have routinely used their teeth like a vise, as my baby sons do when they grip a milk bottle in their teeth and run around with their hands free. Alternatively, Neanderthals may have bitten hides to make leather or wood to make tools.

While a Neanderthal in a business suit or a dress would attract your attention, one in shorts or a bikini would be even more startling. Neanderthals were more heavily muscled, especially in their shoulders and neck, than all but the most avid bodybuilders. Their limb bones, which took the force of those big muscles contracting, had to be considerably thicker than ours to withstand the stress. Their arms and legs would have looked stubby to us, because the lower leg and forearm were relatively shorter than ours. Even their hands were much more powerful than ours; a Neanderthal's handshake would have been bone-crushing. While their average height was only around 5 feet 4 inches, their weight was at least 20 pounds more than that of a modern person of that height, and this excess was mostly in the form of lean muscle.

One other possible anatomical difference is intriguing, although its reality as well as its interpretation are quite uncertain—the fossil evidence so far simply doesn't allow a definitive answer. But a Neanderthal woman's birth canal may have been wider than a modern woman's, permitting her baby to grow inside her to a bigger size before birth. If so, a Neanderthal pregnancy might have lasted one year, instead of nine months.

Besides their bones, our other main source of information about Neanderthals is their stone tools. Like earlier human tools, Neanderthal tools may have been simple hand-held stones not mounted on separate parts such as handles. The tools don't fall into distinct types with unique functions. There were no standardized bone tools, no bows and arrows. Some of the stone tools were undoubtedly used to make wooden tools, which rarely survive. One notable exception is a wooden thrusting spear eight feet long, found in the ribs of a long-extinct species of elephant at an archeological site in Germany. Despite that (lucky?) success, Neanderthals were probably not very good at big-game hunting; even anatomically more modern people living in Africa at the same time as the Neanderthals were undistinguished as hunters.

If you say "Neanderthal" to friends and ask for their first association, you'll probably get back the answer "caveman." While most excavated Neanderthal remains do come from caves, that's surely an artifact of preservation, since open-air sites would be eroded much more quickly. Neanderthals must have constructed some type of shelter against the cold climate in which they lived, but those shelters must have been crude. All that remain are postholes and a few piles of stones.

The list of quintessentially modern human things that Neanderthals lacked is a long one. They left no unequivocal art objects. They must have worn some clothing in their cold environment, but that clothing had to be crude, since they lacked needles and other evidence of sewing. They evidently had no boats, as no Neanderthal remains are known from Mediterranean islands nor even from North Africa, just eight miles across the Strait of Gibraltar from Neanderthal-populated Spain. There was no long-distance overland trade: Neanderthal tools are made of stones available within a few miles of the site.

Today we take cultural differences among people inhabiting different areas for granted. Every modern human population has its characteristic house style, implements, and art. If you were shown chopsticks, a Schlitz beer bottle, and a blowgun and asked to associate one object each with China, Milwaukee, and Borneo, you'd have no trouble giving the right answers. No such cultural variation is apparent for Neanderthals, whose tools look much the same no matter where they come from.

We also take cultural progress with time for granted. It is obvious to us that the wares from a Roman villa, a medieval castle, and a Manhattan apartment circa 1988 should differ. In the 1990s my sons will look with astonishment at the slide rule I used throughout the 1950s. But Neanderthal tools from 100,000 and 40,000 years ago look essentially the same. In short, Neanderthal tools had no variation in time or space to suggest that most human of characteristics, *innovation.*

What we consider old age must also have been rare among Neanderthals. Their skeletons make clear that adults might live to their thirties or early forties but not beyond 45. If we lacked writing and if none of us lived past 45, just think how the ability of our society to accumulate and transmit information would suffer.

But despite all these subhuman qualities, there are three respects in which we can relate to Neanderthals' humanity. They were the first people to leave conclusive evidence of fire's regular, everyday use: nearly all well-preserved Neanderthal caves have small areas of ash and charcoal indicating a simple fireplace. Neanderthals were also the first people who regularly buried their dead, though whether this implies religion is a matter of pure speculation. Finally, they regularly took care of their sick and aged. Most skeletons of older Neanderthals show signs of severe impairment, such as withered arms, healed but inca-

pacitating broken bones, tooth loss, and severe osteoarthritis. Only care by young Neanderthals could have enabled such older folks to stay alive to the point of such incapacitation. After my litany of what Neanderthals lacked, we've finally found something that lets us feel a spark of kindred spirit in these strange creatures of the Ice Age—human, and yet not really human.

Did Neanderthals belong to the same species as we do? That depends on whether we would have mated and reared a child with a Neanderthal man or woman, given the opportunity. Science fiction novels love to imagine the scenario. You remember the blurb on a pulpy back cover: "A team of explorers stumbles on a steep-walled valley in the center of deepest Africa, a valley that time forgot. In this valley they find a tribe of incredibly primitive people, living in ways that our Stone Age ancestors discarded thousands of years ago. Are they the same species as us?" Naturally, there's only one way to find out, but who among the intrepid explorers—male explorers, of course—can bring himself to make the test? At this point one of the bone-chewing cavewomen is described as beautiful and sexy in a primitively erotic way, so that readers will find the brave explorer's dilemma believable: Does he or doesn't he have sex with her?

Believe it or not, something like that experiment actually took place. It happened repeatedly around 35,000 years ago, around the time of the Great Leap Forward. But you'll have to be patient just a little while longer.

Remember, the Neanderthals of Europe and western Asia were just one of at least three human populations occupying different parts of the Old World around 100,000 years ago. A few fossils from eastern Asia suffice to show that people there differed from Neanderthals as well as from us moderns, but too few have been found to describe these Asians in more detail. The best characterized contemporaries of the Neanderthals are those from Africa, some of whom were almost modern in their skull anatomy. Does this mean that, 100,000 years ago in Africa, we have at last arrived at the Great Leap Forward?

Surprisingly, the answer is still no. The stone tools of these modern-looking Africans were very similar to those of the non-modern-looking Neanderthals, so we refer to them as Middle Stone Age Africans. They still lacked standardized bone tools, bows and arrows, art, and cultural variation. Despite their mostly modern bodies, these Africans were still missing something needed to endow them with modern behavior.

Some South African caves occupied around 100,000 years ago provide us with the first point in human evolution for which we have detailed information about what people were eating. Among the bones found in the caves are many of seals and penguins, as well as shellfish such as limpets; Middle Stone Age Africans are the first people for whom there is even a hint that they exploited the seashore. However, the caves contain very few remains of fish or flying birds, undoubtedly because people still lacked fishhooks and nets.

The mammal bones from the caves include those of quite a few medium-size species, predominant among which are those of the eland, an antelope species. Eland bones in the caves represent animals of all ages, as if people had somehow managed to capture a whole herd and kill every individual. The secret to the hunters' success is most likely that eland are rather tame and easy to drive in herds. Probably the hunters occasionally managed to drive a whole herd over a cliff: that would explain why the distribution of eland ages among the cave kills is like that in a living herd. In contrast, more dangerous prey such as Cape buffalo, pigs, elephants, and rhinos yield a very different picture. Buffalo bones in the caves are mostly of very young or very old individuals, while pigs, elephants, and rhinos are virtually unrepresented.

So Middle Stone Age Africans can be considered big-game hunters, but just barely. They either avoided dangerous species entirely or confined themselves to weak old animals or babies. Those choices reflect prudence: their weapons were still spears for thrusting rather than bows and arrows, and—along with drinking a strychnine cocktail—poking an adult rhino or Cape buffalo with a spear ranks as one of the most effective means of suicide that I know. As with earlier peoples and modern Stone Age hunters, I suspect that plants and small game made up most of the diet of these not-so-great hunters. They were definitely more effective than baboons, but not up to the skill of modern Bushmen and Pygmies.

Thus, the scene that the human world presented from around 130,000 years ago to somewhat before 50,000 years ago was this: Northern Europe, Siberia, Australia, and the whole New World were still empty of people. In the rest of Europe and western Asia lived the Neanderthals; in Africa, people increasingly like us in anatomy; and in eastern Asia, people unlike either the Neanderthals or Africans but known from only a few bones. All three populations were still primitive in their tools, behavior, and limited innovativeness. The stage was set for the Great Leap Forward. Which among these three contemporary populations would take that leap?

The evidence for an abrupt change—at last!—is clearest in France and Spain, in the late Ice Age around 35,000 years ago. Where there had previously been Neanderthals, anatomically fully modern people (often known as Cro-Magnons, from the French site where their bones were first identified) now appear. Were one of those gentlemen or ladies to stroll down the Champs Elysées in modern attire, he or she would not stand out from the Parisian crowds in any way. Cro-Magnons' tools are as dramatic as their skeletons; they are far more diverse in form and obvious in function than any in the earlier archeological record. They suggest that modern anatomy had at last been joined by modern innovative behavior.

Many of the tools continue to be of stone, but they are now made from thin blades struck off a larger stone, thereby yielding roughly ten times more cutting edge from a given quantity of raw stone. Standardized bone and antler tools appear for the first time. So do unequivocal compound tools of several parts tied or glued together, such as spear points set in shafts or ax heads hafted to handles. Tools fall into many distinct categories whose function is often obvious, such as needles, awls, and mortars and pestles. Rope, used in nets or snares, accounts for the frequent bones of foxes, weasels, and rabbits at Cro-Magnon sites. Rope, fishhooks, and net sinkers explain the bones of fish and flying birds at contemporary South African sites.

Sophisticated weapons for killing dangerous animals at a distance now appear also—weapons such as barbed harpoons, darts, spear-throwers, and bows and arrows. South African caves now yield bones of such vicious prey as adult Cape buffalo and pigs, while European caves are full of bones of bison, elk, reindeer, horse, and ibex.

Several types of evidence testify to the effectiveness of late Ice Age people as big-game hunters. Bagging some of these animals must have required communal hunting methods based on detailed knowledge of each species' behavior. And Cro-Magnon sites are much more numerous than those of earlier Neanderthals or Middle Stone Age Africans, implying more success at obtaining food. Moreover, numerous species of big animals that had survived many previous ice ages became extinct toward the end of the last ice age, suggesting that they were exterminated by human hunters' new skills. Likely victims include Europe's woolly rhino and giant deer, southern Africa's giant buffalo and giant Cape horse, and—once improved technology allowed humans to occupy new environments—the mammoths of North America and Australia's giant kangaroos.

Australia was first reached by humans around 50,000 years ago, which implies the existence of watercraft capable of crossing the 60 miles from eastern Indonesia. The occupation of northern Russia and Siberia by at least 20,000 years ago depended on many advances: tailored clothing, as evidenced by eyed needles, cave paintings of parkas, and grave ornaments marking outlines of shirts and trousers; warm furs, indicated by fox and wolf skeletons minus the paws (removed in skinning and found in a separate pile); elaborate houses (marked by postholes, pavements, and walls of mammoth bones) with elaborate fireplaces; and stone lamps to hold animal fat and light the long Arctic nights. The occupation of Siberia in turn led to the occupation of North America and South America around 11,000 years ago.

Whereas Neanderthals obtained their raw materials within a few miles of home, Cro-Magnons and their contemporaries throughout Europe practiced long-distance trade, not only for raw materials for tools but also for "useless" ornaments. Tools of obsidian, jasper, and flint are found hundreds of miles from where those stones were quarried. Baltic amber reached southeast Europe, while Mediterranean shells were carried to inland parts of France, Spain, and the Ukraine.

The evident aesthetic sense reflected in late Ice Age trade relates to the achievements for which we most admire the Cro-Magnons: their art. Best known are the rock paintings from caves like Lascaux, with stunning polychrome depictions of now-extinct animals. But equally impressive are the bas-reliefs, necklaces and pendants, fired-clay sculptures, Venus figurines of women with enormous breasts and buttocks, and musical instruments ranging from flutes to rattles.

Unlike Neanderthals, few of whom lived past the age of 40, some Cro-Magnons survived to 60. Those additional 20 years probably played a big role in Cro-Magnon success. Accustomed as we are to getting our information from the printed page or television, we find it hard to appreciate how important even just one or two old people are in preliterate society. When I visited Rennell Island in the Solomons in 1976, for example, many islanders told me what wild fruits were good to eat, but only one old man could tell me what other wild fruits could be eaten in an emergency to avoid starvation. He remembered that information from a cyclone that had hit Rennell around 1905, destroying gardens and reducing his people to a state of desperation. One such person can spell the difference between death and survival for the whole society.

I've described the Great Leap Forward as if all those advances in tools and art appeared simultaneously 35,000 years ago. In fact, different innovations appeared at differ-

ent times: spear-throwers appeared before harpoons, beads and pendants appeared before cave paintings. I've also described the Great Leap Forward as if it were the same everywhere, but it wasn't. Among late Ice Age Africans, Ukrainians, and French, only the Africans made beads out of ostrich eggs, only the Ukrainians built houses out of mammoth bones, and only the French painted woolly rhinos on cave walls.

These variations of culture in time and space are totally unlike the unchanging monolithic Neanderthal culture. They constitute the most important innovation that came with the Great Leap Forward: namely, the capacity for innovation itself. To us innovation is utterly natural. To Neanderthals it was evidently unthinkable.

Despite our instant sympathy with Cro-Magnon art, their tools and hunter-gatherer life make it hard for us to view them as other than primitive. Stone tools evoke cartoons of club-waving cavemen uttering grunts as they drag women off to their cave. But we can form a more accurate impression of Cro-Magnons if we imagine what future archeologists will conclude after excavating a New Guinea village site from as recently as the 1950s. The archeologists will find a few simple types of stone axes. Nearly all other material possessions were made of wood and will have perished. Nothing will remain of the multistory houses, drums and flutes, outrigger canoes, and world-quality painted sculpture. There will be no trace of the village's complex language, songs, social relationships, and knowledge of the natural world.

New Guinea material culture was until recently "primitive" (Stone Age) for historical reasons, but New Guineans are fully modern humans. New Guineans whose fathers lived in the Stone Age now pilot airplanes, operate computers, and govern a modern state. If we could carry ourselves back 35,000 years in a time machine, I expect that we would find Cro-Magnons to be equally modern people, capable of learning to fly a plane. They made stone and bone tools only because that's all they had the opportunity to learn how to make.

It used to be argued that Neanderthals evolved into Cro-Magnons within Europe. That possibility now seems increasingly unlikely. The last Neanderthal skeletons from 35,000 to 32,000 years ago were still full-blown Neanderthals, while the first Cro-Magnons appearing in Europe at the same time were already anatomically fully modern. Since anatomically modern people were already present in Africa and the Near East tens of thousands of years earlier, it seems much more likely that such people invaded Europe rather than evolved there.

What happened when invading Cro-Magnons met the resident Neanderthals? We can be certain only of the result: within a few thousand years no more Neanderthals. The conclusion seems to me inescapable that Cro-Magnon arrival somehow caused Neanderthal extinction. Yet many anthropologists recoil at this suggestion of genocide and invoke environmental changes instead—most notably, the severe Ice Age climate. In fact, Neanderthals thrived during the Ice Age and suddenly disappeared 42,000 years after its start and 20,000 years before its end.

My guess is that events in Europe at the time of the Great Leap Forward were similar to events that have occurred repeatedly in the modern world, whenever a numerous people with more advanced technology invades the lands of a much less numerous people with less advanced technology. For instance, when European colonists invaded North America, most North American Indians proceeded to die of introduced epidemics; most of the survivors were killed outright or driven off their land; some adopted European technology (horses and guns) and resisted for some time; and many of those remaining were pushed onto lands the invaders did not want, or else intermarried with them. The displacement of aboriginal Australians by European colonists, and of southern African San populations (Bushmen) by invading Iron Age Bantu-speakers, followed a similar course.

By analogy, I suspect that Cro-Magnon diseases, murders, and displacements did in the Neanderthals. It may at first seem paradoxical that Cro-Magnons prevailed over the far more muscular Neanderthals, but weaponry rather than strength would have been decisive. Similarly, humans are now threatening to exterminate gorillas in central Africa, rather than vice versa. People with huge muscles require lots of food, and they thereby gain no advantage if less-muscular people can use tools to do the same work.

Some Neanderthals may have learned Cro-Magnon ways and resisted for a while. This is the only sense I can make of a puzzling culture called the Chatelperronian, which co-existed in western Europe along with a typical Cro-Magnon culture (the so-called Aurignacian culture) for a short time after Cro-Magnons arrived. Chatelperronian stone tools are a mixture of typical Neanderthal and Cro-Magnon tools, but the bone tools and art typical of Cro-Magnons are usually lacking. The identity of the people who produced Chatelperronian culture was debated by archeologists until a skeleton unearthed with Chatelperronian artifacts at Saint-Césaire in France proved to be Neanderthal. Perhaps, then, some Neanderthals managed to master some Cro-Magnon tools and hold out longer than their fellows.

What remains unclear is the outcome of the inter-breeding experiment posed in science fiction novels. Did some invading Cro-Magnon men mate with some Neanderthal women? No skeletons that could reasonably be considered Neanderthal-Cro-Magnon hybrids are known. If Neanderthal behavior was as relatively rudimentary and Neanderthal anatomy as distinctive as I suspect, few Cro-Magnons may have wanted to mate with Neanderthals. And if Neanderthal women were geared for a 12-month pregnancy, a hybrid fetus might not have survived. My inclination is to take the negative evidence at face value, to accept that hybridization occurred rarely if ever, and to doubt that any living people carry any Neanderthal genes.

So much for the Great Leap Forward in western Europe. The replacement of Neanderthals by modern people occurred somewhat earlier in eastern Europe, and still earlier in the Near East, where possession of the same area apparently shifted back and forth between Neanderthals and modern people from 90,000 to 60,000 years ago. The slowness of the transition in the Near East, compared with its speed in western Europe, suggests that the anatomically modern people living around the Near East before 60,000 years ago had not yet developed the modern behavior that ultimately let them drive out the Neanderthals.

Thus, we have a tentative picture of anatomically modern people arising in Africa over 100,000 years ago, but initially making the same tools as Neanderthals and having no advantage over them. By perhaps 60,000 years ago, some magic twist of behavior had been added to the modern anatomy. That twist (of which more in a moment) produced innovative, fully modern people who proceeded to spread westward into Europe, quickly supplanting the Neanderthals. Presumably, they also spread east into Asia and Indonesia, supplanting the earlier people there of whom we know little. Some anthropologists think that skull remains of those earlier Asians and Indonesians show traits recognizable in modern Asians and aboriginal Australians. If so, the invading moderns may not have exterminated the original Asians without issue, as they did the Neanderthals, but instead interbred with them.

Two million years ago, several protohuman lineages existed side-by-side until a shakedown left only one. It now appears that a similar shakedown occurred within the last 60,000 years and that all of us today are descended from the winner of that shakedown. What was the Magic Twist that helped our ancestor to win?

The question poses an archeological puzzle without an accepted answer. You can speculate about the answer as well as I can. To help you, let me review the pieces of the puzzle: Some groups of humans who lived in Africa and the Near East over 60,000 years ago were quite modern in their anatomy, as far as can be judged from their skeletons. But they were not modern in their behavior. They continued to make Neanderthal-like tools and to lack innovation. The Magic Twist that produced the Great Leap Forward doesn't show up in fossil skeletons.

There's another way to restate that puzzle. Remember that we share 98 percent of our genes with chimpanzees. The Africans making Neanderthal-like tools just before the Great Leap Forward had covered almost all of the remaining genetic distance from chimps to us, to judge from their skeletons. Perhaps they shared 99.9 percent of their genes with us. Their brains were as large as ours, and Neanderthals' brains were even slightly larger. The Magic Twist may have been a change in only 0.1 percent of our genes. What tiny change in genes could have had such enormous consequences?

Like some others who have pondered this question, I can think of only one plausible answer: the anatomical basis for spoken complex language. Chimpanzees, gorillas, and even monkeys are capable of symbolic communication not dependent on spoken words. Both chimpanzees and gorillas have been taught to communicate by means of sign language, and chimpanzees have learned to communicate via the keys of a large computer-controlled console. Individual apes have thus mastered "vocabularies" of hundreds of symbols. While scientists argue over the extent to which such communication resembles human language, there is little doubt that it constitutes a form of symbolic communication. That is, a particular sign or computer key symbolizes a particular something else.

Primates can use as symbols not just signs and computer keys but also sounds. Wild vervet monkeys, for example, have a natural form of symbolic communication based on grunts, with slightly different grunts to mean *leopard, eagle,* and *snake.* A month-old chimpanzee named Viki, adopted by a psychologist and his wife and reared virtually as their daughter, learned to "say" approximations of four words: *papa, mama, cup,* and *up.* (The chimp breathed rather than spoke the words.) Given this capability, why have apes not gone on to develop more complex natural languages of their own?

The answer seems to involve the structure of the larynx, tongue, and associated muscles that give us fine control over spoken sounds. Like a Swiss watch, our vocal tract depends on the precise functioning of many parts. Chimps are thought to be physically incapable of producing several of the commonest vowels. If we too were limited to just a few vowels and consonants, our own vocabulary would be greatly reduced. Thus, the Magic Twist may have been some modifications of the protohuman vocal tract to give us finer control and permit formation of a much greater variety of sounds. Such fine modifications of muscles need not be detectable in fossil skulls.

It's easy to appreciate how a tiny change in anatomy resulting in capacity for speech would produce a huge change in behavior. With language, it takes only a few seconds to communicate the message, "Turn sharp right at the fourth tree and drive the male antelope toward the reddish boulder, where I'll hide to spear it." Without language, that message could not be communicated at all. Without language, two protohumans could not brainstorm together about how to devise a better tool or about what a cave painting might mean. Without language, even one protohuman would have had difficulty thinking out for himself or herself how to devise a better tool.

I don't suggest that the Great Leap Forward began as soon as the mutations for altered tongue and larynx anatomy arose. Given the right anatomy, it must have taken humans thousands of years to perfect the structure of language as we know it—to hit on the concepts of word order and case endings and tenses, and to develop vocabulary. But if the Magic Twist did consist of changes in our vocal tract that permitted fine control of sounds, then the capacity for innovation that constitutes the Great Leap Forward would follow eventually. It was the spoken word that made us free.

This interpretation seems to me to account for the lack of evidence for Neanderthal-Cro-Magnon hybrids. Speech is of overwhelming importance in the relations between men and women and their children. That's not to deny that mute or deaf people learn to function well in our culture, but they do so by learning to find alternatives for an existing spoken language. If Neanderthal language was much simpler than ours or nonexistent, it's not surprising that Cro-Magnons didn't choose to associate with Neanderthals.

I've argued that we were fully modern in anatomy and behavior and language by 35,000 years ago and that a Cro-Magnon could have been taught to fly an airplane. If so, why did it take so long after the Great Leap Forward for us to invent writing and build the Parthenon? The answer may be similar to the explanation why the Romans, great engineers that they were, didn't build atomic bombs. To reach the point of building an A-bomb required 2,000 years of

Perspectives

After reading "The Great Leap Forward," one can reflect on these human phenomena. The human body has not changed biologically in the last 50,000 years. The Homo sapien brain has the potential of 10 (to 11th power) neurons and (to the 14th power) connectors, which are capable of storing the equivalent of 20,500,000 volumes of information! (Pytlik, 1985, p. 286). With this brain power, humans have the capability to think abstractly, intuitively, and creatively and to project into the unknown "Science and technology have fostered this potential dramatically. Ten thousand years ago humans learned to write and so to store information outside their bodies. This landmark moved the human drama from pure biological evolution to a cultural revolution." (Pytlik, p. 286). Excerpted from

Pytlik, E., Lauda, D., and Johnson, D. (1985). *Technology, Change and Society.* Worcester, Massachusetts: Davis.

technological advances beyond Roman levels, such as the invention of gunpowder and calculus, the development of atomic theory, and the isolation of uranium. Similarly, writing and the Parthenon depended on tens of thousands of years of cumulative developments after the Great Leap Forward—developments that included, among many others, the domestication of plants and animals.

Until the Great Leap Forward, human culture developed at a snail's pace for millions of years. That pace was dictated by the slowness of genetic change. After the Great Leap Forward, cultural development no longer depended on genetic change. Despite negligible changes in our anatomy, there has been far more cultural evolution in the past 35,000 years than in the millions of years before. Had a visitor from outer space come to Earth before the Great Leap Forward, humans would not have stood out as unique among the world's species. At most, we might have been mentioned along with beavers, bowerbirds, and army ants as examples of species with curious behavior. Who could have foreseen the Magic Twist that would soon make us the first species, in the history of life on Earth, capable of destroying all life?

QUESTIONS

1. What three changes took place in human development to distinguish humans from apes?

2. What physical characteristics are evident in the development of *homo erectus?*

3. Approximately when did the Great Leap Forward occur?

4. What misconceptions does the author mention regarding popular beliefs about Neanderthals?

5. What conclusions might be drawn from the fact that tools used 40,000 years ago by Neanderthals differed little from those used by them 100,000 years ago?

6. What human behaviors have been found in the investigation of Neanderthal societies?

7. What are some basic differences between the Neanderthal and the Cro-Magnon of the Great Leap Forward?

8. Speculate why today's society would find a more comfortable link with Cro-Magnon society than with Neanderthal society.

9. Discuss why a possible encounter between Cro-Magnon and Neanderthal societies might have meant an end to Neanderthal life.

DISCUSSION

1. When you read the above paragraph, what are your reflections about the unique development of humankind?

2. After reading "The Great Leap Forward," what are your reflections on the role of technology and language and writing to humankind's development? Discuss the contributions of language and writing and information processing to the continued development of humankind.

31

Putting the Bite on Planet Earth

DON HINRICHSEN

Each year, about 90 million new people join the human race. This is roughly equivalent to adding three Canadas or another Mexico to the world annually, a rate of growth that will swell human numbers from today's 5.6 billion to about 8.5 billion by 2025.

These figures represent the fastest growth in human numbers ever recorded and raise many vital economic and environmental questions. Is our species reproducing so quickly that we are outpacing the Earth's ability to house and feed us? Is our demand for natural resources destroying the habitats that give us life? If 40 million acres of tropical forest—an area equivalent to twice the size of Austria—are being destroyed or grossly degraded every year, as satellite maps show, how will that affect us? If 27,000 species become extinct yearly because of human development, as some scientists believe, what will that mean for us? If nearly 2 billion people already lack adequate drinking water, a number likely to increase to 3.6 billion by the year 2000, how can all of us hope to survive?

The answers are hardly easy and go beyond simple demographics, since population works in conjunction with other factors to determine our total impact on resources. Modern technologies and improved efficiency in the use of resources can help to stretch the availability of limited resources. Consumption levels also exert considerable impact on our resource base. Population pressures work in conjunction with these other factors to determine, to a large extent, our *total* impact on resources.

For example, although everyone contributes to resource waste, the world's bottom-billion poorest and top-billion richest do most of the environmental damage. Poverty compels the world's 1.2 billion bottom-most poor to misuse their environment and ravage resources, while lack of access to better technologies, credit, education, healthcare and family planning condemns them to subsistence patterns that offer little chance for concern about their environment. This contrasts with the richest 1.3 billion, who exploit and consume disproportionate amounts of resources and generate disproportionate quantities of waste.

One example is energy consumption. Whereas the average Bangladeshi consumes commercial energy equivalent to three barrels of oil yearly, each American consumes an average of 55 barrels. Population growth in Bangladesh, one of the poorest nations, increased energy use there in 1990 by the equivalent of 8.7 million barrels, while U.S. population growth in the same year increased energy use by 110 million barrels. Of course, the U.S. population of 250 million is more than twice the size of the Bangladeshi population of 113 million, but even if the consumption figures are adjusted for the difference in size, the slower growing U.S. population still increases its energy consumption six or seven times faster yearly than does the more rapidly growing Bangladeshi population.

In the future, the effects of population growth on natural resources will vary locally because growth occurs unevenly across the globe. Over the course of the 1990s, the Third World's population is likely to balloon by more

Reprinted with permission from National Wildlife Federation, Hinrichsen, D. (1994, Sept./Oct.) Putting the Bite on Planet Earth. *International Wildlife.*

than 900 million, while the population of the developed world will add a mere 56 million. Asia, with 3.4 billion people today, will have 3.7 billion by the turn of the century; Africa's population will increase from 700 million to 867 million; and Latin America's from 470 million to 538 million. By the year 2000, the Third World's total population is expected to be nearly 5 billion; only 1.3 billion people will reside in industrialized countries.

The United Nations estimates that world population will near 11.2 billion by 2100. However, this figure is based on the assumption that growth rates will drop. If present rates continue, world population will stand at 10 billion by 2030 and 40 billion by 2110.

The United Nations Population Fund estimates that to achieve the 11.2 billion projection, the number of couples using family planning services—such as modern contraceptives—in the developing world will have to rise to 567 million by the year 2000 and to 1.2 billion by 2025. In sub-Saharan Africa this means a 10-fold increase by 2025 in the number of people who use family planning. If these measures do not succeed , human population growth could blast the 11.2 billion figure clear out of the ball park.

Perhaps the most ominous aspect of today's unprecedented growth is its persistence despite falling annual population growth rates everywhere except in parts of Africa, the Middle East and South Asia. Annual global population growth stands at 1.6 percent, down from 2 percent in the early 1970s. Similarly, the total fertility rate (the average number of children a woman is likely to have) has dropped from a global average of six only three decades ago to slightly more than three today.

Population continues to grow because of tremendous demographic momentum. China's annual growth rate, for example, is only 1.2 percent. However, the country's huge population base—1.2 billion people—translates this relatively small rate of growth into a net increase in China's population of around 15 million yearly. Clearly, any attempt to slow population growth is a decades-long process affected by advances in medicine, extended life spans and reduced infant, child, and maternal mortality.

The following pages survey the effects of human population growth on a wide range of natural resources.

PLANTS AND ANIMALS: THE SHRINKING ARK

Biologists have catalogued 1.7 million species and cannot even estimate how many species remain to be documented. The total could be 5 million, 30 million or even more. Yet,

we are driving thousands of species yearly to extinction through thoughtless destruction of habitat.

A survey conducted recently in Australia, Asia and the Americas by the International Union for Conservation of Nature and Natural Resources—The World Conservation Union (IUCN) found that loss of living space affected 76 percent of all mammal species. Expansion of settlements threatened 56 percent of mammal species, while expansion of ranching affected 33 percent. Logging and plantations affected 26 percent.

IUCN has declared human population growth the number one cause of extinctions. The 10 nations with the worst habitat destruction house an average of 189 people per square kilometer (250 acres), while the 10 that retain the most original habitat stand at only 29 people per square kilometer.

Future population growth poses a serious threat to wildlife habitat. Every new person needs space for housing, food, travel, work and other needs. Human needs vary widely from place to place, but a UN survey found that the average person requires about 0.056 hectares (a hectare is a standard unit of land measurement equal to about 2.47 acres) of nonfarm land for daily living. To this must be added land for food production. This varies with land quality and available technologies, but each newborn person probably will need at least 0.2 hectare of cropland unless food production per acre increases in the years ahead. This will require the conversion of more and more wildland into cropland. In East Asia, for example, the amount of irrigated, high-yield cropland per person is already near the 0.2 hectare limit.

UN consultant and author Paul Harrison estimates, very conservatively, that each new person will need at least a quarter of a hectare. Thus, every billion people that we add to the planet in the years ahead will require 250 million hectares more of agricultural land. Most of this land will have to come from what is currently wildlife habitat. The UN's projected population of 11.2 billion by 2100 would require creation of roughly 20 million square kilometers (8 million sq. mi.) of new cropland—equivalent to more than 80 percent of all forest and woodland in developing countries today.

Conversion of natural habitat for human use can even reduce the value of remaining wild areas for wildlife. When development chops wild lands into fragments, native species often decline simply because the small remnants do not meet their biological needs. For example, studies of U.S. forest birds indicate that species that prefer to nest in forest interiors are more subject to predation and lay fewer eggs when habitat fragmentation forces them to nest along forest edges. A study in southern California indicated that most canyons lose about half of native bird species dependent on chaparral habitat within 20 to 40 years after the canyons become isolated by development, even though the chaparral brush remains. Biologist William Newmark's 1987 study of 14 Canadian and U.S. national parks showed that 13 of the parks had lost some of their mammal species, at least in part because the animals could not adapt to confinement within parks surrounded by developed land.

Habitat loss in North America and in Latin American tropics has caused declines in many bird species that migrate between those regions. The Breeding Bird Survey, a volunteer group that tabulates nesting birds each June, found that 70 percent of neotropical migrant species monitored in the eastern United States declined from 1978 to 1987. So did 69 percent of monitored neotropical migrants that nest in prairie regions. Declining species include such familiar songbirds as veeries, wood thrushes, blackpoll warblers and rosebreasted grosbeaks. As human population growth continues to push development into wild areas, fragmentation will increase and its effects on wildlife survival will intensity.

LAND LOSS: A FOOD CRISIS

Land degradation, a global problem, is becoming acute in much of the developing world. Population pressures and inappropriate farming practices contribute to soil impoverishment and erosion, rampant deforestation, overgrazing of common lands and misuse of agrochemicals.

Worldwide, an estimated 1.2 billion hectares, an area about the size of China and India combined, have lost much of their agricultural productivity since 1945. Every year, farmers abandon about 70,000 square kilometers (27,000 sq. mi.) of farmland because soils are too degraded for crops.

Drylands, including grasslands that provide rich pastures for livestock, have been hardest hit. Although not as extensive as once thought, desertification—the ecological destruction that turns productive land into deserts—still threatens the Middle East and parts of Africa and Asia.

Because of land degradation, large portions of the Sahel, including Burkina Faso, Chad, Mali, Mauritania, Niger and Senegal, can no longer feed their people. Although annual fluctuations in rainfall may interrupt the trend of cropland loss, the Sahel could suffer agricultural collapse within a decade. Sahelian croplands, as presently farmed, can support a maximum of 36 million people. In 1990 the rural population stood at an estimated 32 million and will exceed 40 million by the end of the decade even if annual population growth slows from the current 3 percent to 2 percent.

Since 1961, food production has matched world population growth in all developing regions except sub-Saharan Africa. In the early 1980s, the UN Food and Agriculture Organization (FAO) predicted that more than half of all developing nations examined in its study of carrying capacity (62 out of 115) may be unable to feed their projected populations by 2000 using current farming technology. Most of the 62 countries probably will be able to feed less than half of their projected populations without expensive food imports.

As a direct result of population growth, especially in developing nations, the average amount of cropland per person is projected to decline from 0.28 hectares in 1990 to 0.17 by 2025.

Three factors will determine whether food production can equal population growth:

1. *New Croplands.* Currently, the amount of new land put into production each year may equal the amount taken out of production for various reasons, such as erosion, salt deposits and waterlogging. Thus, the net annual gain in arable land, despite widespread habitat destruction to create it, may be zero.

2. *New Water Sources.* Agricultural demand for water is expected to double between 1970 and 2000. Already more than 70 percent of water withdrawals from rivers, underground reservoirs and other sources go to crop irrigation.

3. *Agrochemical Use.* Pesticides and fertilizers are boosting crop yields. However, in many areas agrochemicals are too expensive to use, while in other areas they are overused to prop up falling yields. Agrochemicals can pose health hazards, creating another expense for developing nations.

FOREST: THE VANISHING WORLD

The quest for more crop and grazing land has sealed the fate of much of the world's tropical forests. Between 1971 and 1986, arable land expanded by 59 million hectares, while forests shrank by at least 125 million hectares. However, consultant Harrison estimates that during the same period, land used for settlements, roads, industries, office buildings and other development expanded by more than 30 million hectares as a result of growth in urban centers, reducing the amount of arable land in surrounding areas. Consequently, the amount of natural habitat wiped out to produce the 59-million-hectare net in arable land may have exceeded 100 million hectares.

When both agricultural and nonagricultural needs are taken into account, human population growth may be responsible for as much as 80 percent of the loss of forest cover worldwide. Asia produces the highest rate of loss, 1.2 percent a year. Latin America loses 0.9 percent yearly and Africa 0.8 percent.

If current trends continue, most tropical forests will soon be destroyed or damaged beyond recovery. Of the 76 countries that presently encompass tropical forests, only four—Brazil, Guyana, Papua New Guinea and Zaire—are likely to retain major undamaged tracts by 2010, less than a generation away.

Population pressure contributes to deforestation not only because of increased demand for cropland and living space but also because of increased demand for fuelwood, on which half of the world's people depend for heating and cooking. The majority of sub-Saharan Africa's population is dependent on fuelwood: 82 percent of all Nigerians, 70 percent of Kenyans, 80 percent of all Malagasies, 74 percent of Ghanaians, 93 percent of Ethiopians, 90 percent of Somalians and 81 percent of Sudanese.

By 1990, 100 million Third World residents lacked sufficient fuelwood to meet minimum daily energy requirements, and close to 1.3 billion were consuming wood faster than forest growth could replenish it. On average, consumption outpaces supply by 30 percent in sub-Saharan Africa as a whole, by 70 percent in the Sudan and India, by 150 percent in Ethiopia and by 200 percent in Niger. If present trends continue, FAO predicts, another 1 billion people will be faced with critical fuelwood shortages by the end of the decade. Already, growing rings of desolation—land denuded for fuelwood or building materials—surround many African cities, such as Ouagadou-gou in Burkina Faso, Niamey in Niger and Dakar in Senegal. By 2000, the World Bank estimates, half to three-quarters of all West Africa's fuelwood consumption will be burned in towns and cities.

According to the World Bank, remedying the fuelwood shortage will require planting 55 million hectares—an area nearly twice the size of Italy—with fast-growing trees at a rate of 2.7 million hectares a year, five times the present annual rate of 555,000 hectares.

TROUBLED OCEANS: DISAPPEARING RESOURCES

Population and development pressures have been mounting in coastal areas worldwide for the past 30 years, triggering widespread resource degradation. Coastal fisheries are overexploited in much of Asia, Africa and parts of Latin America. In some cases—as in the Philippines, Indonesia, Malaysia, China, Japan, India, the west coast of South America, the Mediterranean and the Caribbean—economically important fisheries have collapsed or are in severe decline. "Nearly all Asian waters within 15 kilometers of land are considered overfished," says Ed Gomez, director of the Marine Science Institute at the University of the Philippines in Manila.

Overfishing is not the sole cause of these declines. Mangroves and coral reefs—critical nurseries for many marine species and among the most productive of all ecosystems—are being plundered in the name of development.

In 1990, a UN advisory panel, the Group of Experts on the Scientific Aspects of Marine Pollution (GESAMP), reported that coastal pollution worldwide has grown worse over the decade of the 1980s. Experts pointed to an overload of nutrients—mainly nitrogen and phosphorus from untreated or partially treated sewage, agricultural runoff and erosion—as the most serious coastal pollution problem. Human activities may be responsible for as much as 35 million metric tons of nitrogen and up to 3.75 million metric tons of phosphorus flowing into coastal waters every year. Even such huge amounts could be dissolved in the open ocean, but most of the pollution stays in shallow coastal waters where it causes massive algal blooms and depletes oxygen levels, harming marine life near the shores.

Although the world still possesses an estimated 240,000 square kilometers (93,000 sq. mi.) of mangrove swamps—coastal forests that serve as breeding grounds and nurseries for many commercially important fish and shellfish species—this represents only about half the original amount. Clear-cutting for timber, fuelwood and wood chips; conversion to fish and shellfish ponds; and expansion of urban areas and croplands have claimed millions of hectares globally. For example, of the Philippine's original mangrove area—estimated at 500,000 to 1 million hectares—only 100,000 hectares remain; 80 to 90 percent are gone.

Some 600,000 square kilometers (230,000 sq. mi.) of coral reefs survive in the world's tropical seas.

Unfortunately, these species-rich ecosystems are suffering widespread decline. Clive Wilkinson, a coral reef specialist working at the Australian Institute of Marine Science, estimates that fully 10 percent of the world's reefs have already been degraded "beyond recognition." Thirty percent are in critical condition and will be lost completely in 10 to 20 years, while another 30 percent are threatened and will be lost in 20 to 40 years. Only 30 percent, located away from human development or otherwise too remote to be exploited, are in stable condition.

Throughout much of the world, coastal zones are overdeveloped, overcrowded, and overexploited. Already nearly two-thirds of the world's population—some 3.6 billion people—live along coasts or within 150 kilometers (100 mi.) of one. Within three decades, 75 percent, or 6.4 billion, will reside in coastal areas—nearly a billion more people than the current global population.

In the United States, 54 percent of all Americans live in 772 coastal counties adjacent to marine coasts or the Great Lakes. Between 1960 and 1990, coastal population density increased from 275 to nearly 400 people per square kilometer. By 2025, nearly 75 percent of all Americans will live in coastal counties, with population density doubling in areas such as southern California and Florida.

Similarly, nearly 780 million of China's 1.2 billion people—almost 67 percent—live in 14 southeast and coastal provinces and two coastal municipalities, Shanghai and Tianjin. Along much of China's coastline, population densities average more than 600 per square kilometer. In Shanghai they exceed 2,000 per square kilometer. During the past few years, as many as 100 million Chinese have moved from poorer provinces in central and western regions to coastal areas in search of better economic opportunities. More ominously, population growth is expected to accelerate in the nation's 14 newly created economic free zones and five special economic zones, all of them coastal.

WATER: DISTRIBUTION WOES

Nearly 75 percent of the world's freshwater is locked in glaciers and icecaps, with virtually all the rest underground. Only about 0.01 percent of the world's total water is easily available for human use. Even this tiny amount would be sufficient to meet all the world's needs if it were distributed evenly. However, the world is divided into water "haves" and "have nots." In the Middle East, north Asia, northwestern Mexico, most of Africa, much of the western United States, parts of Chile and Argentina and nearly all of Australia, people need more water than can be sustainably supplied.

As the world's human population increases, the amount of water per person decreases. The United Nations Educational, Scientific and Cultural Organization (UNESCO) estimates that the amount of freshwater available per person has shrunk from more than 33,000 cubic meters (1.2 million cu. ft.) per year in 1850 to only 8,500 cubic meters (300,000 cu. ft.) today. Of course, this is a crude, general figure. But because of population growth alone, water demand in more than half the world's countries by 2000 is likely to be twice what it was as recently as 1971.

Already some 2 billion people in 80 countries must live with water constraints for all or part of the year. By the end of the 1990s Egypt will have only two-thirds as much water for each of its inhabitants as it has today, and Kenya only half as much. By then, six of East Africa's seven nations and all five nations on the south rim of the Mediterranean will face severe shortages. In 1990, 20 nations suffered water scarcity, with less than 1,000 cubic meters (35,000 cu. ft.) of water per person, according to a study by Population Action International. Another eight experienced occasional water stress. The 28 nations represent 333 million people. By 2025, some 48 nations will suffer shortages, involving some 3 billion residents, according to the study.

China—although not listed as water short because of the heavy amount of rain that falls in its southern region—has, nevertheless, exceeded its sustainable water resources. According to Qu Geping, China's Environment Minister, the country can supply water sustainably to only 650 million people, not the current population of 1.2 billion. In other words, China is supporting twice as many people as its water resources can reasonably sustain without drawing down groundwater supplies and overusing surface waters.

FOSSIL FUELS: ENERGY BREAKDOWN

Human society runs on energy, principally fossil fuels such as oil, gas and coal. These three account for 90 percent of global commercial energy production. Nuclear power, hydro-electricity and other sustainable resources provide the rest.

The industrialized nations, with less than a quarter of the world's people, burn about 70 percent of all fossil fuels. The United States alone consumes about a quarter of the world's commercial energy, and the former Soviet Union about a fifth. In terms of per capita consumption patterns, Canada burns more fuel than any other nation—in 1987 the equivalent of 9 metric tons of oil per person—followed by Norway at 8.9 metric tons of oil per person and the United States at 7.3. By contrast, developing nations on average use the equivalent of only about half a metric ton of oil per person yearly.

Known oil reserves should meet current levels of consumption for another 41 years, up from an estimated 31 years in 1970 thanks to better energy efficiency and conservation measures, along with new oil fields brought into production. Natural-gas reserves should meet current demand for 60 more years, up from 38 years in 1970. Coal reserves should be good for another 200 years.

But our addiction to fossil fuel has resulted in chronic, sometimes catastrophic, pollution of the atmosphere, in some cases far beyond what natural systems or man-made structures can tolerate. A noxious atmospheric cocktail of chemical pollutants is primarily responsible for the death and decline of thousands of hectares of European forests. Acid rain—caused by a combination of nitrogen and sulfur dioxides released from fossil-fuel combustion—has eaten away at priceless monuments and buildings throughout Europe and North America, causing billions of dollars in damages.

Urban air contains a hazardous mix of pollutants—everything from sulfur dioxide and reactive hydrocarbons to heavy metals and organic compounds. Smog alerts are now commonplace in many cities with heavy traffic. In Mexico City, for example, smog levels exceeded World Health Organization standards on all but 11 days in 1991. Breathing the city's air is said to be as damaging as smoking two packs of cigarettes a day, and half the city's children are born with enough lead in their blood to hinder their development.

The only way to stretch fossil fuel reserves and reduce pollution levels is to conserve energy and use it much more efficiently than we do now. Some progress has been made, but the benefits of energy conservation have been realized in only a few industrialized countries.

Recent history has shown what can happen. In the decade following the first oil shock, per capita energy consumption fell by 5 percent in the member states of the Organization for Economic Cooperation and Development (OECD)—consisting of the industrialized countries of Western Europe and North America, plus Japan, Australia and New Zealand—while their per capita gross domestic product grew by a third.

Buildings in the OECD countries use a quarter less energy now than they did before 1973, while the energy efficiency of industry has improved by a third. Worldwide, cars now get 25 percent more kilometers per gallon than they did in 1973. In all, increased efficiency since 1973 has saved the industrialized nations $250 billion in energy costs.

Even more savings could be realized through concerted efforts to conserve energy and improve efficiency. Three relatively simple, cost effective measures could be introduced immediately: 1) making compact fluorescent lamps generally available in homes and offices; 2) tightening up building codes to require better insulation against cold and heat; and 3) requiring lean-burn engines, which get up to 80 kilometers per gallon (50 mpg), in all new compact cars. These three "technical fixes" could save billions of dollars in energy costs.

POLICY: BUILDING A FUTURE

The main population issues—urbanization, rapid growth and uneven distribution—when linked with issues of environmental decline, pose multiple sets of problems for policymakers. The very nature of these interrelated problems makes them virtually impossible to deal with in balkanized bureaucracies accustomed to managing only one aspect of any problem. Population and resource issues require integrated, strategic management, an approach few countries are in a position to implement.

Sustainable-management strategies, designed to ensure that resources are not destroyed by overexploitation, are complicated to initiate because they require the cooperation of ministries or departments often at odds over personnel, budgets and political clout. Most governments lack institutional mechanisms that ensure a close working relationship among competing ministries. Consequently, most sustainable-development initiatives never get beyond words on paper. "We talk about integrated resource management, but we don't do it," admits one Indian official in Delhi. "Our ministries are like fiefdoms, they seldom cooperate on anything."

Fragmented authority yields fragmented policies. Big development ministries—such as industry and commerce, transportation, agriculture, fisheries and forestry—rarely cooperate in solving population and resource problems. Piecemeal solutions dominate, and common resources continue to deteriorate.

The world's population and resource problems offer plenty of scope for timely and incisive policy interventions that promise big returns for a relatively small investment. As little as $17 billion a year could provide contraceptives to every woman who wants them, permitting families throughout the globe to reduce births voluntarily. This approach might produce the same or better results than would government-set population targets, according to one study. Moreover, population specialists recognize that educating girls and women provides a higher rate of return than most other investments. "In fact, it may well be the single most influential investment that can be made in the developing world," says Larry Summers, a former World Bank economist.

But time is at a premium. The decision period for responding to the crises posed by rapidly growing populations, increased consumption levels and shrinking resources will be confined, for the most part, to the next two decades. If human society does not succeed in checking population growth, the future will bring widespread social and economic dislocations as resource bases collapse. Unemployment and poverty will increase, and migrations from poorer to richer nations will bring Third World stresses to the developed world.

QUESTIONS

1. How many people are added to the human race each year? Calculate the percentage of growth annually.

2. List some of the stresses on the earth's environment as a result of unprecedented population growth.

3. What are the difficulties of lowering world population rates even if a country's annual growth rate declines?

4. How much land does a person need to survive? How does this impact wildlife, plants and land use?

5. Discuss the problem of land degradation. How much is the average amount of cropland per person projected to decline from 1990 to 2025? Discuss implications. What are possible solutions?

6. How soon will most tropical forests disappear? What are the implications?

7. What are some of the pressures on oceans? What are the ramifications?

8. How much have freshwater supplies shrunk from 1850? What are the issues here?

9. What is the effect of world reliance on fossil fuels? What are some solutions suggested in the article as well as your own?

10. Name some practical solutions to building a future policy that can sustain world population. Aside from suggestions in the article, also name your own possible solutions and approaches.

Can the Growing Human Population Feed Itself?

JOHN BONGAARTS

Demographers now project that the world's population will double during the next half century, from 5.3 billion people in 1990 to more than 10 billion by 2050. How will the environment and humanity respond to this unprecedented growth? Expert opinion divides into two camps. Environmentalists and ecologists, whose views have widely been disseminated by the electronic and print media, regard the situation as a catastrophe in the making. They argue that in order to feed the growing population farmers must intensify agricultural practices that already cause grave ecological damage. Our natural resources and the environment, now burdened by past population growth, will simply collapse under the weight of this future demand.

The optimists, on the other hand, comprising many economists as well as some agricultural scientists, assert that the earth can readily produce more than enough food for the expected population in 2050. They contend that technological innovation and the continued investment of human capital will deliver high standards of living to much of the globe, even if the population grows much larger than the projected 10 billion. Which point of view will hold sway? What shape might the future of our species and the environment actually take?

Many environmentalists fear that world food supply has reached a precarious state: "Human numbers are on a collision course with massive famines. . . . If humanity fails to

act, nature will end the population explosion for us—in very unpleasant ways—well before 10 billion is reached," write Paul R. Ehrlich and Anne H. Ehrlich of Stanford University in their 1990 book *The Population Explosion*. In the long run, the Ehrlichs and like-minded experts consider substantial growth in food production to be absolutely impossible. "We are feeding ourselves at the expense of our children. By definition farmers can overplow and overpump only in the short run. For many farmers the short run is drawing to a close," states Lester R. Brown, president of the Worldwatch Institute, in a 1988 paper.

Over the past three decades, these authors point out, enormous efforts and resources have been pooled to amplify agricultural output. Indeed, the total quantity of harvested crops increased dramatically during this time. In the developing world, food production rose by an average of 117 percent in the quarter of a century between 1965 and 1990. Asia performed far better than other regions, which saw increases below average.

Because population has expanded rapidly as well, per capita food production has generally shown only modest change; in Africa it actually declined. As a consequence, the number of undernourished people is still rising in most parts of the developing world, although that number did fall from 844 million to 786 million during the 1980s. But this decline reflects improved nutritional conditions in Asia alone. During the same period, the number of people having energy-deficient diets in Latin America, the Near East and Africa climbed.

Many social factors can bring about conditions of hunger, but the pessimists emphasize that population pres-

Reprinted with permission from *Scientific American*, Bongaarts, J. (1994, March). Can the growing human population feed itself? Copyright © (1994, March) by Scientific American, Inc. All rights reserved.

sure on fragile ecosystems plays a significant role. One specific concern is that we seem to be running short on land suitable for cultivation. If so, current efforts to bolster per capita food production by clearing more fertile land will find fewer options. Between 1850 and 1950 the amount of arable land grew quickly to accommodate both larger populations and greater demand for better diets. This expansion then slowed and by the late 1980s ceased altogether. In the developed world, as well as in some developing countries (especially China), the amount of land under cultivation started to decline during the 1980s. This drop is largely because spreading urban centers have engulfed fertile land or, once the land is depleted, farmers have abandoned it. Farmers have also fled from irrigated land that has become unproductive because of salt accumulation.

Moreover, environmentalists insist that soil erosion is destroying much of the land that is left. The extent of the damage is the subject of controversy. A recent global assessment, sponsored by the United Nations Environment Program and reported by the World Resources Institute and others, offers some perspective. The study concludes that 17 percent of the land supporting plant life worldwide has lost value over the past 45 years. The estimate includes erosion caused by water and wind, as well as chemical and physical deterioration, and ranks the degree of soil degradation from light to severe. This degradation is least prevalent in North America (5.3 percent) and most wide spread in Central America (25 percent), Europe (23 percent), Africa (22 percent) and Asia (20 percent). In most of these regions, the average farmer could not gather the resources necessary to restore moderate and severely affected soil regions to full productivity. Therefore, prospects for reversing the effects of soil erosion are not good, and it is likely that this problem will worsen.

Despite the loss and degradation of fertile land, the "green revolution" has promoted per capita food production by increasing the yield per hectare. The new, high-yielding strains of grains such as wheat and rice have proliferated since their introduction in the 1960s, especially in Asia. To reap full advantage from these new crop varieties, however, farmers must apply abundant quantities of fertilizer and water.

Environmentalists question whether further conversion to such crops can be achieved at reasonable cost, especially in the developing world, where the gain in production is most needed. At the moment, farmers in Asia, Latin America and Africa use fertilizer sparingly, if at all, because it is too expensive or unavailable. Fertilizer use in the developed world has recently waned. The reasons for

the decline are complex and may be temporary, but clearly farmers in North America and Europe have decided that increasing their already heavy application of fertilizer will not further enhance crop yields.

Unfortunately, irrigation systems, which would enable many developing countries to join in the green revolution, are often too expensive to build. In most areas, irrigation is essential for generating higher yields. It also can make arid land cultivable and protect farmers from the vulnerability inherent in natural variations in the weather. Land brought into cultivation this way could be used for growing multiple crop varieties, thereby helping food production to increase.

Such advantages have been realized since the beginning of agriculture: the earliest irrigation systems are thousands of years old. Yet only a fraction of productive land in the developing world is now irrigated, and its expansion has been slower than population growth. Consequently, the amount of irrigated land per capita has been dwindling during recent decades. The trend, pessimists argue, will be hard to stop. Irrigation systems have been built in the most affordable sites, and the hope for extending them is curtailed by rising costs. Moreover, the accretion of silt in dams and reservoirs and of salt in already irrigated soil is increasingly costly to avoid or reverse.

Environmentalists Ehrlich and Ehrlich note that modern agriculture is by nature at risk wherever it is practiced. The genetic uniformity of single, high-yielding crop strains planted over large areas makes them highly productive but also renders them particularly vulnerable to insects and disease. Current preventive tactics, such as spraying pesticides and rotating crops, are only partial solutions. Rapidly evolving pathogens pose a continuous challenge. Plant breeders must maintain a broad genetic arsenal of crops by collecting and storing natural varieties and by breeding new ones in the laboratory.

The optimists do not deny that many problems exist within the food supply system. But many of these authorities, including D. Gale Johnson, the late Herman Kahn, Walter R. Brown, L. Martel, the late Roger Revelle, Vaclav Smil and Julian L. Simon, believe the world's food supply can dramatically be expanded. Ironically, they draw their enthusiasm from extrapolation of the very trends that so alarm those experts who expect doom. In fact, statistics show that the average daily caloric intake per capita climbed by 21 percent (from 2,063 calories to 2,495 calories) between 1965 and 1990 in the developing countries. These higher calories have generally delivered greater amounts of protein. On average, the per capita consumption of protein rose from 52 grams per day to 61 grams per day between 1965 and 1990.

According to the optimists, not only has the world food situation improved significantly in recent decades, but further growth can be brought about in various ways. A detailed assessment of climate and soil conditions in 93 developing countries (excluding China) shows that nearly three times as much land as is currently farmed, or an additional 2.1 billion hectares, could be cultivated. Regional soil estimates indicate that sub-Saharan Africa and Latin America can exploit many more stretches of unused land than can Asia, the Near East and North Africa.

Even in regions where the amount of potentially arable land is limited, crops could be grown more times every year than is currently the case. This scenario is particularly true in the tropics and subtropics where conditions are such—relatively even temperature throughout the year and a consistent distribution of daylight hours—that more than one crop would thrive. Nearly twice as many crops are harvested every year in Asia than in Africa at present, but further increases are possible in all regions.

In addition to multicropping, higher yields per crop are attainable, especially in Africa and the Near East. Many more crops are currently harvested per hectare in the First World than elsewhere: cereal yields in North America and Europe averaged 4.2 tons per hectare, compared with 2.9 in the Far East (4.2 in China), 2.1 in Latin America, 1.7 in the Near East and only 1.0 in Africa.

Such yield improvements, the enthusiasts note, can be achieved by expanding the still limited use of high-yield crop varieties, fertilizer and irrigation. In *World Agriculture: Toward 2000,* Nikos Alexandratos of the Food and Agriculture Organization (FAO) of the United Nations reports that only 34 percent of all seeds planted during the mid-1980s were high-yielding varieties. Statistics from the FAO show that at present only about one in five hectares of arable land is irrigated, and very little fertilizer is used. Pesticides are sparsely applied. Food output could drastically be increased simply by more widespread implementation of such technologies.

Aside from producing more food, many economists and agriculturists point out, consumption levels in the developing world could be boosted by wasting fewer crops, as well as by cutting storage and distribution losses. How much of an increase would these measures yield? Robert W. Kates, director of the Alan Shawn Feinstein World Hunger Program at Brown University, writes in *The Hunger Report: 1988* that humans consume only 60 percent of all harvested crops, and some 25 to 30 percent is lost before reaching individual homes. The FAO, on the other hand, estimates lower distribution losses: 6 percent for cereals, 11 percent for roots and 5 percent for pulses. All the same, there is no doubt that improved storage and distribution systems would leave more food available for human nutrition, independent of future food production capabilities.

For optimists, the long-range trend in food prices constitutes the most convincing evidence for the correctness of their view. In 1992–93 the World Resources Institute reported that food prices dropped further than the price of most nonfuel commodities, all of which have declined in the past decade. Cereal prices in the international market fell by approximately one third between 1980 and 1989. Huge government subsidies for agriculture in North America and western Europe, and the resulting surpluses of agricultural products, have depressed prices. Obviously, the optimists assert, the supply already exceeds the demand of a global population that has doubled since 1950.

Taken together, this evidence leads many experts to see no significant obstacles to raising levels of nutrition for world populations exceeding 10 billion people. The potential for an enormous expansion of food production exists, but its realization depends of course on sensible governmental policies, increased domestic and international trade and large investments in infrastructure and agricultural extension. Such improvements can be achieved, the optimists believe, without incurring irreparable damage to global ecosystems.

Proponents of either of these conflicting perspectives have difficulty accepting the existence of other plausible points of view. Moreover, the polarity between the two sides of expert opinion shows that neither group can be completely correct. Finding some common ground between these seemingly irreconcilable positions is not as difficult as it at first appears if empirical issues are emphasized and important differences in value systems and political beliefs are ignored.

Both sides agree that the demand for food will swell rapidly over the next several decades. In 1990 a person living in the developing world ate on average 2,500 calories each day, taken from 4,000 gross calories of food crops made available within a household. The remaining 1,500 calories from this gross total not used to meet nutritional requirements were either lost, inedible or used as animal feed and plant seed. Most of this food was harvested from 0.7 billion hectares of land in the developing world. The remaining 5 percent of the total food supply came from imports. To sustain this 4,000-gross-calorie diet for more than twice as many residents, or 8.7 billion people, living in the developing world by 2050, agriculture must offer 112 percent more crops. To raise the average Third World diet to 6,000 gross calories per day, slightly above the 1990 world average, food production would need to increase by 218 percent. And to bring the average Third World diet to a level comparable with that currently found in the developed world, or 10,000 gross calories per day, food production would have to surge by 430 percent.

A more generous food supply will be achieved in the future through boosting crop yields, as it has been accomplished in the past. If the harvested area in the developing world remains at 0.7 billion hectares, then each hectare must more than double its yield to maintain an already inadequate diet for the future population of the developing world. Providing a diet equivalent to a First World diet in 1990 would require that each hectare increase its yield more than six times. Such an event in the developing world must be considered virtually impossible, barring a major breakthrough in the biotechnology of food production.

Instead farmers will no doubt plant more acres and grow more crops per year on the same land to help augment crop harvests. Extrapolation of past trends suggests that the total harvested area will increase by about 50 percent by the year 2050. Each hectare will then have to provide nearly 50 percent more tons of grain or its equivalent to keep up with current dietary levels. Improved diets could result only from much larger yields.

The technological optimists are correct in stating that overall world food production can substantially be increased over the next few decades. Current crop yields are well below their theoretical maxima, and only about 11 percent of the world's farmable land is now under cultivation. Moreover, the experience gained recently in a number of developing countries, such as China, holds important lessons on how to tap this potential elsewhere. Agricultural productivity responds to well-designed policies that assist farmers by supplying needed fertilizer and other inputs, building sound infrastructure and providing market access. Further investments in agricultural research will spawn new technologies that will fortify agriculture in the future. The vital question then is not how to grow more food but rather how to implement agricultural methods that may make possible a boost in food production.

A more troublesome problem is how to achieve this technological enhancement at acceptable environmental costs. It is here that the arguments of those experts who forecast a catastrophe carry considerable weight. There can be no doubt that the land now used for growing food crops is generally of better quality than unused, potentially cultivable land. Similarly, existing irrigation systems have been built on the most favorable sites. Consequently, each new measure applied to increase yields is becoming more expensive to implement, especially in the developed world and parts of the developing world such as China, where productivity is already high. In short, such constraints are raising the marginal cost of each additional ton of grain or its equivalent. This tax is even higher if one takes into account negative externalities—primarily environmental costs not reflected in the price of agricultural products.

The environmental price of what in the Ehrlichs' view amounts to "turning the earth into a giant human feedlot" could be severe. A large inflation of agriculture to provide growing populations with improved diets is likely to lead to widespread deforestation, loss of species, soil erosion and pollution from pesticides, and runoff of fertilizer as farming intensifies and new land is brought into production. Reducing or minimizing this environmental impact is possible but costly.

Given so many uncertainties, the course of future food prices is difficult to chart. At the very least, the rising marginal cost of food production will engender steeper prices on the international market than would be the case if there were no environmental constraints. Whether these higher costs can offset the historical decline in food prices remains to be seen. An upward trend in the price of food sometime in the near future is a distinct possibility. Such a hike will be mitigated by the continued development and application of new technology and by the likely recovery of agricultural production and exports in the former Soviet Union, eastern Europe and Latin America. Also, any future price increases could be lessened by taking advantage of the underutilized agricultural resources in North America, notes Per Pinstrup-Andersen of Cornell University in his 1992 paper "Global Perspectives for Food Production and Consumption." Rising prices will have little effect on high-income countries or on households possessing reasonable purchasing power, but the poor will suffer.

In reality, the future of global food production is neither as grim as the pessimists believe nor as rosy as the optimists claim. The most plausible outcome is that dietary intake will creep higher in most regions. Significant annual fluctuations in food availability and prices are, of course, likely; a variety of factors, including the weather, trade interruptions and the vulnerability of monocropping to pests, can alter food supply anywhere. The expansion of agriculture will be achieved by boosting crop yields and by using existing farmland more intensively, as well as by bringing arable land into cultivation where such action proves economical. Such events will transpire more slowly than in the past, however, because of environmental constraints. In addition, the demand for food in the developed world is approaching saturation levels. In the U.S., mounting concerns about health have caused the per capita consumption of calories from animal products to drop.

Still, progress will be far from uniform. Numerous countries will struggle to overcome unsatisfactory nutrition levels. These countries fall into three main categories. Some low-income countries have little or no reserves of fertile land or water. The absence of agricultural resources is in itself not an insurmountable problem, as is demonstrated by regions, such as Hong Kong and Kuwait, that can purchase their food on the international market. But many poor countries, such as Bangladesh, cannot afford to buy food from abroad and thereby compensate for insufficient natural resources. These countries will probably rely more on food aid in the future.

Low nutrition levels are also found in many countries, such as Zaire, that do possess large reserves of potentially cultivable land and water. Government neglect of agriculture and policy failures have typically caused poor diets in such countries. A recent World Bank report describes the damaging effects of direct and indirect taxation of agricultures, controls placed on prices and market access, and overvalued currencies, which discourage exports and encourage imports. Where agricultural production has suffered from misguided government intervention (as is particularly the case in Africa), the solution—policy reform—is clear.

Food aid will be needed as well in areas rife with political instability and civil strife. The most devastating famines of the past decade, known to television viewers around the world, have occurred in regions fighting prolonged civil wars, such as Ethiopia, Somalia and the Sudan. In many of these cases, drought was instrumental in stirring social and political disruption. The addition of violent conflict prevented the recuperation of agriculture and the distribution of food, thus turning bad but remediable situations into disasters. International military intervention, as in Somalia, provides only a short-term remedy. In the absence of sweeping political compromise, hunger and malnutrition will remain endemic in these war-torn regions.

Feeding a growing world population a diet that improves over time in quality and quantity is technologically feasible. But the economic and environmental costs incurred through bolstering food production may well prove too great for many poor countries. The course of events will depend crucially on their governments' ability to design and enforce effective policies that address the challenges posed by mounting human numbers, rising poverty and environmental degradation. Whatever the outcome, the task ahead will be made more difficult if population growth rates cannot be reduced.

QUESTIONS

1. As the population of the world will probably exceed 10 billion by 2050, name the two schools of thought and their positions in dealing with the environment and the future of humanity. What is your view before you read this article?

2. How much has food production increased in the Third World between 1965 and 1990? How much has population expanded during this time period? What is the result?

3. What is the argument in the article concerning the amount of arable land in the developed world and undeveloped world?

4. What is the problem with increased food per hectare from the "green revolution" and irrigation systems?

5. Explain the issue of genetic uniformity.

6. According to the optimists, how much additional land could be cultivated?

7. How can yield improvements be achieved according to the optimists?

8. Explain the argument of the optimists concerning waste, storage and distribution losses of food.

9. How can the potential for food expansion be realized?

10. How much would food production have to rise in order to bring the Third World diet to the level found in the developed world?

11. What do you feel is the role of new technology in boosting this production?

12. Explain the author's view in accomplishing increase in food production. Explain your view.

The 1994 International Conference on Population and Development in Cairo

Unless women can manage and control their own fertility, they cannot manage and control their own lives.

KAVAL GULHATI, INDIAN FAMILY PLANNER

In September 1994, 179 delegations of the United Nations, assembled in Cairo, Egypt, at the International Conference on Population and Development, reached a bold and sweeping plan to stabilize world population. (The conference attendees included not only 179 national delegations, but also 2,500 government delegates, 7 heads of state, 5 vice presidents, 20 deputy prime ministers, 1,200 nongovernmental organizations, and 4,000 journalists. The momentum of the population issue, as it climbed to the top of the global agenda with the sheer magnitude and the percentage of the global community that the conference involved, was, in itself, an important accomplishment.) In the preparatory meetings leading up to the conference, delegates rejected the concept of the high trajectory of population growth reaching 11.9 billion people by 2050. Instead, they recommended an ambitious plan to stabilize world population between the low and medium projections of 7.9 to 9.8 billion—by 2050. Their strategy reflected a sense of urgency—that unless population growth can be slowed quickly, it will push human demands beyond the carrying capacity of many countries, causing environmental degradation, economic decline, and social disintegration (Roush, 1994).

The plan calls for active services to provide family planning to the estimated 120 million women in the world who want to limit the number of their children but lack access to the family planning services. It also addresses the underlying causes of high fertility, such as female illiteracy, and strives for universal primary school education for girls, recognizing that as female educational levels rise, fertility levels fall—a relationship that holds across all cultures. The goals of the conference were to achieve gender equality and to permit women to have the right to control and manage their own lives. A major achievement of the Cairo conference was to change the world view of the population issue. From the previous view, which was concerned with demography and contraception, the agenda moved to a focus on the situation of women.

The Cairo conference highlighted the fact that real change in demographic patterns will occur only through fundamental changes in women's lives, such as increases in women's access to education, cash, and credit. Two thirds of the world's illiterate people are women. Women are responsible for over 80% of food production in Africa, but they have very little access to agricultural extension, credit, land titles, or management. Studies show that with better education and an increase of stature in family and community for women, birth rates decline. Ninety-five percent of the growth in population will occur in the developing world, increasing from the current 5–6 billion people to 8.6–12 billion by the year 2100.

There also needs to be an increase in the amount of information regarding reproduction available to adolescents around the world. As there are currently over 1 billion teenagers on the planet and increasingly large numbers of them are unmarried and living in cities, there are radical changes in values of "the young and the restless" as values are beamed around the planet on television, raising questions that do not relate in their culture and the lives of their ancestors (Catley-Carlson, 1994).

The document produced by the conference sets a plan for developing and supportive countries to invest at least 20% of their public expenditures in the social sector. Special focus is placed on a range of population, health, and education programs that would improve the health and status of women. It also requests donor spending on population assistance (currently at about $1 billion) to increase from 1.4% of official development assistance to 4%. One of the prominent themes of the Cairo conference was to try to ensure that the responsibility and funding of this change would be shared by the whole world. These measures, the document states, "would result in world population growth at levels close to the United Nations low [projection] of a global population of 7.8 billion by the year 2050" (Roush, 1994).

The goals set in Cairo will be extremely difficult to achieve, but if the world succeeds in stabilizing human population at 8 or 9 billion, it will satisfy one of the conditions of an environmentally sustainable society. The plan recognizes the earth's natural limits and the urgency to respect those limits. There already is evidence of these limits in the world's declining fish catches, falling water tables, declining bird populations on every continent, rising global temperatures, and the fact that world grain stocks are at their lowest level in 20 years.

In trying to answer the question of how many people the earth can support, we raise the reverse question: What exactly will limit the growth in human numbers? It appears that it is the supply of food that will determine the earth's population carrying capacity. Three natural limits for food production are already slowing food yields: the sustainable yield of oceanic fisheries, the amount of fresh water produced by the hydrological cycle, and the amount of fertilizer that the existing crop varieties can effectively use.

Understanding and respecting the earth's limits is dependent on the urgent recognition of limiting the draw on those resources and translating that recognition into a worldwide population plan for limitation and stabilization. According to the plan developed in Cairo, this can happen best if women have an informed voice in the planning of their families. As the delegates gathered in Cairo debated the plan of action for nine days, the world's population grew by some 2.1 million people (Brown, 1995).

REFERENCES

Brown, L. (1995). Nature's limits. *State of the World,* 1995 (pp. 3–4). New York: W.W. Norton & Co, 1995.

Catley-Carlson, M.: Interview. (1994, October). Cairo Conference Finds Women's Status Central to Slowing Population Growth, Increasing Food Production. *2020 News & Reviews.* Available: http://www.cgiar.org./ifpri/2020/newslet/nv_1094/nv-1094f.htm.

Roush, W. (1994, August 26). Population: The view from Cairo. *Science,* p. 1164. Available: http://www.mit.edu:8001/people/weroush/population.html.

QUESTIONS

1. The Cairo conference plan is ambitious on many counts. Identify five areas in which this plan is especially ambitious.

2. What do you think is the response to the conference's resolutions from nations whose religious and cultural convictions do not support the equal status, education, and self-control of women? What can be done about their response?

3. Some sources have said that this conference was one of the most revolutionary of any U.N. conferences. Why do you think this might be so?

4. Relate the World Birth Control Fact Sheet to the U.N. resolutions from the Cairo conference. Given those facts on birth control, what are some of the biggest obstacles to accomplishing the resolutions? Do you think that the resolutions deal with the core of the problem? Why or why not?

A Bridge to Your 21st Century Understanding

Complete and discuss the following flowchart.

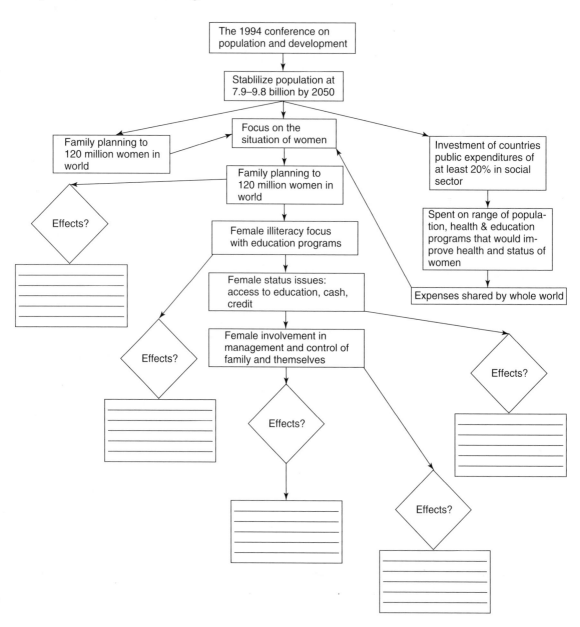

World Birth Control Discussion Issues

SOME WORLD ARGUMENTS AGAINST POPULATION AND BIRTH CONTROL

1. *Progrowth position:* Rapid population growth in a particular country or region is a positive force on grounds of
   (a) economic development (a larger population provides necessary economies of scale and a good labor supply);
   (b) protection of presently underpopulated areas from covetous neighbors;
   (c) differences in fertility among ethnic, racial, religious, or political population segments; and
   (d) military and political power and the strength of a younger age structure.

2. *Revolutionary position:* Population programs are merely a way to hide or placate fundamental social and political contradictions that would lead to a just revolution and therefore could be viewed as inherently counterrevolutionary.

3. *Anticolonial and genocidal position:* The motives of highly developed countries of pushing less developed countries (LDCs) to adopt aggressive population programs are open to suspicion. These more developed countries went through a period of rapid population growth as part of their own development processes, and their current efforts to restrain population growth in the less developed countries are an attempt to keep control and power by retarding the development of these countries. Population limitation could be seen as attempt of the rich countries to "buy development cheaply" or a racist or genocidal attempt to reduce the size of poor and largely nonwhite populations.

4. *Overconsumption by rich countries position:* Population problems are actually resource-scarcity and environmental-deterioration problems that derive from activities of the rich, highly developed countries and not from high fertility in the LDCs. Even if fertility is too high in the LDCs, this is a consequence of their poverty, which, in turn, results from overconsumption of the world's scarce resources by rich countries.

5. *Agricultural and technological improvement position:* Growing population numbers can be accommodated as they have in the past by improvements in agricultural and industrial technologies. Past Malthusian predictions were incorrect, and the same is true of these new Malthusian predictions and solutions. Overpopulation is really underemployment. A humane and well-structured economy can provide employment and subsistence for all people, no matter what the size of the population.

6. *Distribution problem position:* It is not the population numbers themselves that are causing population problems, but their distribution in space. Many areas of the world or countries are underpopulated; others have too many people condensed into too small an area. Instead of efforts to moderate the rate of growth,

Reprinted by permission of *Foreign Affairs,* July 1974. Copyright 1974 by the Council on Foreign Relations, Inc. Teitelbaum, M. (1974, July). Population and Development: Is a Consensus Possible? *Foreign Affairs,* Vol. 52 No. 4.

governments should undertake efforts to reduce more population flows to urban areas and to distribute population on available land.

7. *High mortality and social security position:* High fertility is a response to high death and disease rates; if these levels were reduced, fertility would decline naturally. Living children are the primary way that poor people can achieve security in old age. Therefore, improvements in mortality rates and social security programs would lead to a reduction in fertility.

8. *Status of women position:* High fertility levels are perpetuated by the status and roles of women primarily as procreative agents. As long as women's economic and social status depends largely upon the number of children they bear, there is little possibility that societal fertility levels will decline.

9. *Religious doctrine positions:* The ideology of "Be fruitful and multiply; God will provide" does not recognize population as a serious problem. The other argument holds that while current rates of population growth are a serious problem, the primary instruments to deal with them—modern contraception, surgical sterilization, and abortion—are morally unacceptable; abortion is "murder," and surgical sterilization is "unnatural."

10. *Medical risk position:* Fertility reduction is not worth the medical risks of using the medical means of population programs. Oral contraceptives and intrauterine devices have measurable, if small, short-term risks and possible long-term effects. Sterilization and abortion are operative procedures, both of which have an element of risk, especially when performed outside a hospital.

11. *Progress and development position:* As the social and economic aspects of a society further develop, fertility rates decline. Most of the decline in fertility in LDCs derives from social and economical development than from the success of population control programs. International assistance for development is too heavily concentrated on population programs and is shortchanging the focus on general overall development.

12. *Social justice position:* Fertility will not decline and population programs will not be successful until the basic reasons for high fertility—poverty, ignorance, and fatalism—are eliminated through social policies that result in a redistribution of power and wealth among the rich and poor, both within and among nations.

SOME WORLD ARGUMENTS SUPPORTING POPULATION AND BIRTH CONTROL

1. *Population activist position:* Unrestrained population growth is the principal cause of poverty, malnutrition, environmental disruption, and other social problems. The situation is desperate and necessitates action to restrain population growth, even if coercion is required: "Mutual coercion, mutually agreed upon." Population programs are fine as far as they go, but they are wholly insufficient in scope and strength to meet the urgency and reality of the situation.

2. *Services need position:* According to data and surveys, there is a great unmet demand for fertility control in all countries; therefore, the main problem is to provide modern fertility control to already motivated people. Some proponents also hold that the failure of some population service programs is due to inadequate fertility control technologies and that there is an urgent need for technological improvements in this area.

3. *Human rights position:* According to U.N. propositions, it is a fundamental human right for each person to be able to determine the size of his or her own family. Furthermore, some argue that each woman has the fundamental right to the control of her own bodily processes, a rationale that supports contraception and possibly abortion. Health is also a basic human right that population programs support directly and indirectly and includes the direct medical benefits of increased child spacing for maternal and child health and the indirect effects of reducing numbers of dangerous illegal abortions.

4. *Population control and development position:* Social and economic development programs help, but alone are not sufficient to bring about low mortality and fertility levels. Special population programs are also required. Too rapid population growth intensifies other social and economic problems. Some countries might benefit from larger populations, but would benefit more by moderate rates of growth over longer periods of time than by rapid rates of growth over shorter periods of time. Economic, social, and population programs are all interlinked.

Use of Contraceptive Methods,
by Region
Courtesy of UN Population Fund.

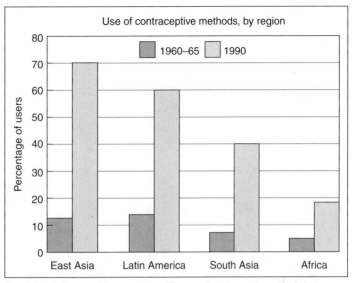

*For all developing countries on average. The use of contraceptive methods has
grown from less than 10% in the 1960s to more than 50% today.*

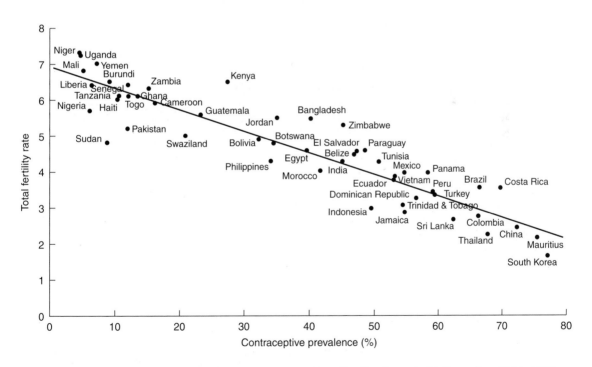

The Relationship between Contraceptive Prevalence and Total Fertility Rates in 50 Countries, 1984–1992.
(From Population Reports 1992.)

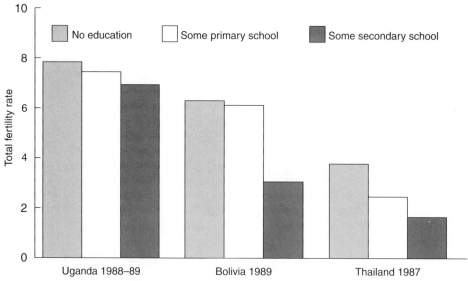

Reprinted with permission from R.V. Short, *Contraceptive Strategies*, p. 325. North American Press.

Total fertility rates in three countries by women's level of education. Before fertility rates start to fall, fertility is high in all groups (Uganda). As educated women are first to reduce their fertility, differences widen amongst groups (Bolivia). Eventually fertility is low in all groups (Thailand). (From Population Reports 1992.)

Percentage Use of Contraceptive Methods Worldwide (1987)

Currently, 390 million couples are using modern contraceptives, about 51% of the total

female sterilization	29
intrauterine devices	20
oral contraceptives	14
condoms	9
male sterilization	8
coitus interruptus	8
rhythm	7
injectables	2
other methods	2
barrier methods	1

Note. Methods with Lowest Failure Rates Predominate in Developing Countries. Methods with Highest Failure Rates Predominate in Developed Countries

Reprinted with Permission from R.V. Short, Contraceptive Strategies, p. 330, North American Press.

BIRTH CONTROL

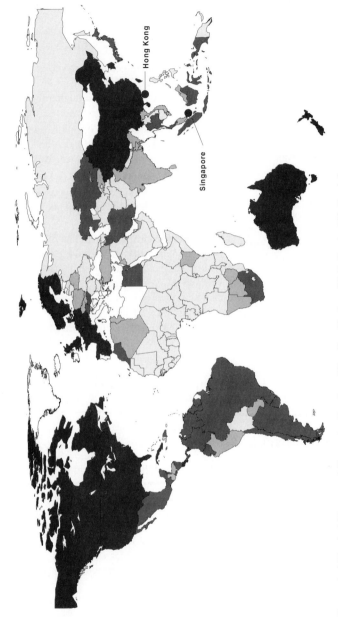

Percent of Married Women of Reproductive Age Who Use Modern Birth Control Methods*

Hong Kong

Singapore

| Under 25 | 25 to 44 | 45 to 64 | 65 or more | No data |

*Includes women in non-marital unions.

Rodger Doyle Copyright 1996

World Birth-Control Use

Source: Based on data compiled by the Population Reference Bureau and Population Crisis International. Data for some countries are estimates. Data for most countries were collected in the late 1980s and early 1990s. Data apply to married women and women in nonmarital unions. Excerpted with permission from Doyle, R (1996, September). World birth-control use. *Scientific American*. Vol. 275, No. 3, p. 34.

WORLD BIRTH CONTROL FACT SHEET

I. In the past 30 years, there has been a marked decline in world fertility rates, particularly in developing countries.

1. Between 1960–1965, women in developing countries averaged six births over a lifetime while in 1990–1995, they averaged 3.4 births over a lifetime.

2. In east Asia for the past 30 years, births fell 65% and are now below replacement rate of 2.1 children.

3. In other parts of Asia, births declined by about a third, Latin America halved, and Africa only by 10%.

II. In the developed countries, the number of births per woman declined by about 40% and are now below replacement level in all these countries, including the United States.

III. Modern contraceptive methods have played a key role in lowering fertility.

1. Among women of reproductive age who are married (or in nonmarital unions), half now depend on the following methods: female sterilization (most popular), male sterilization, hormonal implants (such as Norplant), injectibles (such as Depo-Provera), intrauterine devices (IUD's), birth control pills, condoms, and diaphragms.

2. The first four methods are almost 100% effective, followed by IUD's, the pill, and the male condom in effectiveness. Diaphragms are among the least effective.

3. Condoms (both the male and female type) are the only methods that provide protection against sexually transmitted diseases, such as AIDS.

4. The percentage of women using modern contraception is as follows: 54% in Asia (39% if China is excluded), 53% in Latin America, 30–40% in the Muslim Middle East and North Africa, 48% in the countries of the southern tip of Africa, <10% in middle Africa, 65–75% in the United States and western Europe, in the former Soviet Union <20% because of short supply.

IV. Growth in birth-control use and decline in fertility in developing countries is closely linked to educational opportunities for women (e.g., sub-Saharan Africa, with the highest fertility rates, has the lowest female educational levels).

1. Increased literacy makes it easier for women to receive reliable information on contraception.

2. The demands of education, particularly at the post-secondary level, cause women to delay marriage and childbearing.

V. Some developing countries, such as China and Cuba, are already below the replacement level of 2.1 children, mostly because of modern birth-control methods.

1. Countries such as Brazil, Indonesia, Vietnam, South Africa, Turkey, Egypt, and India should reach this level within the approximate next ten years.

2. Pakistan and Nigeria, where few women use modern contraception, are at the other extreme and are not likely to reach the replacement rate for several decades.

VI. More historical and traditional methods of birth control include the rhythm method, coitus interruptus, and prolonged breast-feeding (which suppresses ovulation). Worldwide, 7% of all married women (or women in nonmarital unions) of reproductive age depend on these methods, which are far less reliable. The rhythm method is prevalent in Peru, and coitus interruptus is the birth control method of choice in Turkey.

Excerpted with permission from Doyle, R. (1996, September). World birth control use. *Scientific American* Vol 275, No. 3, p. 34.

A Day in the Life of the World 1993

100,000,000	Acts of intercourse will take place
910,000	Conceptions will occur
150,000	Abortions will be performed
500	Mothers will die as a result of abortion
384,000	Babies will be born
1,370	Mothers will die or pregnancy-related causes
25,000	Infants in the first year of life will die
14,000	Children aged 1–4 will die
356,000	Adults will get a sexually transmitted disease
250,000	Is the net increase in population of the world

Reprinted with Persmission from North American Press; Short, R.V. *Contraceptive Strategies*, p. 327.

QUESTIONS

To answer these questions refer to all of chapter 34 including graphs and tables.

1. What positions against population and birth control do you think have the most merit? Why?

2. What positions for population and birth control do you think have the most merit? Why?

3. What positions do you think are the most difficult to address concerning the population issue and the 1994 U.N. resolutions made in Cairo?

4. What do these positions for and against population and birth control tell you about other cultures and societies? What do they tell you about their priorities?

5. What plans might you have to address or change the positions you identified in Questions 1 and 2?

6. Do you think that the U.N. resolutions adequately address the concerns of the positions against birth control?

POPULATION CONTROL: A "THIRD WORLD" PERSPECTIVE

During recent years, many scientists and economists have expressed a concern about the population growth and scarcity of resources. The world population has increased from 3.72 billion (1970) to 5.32 billion (1990) and is projected to be 10 billion by the year 2010. Many experts believe that earth's population is about to surpass the planet's "carrying capacity." Contrary to this, World Bank figures show that food prices have declined dramatically to an historic low—reflecting improvements and a worldwide surplus of grain. The population growth rates in the developed countries are low whereas for the developing countries they are high.

To address the question of population control, the United Nations (UN) held a conference in Cairo. The conference adopted a resolution for the population control. The resolution document, called "Proposed Program of Action for the Cairo Conference," calls for adoption of various birth control methods, including abortion for population control. The majority of the Third-World countries with Catholic and Muslim populations believe that the UN has gone too far in making these recommendations. The UN recommendations are in direct contradiction to the religious beliefs of Catholics and Muslims all over the world. The Pope has formed a Catholic-Islamic alliance to oppose UN recommendations for population control.

- Do you think that the UN is justified in recommending population control strategies that contradict people's religious beliefs?

- In many Third-World countries the majority of people are farmers. Due to a lack of technology and resources, they believe in having large families in order to manage their farms. What population control strategy would you recommend to help Third-World farmers?

- Many Third-World nations believe that population is an asset and that the developed countries are trying to control their future potential through the United Nations. Do you agree with this view? Why? Why not?

Case Study: Japan's Recent Lift of Birth Control Pill Ban A Thirty Year Battle

In 1999, Japan's Ministry of Health and Welfare finally announced that oral contraceptives, which had been officially banned for more than 30 years, would be approved. Even though the pill is now approved, doctors say that it will not become popular quickly, as opinion polls regularly show widespread misconceptions from women about the pill and its side effects.

In the past, only women with menstrual disorders could legally use the pill in high and middling doses; however, the number of women who redirected the use of the pill from the treatment of such disorders to birth control are estimated as high as 500,000. Many believe that Japan's abortion rate, one of the highest in the world at 400,000 per year, would be much reduced if pill use became more widespread.

The passage of the law allowing oral contraceptive has been a difficult road. In 1965, pill approval was denied because of fears that it would corrupt sexual morals. In 1986, the government set up a medical study group to examine the feasibility of the use of a low-dose contraceptive pill that concluded in favor of its introduction, but legislation was never forthcoming. Drug companies petitioned the government in 1990, but fears that condom use (the main form of birth control in Japan) would decline, which could lead furthermore to an increase in the spread of HIV, brought about the pill's rejection in 1992. Another possible approval was lost again in late 1995 with publicity surrounding studies on venous thromboembolic disease and the use of combined oral contraceptives.

Many blamed the 30-year ban on a male-dominated bureaucracy that denied women contraceptive freedom. Other important factors that stopped the pill from becoming legal were the profits of condom manufacturers, profits of those who perform abortions and anxiety about Japan's rapidly declining population. There were even debates about concern of polluting effects of hormone-tainted urine to the environment. Without strong feminist and antiabortion movements to keep the issue alive, complacency added to the pill not becoming legalized. In contrast, it should be noted that Japan's Ministry of Health approved Viagra, the pill that helps prevent erectile dysfunction, in six months.

"The resistance to pass the law reflects a general disregard for women and their health needs," says Michael Reich, chair, Department of Population and International Health; Iain Aitken, lecturer, Department of Maternal and Child Health Care; and Aya Gota, a former student of Harvard School of Public Health who now lives in Japan, in a December 8, 1999 article in the *Journal of the American Medical Association* (hsph.harvard, 2000).

Similar to the predictions in spite of the passage, Japanese women have not rushed for pill prescriptions because, as Aitken states, "a lot of residual fears about side effects remain." Other reasons for the slow adoption he said could also be that Japanese society as a whole resist using pharmaceuticals extensively. There is the additional inconvenience for women using the oral contraceptive who must visit their gynecologists every three months, which they have to pay for themselves. Abortion costs, however, are covered by health insurance policies.

International health officials have applauded the Japanese government for finally approving the pill, but emphasize the need to increase focus on women's issues and health concerns which is especially indicated by this long battle of women's reproductive safety, health, and contraceptive options.

REFERENCES

Excerpted with permission from Gutierrez, E. And Netley, G. (1996, September 28). Japan's ban on the pill seems unlikely to change. *The Lancet.* Vol. 346, p. 886.

Birth control pill hard to swallow for some Japanese. (2000, February 25). *Around the School: News and Notices of the Harvard School of Public Health.* Retrieved August 25, 2001 from the World Wide Web: http://www.hsph.harvard.edu/ats/Feb25/feb25_02.html.

QUESTIONS

1. Even though the ban on the pill has been lifted, why is this still an issue in Japan?

2. What does this article imply about the reproductive rights of women in Japan?

3. What does this indicate about problems of passage of laws as a case study for other countries whose laws are striving to support population control and women?

4. What might this possibly indicate about political power and power groups in other countries and the importance of political activism on population control issues?

5. Are oral contraceptives an issue for male political power groups in many countries throughout the world? Why or why not? Discuss from many perspectives including political power, women's rights, cultural, religious, historical, etc.

6. Research other countries' birth control policies and laws and discuss the implications to their population control practices, rights of women, and population growth numbers.

A Bridge To Your 21st Century Understanding

Complete and discuss the following flowchart.

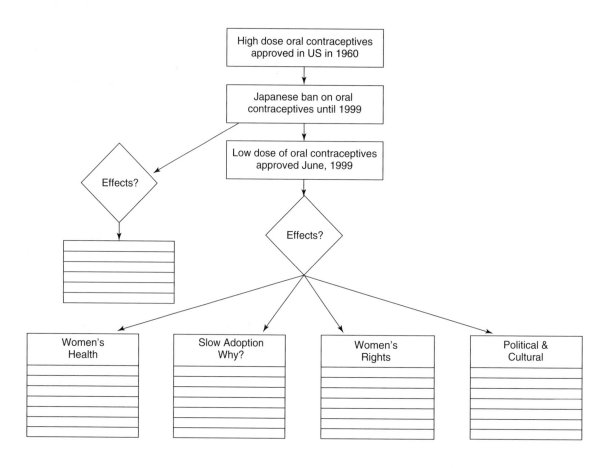

Fertility Rates: The Decline Is Stalling

LINDA STARKE

During the 1970s, one of the encouraging developments in population trends was the reduction in the total fertility rate in several key countries, including the world's two largest nations—China and India. (The fertility rate measures the average number of children born to women in their childbearing years.) In China, the rate dropped precipitously, from 6.4 children per woman in 1968 to 2.2 in 1980. In India, the decline was more modest, but still significant: from 5.8 children per woman between 1966 and 1971 to 4.8 children between 1976 and 1981.

These trends helped slow the rate of world population growth from 2.1 percent between 1965 and 1970 to 1.7 percent between 1975 and 1980. At that point, however, the decline in the number of children that women were having in these two population giants stalled.

In China, despite the most aggressive and least democratic population control program in the world, the fertility rate remained around 2.5 throughout much of the 1980s as couples continued to want to marry young and to have two or more children. In India, the overzealous promotion of family planning by the ruling Congress Party through 1977 apparently backfired after the party's defeat, and progress toward lower birth rates ran out of steam.

One important lesson from these experiences is that governments must do more than just supply contraceptives; they need to lower the demand for children by making fundamental changes that improve women's lives and increase their access to and control over money, credit, and other resources.

Many countries still register fertility rates above replacement level (see Table 36.1), which is generally 2.1 children per woman or basically two children per couple. The total fertility rate for the world as a whole in 1991 was 3.3, ranging from 1.8 in more developed nations to 4.4 in less developed ones (excluding China). In a number of countries, such as Brazil, Egypt, Indonesia, Mexico, and Thailand, fertility rates have been dropping as they did in the 1970s in China and India. At the same time, many developing countries have not yet entered the demographic transition.

The demographic transition occurs when both birth rates and death rates in a country drop from historically high levels to low ones that translate into a stable population—one that merely replaces itself with each new generation. Traditionally, although not always, death rates have declined first, following the spread of sanitation and improved health care overall. Rapid population growth often follows this first phase of the demographic transition, as the gap between fertility and mortality rates widens for a time. Eventually, however, fertility rates fall too.

At the moment, they remain high in a number of countries. The reasons include unequal rights and opportunities for women, as well as inadequate access to birth control. Whatever the reason, the effect is the same: 67 countries, home to 17 percent of the world population, are at best in the early stages of a transition to low fertility rates. Most of them are in Africa and South Asia, and their populations are likely to double in 20 to 25 years.

From the Worldwatch Institute's annual report of key global indicators, *Vital Signs 1993.*

Table 36.1

Population Size, Fertility Rate, and Doubling Time, 20 Largest Countries, 1993

Country	Population (Millions)	Fertility Rate (Average Number of Children Per Woman)	Doubling Time (Years)
Italy	58	1.3	3466
Germany	81	1.4	*
Japan	125	1.5	217
United Kingdom	58	1.8	267
France	58	1.8	169
Russia	149	1.7	990
United States	258	2.0	92
China	1,178	1.9	60
Thailand	57	2.4	49
Indonesia	188	3.0	42
Brazil	152	2.6	46
Turkey	61	3.6	32
Mexico	90	3.4	30
India	97	3.9	34
Viet Nam	72	4.0	31
Philippines	5	4.1	28
Egypt	8	4.6	30
Pakistan	122	6.7	23
Iran	3	6.6	20
Nigeria	5	6.6	23

Source: Population Reference Bureau, 1993 World Population Data Sheet (Washington, D.C.: 1993).
*At its Current Mortality Rate, Germany's Population is Shrinking, and at this Fertility Rate Will Never Double.

This is leading to a two-tiered demographic world that is every bit as worrying as the world of economics haves and have-nots. Countries such as Nigeria and Pakistan are finding it harder to keep up with the demand for food, health care, jobs, housing, and education than countries that are in the middle of the demographic transition.

As Shiro Horiuchi of Rockefeller University notes, "the demographic gap seems to overlap with a growing gap in economic development." This gap has been growing for more than 20 years. In 1970, 34 percent of the world lived in countries with fertility rates below 5.5. Just five years later, thanks to the dramatic declines in India and especially China, the figure was 80 percent. It has not gained much since then, however, reaching 83 percent by 1985. More than three fourths of the significant declines in fertility rates started in the 1965 to 1970 period, and not many have begun since then.

Even when a country does reach replacement-level fertility, its population can continue growing for decades. There is a built-in momentum created by all the people who have yet to enter their childbearing years. Indeed, the decline in the world's population growth rate stalled in the 1980s in part because even in China, India, and other countries where fertility rates had been dropping, a large number of people who had been born in the 1960s reached childbearing age. So even if couples had two or three children instead of five or six, as their parents did, the population would grow substantially.

For the world as a whole, even if replacement-level fertility had been achieved in 1990, the population would continue to grow until it reached 8.4 billion in 2150 because of all the young people already alive.

This built-in momentum obviously limits how quickly any country can stop population growth. Nevertheless, reaching replacement-level fertility is an all-important first step. The 67 countries that have not yet begun the demographic transition—nations in which invariably the government believes fertility levels are too high—could move in the right direction by providing the contraceptive and health care services that would help couples have only the number of children they desire.

QUESTIONS

1. What is the fertility rate?

2. What was the fertility rate as a whole in the world in 1991?

3. How many countries are in the early stages of a transition to low fertility rates? What are the characteristics of this early transition phase? What is the longer term effect of being in the early stages of a demographic transition?

4. Why is this demographic gap overlapping with an economic gap?

5. Why did the decline of the fertility rate stall in the 1980s?

6. Why, if replacement-level fertility had been reached in 1990, would the population continue to grow to 8.4 billion?

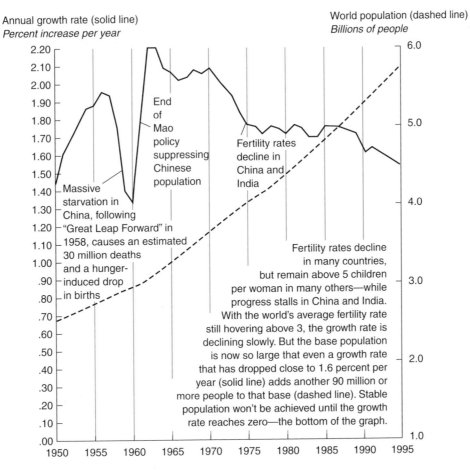

Annual growth rate (solid line)
Percent increase per year

World population (dashed line)
Billions of people

End
of
Mao
policy
suppressing
Chinese
population

Fertility rates
decline in
China and
India

Massive
starvation in
China, following
"Great Leap Forward" in
1958, causes an estimated
30 million deaths
and a hunger-
induced drop
in births

Fertility rates decline
in many countries,
but remain above 5 children
per woman in many others—while
progress stalls in China and India.
With the world's average fertility rate
still hovering above 3, the growth rate is
declining slowly. But the base population
is now so large that even a growth rate
that has dropped close to 1.6 percent per
year (solid line) adds another 90 million or
more people to that base (dashed line). Stable
population won't be achieved until the growth
rate reaches zero—the bottom of the graph.

Figure 36.1
World Population: Even as the Rate of Growth Declines, the Number of Humans Continues
to Climb.

Table 36.2

Projected Population Size at Stabilization, Selected Countries

Country	Population in 1986	Annual Rate of Population Growth	Size of Population at Stabilization	Change from 1986
	(million)	(percent)	(million)	(percent)
Slow Growth Countries				
China	1,050	1.0	1,571	+ 50
Soviet Union	280	0.9	377	+ 35
United States	241	0.7	289	+ 20
Japan	121	0.7	128	+ 6
United Kingdom	56	0.2	59	+ 5
West Germany	61	−0.2	52	− 15
Rapid Growth Countries				
Kenya	20	4.2	111	+435
Nigeria	105	3.0	532	+406
Ethiopia	42	2.1	204	+386
Iran	47	2.9	166	+253
Pakistan	102	2.8	330	+223
Bangladesh	104	2.7	310	+198
Egypt	46	2.6	126	+174
Mexico	82	2.6	199	+143
Turkey	48	2.5	109	+127
Indonesia	168	2.1	368	+119
India	785	2.3	1,700	+116
Brazil	143	2.3	298	+108

Source: World Bank, *World Development Report 1985* (New York: Oxford University Press, 1985).

Table 36.3

Countries That Reached Population Stability by 1995

Country	Population Mid-1995	Birth Rate	Death Rate	Annual Rate of Natural Increase Or Decrease[1]
	(million)	(births per thousand population)	(deaths per thousand population)	(percent)
Austria	8	12	10	+0.1
Belarus	10	11	13	−0.2
Belgium	10	12	11	+0.1
Bulgaria	8	10	13	−0.3
Croatia	4	10	11	−0.1
Czech Republic	10	12	11	0
Denmark	5	13	12	+0.1
Estonia	2	9	14	−0.5

continued

Table 36.3
Countries That Reached Population Stability by 1995, *continued*

Country	Population Mid-1995	Birth Rate	Death Rate	Annual Rate of Natural Increase Or Decrease[1]
	(million)	(births per thousand population)	(deaths per thousand population)	(percent)
Finland	5	13	10	+0.3
France	58	12	9	+0.3
Georgia	5	12	10	+0.2
Germany	82	10	11	−0.1
Greece	10	10	9	0
Hungary	10	12	14	−0.3
Italy	58	9	10	0
Japan	125	10	7	+0.3
Latvia	2	10	15	−0.5
Lithuania	4	13	12	0
Norway	4	14	11	+0.3
Poland	39	12	10	+0.2
Portugal	10	12	11	+0.1
Romania	23	11	12	−0.1
Russia	148	9	16	−0.6
Slovenia	2	10	10	+0.1
Spain	39	10	9	+0.1
Sweden	9	13	12	+0.1
Switzerland	7	12	9	+0.3
Ukraine	52	11	14	−0.4
United Kingdom	59	13	11	+0.2
Yugoslavia[2]	11	13	10	+0.3

[1]Does not Always Reflect Exactly the Difference Between Birth Rates and Death Rates Shown Because of Rounding.
[2]Reflects the New Yugoslavia, Consisting of Serbia Plus Montenegro.
Source: Population Reference Bureau, *1995 World Population Data Sheet* (Washington, D.C.: 1995).

QUESTIONS ABOUT THE TABLES

1. What do you think is meant by "stabilization of population rate"?

2. Explain the information in Table 36.2. Why do the rapid-growth countries experience a much larger change of percentage of growth rate than the slow-growth countries?

3. Compare Table 36.3 with Table 36.2. What are the significant differences between stable-growth countries and the rapid-growth countries?

4. Compare your results from Question 3 with the information from Table P5.7 and Table P5.8 at the beginning of this part. Note specifics like fertility rate per woman of Table P5.7 and Table P5.8 which further support Tables 36.2 and 36.3. Select a country and give three specifics from Tables P5.7 and P5.8 which support Tables 36.2 and 36.3.

37

Population Carrying Capacities: Four Case Studies—Pakistan, India, Ethiopia, and Mexico

THE CONCEPT OF POPULATION CARRYING CAPACITY

Population carrying capacity is an approach that many ecologists use to mark and provide a sustainable limit philosophy that measures needed ecological policy. Economists tend to view carrying capacity with more flexibility due to technological and policy interventions so that the approach has a less fixed value.

Population carrying capacity can be defined as "the number of people that the planet can support without irreversibly reducing its capacity or ability to support people in the future." While this is a global-level definition, it applies at a national level too, albeit with many qualifications, for example, international trade, investment and debt. Furthermore, it is a function of factors that reflect technological change, food and energy supplies, ecosystem services (such as provision of freshwater and recycling of nutrients), human capital, people's lifestyles, social institutions, political structures and cultural constraints among many other factors, all of which interact with each other.

Two points are particularly important: that carrying capacity is ultimately determined by the component that yields the lowest carrying capacity; and that human communities must learn to live off the "interest" of environmental resources rather than off their "principal." Thus, the

concept of carrying capacity is closely tied in with the concept of sustainable development. There is now evidence that human numbers with their consumption of resources, plus the technologies deployed to supply that consumption, are often exceeding the land's carrying capacity already. In many parts of the world, the three principal and essential stocks of renewable resources—forests, grasslands and fisheries—are being utilized faster than their rate of natural replenishment.

Consider a specific example—the Earth's carrying capacity with respect to food production. According to the World Hunger Project, the planetary ecosystem could, with present agro-technologies and with equal distribution of food supplies, satisfactorily support 5.5 billion people if they all lived on a vegetarian diet (the 1991 global population is already 5.4 billion). If people derived 15 percent of their calories from animal products, as tends to be the case in South America, the total would decline to 3.7 billion. If they derived 25 percent of their calories from animal protein, as is the case with most people in North America, the Earth could support only 2.8 billion people.

True, these calculations reflect no more than today's food production technologies. Certain observers protest that such an analysis underestimates the scope for technological expertise to keep on expanding the Earth's carrying capacity. We can surely hope that many advances in agro-technologies will come on stream. But consider the population-food record over the past four decades. From 1950 to 1984, and thanks largely to remarkable advances in Green Revolution agriculture, there was a 2.6-fold

---

Excerpted from *Population Resource and the Environment: The Critical Challenges* United Nations Population Fund.

increase in world grain output. This achievement, representing an average increase of almost 3 percent a year, raised per capita production by more than a third. But from 1985 to 1989, there was next to no increase at all, even though the period saw the world's farmers investing billions of dollars to increase output (fertilizer use alone expanded by 14 percent). These big investments were supported by rising grain prices and by the restoration to production of idled United States cropland. Crop yields had "plateaued"; it appeared that plant breeders and agronomists had exhausted the scope for technological innovation. So the 1989 harvest was hardly any higher than that of 1984. During that same period, there were an extra 440 million people to feed. While world population increased by almost 8.5 percent, grain output per person declined by nearly 7 percent.

To put the case more succinctly, every 15 seconds sees the arrival of another 44 people, and during the same 15 seconds the planet's stock of arable land declines by one hectare. Plainly, this is not to say the first is a singularly causative factor of the second. Many other factors, notably technology, contribute to the linkage. Equally, it is not to deny that there is a strong relationship between the two factors: more people are trying to sustain themselves from less cropland.

Regrettably, there is all too little concise analysis of the concept of carrying capacity. So the evaluation of its nature, and especially of the threat of environmental overloading, must remain largely a matter of judgment. But remember that in a situation of pervasive uncertainty, we have no alternative but to aim for the best assessment we can muster. We cannot defer the question until such time as we have conducted enough research. After all, if we do not derive such conclusions as we can and make explicit planning decisions on their basis, there will be implicit planning decisions taken by large numbers of people who, through their daily lifestyles, are determining the outcome. In other words, decisions on the population-environment nexus will be taken either by design or by default. We must make do with such information and understanding as we have at hand, however imperfect that may be.

To comprehend the concept of carrying capacity in real-world terms, consider the following case studies. Through the experience of four countries, drawn from all three main developing regions, we can discern the strong links between population pressure and the capacity of the natural resource base to support increasing human numbers.

QUESTIONS

1. Define "population carrying capacity."

2. What two points underlie carrying capacity?

3. What is the cause of the difference between earth's carrying capacity estimates of food production of 5.5 billion people, 3.7 billion people, and 2.8 billion people, as mentioned in the article? What number do you think will reflect a sustainable world standard of living and why? Explain.

4. What happens every 15 seconds in the way of adding and subtracting to earth's carrying capacity?

5. What do you think of the validity of the concept of carrying capacity? Do you think that this concept can help us design our future, or will we proceed to our future by default?

PAKISTAN

Pakistan, like its neighbor India, suffers much environmental degradation in the form of deforestation, soil erosion, desertification and depletion of water supplies. This resource decline is leading to agricultural problems; indeed, to a growing incapacity of natural environments to support present human numbers. Yet the country's natural resource base is central to the national economy, with agriculture the largest single development sector. In 1985, agriculture accounted for 32 percent of Pakistan's GDP, employed 55 percent of the population and earned 79 percent of all export revenues.

Pakistan's land area of 887,722 square kilometers is mostly semi-arid or arid. At the time of independence in 1947, the populace totalled 31 million people, and the country was self-sufficient in most staple foods. By 1990, the populace totaled 123 million people, and was growing at an annual rate of 2.9 percent, doubling the population in 24 years. Allowing for a further decline in the death rate and some decline in the birth rate, the population is projected to reach 162 million people by 2000 and 248 million by 2020; more than twice as many people in a mere three decades. But it is questionable whether there will be much fertility decline in the foreseeable future. The average family size today is 7 children (technically expressed as 6.7), the highest in southern Asia except for Afghanistan. This is considered to be due, in part, to the poor social status of women, and the general neglect of women's development needs in this predominantly Muslim country. Only

3 percent of women are educated, and the female to male school enrollment ratio is among the 10 lowest in the world.

Between 1947 and 1985, Pakistan's birth rate fell from 45 to only 43 per 1,000 people. But in 1990, it is estimated to have risen to 44 per 1,000. By contrast, the crude death rate has fallen from over 25 to 13 per 1,000 people, still higher than the rate for India (11 per 1,000) and the average for the region (12 per 1,000). Obviously, there is some scope for a further decline in the death rate. The level of contraceptive use among reproductive-age women is only 8 percent, one seventh that recorded for the region. So the population size in 2100 and 2020 could be rather higher than projected. With only a marginal decline in the birth rate and a continuing decline in the death rate, Pakistan could find itself trying to support twice as many people in less than a generation, and more than 400 million by the time it has enjoyed another 45 years of nationhood.

The 3.7 times increase in human numbers since 1947 has been accommodated mainly by three major advances in agriculture during the 1960s. These were: the Indus Basin Treaty that provided a greater volume of riverwater for irrigation; a sharp increase in the number of tubewells supplying access to more underground water; and the introduction of high-yielding varieties of grain. Today, there appears to be little opportunity to gain much more advantage from the first two, while the third provided scant support during the 1980s when grain yields plateaued.

Meantime, the country's natural resource base has experienced growing stress for at least 30 years. Consider three key sectors: forestry, water and agricultural land. A maximum of 3 percent of national territory is still forested, whereas the government believes it should be at least 15 percent. The Director-General of the Pakistan Forestry Institute, referring to deforested watersheds, concludes the country "already faces a grave situation insofar as water for irrigation and power generation is concerned." Yet deforestation seems set to grow worse if only because of growing demand for fuelwood, which in the mid-1980s accounted for 50 percent of the country's energy requirements and almost 90 percent of total wood consumption. There was an acute shortage of fuelwood as far back as 1978, when the annual growth of wood was supplying only 62 percent of the annual harvest. Yet demand for fuelwood is projected to double during the last two decades of this century. So there appears to be little relief for watershed forests in the foreseeable future.

Second, water is in short supply, especially for agriculture. An irrigation network covers almost a fifth of the country, the largest such system in the world. Some 156,000 square kilometres of croplands are irrigated, a full 70 percent of the cropland expanse; a proportion way ahead of Japan's 63 percent, China's 48 percent and India's 33 percent. These irrigated croplands produce 80 percent of the country's food. Irrigated wheatlands, comprising 38 percent of all irrigated croplands, and four times as large an area as rain fed wheatlands, average over 2,000 kilograms of harvest per hectare a year, rain fed wheatlands only around 850 kilograms.

But water is the scarcest input to agriculture, and supplies are likely to run out well before other resources such as land, energy, high yielding grains, fertilizers and pesticides. On four occasions during the 1980s, Pakistan has experienced critical water shortages, leading to inadequate irrigation-water flows for the main crop-growing season. According to the Director of the Pakistan Water and Power Development Authority, "Water supplies presently available for irrigation are already short of optimum crop water requirements . . . Even if all surface supplies and groundwater reservoirs are exploited, water would still fall far short of future requirements."

Indeed, there is little prospect of increasing the water supply at all. The Indus River and its tributaries could provide perhaps another 25 to 30 percent in volume, but this would lead to saltwater intrusions into the river's lower reaches. Nor is there much scope for recycling irrigation water. Recycled flows are usually only half as useful as original withdrawals, because of grossly declining quality as irrigation channels pick up fertilizers, salts, pesticides and toxic elements from croplands.

Fortunately, there could be much greater efficiency in water use. If water use were better regulated and controlled, the expanse of double-cropped lands could theoretically be tripled. But because of incorrect pricing policies for water, among other factors, wheat yields per hectare/metre (the amount of water needed to cover one hectare to a depth of one metre) have increased from 2.17 tons per hectare a year in 1971–76 to only 2.34 tons in the period 1981–86, despite the use of improved seeds and expanded use of fertilizers. In addition, there is widespread leakage of irrigation water: the first 20 percent is lost from conveyance systems, and 40 percent of the remainder from on-farm water courses, with a further 25 percent attrition from runoff due to inefficient application to crops (fields are often far from level). Because of policy constraints—perverse water pricing has persisted for two decades—there seems little hope that efficiency of water use will be improved within the foreseeable future.

Associated with problems of water supply is the phenomenon of salinization. At least 24 percent of Pakistan's croplands are salinized on the surface, and another 29 percent feature below-surface salinization. Together with waterlogging and alkalization, this means that more than 70 percent of the country's croplands are affected by water-related degradation.

As a result of these constraints on the agricultural resource base, in conjunction with the pressures of population growth, there is pervasive land hunger. To sustain a subsistence income through agriculture (consistent with nutritional needs of 2,400 calories per adult per day), an average-size family of 7 persons needed 2.5 hectares of farmland in 1980. But even a decade ago, when the population was 30 percent smaller than today's, two thirds of farm households did not have enough land to support themselves through this means alone. Whereas the amount of cultivated land per rural inhabitant was a third of a hectare in 1983 (already less than the 0.37 of a hectare considered a minimum with agro-technologies available), it is projected to fall to 0.14 of a hectare in the year 2010. Not surprisingly, there is great and growing pressure from agricultural encroachment on the remaining forests. This, in turn, means that deforestation further aggravates water supply problems for downstream agriculture, reducing the carrying capacity of existing farmlands.

This latter phenomenon of positive feedback between factors of environment and development is paralleled by a reinforcing process between environment and population. Deforestation-derived water deficits affect not only agriculture but public health. Water of sufficient quantity and quality is available to only 23 percent of the rural population. As a result, there is a pervasive problem of water related diseases, especially among young children: infant mortality remains a good deal higher than in adjoining states of India.

All this is not to deny Pakistan's remarkable achievement in agriculture. During much of its 43 years of nationhood, Pakistan has been self-sufficient in most major food staples. But by reason of population growth alone, the 1988 level of wheat demand—16 million tons—is projected to reach 25 million tons by 2000 and 40 million tons by 2010; import requirements are projected to grow to 2 million tons and 5 million tons by each date, respectively. These wheat imports, plus imports of edible oil, are projected to cost 37 billion rupees by 2010 (at 1980 prices), exceeding the value of all projected exports in that year by 25 percent.

On a more positive side, there is much room to step up agricultural productivity. Farmers achieve only 25 to 30 percent of demonstrated potential crop yields. Wheat output per hectare is only 85 percent that of India, corn output 47 percent that of Turkey, and rice output 46 percent that of Egypt. And while the expanse of croplands can scarcely be increased by any more because of soil and climate factors, some 100,000 square kilometers of wastelands, or rather wasted lands, could, through careful management, be brought into production for raising livestock. So theoretically, Pakistan eventually could feed at least three times as many people, or almost 350 million. How, then, does this leave the concept of population carrying capacity?

In Pakistan's case, it appears to reflect three key factors. First, for the country to achieve much advance at all beyond its present meager levels of human welfare (GNP per head $350, with little improvement during the 1980s), it is going to have to tackle the challenge of doing most things right most of the time. In other words, it is going to have to do much better with several leading sectors of development: forestry, agriculture, water, public health and population, to name the more relevant—whether through policy reform, improved planning and programming, institutional adaptation, or more advanced and appropriate technology. This "leap ahead" in socio-economic infrastructure has apparently proved beyond the capacities of the government in a situation of growing constraints and stresses on several fronts. Indeed, the challenge would tax the capacities of a nation with vastly more planning experience and with much more extensive infrastructure.

Secondly, and insofar as carrying capacity must be considered a long-term affair, it is questionable whether Pakistan's development record during its four decades of nationhood has been of a sort to prove sustainable into the foreseeable future. There is much evidence that increases in agricultural output have been accomplished at a cost to sustainable productivity, as evidenced by the rundown of the agricultural resource base via land degradation (salinization, soil erosion) and depletion of resource stocks (forests, surface and groundwater reserves). Technology, in the form of high-yielding grains, fertilizers and pesticides, has enabled food production to keep pace with, or even exceed, population growth for most of the last four decades—a remarkable accomplishment by any standard. But there are signs that the "techno-fix" response to growing human numbers, with their growing aspirations, can no longer maintain its momentum.

The third key factor is that whatever opportunity remains for the nation's carrying capacity to keep on expanding through enhanced agricultural output (sorely limited as this is now turning out to be), it would be taxed in the extreme to keep abreast of population growth at a level of 3 percent a year. Even the most advanced nation would find it difficult to deploy the planning abilities to keep up with a growth rate of that magnitude.

These three factors are crucial to the concept of population carrying capacity for Pakistan. It is difficult to ascertain their individual importance in a conclusive manner, because they have yet to be analyzed (whether in isolation or in combination) to a degree that allows each to be appraised apart from other variables. Nor would a factor-by-factor assessment present the full picture since each interacts with the others in multiple dynamic ways. It may eventually be possible through refined analysis to determine the precise significance of each; to separate out and quantify the carrying-capacity impacts of each, plus their synergistic reinforcing of the other two. Unfortunately, no such analysis has been undertaken.

An operational evaluation of Pakistan's carrying capacity becomes, in essence, a matter of judgment. This judgment should be based upon an informed and considered appraisal of all component factors at issue, together with an assessment of how far the country's problems can be relieved through improved technology, policy changes, institutional innovation and the like in the future. Of course, the future is unknown. But one can usefully speculate in light of the recent past: Pakistan's record during the 1980s surely serves to give warning rather than to inspire hope. Nonetheless, the assessment must remain a matter of judgment—no more and no less.

So while the concept of carrying capacity can still be viewed as something that is putatively expandable in Pakistan's case, there is much evidence to suppose it is practically close to its apparent limits for the foreseeable future—if, indeed, it has not been surpassed already. In these circumstances, it would make sense for Pakistan to respond through two prime strategies: an unprecedented effort to reform agricultural policies together with whatever other measures are necessary to increase food output and to reduce the rate of population growth.

What are the possibilities for population planning? Taking a rather extreme view, were fertility rates to remain constant between 1990 and 2025, total population would grow from approximately 124 million in 1990 to 486 million by 2025. Conversely, were fertility rates to decline to correspond to the United Nations "medium variant" growth projection, total population would rise much less—from approximately 123 million in 1990 to 267 million by 2025. The corresponding reduction in fertility rates to achieve this

decline in total population would require bringing the present birth rate down from 47 births per1,000 population in 1985–90, to 20 births per 1,000 by 2020–25.

Fortunately, Pakistan's present demographic trends are not only a matter of deep national concern, but are mobilizing effective action. Since the 1974 United Nations World Population Conference, the government has consistently viewed its rates of population growth as being "too high" and has sought to support family planning and other population programmes to reduce growth. In addition, the government has consistently viewed its spatial distribution patterns as inappropriate and has sought to decelerate rates of rural-urban migration and to promote a more balanced relationship between population and natural resources.

Balancing resources and human requirements has become a high priority in government planning during the last two five-year national development plans. To achieve its objectives, the government has adopted a population welfare program which utilizes a multi sectoral approach with the specific aim of changing desired family size. This approach encourages active participation of governmental and non-governmental organizations, in conjunction with local communities, and aims at strengthening communication strategies. The government's commitment is reflected in a recent statement by the Minister of Planning and Development that "investment in family planning is a crucial investment for a nation. That is why our population welfare program must rank—and it does—as the top priority in development planning."

A major aim of Pakistan's population welfare program has been to bring about an increase in the small proportion of married women who knew of any contraceptive method in 1984. Only 20 percent of currently married women today know of a source of family planning information or supplies, and only 12 percent have ever used contraception. These figures contrast sharply with Bangladesh where almost 100 percent of married women knew of some contraceptive method in 1985, and 32 percent had used some form of contraception.

Clearly, what is required is the financial and human resources to strengthen and consolidate the existing family welfare program of Pakistan in order to support and expand reproductive health and clinical services; to pursue an integrated clinical training program for family planning and health personnel; to foster community-based maternal and child health/family planning (MCH/FP) services through family welfare centers; and to expand family planning service delivery through the involvement of traditional medical practitioners (hakeems). It also requires the development of research and evaluation capabilities as regards interrelationships between population, environment and sustainable development; the building up of national capacities in policy analysis; the integration of population factors into national development plans; assistance to promote training of nationals to assure self-reliance in population planning in the future; and strategies to promote extensive population education in the formal and non-formal schooling sector, among workers in urban industry, and among male and female workers in farm settings and rural cooperatives.

QUESTIONS

1. What is the main problem for Pakistan's future economic and ecological stability?

2. What are some suggestions that you might have to build Pakistan's natural resource base and economic stability?

3. How much of a component is fuelwood in terms of a source for Pakistan's overall energy consumption? What are some viable alternatives to it?

4. What three factors affect Pakistan's carrying capacity?

5. What conclusions or strategies are drawn?

6. Suppose that you have been given the task for recommending ways to increase Pakistan's agricultural production. What ideas would you propose that would make Pakistan self-sufficient in food production?

INDIA

India features the second largest population on Earth, 853 million people in 1990. With a population growth rate of 2.1 percent in 1990, and little fertility decline in recent years, the total is projected to reach slightly more than a billion in the year 2000 and 1.4 billion by 2020, assuming the United Nations "medium variant" projection. If, however, fertility rates were to remain constant, India could end up with a staggering 2.1 billion people by the year 2100. Yet there are many signs the country has already exceeded its carrying capacity, if only through its flagging efforts to feed vast numbers of people.

As in the case of Pakistan, a good part of India's economy depends on its natural resource base. India's territory comprises 3.29 million square kilometers, of which 1.43 million square kilometers (43 percent) are cultivated, and of these some 470,000 square kilometers (33 percent) are irrigated. It is the irrigated lands that have underpinned India's success in turning itself from a food-deficit nation into a nation with a slight food surplus. But outside the irrigated areas there has been virtually no Green Revolution. The remaining farmlands are rainfed, featuring ultra-low levels of productivity. It is in these lands that most of India's impoverished people live.

Irrigation-water supplies are critically dependent on forested catchments, notably in the Himalayan foothills. Yet, according to the National Remote Sensing Agency of India, forest cover declined from nearly 17 percent of national territory in 1970 to just over 14 percent in 1981. The government believes forest cover should be 25 percent of the land area. According to the late Prime Minister Rajiv Gandhi, deforestation has brought the nation "face to face with a major ecological and socio-economic crisis." Yet, despite the important links between forest cover and irrigated agriculture, investment in forestry and soil conservation programs under the sixth five-year plan has been only 17.9 percent of investment in large scale irrigation projects. The late Prime Minister Gandhi called for 48,000 square kilometres to be reforested each year, but only one third of this goal has been achieved to date.

Much of the deforestation is due to excessive fuelwood gathering as well as agricultural encroachment. In 1982, fuelwood demand was estimated at 133 million tons, whereas remaining forestlands could sustain an annual harvest of only 39 million tons. The gap of 94 million tons was ostensibly met either by over-cutting (thus compromising future forest production) or by burning cow dung and crop residues (which compromises future soil fertility).

A Bridge to Your 21st Century Understanding

Complete and discuss the following flowchart.

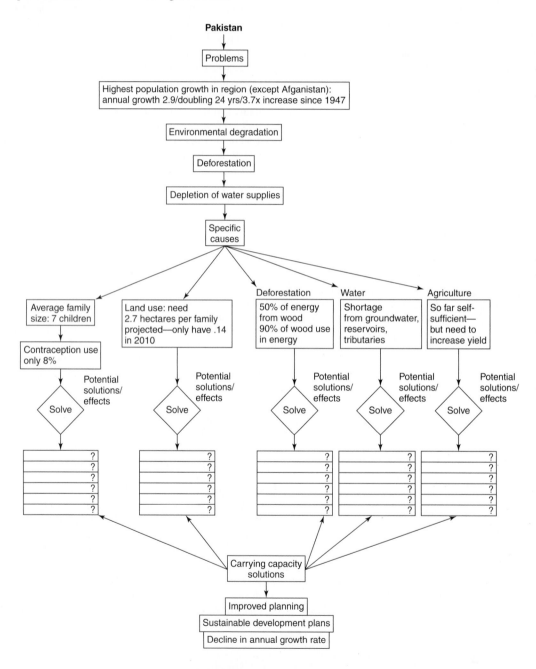

By the end of this century, the gap will be much larger if India's population continues to grow and its forests continue to contract. Indeed, even if all planned plantations are established, there will be a predicted shortfall of 182 million tons of wood, or double the shortfall in 1982.

Deforestation has already disrupted India's watershed systems, causing increased downstream flooding during the monsoon and water shortages during the dry season. In the Ganges River system, dry season water-flows declined by almost a fifth during the 1970s. Much of India's success in attaining food self-sufficiency by the early 1980s has been due to a doubling of its irrigated area since 1960. Yet because of the degradation of watersheds and the disruption of major water-flows, coupled with a surging demand for municipal and industrial water—partly reflecting the sheer rise in human numbers—India faces the prospect of acute water shortages over much of its territory by the middle of this decade.

Moreover, in 1984, the amount of flood-prone lands nationwide totalled 590,000 square kilometers, compared to only 230,000 square kilometers in 1970. In the Ganges Plains alone, flooding annually affects an area averaging 80,000 square kilometers, including 35,000 square kilometers of valuable croplands. The value of crops lost averages more than $250 million a year, while damage to buildings and public utilities averages around $750 million a year.

On top of this is the problem of land degradation. Some 1.75 million square kilometers are losing agricultural productivity through soil erosion, salinization and waterlogging, with a marked decline in soil fertility and crop yields across an area of a million square kilometres. Soil erosion alone causes Indian farmlands to lose 6 million tons a year of nitrogen, phosphorus and potash. Replacing these critical nutrients through chemical fertilizers requires an annual outlay of around $6 billion. In addition, 130,100 square kilometers of irrigated lands suffer from salinization and waterlogging. Much of this land may eventually have to be taken out of production. Land degradation of all sorts is estimated to cause a loss of 30 to 50 million tons of produce every year.

In addition, there is the growing problem of siltation. Twenty eight of the biggest irrigation projects have a total catchment area of 690,000 square kilometers, of which 220,000 square kilometers are critically eroded and fewer than 20,000 square kilometers have been rehabilitated. In 1980, 12 irrigation reservoirs were filling with sediment at between two and five times the initial rates, and a good number of others at between six and 20 times the initial

rates. If only 20 percent of the live-storage capacity of major and medium sized reservoirs is silted up by the end of this century, it will mean a loss of irrigation potential equivalent to about 40,000 square kilometers, a full tenth of the present irrigated area. Whereas irrigated lands were meeting the needs of 56 percent of India's population in 1975, it is planned they will have to meet the needs of 87 percent of the projected population in the year 2000.

All this is not to discount India's remarkable feat in increasing its wheat production from 10 million tonnes in 1964 to more than 45 million tonnes in 1985, and in raising total food grain production to a record 150 million tonnes in 1984. For almost a quarter of a century the nation has maintained a sufficient rate of growth in food output to keep just ahead of population growth (though with hardly any increase in per capita food output, and a slight decline in average consumption of calories, protein and fat). By the end of the century, however, the nation will need at least 230 million tons of food grain each year to feed its growing population, a 53 percent increase over the 1984 output. These needs notwithstanding, there has been a steady decline in the growth rate of agricultural production since 1960 and, since 1984, there has been virtually no increase at all in food-grain output.

The 1960 to 1984 upsurge in total food production was accomplished primarily in irrigated areas such as those of the Punjab and Haryana States with their level terrain. Additional lands for irrigated agriculture are limited. Not only are watershed supplies of irrigation water becoming irregular, but groundwater stocks have been over-exploited to such an extent that in many areas water tables have dropped by at least one metre. So severe are these water problems that India's future irrigation potential is severely constrained. By 1996, India's food-grain production could well plateau, with per capita output holding at around 204 kilograms, only marginally more than the 195 kilograms of 1985. Thereafter, the growth rate for food-grain production could decrease to only 1.9 percent a year during the period 1996 to 2005, only fractionally higher than the projected growth rate for population. There remains much scope to intensify agricultural production. India has about 20 percent more land under paddy rice than China, yet produces 40 percent less. This anomaly is due primarily to the fact that the agricultural resource base is being progressively depleted through land degradation on several fronts.

Thus, profound questions arise about India's capacity to keep feeding its populace which is projected to expand by 190 million, or an additional 22 percent, by the end of the century. Moreover, the challenge is not only to feed extra people, but also to upgrade nutritional levels among the third of the populace that remains malnourished. If these food-deficient people were to be adequately fed through more efficient and equitable grain distribution, India's grain surplus would be instantly eliminated.

Overall, then, it appears that in certain respects India's renowned Green Revolution has been a short-term success accompanied by long-term costs. India's farmers have posted some impressive results since 1964. But in major measure these advances have been achieved through over-working croplands and over-pumping water stocks. So a key question arises: have Indians been feeding themselves today by borrowing against future food productivity?

Recent trends suggest the outlook is not promising. The annual growth rate in agricultural production in India slipped from an average of 2.8 percent during 1965–75 to 2.2 percent during 1981–85, scarcely above the annual rate for population growth. True, there is some scope for India to purchase food from outside sources. But this depends largely on the strength of its economy, burdened as it is with a foreign debt that totalled $37.3 billion in 1987. While this sum appears slight compared with those of certain Latin American nations, debt servicing nevertheless absorbed nearly 19 percent of India's export earnings.

Furthermore, population growth not only means extra mouths to feed. It exacerbates the problem of landlessness. The total number of landless households was 15 million in 1961, rising to 26 million by 1981, and is projected to reach 44 million by the year 2000. As noted, agriculture accounts for 79 percent of employment nationwide. The total workforce is projected to expand from 224 million in 1980 to 376 million in the year 2000, and the level of unemployment and under-employment is not expected to change from its present level of 33 percent. For off-farm employment to absorb all those who are unemployed and underemployed, non-agricultural sectors would have to grow at more than 12 percent a year, or well over twice the rate during the decade 1975–84. So there is the prospect of a fast-growing pool of rural people without land or adequately remunerative work.

To reiterate a basic point, these food-population issues place a premium on India's continuing access to water in sufficient amounts at appropriate times of the year. While there is often too much water during the monsoon season, there is often too little during the subsequent dry season. According to a number of reports and analyses summarized in *The State of India's Environment, 1984–1985,* several parts of India could face severe seasonal water deficits

by the turn of the century when irrigation demand is projected to be half as high again as it is today and industry's demand to be at least twice as high. This raises the prospect of strife and conflict over water supplies.

Resource-sharing conflicts over the waters of the Ganges River are also a reality. Rising in Nepal, where watershed degradation is pervasive and pronounced, the river runs for 2,700 kilometers before reaching the Bay of Bengal in the Bangladesh delta. Within its valleylands live more than 500 million Indian and Bangladeshi farmers. Bangladesh, occupying the lower reaches of the Ganges River, makes ever greater demands on the river's flows— and the present deteriorating situation arises from a river basin that is projected to feature almost 200 million more people by the end of the century.

Overall hangs the prospect of climatic change, with potentially critical impacts. Scientists predict that a planetary warming is surely on its way, caused by a buildup of carbon dioxide and other greenhouse gases in the atmosphere. As a result of global warming, monsoon systems may be disrupted which could prove critical insofar as India receives 70 percent of its precipitation from the monsoon. Global warming could also bring on a greater incidence of typhoons and coastal storms along India's coastline, possibly 40 to 50 percent more destructive than today's due to the increased temperature of the ocean surface.

The other side of the carrying capacity equation is population growth. As noted previously, if population growth were to follow the path of the United Nations "constant fertility variant," it will grow from approximately 856 million in 1990 to more than 2 billion by 2025 (implying a constant crude birth rate of about 32 per 12,000 population). Achieving the United Nations "medium variant" projection would result in considerably lower population growth, amounting to about 1.4 billion in 2025; and with the "low variant," an even lower population of 1.3 billion. Achieving the "low variant" would require a drop in crude birth rates from 32 per 1,000 in 1990 to about 13 per 1,000 by 2025.

The government of India has recognized population growth rates as "too high" since the 1974 United Nations Conference on Population and it favours expanded support to increase contraceptive use. Increasingly, policy statements indicate the government considers its population problem to be extremely serious, particularly in relation to alleviating poverty. The basic aim of the government's population policy is to reduce fertility by influencing social variables known to promote fertility reduction, such as health, education and literacy, as well as by implementing a nationwide family planning program. The government also perceives the population's spatial distribution to be inappropriate and it has adopted policies to reduce imbalances in population distribution in relation to natural resources, as well as congestion in urban areas.

Under the fifth five-year national development plan (1980–84), family planning continued to be accorded high priority, with emphasis on program integration and coordination of activities involving all ministries and departments. During the sixth plan, the government sought to extend the delivery of services through the strengthening of health and family planning infrastructure, particularly rural infrastructure, with the aim of making family planning services available on a wider scale and at all levels.

Though statistics from contraceptive prevalence surveys are dated, they reveal that whereas 95 percent of married women knew of some contraceptive method in the early 1980s, only 34 percent were using any method at all. Again, in a country as populous and poor as India, the success of its population policies and family planning programs will be highly contingent on financial and human resources needed to extend its present services. Priorities include upgrading the service delivery capacity of the health care and family welfare network through intensive area development projects; strengthening the data collection and analysis system on interrelationships between population, environment and sustainable development; and strengthening the managerial capability at all levels of India's population program through training, research, and the development and improvement of management information and evaluation systems.

A great deal needs to be done to support India's population education program, both in and out of schools; and to produce additional audio-visual and printed materials for motivation campaigns. Most importantly, additional support for health and family welfare programs aimed specifically at women should be developed, through innovative projects which emphasize the positive role women play as managers of reproduction as well as of natural resources required in everyday household use, notably fuelwood, water, and food security.

QUESTIONS

1. What is the evidence that India has exceeded its carrying capacity?

2. What is the cause of much of the deforestation in India?

3. How did food production in India increase from 1960 to 1984? Why are projections of food production not expected to increase? Has the Green Revolution been a success? Why or why not?

4. Compose some creative ideas to increase land irrigation projects without causing further land degradation and water shortages.

5. Explain the problem of landlessness.

6. What is India's problem of water supply?

7. Imagine for a moment that you have been given the job of proposing ways to reduce India's population. What ideas would you propose that would reduce the population without disturbing the nation's culture?

A Bridge to Your 21st Century Understanding

Complete and discuss the following flowchart.

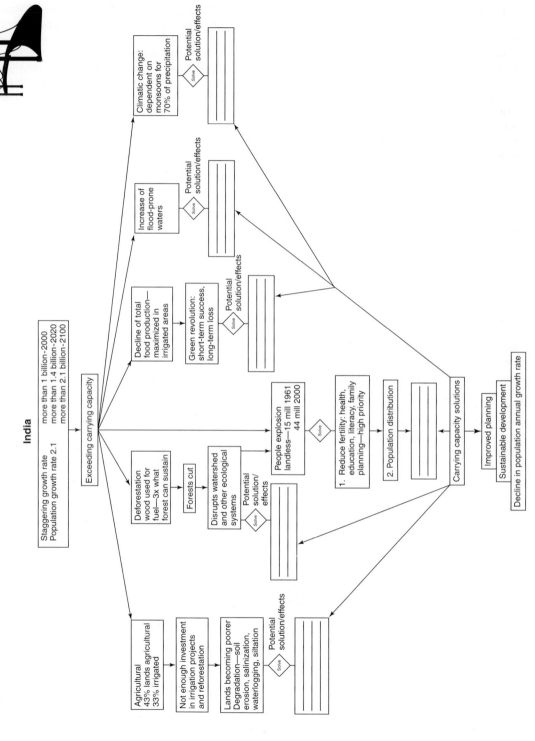

India

Staggering growth rate
Population growth rate 2.1

more than 1 billion–2000
more than 1.4 billion–2020
more than 2.1 billion–2100

Exceeding carrying capacity

Climatic change: dependent on monsoons for 70% of precipitation
Solve — Potential solution/effects

Increase of flood-prone waters
Solve — Potential solution/effects

Decline of total food production—maximized in irrigated areas
Green revolution: short-term success, long-term loss
Solve — Potential solution/effects

Deforestation wood used for fuel—3x what forest can sustain
Forests cut
Disrupts watershed and other ecological systems
Solve — Potential solution/effects

People explosion landless—15 mill 1961 44 mill 2000
Solve

1. Reduce fertility: health, education, literacy, family planning—high priority
2. Population distribution

Carrying capacity solutions
Improved planning
Sustainable development
Decline in population annual growth rate

Agricultural 43% lands agricultural 33% irrigated
Not enough investment in irrigation projects and reforestation
Lands becoming poorer Degradation—soil erosion, salinization, waterlogging, siltation
Solve — Potential solution/effects

ETHIOPIA

Ethiopia and Kenya are symptomatic of many problems suffered by sub-Saharan Africa as a whole, even though Ethiopia is one of the most impoverished nations, while Kenya is frequently regarded as something of a development success story in the region. Both display many of the adverse repercussions of rapid population growth. But before examining these two countries in detail, and in order to supply a conceptual context for the entire region, let us take stock of some pervasive factors afflicting sub-Saharan Africa as a whole.

Of the region's nearly 450 million people in 1985, some 250 million were chronically malnourished; 150 million were subject to acute food deficits; and 30 million were actually starving. In much if not most of the region, per capita food production has been declining for a full two decades. At least 62 percent of the entire populace endures absolute poverty. As much as 80 percent of croplands and 90 percent of stock-raising lands are affected to some degree by land degradation processes. Yet the population of sub-Saharan Africa, with an annual average growth rate of 3 percent in 1985–90, is projected to reach 722 million by the year 2000 and almost 1.4 billion by 2025, according to the United Nations "medium variant" projection. This ultra-rapid rate of growth operates in conjunction with adverse climatic conditions, characterized by erratic rainfall patterns, frequent droughts and increasing desertification, that have affected the region for most of the past two decades.

Sub-Saharan Africa has continued to feature a fertility rate higher than anywhere else in the world; and since 1960, the region has been growing poorer and hungrier in absolute as well as relative terms. Today's average per capita income of roughly $250 is only 95 percent of real income in 1960; of the 36 poorest countries in the world, 29 are in Africa. Worse, average per capita agricultural production has declined by an average of 2 percent annually since 1970. And the World Bank estimates that production is unlikely to grow at more than 2.5 percent a year for at least the next two decades, even while population growth remains at 3 percent or more a year. As a result, food output per head, which has declined by 20 percent since 1970, is scheduled to decline by a further 30 percent during the next 25 years. The region serves as a prime example of an "adverse outlook" scenario.

In recent years there has been some respite, thanks to better rains. But because of unpromising baseline conditions generally, and particularly in respect to harsh climate, together with widespread environmental degradation

and poor agricultural policies, the return of only moderately adverse weather conditions could quickly trigger a renewed onset of broad-scale famine. If these adverse conditions persist—aggravated, perhaps, by the climatic vicissitudes entrained by the greenhouse effect—the number of chronically malnourished, which totaled 30 million in 1985, could well increase to 130 million by the end of this century. This means that the proportion of starving people would expand from less than 7 percent of the region's population in 1985 to 18 percent by the year 2000.

It is within the context of sub-Saharan Africa as a whole that we should examine the case of Ethiopia. It epitomizes those nations with high population growth rates that must work exceptionally hard to satisfy their basic needs—food, water, shelter, health care, employment and education. When a nation is economically impoverished, it may well find the task all but beyond its means. So human needs overtake the pace of development, until eventually they threaten the very structure and stability of the nation. In Ethiopia, the essentials of everyday life, in terms of food supplies alone, are increasingly maintained by outside agencies rather than by the government. The country's output of cereal grain in 1980, 5.1 million tons, has been theoretically projected to expand to 7.3 million tons in 1991—though in fact, food production declined by an average of 1 percent a year during the 1980s. Despite this projected increase in cereal—grain output in 1991 to 7.3 million tons, Ethiopia's need for cereal imports jumped from 214,000 tons in 1980 to more than 3 million tons by 1991.

As a consequence of the severe imbalance between population and food supplies, plus associated political upheavals, there are now at least 5 million people facing starvation, and a total of almost 15 million people—nearly a third of the population—chronically undernourished. Furthermore, there are now at least 3 million displaced people within Ethiopia and another half million in refugee camps in the Sudan (where they are explicitly recognized as environmental refugees rather than political refugees).

How did Ethiopia get to this state? By the early 1970s, as much as 470,000 square kilometers of Ethiopia's traditional farming areas in the highlands—home to 88 percent of the population—were severely eroded. These formerly fertile upland areas were losing an estimated billion tons of topsoil a year (a more recent and refined estimate puts the loss at 1.5 to 3.5 billion tons a year). This massive soil erosion was due partly to rudimentary agricultural practices, partly to inequitable land-tenure systems, and partly to pressures generated by a population that increased from 20 million in 1950 to 31 million by 1970. The results included a marked fall-off in agricultural production accompanied by food shortages in cities, with ensuing disorders that precipitated the overthrow of Emperor Haile Selassie in 1974.

The Dergue regime did not move fast enough to restore agricultural production. Primarily for this reason, throngs of impoverished peasants started to stream into the country's lowlands, including the Ogaden zone that borders Somalia—a zone of long-standing conflict between the two nations. In Somalia, too, steadily increasing human numbers, together with inefficient agricultural practices, had led to much over-taxing of traditional farmlands. Largely for these reasons (plus some ethnic complications), there was a migration into the Ogaden from the Somali side as well. The result was a clash between the two sides, with an outbreak of hostilities in 1977.

Primarily as a result of a regional arms race and internal conflicts, Ethiopia in 1981 spent $447 million on defense, and Somalia $105 million. Added to the outlays of previous years, the total sum expended in the Horn of Africa because of the Ogaden conflict can be estimated at well over $1 billion during a five-year period. If only a small part of that sum had been allocated ahead of time to reforestation, soil safeguards and associated aspects of restoring the agricultural resource base in the two countries—estimated by the United Nations Anti-Desertification Plan to cost no more than $50 million a year for 10 years—the disastrous outcome could well have been avoided.

According to the United Nations "constant fertility variant" projection, if crude birth rates were to remain at 1990 levels over the next 45 years, total population would increase from 49 million in 1990 to 161 million by 2025. Were crude birth rates brought more into line with the UN "medium variation" projection—requiring a birth-rate decline from 49 per 1,000 in 1985–90 to about 30 per 1,000 in 2020–25 the total population would grow far less, to about 127 million by 2025.

Between 1974 and 1980, the government of Ethiopia perceived its rates of population growth and fertility as "satisfactory," requiring no intervention. Furthermore, it provided little support for family planning services. More recently, the government has changed its position, now viewing population growth and fertility levels as being "not satisfactory" because they are "too high" in relation to family well-being and in particular to maternal and child health. While family planning services were initiated by the Ministry of Health in 1981, only 23 percent of the country's 2,500 health stations, centres and hospitals provided integrated maternal, child health and family planning services. In 1981, the contraceptive prevalence rate was estimated to be only 2 percent; more recent estimates suggest rates of 2 to 8 percent in urban areas and from 1 to 2 percent in rural areas.

Government priorities include the formulation of a more focused population policy which rationalizes current population growth patterns relative to the country's economic and environmental carrying capacity. Furthermore, in view of the extensive poverty in the country, and limited resources available to the government for population activities, various forms of assistance are urgently needed to strengthen Ethiopia's existing population program. These include strengthening the national maternal and child health care program as part of the government's family welfare goals; greatly expanded population education on relationships between population, environment and sustainable development in primary and secondary schools, as well as in non-formal education; strengthening national capacities for population analysis, studies and training in Ethiopia's major universities and research-oriented non-governmental organizations; and programs to support and encourage the potential of women in their role as natural resource managers.

The Dergue government, like the Selassie regime before it, has been unable to resolve the problem of growing human numbers seeking to survive off an impoverished natural resource base. Even though Ethiopia has had one of the lowest population growth rates in sub Saharan Africa—around 2 percent a year for most of the time between 1950 and 1980, rising to 2.9 percent in 1990—Ethiopia's population had increased from 20 million in 1950 to about 49 million by 1990.

QUESTIONS

1. What is the profile of sub-Saharan Africa as a whole in terms of food production, poverty, and population? What figures does the article provide to support your view?

2. What is the reason that the number of chronically malnourished Ethiopian people of 30 million in 1985 could increase to 130 million by 2000?

3. How did Ethiopia get to the state of being unable to serve its population's basic needs?

4. How did defense spending affect the current population and environmental situation in Ethiopia?

5. What plan might you have to increase population planning participation among Ethiopian people?

A Bridge to Your 21st Century Understanding

Complete and discuss the following flowchart.

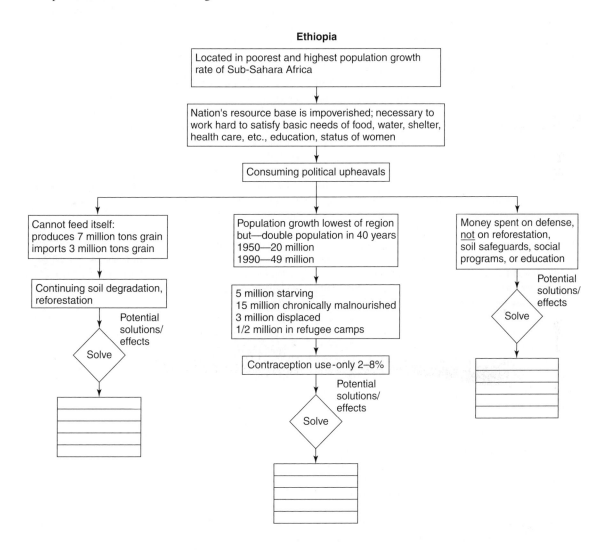

Ethiopia

Located in poorest and highest population growth rate of Sub-Sahara Africa

Nation's resource base is impoverished; necessary to work hard to satisfy basic needs of food, water, shelter, health care, etc., education, status of women

Consuming political upheavals

Cannot feed itself: produces 7 million tons grain imports 3 million tons grain

Continuing soil degradation, reforestation

Potential solutions/ effects

Solve

Population growth lowest of region but—double population in 40 years
1950—20 million
1990—49 million

5 million starving
15 million chronically malnourished
3 million displaced
1/2 million in refugee camps

Contraception use-only 2–8%

Potential solutions/ effects

Solve

Money spent on defense, <u>not</u> on reforestation, soil safeguards, social programs, or education

Potential solutions/ effects

Solve

MEXICO

Mexico is more of a middle-income country than the others considered thus far, with a current per capita income of $1,800. But it still illustrates the problems of population growth with respect to environment, albeit problems of a rather different sort from those already reviewed.

Mexico was the first developing country to engage in Green Revolution agriculture, expanding its grain production four-fold between the mid-1950s and the mid-1980s. But its population growth has also been among the most rapid anywhere, increasing from 28 million in 1950 to 89 million by 1990 (a 218 percent increase), making it the second largest populace in Latin America. As a result of the upsurge in human numbers and nutritional needs, plus an increase in environmental degradation, the benefits of the Green Revolution breakthroughs have all but dwindled away in terms of per capita consumption. Mexico has once again become a net food importer.

Because much of the country is dry, it possesses limited agricultural land. In more than 70 percent of these lands, soil erosion is significant: 60 percent medium and growing worse, 12 percent severe. Outright desertification claims 2,250 square kilometres of farmlands each year. At least one tenth of irrigated areas have become highly salinized, and 10,000 square kilometers need urgent (and expansive) rehabilitation if they are to be restored to productivity.

Moreover, the country's remaining forests total only 120,000 square kilometers, and are giving way to large-scale ranchers and small-scale peasants alike. The loss of the forests' "sponge effect" disrupts river flows: in two thirds of arable lands, water supplies constitute the main factor that is limiting agricultural productivity. Because of dwindling water supplies, together with soil erosion, at least 1,000 square kilometers of farmlands are abandoned each year. From the late 1950s to the early 1970s, the area under staple crops—corn, wheat, rice and beans—expanded by almost half, extending into marginal lands with highly erodible soils. At least 45 percent of all farmers occupy those 20 percent of croplands located on steep slopes. By the mid-1970s, these newly opened-up areas were starting to feature declining crop yields—precisely at the time when there was a peak in population growth. Domestic calorie production as a percentage of total supply reached 105 percent in 1970 but, by 1982, it had slipped to 94 percent.

The distribution of agricultural lands is becoming ever more skewed as large farmers buy out small farmers and engage in capital-intensive cash cropping for export. But growing numbers of smallholder peasants impose still greater strains on over-worked croplands, reducing their carrying capacity; and migrating throngs of marginalized farmers are pushed into those lands most susceptible to degradation.

As a result of this agricultural squeeze, there has been a recent upsurge in migration from Mexico's rural areas. Many migrants head for Mexico's cities which, after growing at 5 percent or more a year since 1960, are less and less capable of absorbing new arrivals. For every two rural Mexicans who migrate to the city, one now crosses the border into the United States.

What are some likely future population pressures in Mexico? Despite a remarkable economic performance for much of the period 1960 to 1980, average real wages in 1988 were below the level of 1970; and the economy has generally stagnated or even contracted since 1980 while the population has grown by more than a quarter. Nor does the economic outlook presage much improvement. A realistic prognosis is that within a decade Mexicans could well be poorer than they are today. Just to keep pace with a swelling work force, Mexico will have to create jobs at half the rate the United States achieves, while doing it within an economy only one thirtieth the size, and at an investment cost that could conceivably reach $500 billion by the year 2000. This is an unrealistic expectation. In 1986, a year with poor economic performance because of declining oil prices and growing debt burden, the jobs total actually decreased. Not surprisingly, then, it is an optimistic estimate that foresees "only" 20 million Mexicans without proper employment by the turn of the century. At least half of these will be living in rural areas, where there could be an additional 2 million, perhaps even 4 million, landless peasants. According to Professor J.G. Castañeda of the National University of Mexico, "The consequences of not creating nearly 15 million jobs in the next 15 years are unthinkable. The youths who do not find them will have only three options: the United States, the streets or revolution."

The biggest issues confronting Mexico appear to be population growth and urbanization. The population could more than double between 1990 and 2025 if fertility rates were to remain unchanged. Realizing the United Nations "medium variant" projection—implying a drop in the crude birth rate from 28 per 1,000 in 1990 to about 18 per 1,000 in 2025—would still result in a population of 150 million by 2025, up from 89 million in 1990. Bolder reductions, to achieve the UN "low variant" projection, would imply a further drop in the birth rate to 15 per 1,000 in 2025, resulting in a total population of 132 million.

Since 1974, the government of Mexico has considered its rates of population growth to be "too high," and has promoted policies to lower growth rates, chiefly through fertility reduction by providing support to contraceptive information, education, and supplies. The government has consistently expressed dissatisfaction with imbalances in the growth and distribution of population (including migration patterns), in relation to the distribution of natural resources. In view of these policies, the government has taken steps to integrate specific demographic goals into its overall development planning. In 1974, Mexico adopted a General Population Law with the aim of regulating factors affecting population structure, volume, dynamics and distribution, and it established the National Population Council (CONAPO), which is responsible for formulating plans on population and promoting their integration into economic and social programs. In 1977, the country approved a national family planning policy with the aim of reducing the population growth rate; and in 1979, a regional population policy was adopted, addressed to the specific problems of interregional migration and the geographical distribution of the population.

By 1987, approximately 91 percent of married women knew of at least one contraceptive method, while 53 percent were actually using one method or another. On the one hand, this indicates considerable progress since the contraceptive prevalence rate was only about 30 percent in 1976, grew to 48 percent by 1982 and has now passed 50 percent. On the other hand, approximately 45 percent of married women have yet to be reached by family planning programs—an important goal in view of population and environmental problems.

Recognizing the urgency of the problems at hand, the government is now embarking on several aggressive strategies and programs to enhance the effectiveness of its population policies. For example, it seeks to reduce the population growth rate to 1.8 percent by 1994 and to 1 percent by the year 2000. In order to achieve a better distribution of population, more in line with the development potential of different regions, the government aims to lessen the relative size of the great metropolitan zones and to encourage the growth of small and intermediate cities. The government also seeks to promote the integration of demographic objectives and criteria more explicitly into the country's economic and social planning; and it aims to raise the socio-economic status and role of women as contributors to the economy and as environmental managers.

QUESTIONS

1. What is Mexico's current per-capita income?

2. Why hasn't Mexico's Green Revolution grain production growth produced food excess?

3. How would you describe the following events taking place in Mexico: desertification and the sponge effect?

4. What is the effect of large farmers buying out small farmers and using capital-intensive farming approaches rather than the labor-intensive techniques used by small farmers?

5. What will be the consequence of Mexico's being unable to create nearly 15 million jobs in the next 15 years?

6. What are the biggest issues confronting Mexico?

7. How does Mexico plan to make its family planning programs more effective?

8. How can Mexico deal with the high growth of urban areas and the negative impact of such growth? What are short-term and longer term solutions to this problem?

A Bridge to Your 21st Century Understanding

Complete and discuss the following flowchart.

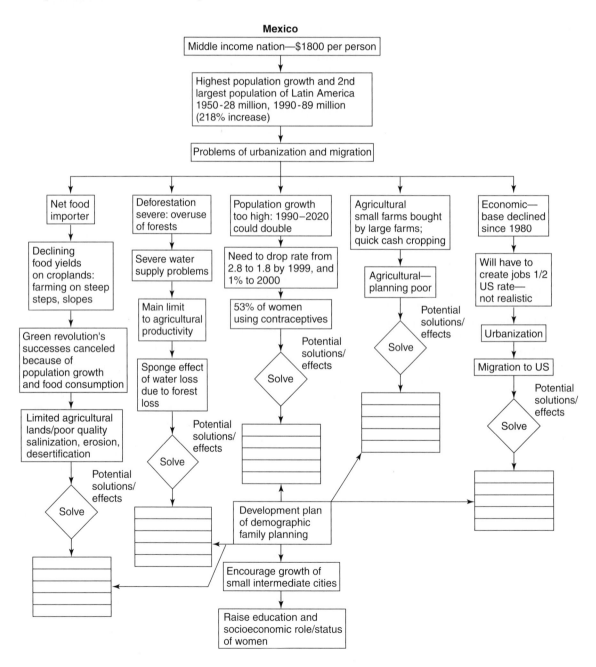

The Humanitarian and Technical Dilemmas of Population Control in China

SU XIAOKANG

TRANSLATED BY YUAN XUE

An important experiment for mankind this century is the control of population growth by state-administered means. The necessity for some controls is beyond dispute from the point of view of global interests as a whole. But the outcome of their implementation and the resulting ethical dilemma in any one particular country is an issue which ultimately concerns values reaching beyond mere economic benefits. It becomes an issue we mankind cannot ignore.

Population controls have created two basic dilemmas. First, the ethics of life is being challenged. This differs from those ethical dilemmas, such as genetic manipulation, facing Western developed countries; instead, it is the conflict between unnatural interference in human reproduction, on the one hand, and traditional values and basic human rights, on the other. Second, the malady that results from applying artificial economic planning to population control shows that the role of the state in population control is suspicious.

An important implication of these two dilemmas is that the assumption of international organizations, such as the United Nations and the World Bank, for instance, and of governments of the Western countries may be turned into a mere illusion. Namely, this is the assumption that by trusting or supporting the population policy of the Chinese

Published by permission of the *Journal of International Affairs* and The Trustees of Columbia University in the City of New York. Xiaokang, Su (Winter, 1996). The humanitarian and technical dilemmas of population control in China. *Journal of International Affairs,* vol. 49, no 2, pp. 343–347. © The Trustees of Columbia University in the City of New York.

government, due to the unavailability of alternatives, the explosion of the Chinese population—the hidden danger of the world—can be solved.

Let us address the first point. China does not have population theories like those in the West, but only a more traditional concept of child bearing that has become a custom for a thousand years or so. This custom regards carrying on the ancestral line as being the first priority of life, thus it is said, "of the three kinds of unfilial acts, having no descendants is the worst." Although it has been seriously challenged since China entered the modern period when traditional families began to disintegrate, this concept still remains today a basic value widely held among the people. The natural economy in the Chinese countryside sustains this basic value; that is, underdevelopment has been maintained mainly by the increase of labor. Therefore, both the rural economy and the concepts of the patriarchal family system stress to a great extent traditional customs such as having an extended family, valuing only the male child and raising children for old-age security. A thorough change of ethical practices before being able to change the rural economy will undoubtedly cause strong social instability.

The unprecedented ethical dilemma in human child-bearing behavior that exists in China stems from the following:

- people make the choice to bear children based on fear;
- for the first time in history, child-bearing has become an illegal action that makes many pregnant women flee from home in large numbers and hide in wild places to escape forced abortions;

- artificial distortions in demographics, that is, the practice of drowning infant females rages on, causing the numbers of female children proportionately less vis-à-vis those of males.

Officially sanctioned data, i.e., the 1987 population sample survey, shows that the death rate of female infants was 47.15 percent of the total infants of the same age, lower than the 48.33 percent female rate of the total population, and the sex ratio of male and female infants was 112.1 to 100. But according to private estimates by population experts with good knowledge of the situation in the countryside, the female infants that are killed plus the *chaosheng yinge* (infants born beyond the official quota) is expected to reach millions every year in China.

All the above is widely known, but there is more seldom known to the outside world:

- the medical ethics crisis faced by doctors, including those who are being forced to induce abortion for pregnant women and even to kill with drugs some infants more than seven months old that have survived the induced abortion;
- as a consequence of the massive number of induced abortions, dead infants are commonly used in autopsies, organ transplants and in medicine and nutriment research;
- those born without designated official sanction, privately called black children, lose basic living rights, including their grain ration.

Their families often are compelled to leave their home villages and wander destitute, emerging in the outskirts of many Chinese cities in *heicun* and *heitun* (both meaning black villages) where those escaping forced birth control live in close quarters.

The second dilemma is the effect of population control after making people pay such costs. In China, the role of the state in population control has changed drastically. After the Communist takeover in 1949, the regime initially carried out an open and conniving population policy; Mao Zedong believed that "more people make the job easier." As a result, for twenty years, China experienced a continuous population growth rate of over two percent. Its population doubled from 500 million to 1 billion in 38 years, forming a population-reproduction typology of high birth rate and high natural growth rate unmatched in the world. This process continued until the 1970s, when it was abruptly ended by the sudden appearance of severe measures: the *yitaihua* (one family, one child) policy carried out by the Chinese government through coercion, and the so-called national policy that stated that the growth rate of the population could not go beyond 1.2 percent per year. The question became whether an irrational natural process can be reversed successfully by another kind of extremely radical artificial measure.

China's central leadership has been confronted with a basic difficulty in the practical application of initiating population control within a planned economy. It is a policy applying flawed math to manage a whole population. This results from the following:

- inaccuracy in the vital statistics used as the basis of planning;
- the impossibility of replacing the natural speed of population growth with human-imagined figures;
- the spuriousness of deriving the national population growth rate by adding up that of each province as rigidly laid down by the central government.

The Chinese government continues to carry out one Five-Year Plan after the next. The Seventh Five-Year Plan of Population Control formulated in 1985 required that the population be kept at 1.113 billion. But the vital statistics of 1990 show that the Plan was surpassed by 30 million. Analyzing the reasons for the failure, Peng Peiyun, Director of Chinese Family Planning Committee, said that at the time of formulating the Seventh Five-Year Plan, about 13 million people went without being counted. She also said that the Eighth Five-Year Plan of Population Control and the ten-year population program formulated in 1991 require that the population in China be kept under 1.3 billion by the end of this century. But, again, according to the announcement of the survey issued by the Chinese government in August 1995, the population has already reached 1.2 billion.

There are two questionable factors in Peng's statement. First, it is likely that the publicized figure of 1.2 billion is still an underestimation. Secondly, the Eighth Five-Year Plan is a continuation of the thinking of the Seventh Five-Year Plan with extremely uncertain consequences. There is one element, however, that is known for sure. During this period, women in their prime of child-bearing age have reached 120 million or more. No matter how severe the control, it can not suppress the figure of those born unplanned; furthermore, a harsh and rigid control means an even greater disaster for the rights and interests of women in mainland China, as well as even greater repercussions. This is another reason for the failure of the Seventh Five-Year Population Plan about which Peng Peiyun spoke.

Meanwhile, such severe controls do not take into consideration standards of education and basic knowledge of birth control. Facing an audience of 260 million illiterate or semi-illiterate (22.27 percent of the total population, mostly women in the countryside), the usefulness of the propaganda and preaching of the Chinese government is next to nothing. This factor is both the reason for the masses to resist the birth control policy (the fundamental driving force of the resistance comes from not only the strong and invisible traditional custom and also a more primitive child-bearing impulse), and, at the same time, the major factor for the rudeness in the policy implementation. Such a rudeness is expressed by the cadres at the grass-roots level carrying out the policy and the so-called "birth control work teams" dispatched into villages to impose on those policy-violators forced abortions and economic punishment and, on all women of child-bearing age, preventative oviduct legation. Deviations in policy implementation and excessive retaliations are common occurrences.

In short, to control child-bearing behavior through state planning, as if it were economic behavior, will not only fail like the past economic plans in China, but also expose serious inhumanity. Its limited control effect is gained at the great expense of deprivation of basic human rights, damage to traditional values and sacrifice of innumerable lives. In recent years in China, the newborn population has been approximately 21 million per year with a net increase of 14 million; this growth momentum is expected to continue towards the end of this century, then reaching 17 million (equivalent to a medium-sized country) per year. According to the estimation made by the Chinese government itself, surpassing a population of 1.3 billion by the end of this century will mean China's total population in the middle of the next century will reach between 1.5 and 1.6 billion, or possibly even 1.8 billion if it gets out of control. At that time, the Chinese population issue will truly become a thorny problem for the entire world.

QUESTIONS

1. What are the two dilemmas of population control in China?

2. What is the customary philosophy in China on having children? What supports this philosophy?

3. What are some of the ethical implications of state-imposed birth quotas?

4. What are the reasons that China reached the fastest natural growth rate in the world?

5. What are some of the reasoning flaws that China used in its initiation of population control?

6. What are some of the reasons that the population control plan in China is not succeeding as expected?

7. Faced with the exploding population of China, do you think that a government-imposed population quota law is the most effective solution? Why or why not? What approach would you suggest?

39

Earth's Carrying Capacity: Not Quite So Easy When Applied to Humans

We face a real crunch. At issue is whether we can do what is totally unprecedented—feed, house, nurture, educate, and employ as many more people in the space of four decades as are alive today.

GEOGRAPHER ROBERT W. KATES OF ELLSWORTH, MAINE

With the current worldwide population increasing by 90 million people annually and the last billion people added just in the last 12 years, people over forty years old have actually lived through a doubling of the human population—the first time this has ever happened. This unprecedented growth evokes the concern of how many people the earth can sustain. In four decades, we again will have a doubling of the world's population—the most difficult challenge for mankind yet.

EARTH'S CARRYING CAPACITY DOES NOT ACCURATELY APPLY TO HUMANS

Scientists, in analyzing the demands placed on the earth by human population, usually refer to the "earth's carrying capacity"—the maximum population its ecosystems can sustain indefinitely. This concept works well when defining how many animals a habitat can support, but it breaks down when applied to people, since humans' variety of habits and environmental patterns have a wide variation of resource consumption and waste generation. The typical New Yorker, for example, uses 10 to 1,000 times more resources daily than the average Chilean or Ghanaian. A second reason that the carrying capacity formula doesn't work well is that humans have a great ability to alter their

physical environment or culture for survival. Many scientists are beginning to argue that they can't predict how many people the earth can sustain until there exists a better way to account for the various social, cultural, and demographic habits and effects of people.

Wolfgang Lutz, a scientist in Laxenburg, Austria, has a different approach. The formula he uses does not just account for the *number* of people; it estimates environmental impact *(I)* as the product of population *(P)* times affluence *(A)* times the technological efficiency *(T)* of a culture *(I = PAT)*. The formula $I = PAT$ was first developed by scientists Paul Ehrlich and John Holdren to examine the environmental impacts of population in general. Some scientists use this formula to calculate the economic and environmental costs associated with each new birth. There was a problem with the use of this formula in calculations of the impact of CO_2 emissions, however: The formula better correlated with the number of *households* rather than individuals. Within the last couple of decades, there has been a trend toward the existence of more households with fewer members caused by the patterns of divorces, young adults setting up their own households, the postponement of marriage, an increase in life expectancy, and a growing sense of privacy. The trend of more, but smaller, households is most prevalent in the nations that already use the most resources, and this trend could cause substantial impacts. When calculated using households, rather than individuals, in the $I = PAT$ formula, CO_2 emissions were 50% higher than those suggested by the use of individuals in the formula.

Reprinted with permission from "The Human Numbers Crunch" by J. Raloff, pp. 396–397, *Science News,* the weekly news magazine of science, Copyright © 1996, 1997 by Science Service.

FOOD PRODUCTION AS AN EXAMPLE

Population growth impacts food production. Since 1955, nearly one third of the world's cropland (an area larger than China and India combined) has been abandoned because its overuse has led to soil loss, depletion, or degradation. The search for cropland substitute accounts for about 60 to 80% of the world's deforestation; however, the areas being deforested are poorly suited for agriculture and are becoming more difficult to find.

Additionally, about 87% of the world's freshwater is being consumed by agriculture—far more than is consumed by any other human activity. Farmers are using surface water and are mining underground aquifers. According to Population Action International, "farmers are literally draining the continents, just as they are allowing billions of tons of topsoil to erode."

Although at present there is enough of a supply of land and water to feed the earth's existing population, the per-capita availability of cereal grains, the source of 80% of the world's food, has been declining for 15 years. Most nations today rely on imports of surplus grains from the resources of only a few nations, which account for 80% of all cereal exports, the United States providing the largest share. If the population of the United States doubles in 60 years as projected, then U.S. food and cereal resources will need to be used to feed 520 million Americans. Meanwhile, if China's population continues to grow to another 500 million people and if China continues to experience loss of cropland to erosion and unabated industrialization, its demands for imports by 2050 might be more than the world's entire 1993 grain exports. Lester Brown of the Worldwatch Institute says, "Who could supply grain on this scale? No one."

Considering these food demands still does not answer the question of how many people earth can sustain. Joel Cohen, head of the Laboratory of Populations at Rockefeller University, states that there is no single answer. It will depend, he says, "on the number of people that are willing to wear clothes from cotton (a renewable resource) versus polyester (a nonrenewable petroleum-based product), how many will eat beef versus bean sprouts, how many will want parks versus parking lots, how many will want Jaguars with a capital J and how many will want jaguars with a small j. These choices will change in time, and so will how many people the Earth can support." If the U.S. population doubles by 2050, the share of its diet from animal products will drop by about half, to perhaps 15%, and food would cost up to 50% of each person's paycheck,

which is about what Europeans spend on food now—but this percentage is far more than what U.S. residents typically pay.

OTHER APPROACHES FOR A SUSTAINABLE PLAN

Another idea that addresses the problem of limited resources for a growing population is the free-market economic approach. If world markets were free, the price of goods would rise as the amount of materials grew limited. Rising prices would alert entrepreneurs to the approaching scarcity and provide an entrepreneurial award for new products and creativity. Thomas Lambert of the Center of the Study of American Business feels that resource scarcity will lead to technological developments that leave everyone better off than before. "The entire history of humanity indicates that with enough economic freedom, overpopulation relative to natural resources or energy will not occur."

The problem with this free-market argument, according to Thomas Homer-Dixon of the Peace and Conflict Studies Program at the University of Toronto, is that not all things are equal. In areas where young children receive inadequate nutrition, stimulation, or education, many will not mature adequately to meet the challenge of innovation. Other people face social barriers to innovation such as business corruption, banking difficulties and lack of monetary support, politics and poverty, and power group considerations. Mr. Homer-Dixon says that as resource scarcity and population stresses rise, "the smarter we have to be socially and technically just to maintain our well-being—the more dire the problem, the more quickly we're going to have to respond." Mr. Homer-Dixon is an optimist in that he feels research institution in developing countries will increase the supply of ingenuity. Most of the creativity that is useful in solving scarcity and food production problems is going to be generated locally by people familiar with local geography, social relations, and resources. To stop the flow of local talent to industrialized nations where greater research opportunities exist, there needs to exist funding of local research by providing scientists in the developing world with monetary and technical resources and with computers, modems, and fax machines—a communication network to peers around the globe.

Global Research on the Environment and Agricultural Nexus (GREAN) is such a proposed organization. GREAN would set up a partnership among U.S. universities and 16 international centers that belong to the Consultative

Group on International Agricultural Research. Since the demand for food in developing countries is expected to double by 2025 and triple by 2050, GREAN wants to involve the outstanding scientists of the world (including those in developing countries) in projects aimed at developing sustainable, environmentally friendly food production by and for the world's poorest people (Raloff, 1996, pp. 396–7).

Another approach is to link population to the environment more closely. The 1992 Earth Summit Conference in Rio targeted many environmental issues but largely overlooked population. The 1994 Conference on Population in Cairo led to a comprehensive program of action but overlooked the environment. Establishing such a link helps to balance the intensity and sensitivity of developing countries with the highest population growth rates with the developed countries in terms of their consumptive use of earth's resources—and therefore better designs a whole-global cooperative action plan. This type of link could measure use of global materials and resources as a means to reduce CO_2 emissions and other patterns directly related to quality of life and patterns of population. Policies on population aim to decrease fertility rates in a noncoercive fashion and focus on improving the well-being of women. This can be done by helping women avoid unwanted pregnancies, improving girls' education, aiming to achieve greater gender equality, and reducing child mortality, among other methods. Tying environmental actions to population can develop an all-beneficial picture of the improvement of environmental habits and the abatement of consequences that stem from population growth in an equally noncoercive, comprehensive fashion. This process then becomes an all-encompassing and realistic method not only of reducing population, but also of improving the quality of life around the world, with all countries as equal participants in the solution (Gaffin, 1996, p. 7).

As we approach these challenges, Mr. Cohen believes that "neither panic nor complacency is in order. Earth's capability to support people is determined both by natural constraints and by human choices. In the coming half century, we and our children are less likely to face absolute limits than difficult trade-offs" such as population size, environmental quality, and lifestyle. Many people have already been questioning the consumerism of the industrialized countries since the 1950s. Choices will begin to occur when people begin to question what they are buying

with their hard-earned money and the real long-term value of these items. For example, environmentalist mediator Paul Wilson, a lawyer in Portland, Oregon, has spent seven years simplifying his life in terms of his diet and aspirations after realizing that the products he was buying were not worth the hours of work needed to pay for them. His sense is that a transition will occur in which people and their organizations will perceive the imbalance of values in their own way until simple values predominate. Earth's carrying capacity for humans will have to be determined by their own well-chosen choices and strategies (Raloff, 1996, p. 397).

REFERENCES

Gaffin, S. R., O'Neill, B. C., and Bongaarts, J. (1996, November). Population growth could affect global warming. *Environmental Defense Letter.* Vol. 27, No. 6, p. 7.

Raloff, J. (1996, June 22). The human numbers crunch: The next half-century promises unprecedented challenges. *Science News,* Vol. 149, No. 25, pp. 396–397.

QUESTIONS

1. Why doesn't the concept of carrying capacity apply so easily when it comes to people?

2. What is $I = PAT$?

3. What was the problem in using this formula? How was it better correlated?

4. How can this formula be used to project intensity of resource use and carrying capacity? How can it be used to better predict the availability of future resources?

5. How are we going to resolve the inequalities of resource use, according to the article?

6. What are some suggestions for helping unindustrialized countries control accelerating populations and future quality of life issues?

7. What is the free-market approach? What is the problem with using this type of approach?

8. What kind of a population reduction plan can involve industrialized nations as well as unindustrialized or developing countries? Develop an outline of such a plan and present it to your class.

Conclusion

The world is in the thrust of an unprecedented expansion of population. It took hundreds of thousands of years for humans to arrive at a population level of 10 million, which occurred only 10,000 years ago. This number expanded to 100 million people about 2,000 years ago and to 2.5 billion people by 1950. Within the span of a single lifetime, it has more than doubled to 5.5 billion in 1993. The implications of this growth are enormous: some very positive, some extremely threatening, and certainly *all* needing planning, awareness, a world community that works together, and wise choices. It has been stated in this part of the book that we have only a few decades to determine the future path of population growth and all the effects that this path brings with it. The text of this part—from discussions of the effects of this population growth to discussions of the earth's natural limits, from a presentation of the proceedings of the 1994 Cairo conference and the information contained in case studies of Pakistan, India, Ethiopia, Mexico, and China to a presentation of details of some possible plans for alleviating problems of population—alludes to some of the problems and solutions. The problems and solutions are very complex, but it is hoped that this part has begun to sort the issues and makes the reader aware of and involved in the important task of finding and implementing the best course.

Already, progress is being made in many countries of the world, with the reduction of birth rates, an increase in education, and the provision of more medical aid. However, much still has to be accomplished in dealing with the key determinants of rapid population growth of poverty, high childhood mortality, low status and education of women, low adult literacy, and various other negative cultural influences on progress. The common goal is to improve the quality of life for all people and future generations, protecting human social, economic, and personal well-being. This goal involves the participation of all the people on earth in using science and technology wisely, finding substitutes for wasteful practices, educating others about the protection of earth's resources, and actively protecting the natural environment. What happens in Africa or Asia or Puerto Rico affects us all . . . three more of us every second.

Our Population: A Social–Economic View

Environmental degradation is generally the effect of attempts to secure improved standards of food, clothing, shelter, and comfort for growing numbers of people. The impact of the threat to the ecosystem is linked both to human population size and resource use per person. The issues, then, of resource use, waste production, and environmental degradation are accelerated by factors of population growth and further exacerbated by consumptive habits, certain technological developments, social habits, and poor resource management.

319

As our population continues to increase, the chance of the occurrence of irreversible changes of magnitude also increases. Already, there are indicators of severe environmental stress caused by the growing loss of biodiversity; increase of greenhouse gases; deforestation; ozone depletion; acid rain; loss of topsoil; and world shortages of water, food and fuelwood. The developed countries, with 85% of the gross world product but only 23% of the population, account for the largest use of these resources (Population Summit, 1993).

As the developing countries seek to adopt standards of living and resource patterns from the developed world, the people of the world are further intensifying the already unsustainable demands on the biosphere. For the future, it is important that economic value be placed on the use of earth's resources in a manner similar to that proposed by Green Net National Production or the net national product. Some equitable measures have to be placed on the value of the resources that can sustain the quality of life and the breath of life in each one of us. Productive activities and models to create a sustainable use of our natural resources must happen globally and coordinately, developed world and developing world alike. The placement of monetary values on natural resources would point us in a positive direction toward conservation and somewhat equalize national population imbalances and provide for intelligent resource use.

Bridge to the Future

As we attend to the challenges of increasing food supplies, limiting population growth, and educating societies to 21st-century imperatives, some special progress has been made. Some of these new approaches are alluded to in the part of the book on the future, Part 8. These methods will receive more attention and refinement as we begin to tackle the population problem on the bridge to the 21st century.

INTERNET EXERCISE
1. Use any of the Internet search engines (e.g., Alta Vista, Yahoo, Infoseek, etc.) to research the following topics:
    a. The various population time frames ("popclocks") given in this chapter. Find other resources describing issues of population. Compare the various pieces of data and determine the reasons for any discrepancies.
    b. Malthus and his predictions.
    c. The three-stage demographic theory.
    d. Lester R. Brown and his work with world food projections.
    e. Other world demographic charts not given in this chapter.
    f. Future predictions of world population growth.
    g. The United Nations Population Fund's work.
    h. The International Union for Conservation of Nature and Natural Resources (IUCN) and its work and predictions.

Some ideas taken from the following source: Graham-Smith, F. (1993). Population Summit. *Population–The Complex Reality.* Cambridge, England: Cambridge Universituy Press.

   i. The Group of Experts on the Scientific Aspects of Marine Pollution (GEASMP) and its work.
   j. Figures available from the United Nations Educational, Scientific and Cultural Organization (UNESCO) on population growth, food supply, and education.
   k. The Population Council in New York City and its work.
   l. Anne H. Ehrlich and Paul R. Ehrlich's predictions and their work.
   m. The Club of Rome's Project on the Predicament of Mankind.
   n. More information on the 1994 International Conference on Population and Development in Cairo, Egypt.
   o. Any country in the world not discussed in this chapter. Research its population growth and find what information you can on its cultural attitudes about population growth and its socioeconomic positions.
   p. More information about the pros and cons of birth control.
   q. The latest developments on birth control. Also, find information on the latest developments on birth control in relation to various religions and countries.
   r. The carrying capacity of earth.
   s. Global Research on the Environment and Agricultural Nexus (GREAN).
   t. $I = PAT$.
   u. The status and education of women and birth control.
   v. Abortion rates throughout the world in various countries and the countries' cultural views on abortion.
   w. Poverty and population rates around the world.
   x. Green Net National Production (NNP).

2. Use any of the Internet search engines (e.g., Alta Vista, Yahoo, Infoseek, etc.) to research the following topics.
   a. Birth rate.
   b. Death rate.
   c. Stabilization of population.
   d. Demography.
   e. Developed countries.
   f. Developing countries.
   g. Growth rate.
   h. Infant mortality rate.
   i. Projections.
   j. Vital events.
   k. Vital rates.
   l. Fertility rate.
   m. Exponential growth.
   n. Sink (the process of a system).
   o. Birth control devices and approaches.
   p. Carrying capacity.
   q. Environmental degradation.
   r. Sustainability.

USEFUL WEB SITES

URL	Site Description
http://www.sunsite.unc.edu/lunarbin/worldpo	World population estimates
http://www.census.gov/cgi-bin/ipc/popclockw	World population estimates—U.S. Census Bureau
http://www.census.gov.cgi-bin/popclock	U.S. Census Bureau population estimates
http://popindex.princeton.edu	Extensive population database
http://www.cpc.unc.edu	Population Research Institute
http://www.uscusa.org/environment/pop. faq.html	FAQ's about population growth
http://www.overpopulation.com/faq/	Population Issues and Statistics
http://www.worldvillage.org/index2.html	World Village Project
http://www.overpopulation.com/population_ _news/1998/popnews_01_05_1998html	Population newsletter
http://www.carnell.com	Population newsletter and information

Boundless Humanity. Photo
Courtesy of United Nations.

PART

VI

Health and Technology

A DNA Double Helix Near Genetic Material.

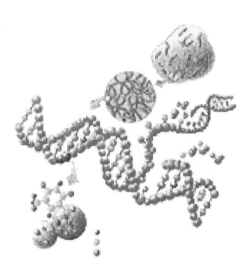

OBJECTIVES

This part will help you to:

- Be able to identify antibiotic resistance and know how to prevent it.
- Look at all sides of the cloning controversy.
- Determine if genetically engineered foods are safe or toxic.
- Understand the impact that AIDS has had on the world.
- Be aware of the financial and psychological impact that dying has on families.
- Understand euthanasia.

INTRODUCTION

You have probably heard the saying, "If you have your health, you have everything." Most would agree that our health is vital to the quality of our lives; this part of the book explores the ethical considerations that the doctor and the patient need to examine when trying to create a life of health. In this part, the interplay between health, medicines, and medical interventions is explored within the foundation of bioethics.

When you get sick, do you go to the doctor and receive an antibiotic? Do you finish all of your medicine? Do you sometimes use a friend's, child's, or spouse's antibiotic rather than going to get your own? Both of these latter practices could relate to antibiotic ineffectiveness or resistance. How to use antibiotics more effectively will be discussed in the chapter "Antibiotic Resistance."

Another issue that is highly relevant to the topics of health and technology is the controversial question of euthanasia. Should individuals who are critically ill and who have no chance for recovery be allowed to die, or should medical heroics keep them alive? When should the medical technology keeping such people alive be deimplemented to allow those people to die? Should the infamous Dr. Kervorkian be allowed to help people who are in chronic pain or who are technology dependent (e.g., dependent on ventilators, feeding tubes, etc.) die with the aid of his "death machine"? These are questions that are explored in the "Physician-Assisted Suicide" article.

Modern medical technology has increased the average life span of people and, in many cases, the quality of life. Yet, it is important to analyze and study all sides of bioethical questions relating to medical care. This chapter just touches the surface of these issues, but it should provide enough information for you to look differently at medical care from this day forward.

Baby, It's You! and You, and You . . .

NANCY GIBBS

RENEGADE SCIENTISTS SAY THEY ARE READY TO START APPLYING THE TECHNOLOGY OF CLONING TO HUMAN BEINGS. CAN THEY REALLY DO IT, AND HOW SCARY WOULD THAT BE?

Before we assume that the market for human clones consists mainly of narcissists who think the world deserves more of them or neo-Nazis who dream of cloning Hitler or crackpots and mavericks and mischief makers of all kinds, it is worth taking a tour of the marketplace. We might just meet ourselves there.

Imagine for a moment that your daughter needs a bone-marrow transplant and no one can provide a match; that your wife's early menopause has made her infertile; or that your five-year-old has drowned in a lake and your grief has made it impossible to get your mind around the fact that he is gone forever. Would the news then really be so easy to dismiss that around the world, there are scientists in labs pressing ahead with plans to duplicate a human being, deploying the same technology that allowed Scottish scientists to clone Dolly the sheep four years ago?

© 2001 Time Inc. Reprinted by permission. —*Reported by David Bjerklie and Andrea Dorfman/New York, Wendy Cole/Chicago, Jeanne DeQuine/Miami, Helen Gibson/London, David S. Jackson/ Los Angeles, Leora Moldofsky/Sydney, Timothy Roche/Atlanta, Chris Taylor/San Francisco, Cathy Booth Thomas/Dallas and Dick Thompson/Washington, with other bureaus.*

All it took was that first headline about the astonishing ewe, and fertility experts began to hear the questions every day. Our two-year-old daughter died in a car crash; we saved a lock of her hair in a baby book. Can you clone her? Why does the law allow people more freedom to destroy fetuses than to create them? My husband had cancer and is sterile. Can you help us?

The inquiries are pouring in because some scientists are ever more willing to say yes, perhaps we can. Last month a well-known infertility specialist, Panayiotis Zavos of the University of Kentucky, announced that he and Italian researcher Severino Antinori, the man who almost seven years ago helped a 62-year-old woman give birth using donor eggs, were forming a consortium to produce the first human clone. Researchers in South Korea claim they have already created a cloned human embryo, though they destroyed it rather than implanting it in a surrogate mother to develop. Recent cover stories in *Wired* and the *New York Times Magazine* tracked the efforts of the Raelians, a religious group committed to, among other things, welcoming the first extraterrestrials when they appear. They intend to clone the cells of a dead 10-month-old boy whose devastated parents hope, in effect, to bring him back to life as a newborn. The Raelians say they have the lab and the scientists, and—most important, considering the amount of trial and error involved—they say they have 50 women lined up to act as surrogates to carry a cloned baby to term.

Given what researchers have learned since Dolly, no one thinks the mechanics of cloning are very hard: take a donor egg, suck out the nucleus, and hence the DNA, and fuse it with, say, a skin cell from the human being copied. Then, with the help of an electrical current, the reconstituted cell should begin growing into a genetic duplicate. "It's inevitable that someone will try and someone will succeed," predicts Delores Lamb, an infertility expert at Baylor University. The consensus among biotechnology specialists is that within a few years—some scientists believe a few months—the news will break of the birth of the first human clone.

At that moment, at least two things will happen—one private, one public. The meaning of what it is to be human—which until now has involved, at the very least, the mysterious melding of two different people's DNA— will shift forever, along with our understanding of the relationship between parents and children, means and ends, ends and beginnings. And as a result, the conversation that has occupied scientists and ethicists for years, about how much man should mess with nature when it comes to reproduction, will drop onto every kitchen table, every pulpit, every politician's desk. Our fierce national debate over issues like abortion and euthanasia will seem tame and transparent compared with the questions that human cloning raises.

That has many scientists scared to death. Because even if all these headlines are hype and we are actually far away from seeing the first human clone, the very fact that at this moment, the research is proceeding underground, unaccountable, poses a real threat. The risk lies not just with potential babies born deformed, as many animal clones are; not just with desperate couples and cancer patients and other potential "clients" whose hopes may be raised and hearts broken and life savings wiped out. The immediate risk is that a backlash against renegade science might strike at responsible science as well.

The more scared people are of some of this research, scientists worry, the less likely they are to tolerate any of it. Yet variations on cloning technology are already used in biotechnology labs all across the country. It is these techniques that will allow, among other things, the creation of cloned herds of sheep and cows that produce medicines in their milk. Researchers also hope that one day, the ability to clone adult human cells will make it possible to "grow" new hearts and livers and nerve cells.

But some of the same techniques could also be used to grow a baby. Trying to block one line of research could impede another and so reduce the chances of finding cures for ailments such as Alzheimer's and Parkinson's, cancer and heart disease. Were some shocking breakthrough in human cloning to cause "an overcompensatory response by legislators," says Rockefeller University cloning expert Tony Perry, "that could be disastrous. At some point, it will potentially cost lives." So we are left with choices and trade-offs and a need to think through whether it is this technology that alarms us or just certain ways of using it.

By day, Randolfe Wicker, 63, runs a lighting shop in New York City. But in his spare time, as spokesman for the Human Cloning Foundation, he is the face of cloning fervor in the U.S. "I took one step in this adventure, and it took over me like quicksand," says Wicker. He is planning to have some of his skin cells stored for future cloning. "If I'm not cloned before I die, my estate will be set up so that I can be cloned after," he says, admitting, however, that he hasn't found a lawyer willing to help. "It's hard to write a will with all these uncertainties," he concedes. "A lot of lawyers will look at me crazy."

As a gay man, Wicker has long been frustrated that he cannot readily have children of his own; as he gets older, his desire to reproduce grows stronger. He knows that a clone would not be a photocopy of him but talks about the traits the boy might possess: "He will like the color blue, Middle Eastern food and romantic Spanish music that's out of fashion." And then he hints at the heart of his motive. "I can thumb my nose at Mr. Death and say, 'You might get me, but you're not going to get all of me,'" he says. "The special formula that is me will live on into another lifetime. It's a partial triumph over death. I would leave my imprint not in sand but in cement."

This kind of talk makes ethicists conclude that even people who think they know about cloning—let alone the rest of us—don't fully understand its implications. Cloning, notes ethicist Arthur Caplan of the University of Pennsylvania, "can't make you immortal because clearly the clone is a different person. If I take twins and shoot one of them, it will be faint consolation to the dead one that the other one is still running around, even though they are genetically identical. So the road to immortality is not through cloning."

Still, cloning is the kind of issue so confounding that you envy the purists at either end of the argument. For the Roman Catholic Church, the entire question is one of world view: whether life is a gift of love or just one more industrial product, a little more valuable than most. Those who believe that the soul enters the body at the moment of conception think it is fine for God to make clones; he does it about 4,000 times a day, when a fertilized egg splits into identical twins. But when it comes to massaging a human life, for the scientist to do mechanically what God does naturally is to interfere with his work, and no possible benefit can justify that presumption.

On the other end of the argument are the libertarians who don't like politicians or clerics or ethics boards interfering with what they believe should be purely individual decisions. Reproduction is a most fateful lottery; in their view, cloning allows you to hedge your bet. While grieving parents may be confused about the technology—cloning, even if it works, is not resurrection—their motives are their own business. As for infertile couples, "we are interested in giving people the gift of life." Zavos, the aspiring cloner, told *Time* this week. "Ethics is a wonderful word, but we need to look beyond the ethical issues here. It's not an ethical issue. It's a medical issue. We have a duty here. Some people need this to complete the life cycle, to reproduce."

In the messy middle are the vast majority of people who view the prospect with a vague alarm, an uneasy sense that science is dragging us into dark woods with no paths and no easy way to turn back. Ian Wilmut, the scientist who cloned Dolly but has come out publicly against human cloning, was not trying to help sheep have genetically related children. "He was trying to help farmers produce genetically improved sheep," notes Hastings Center ethicist Erik Parens. "And surely that's how the technology will go with us too." Cloning, Parens says, "is not simply this isolated technique out there that a few deluded folks are going to avail themselves of, whether they think it is a key to immortality or a way to bring someone back from the dead. It's part of a much bigger project. Essentially the big-picture question is, To what extent do we want to go down the path of using reproductive technologies to genetically shape our children?"

At the moment, the American public is plainly not ready to move quickly on cloning. In a Time/CNN poll last week, 90% of respondents thought it was a bad idea to clone human beings. "Cloning right now looks like it's coming to us on a magic carpet, piloted by a cult leader, sold to whoever can afford it," says ethicist Caplan. "That makes people nervous."

And it helps explain why so much of the research is being done secretly. We may learn of the first human clone only months, even years, after he or she is born—if the event hasn't happened already, as some scientists speculate. The team that cloned Dolly waited until she was seven months old to announce her existence. Creating her took 277 tries, and right up until her birth, scientists around the world were saying that cloning a mammal from an adult cell was impossible. "There's a significant gap between what scientists are willing to talk about in public and their private aspirations," says British futurist Patrick Dixon. "The law of genetics is that the work is always sig-

nificantly further ahead than the news. In the digital world, everything is hyped because there are no moral issues—there is just media excitement. Gene technology creates so many ethical issues that scientists are scared stiff of a public reaction if the end results of their research are known."

Of course, attitudes often change over time. In-vitro fertilization was effectively illegal in many states 20 years ago, and the idea of transplanting a heart was once considered horrifying. Public opinion on cloning will evolve just as it did on these issues, advocates predict. But in the meantime, the crusaders are mostly driven underground. Princeton biologist Lee Silver says fertility specialists have told him that they have no problem with cloning and would be happy to provide it as a service to their clients who could afford it. But these same specialists would never tell inquiring reporters that, Silver says—it's too hot a topic right now. "I think what's happened is that all the mainstream doctors have taken a hands off approach because of this huge public outcry. But I think what they are hoping is that some fringe group will pioneer it and that it will slowly come into the mainstream and then they will be able to provide it to their patients."

All it will take, some predict, is that first snapshot. "Once you have a picture of a normal baby with 10 fingers and 10 toes, that changes everything," says San Mateo, Calif., attorney and cloning advocate Mark Eibert, who gets inquiries from infertile couples every day. "Once they put a child in front of the cameras, they've won." On the other hand, notes Gregory Pence, a professor of philosophy at the University of Alabama at Birmingham and author of *Who's Afraid of Human Cloning?,* "if the first baby is defective, cloning will be banned for the next 100 years."

"I wouldn't mind being the first person cloned if it were free. I don't mind being a guinea pig," says Doug Dorner, 35. He and his wife Nancy both work in health care. "We're not afraid of technology," he says. Dorner has known since he was 16 that he would never be able to have children the old-fashioned way. A battle with lymphoma left him sterile, so when he and Nancy started thinking of having children, he began following the scientific developments in cloning more closely. The more he read, the more excited he got. "Technology saved my life when I was 16," he says, but at the cost of his fertility. "I think technology should help me have a kid. That's a fair trade."

Talk to the Dorners, and you get a glimpse of choices that most parents can scarcely imagine having to make. Which parent, for instance, would they want to clone? Nancy feels she would be bonded to the child just from carrying him, so why not let the child have Doug's genetic material? Does it bother her to know she would, in effect, be raising her husband as a little boy? "It wouldn't be that different. He already acts like a five-year-old sometimes," she says with a laugh.

How do they imagine raising a cloned child, given the knowledge they would have going in? "I'd know exactly what his basic drives were," says Doug. The boy's dreams and aspirations, however, would be his own, Doug insists. "I used to dream of being a fighter pilot," he recalls, a dream lost when he got cancer. While they are at it, why not clone Doug twice? "Hmm. Two of the same kid," Doug ponders. "We'll cross that bridge when we come to it. But I know we'd never clone our clone to have a second child. Once you start copying something, who knows what the next copies will be like?"

In fact the risks involved with cloning mammals are so great that Wilmut, the premier cloner, calls it "criminally irresponsible" for scientists to be experimenting on humans today. Even after four years of practice with animal cloning, the failure rate is still overwhelming: 98% of embryos never implant or die off during gestation or soon after birth. Animals that survive can be nearly twice as big at birth as is normal; or have extra-large organs or heart trouble or poor immune systems. Dolly's "mother" was six years old when *she* was cloned. That may explain why Dolly's cells show signs of being older than they actually are—scientists joked that she was really a sheep in lamb's clothing. This deviation raises the possibility that beings created by cloning adults will age abnormally fast.

"We had a cloned sheep born just before Christmas that was clearly not normal," says Wilmut. "We hoped for a few days it would improve and then, out of kindness, we euthanized it, because it obviously would never be healthy." Wilmut believes "it is almost a certainty" that cloned human children would be born with similar maladies. Of course, we don't euthanize babies. But these kids would probably die very prematurely anyway. Wilmut pauses to consider the genie he has released with Dolly and the hopes he has raised. "It seems such a profound irony," he says, "that in trying to make a copy of a child who has died tragically, one of the most likely outcomes is another dead child."

That does not seem to deter the scientists who work on the Clonaid project run by the Raelian sect. They say they are willing to try to clone a dead child. Though their outfit is easy to mock, they may be even further along than the competition, in part because they have an advantage over other teams. A formidable obstacle to human cloning is that donor eggs are a rare commodity, as are potential surrogate mothers, and the Raelians claim to have a supply of both.

Earlier this month, according to Brigitte Boisselier, Clonaid's scientific director, somewhere in North America, a young woman walked into a Clonaid laboratory whose location is kept secret. Then, in a procedure that has been done thousands of times, a doctor inserted a probe, removed 15 eggs from the woman ovaries and placed them in a chemical soup. Last week two other Clonaid scientists, according to the group, practiced the delicate art of removing the genetic material from each of the woman's eggs. Within the next few weeks, the Raelian scientific team plans to place another cell next to the enucleated egg.

This second cell, they say, comes from a 10-month-old boy who died during surgery. The two cells will be hit with an electrical charge, according to the scenario, and will fuse, forming a new hybrid cell that no longer has the genes of the young woman but now has the genes of the dead child. Once the single cell has developed into six to eight cells, the next step is to follow the existing standard technology of assisted reproduction and gingerly insert the embryo into a woman's womb and hope it implants. Clonaid scientists expect to have implanted the first cloned human embryo in a surrogate mother by next month.

Even if the technology is basic, and even if it appeals to some infertile couples, should grieving parents really be pursuing this route? "It's a sign of our growing despotism over the next generation," argues University of Chicago bioethicist Leon Kass. Cloning introduces the possibility of parents' making choices for their children far more fundamental than whether to give them piano lessons or straighten their teeth. "It's not just that parents will have particular hopes for these children," says Kass. "They will have expectations based on a life that has already been lived. What a thing to do—to carry on the life of a person who has died."

The libertarians are ready with their answers. "I think we're hypercritical about people's reasons for having children," says Pence. "If they want to re-create their dead children, so what?" People have always had self-serving reasons for having children, he argues, whether to ensure there's someone to care for them in their old age or to relive their youth vicariously. Cloning is just another reproductive tool; the fact that it is not a perfect tool, in Pence's view, should not mean it should be outlawed altogether. "We know there are millions of girls who smoke and drink during pregnancy, and we know what the risks to the fetus are, but we don't do anything about it," he notes.

"If we're going to regulate cloning, maybe we should regulate that too."

Olga Tomusyak was two weeks shy of her seventh birthday when she fell out of the window of her family's apartment. Her parents could barely speak for a week after she died. "Life is empty without her," says her mother Tanya, a computer programmer in Sydney, Australia. "Other parents we have talked to who have lost children say it will never go away." Olga's parents cremated the child before thinking of the cloning option. All that remains are their memories, some strands of hair and three baby teeth, so they have begun investigating whether the teeth could yield the nuclei to clone her one day. While it is theoretically possible to extract DNA from the teeth, scientists say it is extremely unlikely.

"You can't expect the new baby will be exactly like her. We know that is not possible," says Tanya. "We think of the clone as her twin or at least a baby who will look like her." The parents would consider the new little girl as much Olga's baby as their own. "Anything that grows from her will remind us of her," says Tanya. Though she and her husband are young enough to have other children, for now, this is the child they want.

Once parents begin to entertain the option of holding on to some part of a child, why would the reverse not be true? "Bill" is a guidance counselor in Southern California, a fortysomething expectant father who has been learning everything he can about the process of cloning. But it is not a lost child he is looking to replicate. He is interested in cloning his mother, who is dying of pancreatic cancer. He has talked to her husband, his siblings, everyone except her doctor—and her, for fear that it will make her think they have given up hope on her. He confides, "We might end up making a decision without telling her."

His goal is to extract a tissue specimen from his mother while it's still possible and store it, to await the day when—if—cloning becomes technically safe and socially acceptable. Late last week, as his mother's health weakened, the family began considering bringing up the subject with her because they need her cooperation to take the sample. Meanwhile, Bill has already contacted two labs about tissue storage, one as a backup. "I'm in touch with a couple of different people who might be doing that," he says, adding that both are in the U.S. "It seems like a little bit of an underground movement, you know—people are a little reluctant that if they announce it, they might be targeted, like the abortion clinics."

If Bill's hopes were to materialize and the clone were born, who would that person be? "It wouldn't be my mother but a person who would be very similar to my mother, with certain traits. She has a lot of great traits: compassion and intelligence and looks," he says. And yet, perhaps inevitably, he talks as though this is a way to rewind and replay the life of someone he loves. "She really didn't have the opportunities we had in the baby-boom generation, because her parents experienced the Depression and the war," he says. "So the feeling is that maybe we could give her some opportunities that she didn't have. It would be sort of like we're taking care of her now. You know how when your parents age and everything shifts, you start taking care of them? Well, this would be an extension of that."

A world in which cloning is common-place confounds every human relationship, often in ways most potential clients haven't considered. For instance, if a woman gives birth to her own clone, is the child her daughter or her sister? Or, says bioethicist Kass, "let's say the child grows up to be the spitting image of its mother. What impact will that have on the relationship between the father and his child if that child looks exactly like the woman he fell in love with?" Or, he continues, "let's say the parents have a cloned son and then get divorced. How will the mother feel about seeing a copy of the person she hates most in the world every day? Everyone thinks about cloning from the point of view of the parents. No one looks at it from the point of view of the clone."

If infertile couples avoid the complications of choosing which of them to clone and instead look elsewhere for their DNA, what sorts of values govern that choice? Do they pick an uncle because he's musical, a willing neighbor because she's brilliant? Through that door lies the whole unsettling debate about designer babies, fueled already by the commercial sperm banks that promise genius DNA to prospective parents. Sperm banks give you a shot at passing along certain traits; cloning all but assures it.

Whatever the moral quandaries, the one-stop-shopping aspect of cloning is a plus to many gay couples. Lesbians would have the chance to give birth with no male involved at all; one woman could contribute the ovum, the other the DNA. Christine DeShazo and her partner Michele Thomas of Miramar, Fla., have been in touch with Zavos about producing a baby this way. Because they have already been ostracized as homosexuals, they aren't worried about the added social sting that would come with cloning. "Now [people] would say, 'Not only are you a lesbian, you are a cloning lesbian,'" says Thomas. As for potential health problems, "I would love our baby if its hand was attached to its head," she says. DeShazo adds, "If it came out green, I would love it. Our little alien . . ."

Just as women have long been able to have children without a male sexual partner, through artificial insemination, men could potentially become dads alone: replace the DNA from a donor egg with one's own and then recruit a surrogate mother to carry the child. Some gay-rights advocates even argue that should sexual preference prove to have a biological basis, and should genetic screening lead to terminations of gay embryos, homosexuals would have an obligation to produce gay children through cloning.

All sorts of people might be attracted to the idea of the ultimate experiment in single parenthood. Jack Barker, a marketing specialist for a corporate-relocation company in Minneapolis, is 36 and happily unmarried. "I've come to the conclusion that I don't need a partner but can still have a child," he says. "And a clone would be the perfect child to have because I know exactly what I'm getting." He understands that the child would not be a copy of him. "We'd be genetically identical," says Barker. "But he wouldn't be raised by my parents—he'd be raised by me." Cloning, he hopes, might even let him improve on the original: "I have bad allergies and asthma. It would be nice to have a kid like you but with those improvements."

Cloning advocates view the possibilities as a kind of liberation from travails assumed to be part of life: the danger that your baby will be born with a disease that will kill him or her, the risk that you may one day need a replacement organ and die waiting for it, the helplessness you feel when confronted with unbearable loss. The challenge facing cloning pioneers is to make the case convincingly that the technology itself is not immoral, however immorally it could be used.

One obvious way is to point to the broader benefits. Thus cloning proponents like to attach themselves to the whole arena of stem-cell research, the brave new world of inquiry into how the wonderfully pliable cells of seven-day-old embryos behave. Embryonic stem cells eventually turn into every kind of tissue, including brain, muscle, nerve and blood. If scientists could harness their powers, these cells could serve as the body's self-repair kit, providing cures for Parkinson's, diabetes, Alzheimer's and paralysis. Actors Christopher Reeve, paralyzed by a fall from a horse, and Michael J. Fox, who suffers from Parkinson's, are among those who have pushed Congress to overturn the government's restrictions on federal funding of embryonic stem-cell research.

But if the cloners want to climb on this train in hopes of riding it to a public relations victory, the mainstream scientists want to push them off. Because researchers see the potential benefits of understanding embryonic stem

cells as immense, they are intent on avoiding controversy over their use. Being linked with the human-cloning activists is their nightmare. Says Michael West, president of Massachusetts-based Advanced Cell Technology, a biotech company that uses cloning technology to develop human medicines: "We're really concerned that if someone goes off and clones a Raelian, there could be an overreaction to this craziness—especially by regulators and Congress. We're desperately concerned—and it's a bad metaphor—about throwing the baby out with the bath water."

Scientists at ACT are leery of revealing too much about their animal-cloning research, much less their work on human embryos. "What we're doing is the first step toward cloning a human being, but we're not cloning a human being," says West. "The miracle of cloning isn't what people think it is. Cloning allows you to make a genetically identical copy of an animal, yes, but in the eyes of a biologist, the real miracle is seeing a skin cell being put back into the egg cell, taking it back in time to when it was an undifferentiated cell, which then can turn into any cell in the body." Which means that new, pristine tissue could be grown in labs to replace damaged or diseased parts of the body. And since these replacement parts would be produced using skin or other cells from the suffering patient, there would be no risk of rejection. "That means you've solved the age-old problem of transplantation," says West. "It's huge."

So far, the main source of embryonic stem cells is "leftover" embryos from IVF clinics; cloning embryos could provide an almost unlimited source. Progress could come even faster if Congress were to lift the restrictions on federal funding—which might have the added safety benefit of the federal oversight that comes with federal dollars. "We're concerned about George W.'s position and whether he'll let existing guidelines stay in place," says West. "People are begging to work on those cells."

That impulse is enough to put the Roman Catholic Church in full revolt; the Vatican has long condemned any research that involves creating and experimenting with human embryos, the vast majority of which inevitably perish. The church believes that the soul is created at the moment of conception, and that the embryo is worthy of protection. It reportedly took 104 attempts before the first IVF baby, Louise Brown, was born; cloning Dolly took more than twice that. Imagine, say opponents, how many embryos would be lost in the effort to clone a human. This loss is mass murder, says David Byers, director of the National Conference of Catholic Bishops' commission on science and human values. "Each of the embryos is a human being simply by dint of its genetic makeup."

Last week 160 bishops and five cardinals met for three days behind closed doors in Irving, Texas, to wrestle with the issues biotechnology presents. But the cloning debate does not break cleanly even along religious lines. "Rebecca," a thirty something San Francisco Bay Area resident, spent seven years trying to conceive a child with her husband. Having "been to hell and back" with IVF treatment, Rebecca is now as thoroughly committed to cloning as she is to Christianity. "It's in the Bible—be fruitful and multiply," she says. "People say, 'You're playing God.' But we're not. We're using the raw materials the good Lord gave us. What does the doctor do when the heart has stopped? They have to do direct massage of the heart. You could say the doctor is playing God. But we save a life. With human cloning, we're not so much saving a life as creating a new being by manipulation of the raw materials, DNA, the blueprint for life. You're simply using it in a more creative manner."

A field where emotions run so strong and hope runs so deep is fertile ground for profiteers and charlatans. In her effort to clone her daughter Olga, Tanya Tomusyak contacted an Australian firm, Southern Cross Genetics, which was founded three years ago by entrepreneur Graeme Sloan to preserve DNA for future cloning. In an e-mail, Sloan told the parents that Olga's teeth would provide more than enough DNA—even though that possibility is remote. "All DNA samples are placed into computer-controlled liquid-nitrogen tanks for long-term storage," he wrote. "The cost of doing a DNA fingerprint and genetic profile and placing the sample into storage would be $2,500. Please note that all of our fees are in U.S. dollars."

When contacted by *Time,* Sloan admitted, "I don't have a scientific background. I'm pure business. I'd be lying if I said I wasn't here to make a dollar out of it. But I would like to see organ cloning become a reality." He was inspired to launch the business, he says, after a young cousin died of leukemia. "There's megadollars involved, and everyone is racing to be the first," he says. As for his own slice of the pie, Sloan says he just sold his firm to a French company, which he refuses to name, and he was heading for Hawaii last week. The Southern Cross factory address turns out to be his mother's house, and his "office" phone is answered by a man claiming to be his brother. David—although his mother says she has no son by that name.

The more such peddlers proliferate, the more politicians will be tempted to invoke prohibitions. Four states—California, Louisiana, Michigan and Rhode Island—have already banned human cloning, and this spring Texas may become the fifth. Republican state senator Jane Nelson has introduced a bill in Austin that would impose a fine of as much as $1 million for researchers who use cloning technology to initiate pregnancy in humans. The proposed Texas law would permit embryonic-stem-cell research, but bills proposed in other states were so broadly written that they could have stopped those activities too.

"The short answer to the cloning question," says ethicist Caplan, "is that anybody who clones somebody today should be arrested. It would be barbaric human experimentation. It would be killing fetuses and embryos for no purpose, none, except for curiosity. But if you can't agree that that's wrong to do, and if the media can't agree to condemn rather than gawk, that's a condemnation of us all."

QUESTIONS

1. What kind of ethical issues should be taken into consideration when deciding to create a clone?

2. What are the risks and benefits of cloning?

3. Should human cloning be allowed? Support your answer with three explanations using examples from the article.

4. Should human cloning be outlawed globally? Support your answer with three explanations using examples from the article.

5. Explain how cloning could be a personal, social, political and ethical issue. Provide an example for each issue.

Antibiotic Resistance

LINDA STEVENS HJORTH

"Before antibiotics were available, isolation was the only way to prevent the spread of infectious diseases. Children with tuberculosis were isolated on ferry boats during the 1920s in New York City's harbor. With the indiscriminate and reckless use of antibiotics in recent times, more and more bacteria are becoming resistant to drugs, including the microorganism that causes tuberculosis. If the trend continues, isolation may once again become necessary."

AMABILE-CUEVAS, ET. AL, 1995

Have you ever gone to the doctor hoping to get an antibiotic to get rid of that hacking cough, sore throat, running nose and fever? Was your doctor's response something like this: "Go home, drink plenty of liquids and rest"? This cautious response by doctors is because they are correctly concerned that the desire to quickly eliminate your ills through the use of antibiotics can be a dangerous practice.

If a patient expects to be "cured" or free from symptoms, doctors may find it difficult to turn that patient down and not provide antibiotics; the goal for most doctors is to treat, heal, cure and relieve pain. The doctors may feel that they are not healing the patient, and the patient may feel frustrated because they believe their discomfort should dissipate quickly with a little pill.

The reality is that doctors have to be cautious when prescribing antibiotics so that their patients will not develop antibiotic resistance (microbial resistance). According to the Food and Drug Administration, microbial resistance can be caused by . . .

1. Administering antibiotics to patients in larger doses than recommended by the healthcare and federal organizations.

2. Patients who do not finish the entire bottle of medication prescribed to them. When patients do not take the entire amount of pills prescribed, the bacterial strain becomes stronger and more resistant to antibiotics.

3. Administering antibiotics for viral infections. Viral infections do not respond to antibiotics.

4. Administering antibiotic drugs to animals that are eaten by humans. The drugs are given to prevent the animal from contracting diseases and to increase production. Potentially those humans who eat the animal that has been treated with the antibiotics could be resistant to medications used to treat human illness. (FDA 2002)

Some turkey and chicken products contain campylobacter which increases the risk of humans being infected with a bacteria that medications on the market will not easily kill. According to the Centers for Disease Control and Prevention, campylobacter is the most common bacterial cause of diarrheal illness in the United States; over 2 million people every year are affected by it. Fever, diarrhea and abdominal cramps can be produced in humans who eat chicken that contains campylobacter, and it can be life threatening for those with weakened immune systems. (Bren, 2001)

Anyone can experience antibiotic resistance. The struggle is that antibiotic resistance makes cures for diseases difficult to administer. For example, parents have learned that after repeatedly using amoxicillin for their children's sore throats and earaches, the antibiotic loses its effectiveness and stronger antibiotics are needed. The struggle lies within a heartfelt fear: if my eight year old already needs a stronger antibiotic, what will my child have in store for him when he needs antibiotic treatments in 25 years?

The Food and Drug Administration states that "about 70 percent of the bacteria that cause infections in hospitals are resistant to at least one of the drugs most commonly used to treat infections." (FDA 2002) It seems that the problem of antibiotic resistance continues to challenge the patients who suffer and the doctors who simply want to care for those patients.

What can be done about this problem? The National Center for Infectious Diseases has created a strategic plan to address the problems of antibiotic resistance. What follows is a short summary of their goals:

- To collect data that can "prevent early warnings of outbreaks or identify changing patterns of resistance."

- "To assess trends in drug use and understand the relationship between drug use and the creation of infections that are antibiotic resistant."

- "To develop and evaluate new lab tests that can improve the accuracy and timeliness of antimicrobial resistance detection in clinical settings."

- "To find ways to decrease the emergence and spread of drug resistance by educating the patients about the 'right' way to use medications."

- "To evaluate a vaccine that could be used in preventing drug-resistant infections." (CDS 2001)

There needs to be a heightened awareness among physicians, pharmacists, pharmaceutical companies and patients so that antibiotic resistance can be reduced. Antibiotics should only be used when necessary, but this will only happen when everyone becomes educated about the ramifications of antibiotic misuse. It is hoped that after reading this article, you will question your own need for antibiotics as well as questioning the need for the prescription in the first place. Pass the word about antibiotic resistance to others. By educating others you are reducing the likelihood of the development of a medical problem that could have been prevented.

REFERENCES

Centers for Disease Control and Prevention. *Target Area Booklet: Addressing the Problem of Antimicrobial Resistance* "Emerging Infectious Diseases: A strategy for the 21st Century." 2001 http://www.cdc.gov/ncidod/emergplan/antiresist/.

Cuevas, Amabile, Maura Cardenas-Garcia, and Mauricio Ludgar. "Antibiotic Resistance." *American Scientist,* 83 (1994); 320–329.

Bren, Linda. "Antibotic Resistance from Down on the Chicken Farm." *FDA Consumer Magazine.* January–February 2001.

U.S. Food and Drug Administration. "Antibiotic Resistance." http://www.fda.gov/oc/opacom/hottopics/anit_resist.html.

QUESTIONS

1. How would you define antibiotic resistance?

2. What causes antibiotic resistance?

3. Describe a conversation that you might have with your doctor to express your educated concerns about antibiotic resistance.

4. What is government doing to stop antibiotic resistance?

The Politics of Life and Death: Global Responses to HIV and AIDS

MARY CARON

●●

If effectively fighting HIV means openly getting condoms to teenagers or clean needles to addicts, or candidly discussing the prevalence of prostitution in their communities, many politicians would rather avoid the subject altogether—even if it means allowing an epidemic to flourish. Where leaders have lifted their heads from the sand, however, millions of lives have been saved.

Both Rajesh and his wife—who prefers not to give her name for fear of being ostracized by neighbors in their Bombay community—are infected with HIV. With the help of money quietly contributed by relatives, they are among the few families in India who can pay for life-prolonging anti-viral drugs—but only for Rajesh. Other couples in India are finding themselves in a similar situation. "It is the woman who is stepping back" so her husband can get treatment, said Subhash Hira, director of Bombay's AIDS Research and Control Center in a recent *Associated Press* story. "She thinks of herself as dispensable."

In Zimbabwe, where 200 people are dying every day from AIDS, life insurance premiums have quadrupled to keep up with rising costs. It would take roughly two years for the average Zimbabwean to pay for one month of treatment at U.S. rates.

In the United States, of course, incomes are much higher—about 74 times those in India, and 46 times those in Zimbabwe. Yet, even here, nearly half of HIV positive patients in a recent national study had annual incomes less than $10,000, whereas the annual costs of their care and treatment came to about $20,000. Two of every three patients in the study had either no insurance or only public health insurance—which may not adequately cover their needs.

And then there is the Central African Republic, where I worked as a Peace Corps volunteer a few years ago. The

Mary Caron is press director for the Worldwatch Institute. Permission granted by World Watch. *World Watch* May/June 1999, pp. 30–38. www.worldwatch.org

five-bedroom house where I lived was owned by my neighbor, Victor, who rented it to me while living in a more modest mud-brick house next door. He used the income from rent, plus whatever his eldest daughter could earn selling food in the market, to support ten people. One of them was a little girl of five or six, who used to come over and sing me songs. I didn't realize that Victor was her uncle, not her father, until someone explained that her mother and father had died following "a long illness." There are many households like Victor's. Worldwide, more than 41 million children will have lost one or both parents by 2010, mostly as a result of AIDS. The grandparents or other surviving family members, who often have their own difficulties making ends meet, may find themselves taking care of up to a dozen children.

Clearly, AIDS is a disease that the world cannot afford. And yet, the relentless spread of the virus forces us to confront painful life-and-death choices about allocating resources. Communities, nations, and international donors are all struggling to care for a growing number of the sick; to invest in prevention that can avert millions of future infections; to fund research that can yield life-prolonging treatments; and ultimately to develop a vaccine. To do all of these things at once seems an almost impossible task. But experience in the field demonstrates that there are reasons for hope, even in relatively poor countries where HIV is already a serious problem.

While other diseases target children or the elderly, HIV often strikes otherwise strong and healthy people—those most likely to be taking care of children and contributing to the economy. And it does so in a way that has repeatedly caught societies unprepared. HIV does not kill within a matter of days or weeks, like other infectious diseases; rather it gives death a kind of "rain check." The asymptomatic period may last 10 years or longer in a country like the United States, though the infection can progress to AIDS in as little as two to three years in a country like Zimbabwe or India, where the percentage of people who can get full treatment and care is much smaller. An infected person may be ill off and on for years, requiring extended care from family or community members. And HIV, while relatively slow to develop in the body, can spread rapidly within a population. About 75 percent of HIV transmission worldwide is through unprotected sex. The rest occurs mainly through sharing of unsterilized needles, through childbirth or breastfeeding from an infected mother to her child, and from the use of infected blood in transfusions.

AIDS now rivals tuberculosis as the world's most deadly infectious disease. Every day last year, 16,000 people were infected with HIV—11 people per minute. Women now account for 43 percent of all adults with HIV/AIDS. And about half of all new infections are in 15- to 24-year-olds. Since AIDS was first recognized in 1981, more than 47 million people have become infected and nearly 14 million have died. The epidemic has taken its heaviest toll in Africa, which has just 10 percent of the world's population but 68 percent of the HIV/AIDS cases—most of them in the sub-Saharan region. In some southern African countries, one in four adults are HIV positive. Thus, the world's poorest countries are staggering under the burden of the world's most unaffordable disease.

As HIV wears down the body's defenses and the infected person becomes increasingly ill from opportunistic infections, the costs of providing care and treatment mount. Worldwide, about 63 percent of the $18.4 billion spent on HIV/AIDS in 1993 went to care, according to a 1996 study by Harvard researchers Daniel Tarantola and the late Jonathan Mann.* Another 23 percent was spent on research and just 14 percent on prevention. Moreover, only 8 percent of global spending took place in "low economy" countries of the developing world, yet more than 95 percent of all HIV-infected people live in developing countries.

Preventing an HIV infection costs much less than caring for an infected individual. And the benefit of preventing an HIV infection costs much less than caring for an infected individual. And the benefit of prevention is compounded, since preventing one person from getting HIV keeps that person from spreading it to others. If a man has sex with three different women in a year, shielding that man from infection also shields his three partners and any children they may have.

However, global resolve to protect the uninfected may be overwhelmed by the challenge of providing for the staggering number already infected. AIDS already rivals the horror of the smallpox epidemic which decimated Native American populations in the 16th century, and the Black Death which wiped out a quarter of Europe's population in the 14th century. If one of every four adults is already infected in Botswana and Zimbabwe, what hope is there for those countries and their neighbors? Many people may now be under the impression that Africa is a continent essentially lost to AIDS, and that the rest of the developing world may soon follow.

Look more closely, however, and two signs make it clear that the situation is far from hopeless.

First, more than half of the populations of developing countries—about 2.7 billion people—live in areas where HIV is still low, even among high-risk groups. Another third live in areas where the epidemic is still concentrated in one or more high-risk groups, yet "HIV prevalence"—the proportion of a population infected at a given time—is still below 5 percent in the general population. Even in hard-hit Africa, there are at least a few countries—such as Benin, Senegal, Ghana, and Guinea—where adult infections are still under 3 percent. These areas of relatively low HIV prevalence present a one-time opportunity—and one that will not last long—for policy makers to implement solid strategies for keeping HIV at bay.

Second, even in places where the epidemic has taken hold, campaigns to stop further escalation have proven successful in both the early *and* the later stages of an epidemic. It's instructive to see how that was done in each case—first in Thailand, where an effective campaign was able to ward off an incipient epidemic and to keep prevalence relatively low in the general population, and then in Uganda, where a high percentage of the population was already infected.

By taking action relatively early, Thailand was able to keep HIV infections from spiraling out of control in the general population. In early 1988, officials were alarmed by reports from an ongoing survey at a Bangkok hospital showing that infections among drug addicts who use needles, or "injecting drug users" (IDUs), had jumped from 1 percent to 30 percent in the preceding 6 months. In response, the Thai Ministry of Public Health set up a system to collect data on

---

*Mann and his wife, Mary Lou Clements-Mann, died in the crash of Swissair flight 111 last September. Mann founded and headed the World Health Organization's Global Program on AIDS.

HIV infection at selected sites throughout the country. These "sentinel surveys," as they are called, revealed even more alarming news. By mid-1989, HIV was present in all 14 provinces surveyed. In the northern city of Chiang Mai, 44 percent of the prostitutes were infected. And HIV was also found in some pregnant women, who are considered representative of the general population.

Concerned about the possibility of a general epidemic, the Thai government then conducted a national survey to identify behaviors that might be driving the spread of the virus. What it found was that more than one-fourth of the country's men were having sex with prostitutes, both before and outside marriage. In 1991, Prime Minister Anand Panyarachun assumed personal leadership of the National AIDS Committee and aggressively escalated the government's response. Official spending on HIV/AIDS was pushed from $2.6 million in 1990 to $80 million in 1996.

The Thai effort mobilized sectors of the population ranging from prostitutes to teachers to monks. In the commercial sex industry, which accounts for an estimated 14 percent of Thailand's GDP, brothel owners and employees now require every male customer to use a condom. Government STD clinics hand out about 60 million free condoms a year, and encourage their use. Several monasteries in northern Thailand are providing counseling services for HIV-affected people, and helping them find employment. Schools are teaching children how to reduce sexual risk-taking.

Within three years after this heightened response got underway, there were signs that it might be working. A second national behavior survey showed that between 1990 and 1993, the percentage of 15–49 year old men reporting sex outside of marriage had dropped from 28 percent to 15 percent. Among men who continued to engage prostitutes, the percentage reporting that they always used a condom doubled. Condom sales rose and sexually transmitted disease declined throughout the country. HIV infection also declined. Annual testing of 21 year-old military conscripts, which had found 0.5 percent of them infected in 1989, showed prevalence peaking at 3.7 percent in mid-1993 (reflecting a predictable lag between risky behavior and evidence of infection), then declining to 1.9 percent in 1997. Similarly, testing of pregnant women in all 76 provinces found HIV infection at 0.5 percent in 1990, increasing to 2.4 percent in 1995, then declining to 1.7 percent in 1997.

The health and social costs of this disease still inflict a heavy burden on the Thai economy, and the costs continue to grow. The Asian financial crisis, which began in Thailand in 1997, has forced cuts in the national AIDS budget and has put greater strains on affected families. And Thailand will have to remain vigilant to keep prevalence low. While condom use has increased in the country as a whole, it remains much lower in rural areas among people with limited education, and among those who engage in casual sex. A 1995 survey also showed that many drug users were reverting to sharing needles. Nonetheless, by acting quickly and aggressively, Thailand may have averted a full-blown HIV epidemic.

Uganda, unlike Thailand, launched its prevention campaign at a time when a high percentage of the population had already been infected. By 1999, in a population of less than 21 million, 1.8 million Ugandans have already died and 900,000 more have HIV. Moreover, Uganda has far less financial capability than does Thailand, with a per-capita GNP of just $300 compared with Thailand's $2,960. Yet Uganda's success in bringing down high HIV prevalence provides evidence that fighting HIV is not impossible, even when the situation at the outset looks dire. When Yoweri Museveni became president in 1986, HIV was already a serious problem. Museveni quickly implemented a national plan, enlisting both government agencies and non-governmental organizations (NGOs) to join the fight. Uganda established the first center in sub-Saharan Africa where people could go for voluntary and anonymous HIV testing and counseling.

Like the Thai campaign, the Uganda one succeeded by mobilizing a wide spectrum of groups. A student heading home after class at Makerere University in Kampala, for example, may well get an update of the latest information on avoiding HIV—courtesy of her "boda boda" bicycle taxi driver, who has been trained by the Community Action for AIDS Prevention project. Or if you live in the Mpigi district, the local Muslim spiritual leader, or Imam, may stop by for a discussion of AIDS and Islam. Trained by the Family AIDS Education and Prevention Through Imams project of the Islamic Medical Association of Uganda, some 850 of these leaders have taken HIV prevention messages directly to the homes of more than 100,000 families throughout the country.

The Ugandan government has conducted regular surveys of sexual behavior, and these studies show signs of substantial change from 1989 to 1995. The share of 15–19 year-olds who report never having had sex has increased from 26 to 46 percent for girls, and from 31 to 56 percent for boys. The share of people reporting that they had used a condom at least once rose from 15 to 55 percent for men and from 6 to 39 percent for women.

HIV prevalence has also dropped, most notably among young people between the ages of 13 and 24. Between 1991 and 1996, the percentage of pregnant women testing positive for HIV in some urban areas dropped by one half, from about 30 percent to 15 percent.

So it is apparently possible to keep HIV in check. But it's not easy, whether the problem is caught in its early stages as it was in Thailand, or has become full-blown as it was in Uganda. It's difficult to change people's behavior, especially when it means challenging highly sensitive—and very personal—questions about sex, prostitution, infidelity, and drug dependence. "We have to stop thinking that HIV/AIDS is only a health problem. It is a development problem," says the World Bank's HIV/AIDS coordinator Debrework Zewdie. To stop it will take "a commitment from governments in developed and developing countries. Zoom-in-and-zoom-out programs are not going to work; we have to build local capacity."

There is no single formula for building that capacity, although the most innovative and appropriate solutions often come from within communities. Wherever initiatives are taken, however, there are some basic policy principles that seem to apply.

- **Early and aggressive action:** Worldwide, we already spend nearly $5 on HIV/AIDS treatment and care for every $1 spent on prevention. Implementing prevention measures before even the first case of AIDS is reported can reduce that ratio and thereby greatly reduce the overall costs of care. On the other hand, if governments avoid thinking about prevention until AIDS cases start to burden the health system, the epidemic may already have invaded large parts of the population. Yet because symptoms of AIDS typically do not show up until several years after infection, the threat may be largely invisible until large numbers of people have been doomed.

- **Communities:** Mobilizing business, religious, and civic leaders can galvanize broad support for raising public awareness of the risks of HIV and reducing the stigmatization of those infected. In Zimbabwe, for example, the Commercial Farmers Union recruited and sponsored farm owners to participate in a Family Health International-supported program that trained more than 2 million farm employees and family members in a nationwide HIV/AIDS prevention effort.

- **Political leadership:** In both Thailand and Uganda, HIV/AIDS prevention moved from simply a public

health concern to a national priority. Prevention campaigns can succeed when political leaders put them at the top of the national agenda, use their public platform to encourage safer behavior, ask communities and NGOs to join the fight, and work to change laws that prohibit such effective tools of prevention as condom advertising and needle purchases.

- **Data collection and dissemination:** HIV is a stealthy attacker that can infiltrate an unsuspecting community and spread rapidly. It is therefore important to collect infection data from health clinics and to assess behavior trends. By publicizing the results of its sentinel and behavior surveys, Thailand made its population aware of the extent of risk in the country.

- **Low-cost, high-quality condoms:** Mr. Lover Man, a human-size condom mascot, can now be seen cruising the streets, attending soccer games—and, of course, passing out condoms—in several South African cities. In Portland, Oregon, teenagers have been given discreet access to protection via 25-cent condom vending machines in public rest rooms. Using new variants of old marketing techniques, organizations like Population Services International (PSI) have dramatically increased the worldwide distribution of HIV prevention information and reliable low-cost condoms. In Zaire, PSI "social marketing" programs helped condom sales to rise from 900,000 in 1988 to 18.3 million in 1991, averting an estimated 7,200 cases of HIV.

- **Targeting interventions to high risk groups:** HIV usually gains a foothold in one or more groups whose behavior puts them at higher risk: prostitutes, IDUs, people with another sexually transmitted disease, young military recruits, migrant workers, truck drivers, or homosexual men. The virus can spread rapidly within the group and, once established, can move to those at lower risk of infection through people who act as a bridge between high and low risk groups—for example, men who have visited prostitutes and then bring the disease home to their wives.

A World Bank report, *Confronting AIDS,* suggests that countries can keep HIV at bay by targeting these high-risk groups with HIV prevention. It is important to note, though, that such efforts, if not managed with particular care, can trigger unintended public reactions. Singling out particular groups may inadvertently raise perceptions that HIV is a problem only for "those" people. Some public health experts have also noted that programs to give prostitutes a regular monthly course of antibiotics, for example, may reduce STDs but are harmful to overall health. And when HIV is present in the general population, questions of equitable distribution of resources arise as well.

Preventing HIV infection in someone with a high rate of partner change can avert many more future infections than preventing infection in a person with low-risk behavior, says the World Bank report. For example, compare two prevention programs. The first, in Nairobi, Kenya, provided free condoms and STD treatment to 500 prostitutes, of whom 400 were infected. Each of the women had an average of four partners per day. Under the program, condom use rose from 10 to 80 percent. A calculation based on the estimated rate of transmission, number of partners, condom effectiveness, and secondary infections, shows that this program averted an estimated 10,200 new cases of HIV infections each year among the prostitutes, their customers, and their customers' wives. If the same program had instead targeted a group of 500 men, who had an average of four partners per year, 88 new cases of HIV would have been prevented. The second program would have saved fewer than 1 percent as many people as the first.

When IDUs share needles contaminated with blood, HIV can sweep through their population even more rapidly than it does among prostitutes, because the risk of transmission per contact is higher. In January 1995, HIV prevalence among such drug users in the Ukraine was under 2 percent. Eleven months later, it had shot up to 57 percent. As of December 1997, 66 percent of HIV infections in China and 75 percent in Kaliningrad, Russia resulted from shared needles. Half of all new HIV infections in the United States occur among intravenous drug users, even though less than half of 1 percent of the U.S. population injects drugs frequently. And as with prostitution, HIV can spread from this high-risk group to the population at large.

Needle exchange programs aim to reduce the transmission of bloodborne infections, including HIV, by providing sterile syringes in exchange for used, potentially contaminated syringes. After the U.S. state of Connecticut made needles available from a pharmacy without a prescription, the percentage of IDUs who share needles dropped from 71 percent to 15 percent in three years. A review of studies conducted between 1984 and 1994 showed that HIV prevalence among IDUs increased by 5.9 percent per year in 52 cities that did not have needle exchange programs, but declined by 5.8 percent per year in 29 cities that did.

The experience of the past two decades has given us a set of policies that are proven to work, at least at mobilizing communities to keep HIV in check. Such policies should be in place in every country in the world. Yet, proven policies aren't always enough. Even when faced with the specter of an ever more devastating human and economic toll, people in positions of political power too often ignore—or thwart—the most effective HIV-fighting strategies. If confronting AIDS means talking about such potentially explosive topics as distribution of condoms to teenagers, or of needles to addicts, or the prevalence of prostitution in their communities, many politicians would rather avoid the subject altogether.

In Kenya, where tourism brings in more money than exports of tea, coffee, or fruit, officials—perhaps leery of scaring off tourists—declared the country AIDS-free, even when studies among Kenyan prostitutes showed 60 percent of them to be HIV-infected. The government did not admit the scope of the epidemic until late in 1997. By then, more than a million Kenyans were infected. The country is belatedly taking steps to implement some HIV-fighting programs, such as an awareness campaign for students. Muslim and Catholic religious leaders, however, object to sex education in schools, saying it would corrupt students' morals. By now, the number of infected Kenyans has passed 1.6 million—about 12 percent of the adult population.

Refusal to pay serious attention has been a common failing in these battles, in which the invasion is so stealthy and the victims are often socially marginalized. Even in Thailand, where the government eventually roused itself to lead an aggressive anti-HIV campaign, there was an initial period of denial in the late 1980s, when infections were burgeoning among prostitutes in the Northern provinces, particularly in the Chiang Mai area. Given the large role of commercial sex in the Thai economy, officials may at first have been more concerned about the possible loss of tourism dollars than about the risk of an epidemic. Fortunately, they did not continue to ignore the problem.

In the United States, where about half of all new HIV infections are spread through shared needles or to sexual partners of IDUs, the government bans the use of federal funds for needle exchange programs. Last April, after carefully reviewing research on needle exchange programs, U.S. president Bill Clinton declared that these programs curb AIDS without promoting increased illegal drug use. Yet, in the same announcement, he declined to lift an existing ban on federal funding for needle exchange, which applies to all domestic and overseas programs. Senator Paul Coverdell of Georgia introduced a bill that would prevent the ban from ever being lifted, and Representative Todd Tiahrt of Kansas authored a provision in the federal budget that bans the use of federal and city funding for needle exchange in the nation's capital. Politicians do not want to appear "soft on drugs" by helping drug users, who are often perceived as a criminal element—and who some cynically believe would die of drug overdoses anyway, even if they didn't die of AIDS. Even the more self-interested argument that preventing HIV among drug users could prevent it from spreading to the general population is largely ignored.

Similarly, U.S. officials were slow to act when HIV was first recognized in the early 1980s among homosexual men. Condemnation of the gay community was widespread, and some people went so far as to suggest AIDS was a heavenly retribution for worldly sins (i.e., homosexual sex). Fortunately for the U.S. population as a whole, as well as for those segments most at risk, members of the gay community launched their own aggressive and highly organized campaign to prevent HIV. Between the 1980s and the 1990s, AIDS was turned from a marginalized problem of "those people" to a high-profile national health threat. And while about half of those infected are still not in ongoing care, prevalence has been kept low.

In less politically or economically stable countries than the United States, however, leaders are sometimes overwhelmed by social and economic upheaval that may fatally distract them from the threat of HIV. As apartheid was ending in South Africa, for example, an influx of commercial trade and migrant workers from neighboring countries opened up a kind of viral superhighway for the epidemic. Legislators, grappling with the momentous political and social changes at hand, failed to foresee that these changes might also bring deadly consequences, and no proper prevention strategy was put in place. Forced displacement of black people under apartheid, and the deploying of workers far from their families, had also led to higher rates of extramarital sex and prostitution. Today, more than 3 million South Africans—one of every eight adults—have HIV. In a country of just 43 million, 1,500 people are infected every day.

The political and social climate in South Africa has been slow to change. The government has been accused of stifling non-governmental action with bureaucratic restrictions. Social stigmatization runs very high. A woman who had just publicly declared her HIV-positive status as a means of helping others to fight discrimination was beaten to death just after Christmas last year by a mob of neighbors who stoned her, kicked her, and beat her with sticks.

After a long period of rarely addressing the issue, departing president Nelson Mandela declared in March that "the time for such silence is now long past. The time

has come to teach our children to have safe sex, to have one partner, to use a condom."

The painful lessons learned in South Africa and other AIDS-ravaged countries can now be brought to bear on the world's two largest countries, where the future health of a large portion of humanity lies at stake. The choices that Chinese and Indian leaders make about fighting HIV in the next few years will affect the course of the epidemic for one-third of the world's people. India and China both have relatively low HIV prevalence, but alarming signs of increasing infections among some groups, coupled with known risk factors, make both countries precariously susceptible. If HIV prevalence in China and India were to reach the levels now seen in some southern African countries, up to 300 million people would be infected. The magnitude of the impact—on economic productivity, social and political stability, psychological health, and the human spirit worldwide—is almost unimaginable.

In India, at present, less than 1 percent of adults are infected. Still, with an adult population of almost 500 million, that comes to 4 million HIV positive people—in absolute numbers, more than any other nation. Prevalence is highest among prostitutes, truckers, and IDUs, and there are signs that HIV is also gaining a foothold in the general population. A study conducted between 1993 and 1996 in the city of Pune, south of Bombay, showed that close to 14 percent of the city's monogamous married women had been infected.

In Bombay, by now, more than 50 percent of the city's 50,000 "sex workers" are HIV-positive, as compared with just 1.6 percent in 1988. Prevalence has also jumped into the double digits among prostitutes in the cities of Pune, Vellore, and Chennai (Madras). By 1993, about 70 percent of the 15,000 IDUs in India's Manipur state, located near the "Golden Triangle" of Myanmar and China, were HIV-positive. And more recently, a random survey in Tamil Nadu indicated that some 500,000 of that state's 25 million people are now infected. The epidemic has also spread among people who live and work along the major north-south truck corridor. In short, the evidence suggests that India's AIDS situation is on the verge of exploding, if the country's leaders don't mobilize quickly enough to stop it. Moreover, in a country with 16 major languages, more than 1,600 dialects, and six major religions, such mobilization will require exceptionally skillful coordination and organization.

The Indian government has made a commitment to fighting HIV and is working with donors to coordinate prevention and care efforts. The question now is whether it can mobilize quickly enough. Last December, Prime

Minister Atal Behari Vajpayee declared HIV and AIDS to be the country's most serious public health challenge. With financial assistance from the World Bank, the government is implementing a National AIDS Control Program. It aims to give autonomy and financial support to the country's 25 states in order to upgrade their health service delivery infrastructures and carry out HIV prevention and care targeted to high risk groups. The state of Tamil Nadu already has a system in place to give financial and technical support to NGOs and has set a precedent for an effective decentralized anti-HIV campaign.

Greater reason for hope, though, lies in India's active local communities and a thriving network of NGOs. Following on the Gandhian legacy of grassroots resistance to British colonialism, local groups are emerging throughout India to tackle HIV. In 1992, for example, representatives from SANGRAM, a rural women's group in Maharashta, went to a local red light district and began passing out condoms, telling prostitutes, "This will save your life and mine." Some prostitutes, resentful of mainstream disdain for them, did not appreciate outsiders coming in to tell them what to do. "In the beginning, it was difficult; they even threw stones at us," said SANGRAM General Secretary Meena Seshu. Eventually, though, a small group of prostitutes took over the condom distribution and began educating their peers on how to avoid STDs and HIV. Since then, some 4,000 prostitutes in seven districts have formed their own collective, called the Veshya AIDS Muquabla Parishad (VAMP). The women attend training sessions on personal health, sexuality, STDs and superstition, negotiating condom use with clients, and how to be counselors for the infected and their families. Seshu notes that, in addition to lowering STDs and pregnancies, the collective has given the women strength to tackle difficult issues that might previously have been neglected. Whereas their health needs were often overlooked in the past, for example, the prostitutes are now demanding that doctors examine them and treat STDs properly. Organizations like SANGRAM and VAMP are gaining strength in several regions of India, and as they grow they are using their programs as a basis for advocating improved AIDS-prevention policies throughout the country.

In China, as far as we know, there is not yet a large-scale HIV epidemic. However, the potential for an enormous epidemic is becoming evident. China—shades of South Africa—is relaxing once-stringent economic constraints and opening previously closed doors to the outside world. These economic policy shifts are driving rapid social change, and may also be paving the way for HIV.

Once confined to foreign visitors and small groups of injecting drug users in Yunnan province, HIV has entered a phase of "fast growth" throughout the country according to a recent report by the Chinese Ministry of Health. Left unchecked, HIV infections could exceed 1 million by the year 2000 and 10 million by 2010. The World Health Organization's most recent estimate puts the number infected in China at 600,000.

China eradicated open prostitution in 1949. Since the 1980s, however, commercial sex has resurfaced and seems to be growing. Girls, lured by money in China's burgeoning cities, are moving from rural areas and are often drawn into prostitution. Economic expansion is also increasing the number of migrant workers, who may now represent up to 15 percent of the total labor force. Often young, unmarried or living away from their spouses, migrants may be more likely to have casual sex or sex with prostitutes, greatly increasing their risk of infection. And here, as elsewhere, the sharing of needles by drug addicts spreads HIV even faster than prostitution. Among injecting drug users in Yunnan province, almost 86 percent are now infected.

The Chinese government apparently recognizes the magnitude of threat to its more than 1.2 billion people. A national program for HIV/AIDS control has been approved by the State Council, China's highest governing body. Hypodermic syringes and needles are available for sale at all pharmacies throughout the country. More than a billion condoms were produced by Chinese manufacturers in 1998 and distributed by the State Family Planning Commission. At a time when other ministries were facing stiff cutbacks in staff and resources, the National Center for HIV Prevention and Control was set up last July to study the epidemiology of HIV, to develop health education, and to conduct clinical work for pharmaceuticals. The Chinese Railways Administration distributed AIDS prevention information to its staff of 6 million workers and among railway passengers, many of them migrants.

China's past performance with public health management offers additional reason to hope the country can keep HIV in check. China has a unique history of bringing about swift social changes to improve health. As part of the "barefoot doctors" program in the 1970s, village representatives throughout the country were given basic public health training. Their efforts to provide basic health care and convey preventive health messages to their communities brought significant declines in infectious disease and child mortality in China. Today, China's health indicators are much closer to those of industrialized nations than to those of the developing world.

An emphasis on preventing HIV infection is essential to stemming this global health catastrophe. But even though behavior changes can dramatically reduce the spread of infection, they will never eradicate HIV. And while scientific advances have greatly improved treatment, no drug therapy has yet been able to fully rid the human body of the virus. Furthermore, anti-viral treatment is out of reach for all but a small fraction of the more than 33 million infected.

Ultimately, successful containment and eventual eradication of HIV will require a safe, effective, and affordable vaccine. Many scientists think we can eventually develop such a vaccine—despite some significant hurdles. HIV is very efficient at making copies of itself, a replication that leads to disease despite a vigorous immune response. It also mutates rapidly, and has produced many different strains of itself, which means an effective vaccine would have to be able to recognize and fight off each nuance of the virus. Nevertheless, candidate vaccines have already been able to stimulate some immune response in human volunteers, and seem to be safe.

Even under the most optimistic scenarios, however, the development of an effective vaccine will take years—and the risks to humanity will continue to escalate if not powerfully addressed. In the last decade or so, 25 experimental vaccines have been tested in studies involving small numbers of volunteers but only one has advanced to larger scale "efficacy trials." "Unless there's a major breakthrough," says Dr. Seth Berkley of the International AIDS Vaccine Initiative, "it's unlikely we'll have a vaccine within the next decade."

Meanwhile, even the testing poses formidable challenges. For example, some standard preparation strategies used for other vaccines cannot be used for fear that a weakened form of the live virus or a whole killed virus might cause HIV infection in the person vaccinated. Even after the basic research on safety and effectiveness has been conducted, private pharmaceutical companies will still need to develop a commercial product—a process that takes an average of 10 years and costs at least $150–250 million. Because HIV has hit developing countries hardest, a vaccine that offers any real hope of eradication will need to be inexpensive, easy to transport and administer, require few if any follow-up inoculations, and protect against any strain or route of transmission of the virus.

Getting adequate funding, too, has been an uphill battle. Five years ago, the National Institutes of Health (NIH) decided not to fund large-scale efficacy trials of any leading AIDS vaccines—at the time causing a serious setback to the research. This year, NIH increased vaccine research funding by 79 percent. But if it takes over 10 years to develop a vaccine as Dr. Berkley expects, it could be several decades before the vaccine allows us to put HIV on the road to eradication.

Sitting on the edge of a big wooden chair in my living room, Amélia—the young niece of Victor, next door—used to sing hymns, waiting for lunchtime, while I did Saturday cleaning. She was a skinny kid so I liked giving her a good meal. But she got thinner and thinner. And then, her grandmother had to tie an old towel around her head to hide the open sores. But she couldn't cover up Amé's nose, which had somehow melted away. My neighbor came one day to tell me that Amé was no longer allowed to eat at my house, for fear she would infect me.

Amé died about a year later. I could tell, by the look in her eyes, that she knew what was happening to her. I didn't do anything to take her pain away. I didn't go over to my neighbors and insist that they let her come back. I didn't explain to them that Amé couldn't possibly infect me by eating from my dishes. I was immobilized—perhaps because I didn't know how to answer the unspoken questions of a little girl who didn't understand why this was happening to her.

This article has been based on numbers, but for every number that adds up incrementally to give us the startling statistics we now have, there is a human tragedy. Facing up to the scale of the suffering that lies behind those rising numbers can become an overwhelming challenge—one that we'd rather not think about. Yet, if we don't think about it clearly enough to find a way of stopping HIV in its tracks, a great many more communities, on every continent, will lose those men and women who are in the prime of life, as Amélia's parents were. And then they will lose their Amélias.

QUESTIONS

1. Summarize the main impact that AIDS has had on adults and children around the world.

2. List and explain several statistics in this article that seem very surprising to you. Explain your surprise.

3. What are two signs that indicate the AIDS epidemic is not hopeless.

4. How did the Thai government deal with AIDS in their country? Provide at least three examples.

5. What are basic policy principles that seem to keep HIV in check?

Physician-Assisted Suicide

What I do is relieve suffering. . . . Unfortunately, death is the result.
DR. JACK KERVORKIAN

Dr. Jack Kervorkian is a delicensed pathologist often referred to as "Dr. Death" because he participates in doctor-assisted suicide for terminally ill patients. Dr. Kervorkian has assisted in at least 50 suicides. He has been involved in three trials and three acquittals and shoulders the responsibility of being the "standard-bearer" for ethical doctor-assisted suicide. (Worthington, August 1993)

In Illinois, hospitals have been turning off ventilators or pulling out the feeding tubes of terminally ill patients at relatives' requests since September 1991, when the Health Care Surrogate Act went into effect. In addition, there are 43 other states that allow medical proxy laws that give a person permission to dictate before the onset of a catastrophic illness that in the event of a hopeless medical situation, his or her life would not be prolonged (Hilkevitch, 1993).

"Pulling the tubes," "euthanasia," and "physician-assisted suicide" are terms that are used interchangeably when discussing the ultimate death of an individual who is terminally ill and wants to die. (This wish is either verbally stated by the patient to his or her doctor or family or via a living will.) Historically, euthanasia was defined as a peaceful death with minimal suffering, regardless of whether the death is natural or deliberate. Recently, however, the meaning has been changed to refer to the deliberate killing of people with chronic illnesses or disabilities (Sosby, 1994). Dr. Kervorkian is quick to point out that euthanasia and physician-assisted suicide are the same thing. He states that euthanasia is a Greek word and means "easy death" (Laurence, 1997). Whatever words are used, however, are only symbols of the emotional process of

making the decision to die or the decision to allow a loved one to die. It is an ethical dilemma that individuals and families face when they encounter a terminal illness.

One ethical question often asked is that of what point a terminally ill patient should end his or her life? When does life become hopeless enough to ask for Dr. Kervorkian to bring his "Mercitron" suicide machine so that the patient can kill him or herself? This particular machine lets the patient release a clamp that injects a lethal combination of anesthesia and poison into him or herself. When does life

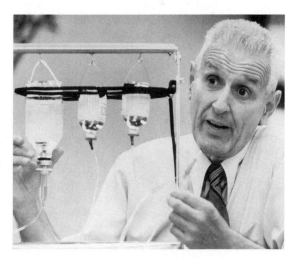

Dr. Jack Kervorkian and His Death Machine.
Photo Courtesy of Gamma-Liaison, Inc.

Scenario I

Gene Theraphy
Consider the following hypothetical news broadcast:

January 20XX New York, NY
Excerpted from the NBC six o'clock news:

> Researchers reported today that they can replace genes that cause disease by using artificial human chromosomes. This treatment will "fix" genes that are not working right and will replace missing genes; the concern is, however, that there could be serious medical risks. There is still hope that artificial chromosomes will be sold soon.

Discussion
How can artificial human chromosomes replace genes? What are the medical risks of such replacement? What ethical issues relate to the sale of artificial human chromosomes? Can artificial chromosomes be used to treat disorders caused by a single gene?

Scenario II

Xenotransplants
Consider the following scene:

October 10, 20XX Chicago, IL

John's kidney has become diseased. He has been on a donor list for months, waiting impatiently for a new kidney. The outlook for the transplant or for returning to a healthy lifestyle seems grim. However, John does not want to give up hope. As he is waiting in the examining room for the doctor, he reviews two detailed reports about xenotransplants and quickly scribbles a list of questions.

John is sitting on the examining table as the doctor walks in. Before the doctor can say a word, John says, "Dr. Michaels, I have been reading reports on xenotransplants, and I want one."

"John, let's discuss this idea of yours a little further," Dr. Michaels replies.

Discussion
Select a year for this scenario to occur (20??). Based on the time period that you choose, answer the following questions: What kind of animal should be used for John's kidney transplant? What complications might occur? How would researchers prevent animal viruses from being transferred to John? Could animals be kept in pathogen-free environments in order to prevent virus transfer? Should there be concern for the possibility that animals carry as-yet-unknown viruses that are fatal to man? If you were John, what five questions would you want Dr. Michaels to answer for you (not including the previously listed questions)?

become hopeless enough that parents make the decision to pull life-support systems from a six-year-old child who has already been comatose for four weeks? When does life become hopeless enough that the children of a 77-year-old, paralyzed stroke patient adhere to her request that a feeding tube not be provided, an act that will ultimately kill her (Worthington, October 1993)?

As if these questions are not tough enough, there are others to consider when contemplating physician-assisted suicide. If a patient is in his or her "right mind," is terminally ill, and wants to die with Dr. Kervorkian's aid, is his or her death to be counted as murder? How would you even know if the patient was in his or her right mind or if he or she was exercising good judgment in the midst of such great physical and emotional suffering? What part should the law play in this decision, or should the decision be a private doctor-patient decision? Should religion be integrated in the final decision of doctor-assisted suicide? Should pain management be used instead of death? Should doctor-assisted suicide happen only when there is intolerable, unrelenting pain and suffering with no chance for recovery? If so, how would you know definitively that there *really* is no chance for recovery? How would courts of law measure "suffering" or "unrelenting pain" or the determination of "no chance for recovery" if the cases went to court (Laurence, 1997)?

Dr. Kervorkian has brought to light a dilemma that many families, medical professionals, and counselors have struggled with in private. It is a difficult issue that should be contemplated, struggled with, and understood by each human prior to finding him or herself in that type of situation.

REFERENCES

Hilkevitch, J. (1993, February 15). Euthanasia Debate is Heating Up. *Chicago Tribune.*

Laurence, C. (1997, April 27). Some Difficult Questions for "Dr. Death." *Chicago Sun-Times.*

Sosby, D. (1994, September 30). *An Issue of Autonomy: Questioning Physician-Assisted Suicide and Euthanasia.* The Special Senate Committee on Euthanasia and Assisted Suicide, Winnipeg, University of Alberta Developmental and Disabilities Center.

Worthington, R. (1993, August 18). Suicide Doctor Finally to Have Day in Court. *Chicago Tribune.*

Worthington, R. (1993, October 23). Kervorkian Fires up Ethics Debate: 19th Suicide Disturbs Even Some of His Sympathizers. *Chicago Tribune.*

44 Future Genetic Testing

DO YOU REALLY WANT THE RESULTS OF YOUR GENETIC TEST?

Genetic testing brings about many ethical, legal, moral, and personal issues that need to be resolved. For example, what would happen if an insurance company knew that you were predisposed to certain diseases because of the results of your genetic tests? Would the company cancel your insurance or not offer you insurance in the first place? Would companies refuse to hire you on the basis of your genetic tests for fear that you would cost them money in sick days, insurance costs, or lack of productivity? What if you had to have a genetic test prior to being hired by a new company. Would you refuse? Could a company demand that you either take a genetic test or be fired?

Other important issues revolve around the availability versus privacy of genetic test results. Would the results of a person's genetic tests be available to anyone who asked for them? Does genetic testing correlate to an invasion of privacy, and would it stigmatize anyone who is found to have cellular deficiencies?

The role of genetic testing in relation to reproduction and children must also be considered. Could prenatal genetic testing lead to an increase in abortions? Should children be tested so that parents know what diseases they will have? Would the genetic testing of children help parents prevent their children from getting diseases, or would it cause psychological trauma to the children instead? What would be the psychological and legal effects of incorrect genetic test results?

Many health professionals believe that counseling provided by health professionals before and after genetic tests are taken will soon be medicine's fastest growing specialty. The prediction is that doctors will be required to take courses in how to explain the effects of genetic predictions to patients. Explaining a genetic prognosis is a complicated process; yet, health professionals must learn how to do it so that the great advances in genetics will not be surpassed by unnecessary anguish, dispute, or lawsuits.

REFERENCE

Jaroff, L. (1996, Fall). Keys to the Kingdom. *Time,* pp. 24–29.

GENETICALLY MODIFIED FOODS EXERCISE

Locate research that compares and contrasts the use of genetically modified foods. Answer the following questions based on your research.

1. Are genetically engineered foods safe or unsafe? Support your answer with three examples from your research

2. What are the benefits of genetically engineered foods? What are the problems?

3. List and explain the positive and negative effects of genetic engineering on humans and the environment.

4. What is the process that is used to make a genetically modified plant?

5. Based on your research, do you believe that labeling genetically modified foods will help consumers make more informed decisions about the foods they eat?

DNA TESTING EXERCISE

Find research articles that discuss the use of DNA testing in criminal cases. Decide if a criminal is innocent or guilty. Then answer the following questions using your personal opinion and your research. Remember to include the way that ethics intertwines in your answers.

1. In Britain, police took a DNA sample from a robbery suspect and matched his DNA on a database with a rape that had occurred nine years earlier. (Possley, Ferkenhoff, 2002) Do you believe that he should be charged with both crimes? Do you think that it is unethical to charge someone with a crime that was done that long ago? Do you feel that there should be a time limit attached to crimes that are identified by DNA testing?

2. Do you believe that DNA samples should be taken at the time of the arrest or should they be taken after conviction? Explain your answer using research and your opinion.

3. Should international, federal, state and local law enforcement officers be able to share DNA information in order to convict criminals?

4. What would be the economic issues related to using more DNA testing with alleged criminals?

Possley M., Ferkenhoff, E. (2002, February 24). "Wide DNA Tests Far Off-If Ever." *Chicago Tribune.*

Conclusion

Between 1980 and 1995, life expectancy has increased globally by 4.6 years: by 4.4 years for males and by 4.9 years for females. Leprosy has been reduced by 82% worldwide in the last 11 years. There are currently 380 million people aged 65 and above (Subramamian 1997). As of December 18, 1998 there were 63,833 people waiting for transplants. The number waiting breaks down as follows: 41,982 kidney transplants, 11,814 liver transplants, 4,161 heart transplants, and 3,142 lung transplants (Gift of Life Foundation, 1999). These facts speak strongly to the positive effects of medical technology. By reading this part of the book and examining the history of technological advancements in medicine, you now have an enhanced awareness of where medical technology has been (e.g., the invention of antibiotics, the irradication of smallpox, etc.) and where it is headed (e.g., tissue and genetic engineering, antibiotic resistance, and organ transplants).

What about the future? You have been asked in this part to access Web sites to find out additional information about medical technology in the future. What did you find? Did you find that xenotransplants may be a part of medical technologies in the future? There have been approximately 30 experimental xenotransplants since the turn of the century. According to Rebecca Williams, in an article entitled "Organ Transplants from Animals: Examining the Possibilities,"

> [I]n April 1995, with FDA permission, doctors at Lahey Hitchcock Medical Center in Burlington, Mass., injected fetal pig brain cells into the brains of patients with advanced Parkinson's disease. The hope was that the fetal tissue would produce dopamine, which the patient's brains lack. The experiments were done primarily to test the safety of the procedures. Other xenotransplant experiments have involved implanting animal hearts, livers and kidneys into humans.

The major concern in conducting xenotransplants is that the body may reject the transplants from other animals. One way to counteract this problem is to prescribe very strong drugs that suppress the immune system—and thus, its negative response to the transplanted organ. Even if the body did not reject the new organ, the drugs may cause the organ recipient to be vulnerable to infections. The hope is that genetic engineering will pave the way for the human body to accept xenotransplants more readily in the future.

Throughout this part of the book, many ethical and social issues were outlined that should cause you to stop, ponder, wonder, and reevaluate the social implications of medical technologies. It is impossible to read this part and not wonder about the amazingly positive effects of medical technology. Yet, this part may also force you to ponder quality-of-life issues, ethical considerations in medical treatment, the choices you have as a patient, and what you will do if you ever have to decide about euthanasia.

The next part of the book will discuss the effects of technology on Third-World countries. It will provide a clearer understanding of the part that technology plays in the development of these countries.

REFERENCES

Subramamian, Director of the Office of the World Health Reporting. "Facts from the World Health Report 1997." Available: http://www.who.ch/whr/1997/facts.htm. Access Date: 2/18/99.

Gift of Life Foundation. (1999) "Gift of Life-Statistics." Available: http://www.giftoflife-sc.org/statistics.asp. Last updated: 1/26/99.

Williams, R. (1996) "Organ Transplants from Animals: Examining the Possibilities." Available: http://www.fda.gov/fdac/features/596_xeno.html.

INTERNET EXERCISE

Using any of the Internet search engines (e.g., Alta Vista, Yahoo, Infoseek, etc.) answer the following questions:

1. Internet questions
    a. What constraints and rules must researchers follow today when working with animals and humans? What is "informed consent"?
    b. What are the legal ramifications of doctors using clients for research without obtaining informed consent?
    c. Find information on current research on the misuse of antibiotics. What are the current medical concerns about patients who fail to finish antibiotic prescriptions or who use others' antibiotics without a doctor's prescription?
    d. What is the current status of the Human Genome Project?
    e. What is the current trend in the field of organ transplants? In what geographical locations are organ transplants done the most? The least? What age groups usually receive the most organ transplants?
    f. Define the following terms:
        colonization
        tissue engineering
        bioengineered tissue
        cell transplant
        euthanasia
        biotechnology

2. Xenotransplants questions
    a. What is a xenotransplant?
    b. Describe the first animal-to-human transplant. When was it? Who did the surgery? What kind of animals were used? Did the recipient live? Did the animals live?
    c. Why is it difficult to keep patients who have had xenotransplants alive? Does the body usually reject the transplanted organ? What effect do immunosuppressive drugs have on the human body?

d. Why do doctors not want to use baboons for xenotransplants? Why, medically, might baboons be the best animal to use for xenotransplants?

e. Ethically, why might pigs be better to use for xenotransplants than baboons?

3. Tissue engineering questions

a. What parts of the body can be re-created through tissue engineering? Can a heart valve, liver, uterus, bone, skin, eye, or brain be recreated? It is doubtful that eyes could be grown in cultures, but some researchers hope to create artificial vision systems instead. What research can you find on each topic?

b. Some biotech companies are waiting to gain approval from the Food and Drug Administration for skin substitutes that would help patients suffering from burns and serious wounds. Until recently, doctors used cadaver tissue to replace dead tissue, but the cadaver tissue is sometimes rejected by the patients' immune system. By growing skin in the lab, the need for cadaver skin is eliminated. What research can you find on this technological phenomenon?

c. Tissue engineering today may start with neonatal foreskin; cow tendons may be used as replacement connective tissues. Try to find more details on this type of tissue engineering.

USEFUL WEB SITES

URL	Site Description
http://www.ornl.gov/TechREsources/Human_Genome/home.html	Human Genome Project
http://www.en./com/users/ddavisfrmain55.htm	Transplant Survival Rates UP
http://www.vparker.home.texas.net/Thinkquest/Manipulating/ Ethics/geneng.htm	Ethics of Genetic Engineering
http://www.Euthansia.com	Euthansia
http://www.ri.bbsrc.ac.uk/library/research/cloning/ readinglist.html#wilmut	Cloning Bibliography
http://www.acmi.canoe.ca/Health0012/07_suicide-ap.html	Dr. Kervorkian Statistics
http://www.safe-food.org	Genetically Engineered Crops
http://www3.utsouthwestern.edu/library/consume/sickbldg.html	Is Your Building Sick?
http://www.who.ch/whr/1997/factse.htm	Medical facts
http://www.who.ch/whr/1997/fig2e.gif	Global causes of death
http://www.giftoflife-sc.org/statistics.asp	Gift of Life statistics

Technology and the Third World

The first world...the third world
...the rich....the poor
...where are the
boundaries...?

One Planet...One People

Photo Courtesy of NASA Headquarters

OBJECTIVES

This chapter will help you to:

- Recognize the historical background and major issues of the developing world
- Distinguish between the developed and the developing world's problems and viewpoints
- Recognize the problems, urgencies and implications of the lack of technology in the developing world
- Analyze and use tables and graphs to promote a better understanding of the techno-economic gap between the first and the third world
- Realize the magnitude of the intrinsic and extrinsic problems of the developing world
- Determine appropriate pathways and policies for the sustainable economic growth in the third world
- Appreciate the importance of the use of technology in solving the economic and environmental problems of the developing world

INTRODUCTION

> *One billion people in the developing world still lack access to clean water . . . nearly 2 billion lack adequate sanitation . . . electric power has yet to reach 2 billion people.*
> WORLD DEVELOPMENT REPORT, WORLD BANK, 1994

As we approach the end of the 20th century, the advances in information technologies have transformed the world into a global village. People of this global village find themselves so close as they witness world events unfolding on the television screens and computer monitors, and yet they still remain oceans apart in terms of their economic conditions and living standards. The gap between the world's poor and its rich is wide and has been getting even wider. The following facts illustrate the dilemma:

- According to the United Nations Development Program (UNDP), in 1991 the richest 20% of the world's people earned 61 times more income than the poorest 20%.

- Thirty years ago, the richest 20% of the world's people received 30 times more than the poorest 20% did—half as wide a gap as today. Since then, the poorest fifth saw its share of global income drop from 2.3 to 1.4%, while that of the richest group rose from 70 to 85%.

- Of the $23 trillion gross world product (GWP) in 1993, $18 trillion was generated in the industrial world, and only $5 trillion circulated in the developing world.

- The gap in per-capita income between the industrial and developing worlds tripled over the past three decades—from $5,700 in 1960 to $15,400 in 1993.

- Economic conditions and growth have been failing during the last 15 years in about 100 countries, home to a quarter of the world's population. The income of 1.5 billion people living in these countries was lower in the 1990s than in earlier decades.

- One fifth of the world's population—those people living in developed nations—now accounts for 70% of the world's energy use.

- More than 80% of the world's population resides in developing countries.

- According to the president of the World Bank, James Wolfensohm, the number of people living in acute poverty could grow from 3 billion to 5 billion in the next 30 years if world leaders do not act now to prevent the trend.

Many developing nations gained independence from the colonial powers after the second World War. But their dreams and aspirations of economic growth and prosperity have yet to materialize. People in developing countries wonder if the colonial era has really ended or if it has merely been replaced by an economic system in which international financial institutions and multinational corporations rule over developing nations.

This part of the book presents an array of issues and challenges that the Third World faces today. The chapters in this part will enable readers to gain insights to the problems of the developing world and their potential solutions. As the developed world prepares itself to enter the 21st century and a new millennium, people in many developing countries, due to their lagging technological level, are still struggling to enter the 18th, 19th, and 20th centuries.

One Planet: Many Worlds

AHMED S. KHAN

The Earth is one but the world is not. We all depend on one biosphere for sustaining our lives.
Yet each community, each country, strives for survival and prosperity with little regard for its
impact on others.
   WORLD COMMISSION ON ENVIRONMENT AND DEVELOPMENT (WCED), OUR COMMON FUTURE

As we approach the new millennium, the great technological accomplishments of the 20th century in the areas of telecommunications, computers, energy, agriculture, materials, medicine, genetic engineering, and defense have transformed the world and brought people closer, yet millions of people worldwide still go to bed hungry at night. These technological advances have enabled us to design advanced early warning systems to warn against missile attacks, but we have failed to develop an advanced early warning system to warn against and prevent global famine or the spread of disease. Thanks to state-of-the-art technologies, we are able to design spaceships to explore life on other planets, yet we have failed to preserve life on planet Earth. Millions of children worldwide continue to die due to malnutrition, disease, and poverty.

In this information age, thanks to the Internet, the world has been transformed into a global village. But in this global village, the economic gulf between the developed and the developing countries, commonly known as the Third World, is wide and growing. This economic disparity is one of the most challenging issues in world affairs.

In 1952, when the French demographer Alfred Sauvey coined the phrase "Third World," he observed that the aspirations of the nations gaining independence from the colonial powers were similar to those of the Third Estate in France at the time of the French Revolution. Like the French commoners, who in the later part of the 18th century struggled against the clergy and the elite, these nations wanted to become the masters of their own destinies (Grimm, 1990). The usage of the phrase "Third World" led to the coinage of the phrases "First World" and "Second World." The "First World" represents mainly economically developed western countries. The "Second World" was meant to represent the communist countries of eastern Europe during the Cold War era, but the term never really caught on. Today, in the post-Cold War era, these countries are referred to as former Soviet-bloc countries. Some of these countries are economically worse off than Third-World countries, but they are never called Third-World countries. Today, the phrase "Third World" is exclusively used to describe a group of nations that are struggling to feed, house, clothe, and educate their people while restrained by poverty, climate, oppression, war, the aftereffects of colonialism, and the economic bondage of the IMF and the World Bank. The phrase "Third World" is also used in a discriminatory manner. According to William J. Grimm, associate editor of Maryknoll Magazine, the Third World is "any nonwhite country whose economy is as bad as or worse than Poland's. No matter how desperate the economic situation may be in eastern Europe or on parts of the Iberian Peninsula, those places are never spoken of as being the Third World, which is a world of brown-, black-, or yellow-skinned people. No white nation is described with the term, though the country may be economically worse off than, for example, Thailand. When whites ruled Rhodesia, it was not the Third World. Now that it is black-ruled Zimbabwe, it is" (Grimm, 1990).

Many African and Asian nations gained independence from the colonial powers after the second World War. But their dreams and aspirations of economic growth and

355

prosperity have yet to materialize. People in developing countries wonder if the colonial era has really ended or if it has merely been replaced by an economic system in which international financial institutions and multinational corporations rule over developing nations. As the developed world prepares itself to enter the 21st century and a new millennium, people in many developing countries, due to their lagging technological level, are still struggling to enter the 18th, 19th, and 20th centuries.

Development in the Third World is inhibited due to a number of extrinsic and intrinsic factors. The extrinsic factors include political interference by the superpowers during the Cold War era, political and economic instability caused by international financial institutions, and economic exploitation of resources by multinational companies. The intrinsic factors are illiteracy; poverty; social injustice; corruption; lack of resources; population pressures; and power-hungry political, feudal, and military elite.

The developed nations have portrayed themselves to the Third World as role models of development and modernization while simultaneously producing a disproportionate share of pollutants emitted into the atmosphere through the burning of fossil fuels relative to the size of their population as compare to the population of the Third World. The Third-World nations confront numerous environmental problems of their own, with water, soil, and air pollution being among the most common. Other problems include desertification, deforestation, the extinction of animal and plant species, international trade in wildlife, the export of toxic waste by developed countries to the Third World, and the disposal of garbage and human waste (Myers, 1990; Mennerick and Mehrangiz, 1991).

Efforts to industrialize by Third-World countries are often hindered by the developed countries in the name of reducing pollution. How would the developed countries respond if the power relationship were reversed? How would they respond if Third-World nations had the economic and technological power to exert pressure on these countries to reduce, for example, the use of fossil fuels and curb industrial processes in order to reduce pollution?

Despite the fact that the average incomes for both the developed and the developing countries have increased, the gap between them has widened during the most recent years. Since 1960, the developed countries have become richer by an average of $5,400 in gross national product (GNP) per capita, while the developing countries have advanced by only $800. Latin American citizens are, on average, poorer than they were in 1970. Countries like

Table 45.1

Development Indicators for the Five Most Populous Low-Income Countries

	China	India	Pakistan	Bangladesh	Nigeria
Population (1995) in millions	1,200	929	130	120	111
Percent of population with access to safe water	83	63	60	83	43
Infant mortality rate per 1,000 births	34	68	90	79	80
Fertility rate (1995)	1.9	3.2	5.2	3.5	5.5
Energy use in (oil eq.) kg	664	248	254	64	162
Carbon dioxide emission per capita	2.3	0.9	0.6	0.2	0.9
Adult illiteracy					
% Female	27	62	76	74	53
% Male	10	35	50	51	33
1990–1995 annual growth rate (GDP) %	12.8	4.6	4.6	4.1	1.6
Exports (million $)	148,797	30,674	7,992	3,173	11,670
Imports (million $)	129,113	34,522	11,461	6,496	7,900
External debt (million $) (1993)	118,090	93,766	30,152	16,370	35,005
External debt as % of GNP (1995)	17.2	28.2	49.5	56.3	140.5
Debt service as % of exports of goods and services (1995)	77.3	201.2	257.9	298.2	274.5

Source: Worldbank, World Development Indicators (http://www.worldbank.org/html/iecdd/products.htm)

Burundi, Somalia, Burkina Faso, Chad, Bangladesh, and Sri Lanka have the lowest per-capita GNP (less than $610), whereas in most developed countries it is more than $14,000 (Mansbach, 1994; Gardner, 1995).

The debt of developing countries is estimated to have grown to over $1.8 trillion in 1994, up from $1.77 trillion in 1993. The worst debt is that of sub-Saharan Africa, excluding South Africa. Collectively, the region's debt amounts to $180 billion, three times the 1980 total, and is 10% higher than the region's entire output of goods and services. Many developing nations have committed a third to a half of their foreign currency earnings to pay the interest and principal, and it is not enough. In Latin America, all of the major debtors, especially Brazil, Argentina, Mexico, Venezuela, Peru, and Chile, are in default and have had to reschedule their repayments (Gardener, 1995).

The developing countries also lag behind in areas like health and education. In the developing countries, the infant mortality rate declined from 109 per 1,000 live births in 1970 to 71 per 1,000 in 1991. Yet, these rates are still much higher than in the developed countries. In 1970, the infant mortality rate in developed countries was 20 per 1,000 live births and declined to about 8 per 1,000 by 1991.

Life expectancy rose from an average of about 50 years in 1965 to slightly more than 62 in 1991 in developing countries, but remained far behind the figure for most developed countries (Mansbach, 1994). The story is the same for adult literacy rates, which improved in poor countries but remained far behind those of developed countries. (See Tables 45.1 through 45.4 and Figures 45.1 through 45.3.)

Today the developed countries have set the membership standards for the First and the Third Worlds. Basically, these standards are economic (Grimm, 1990). Many people in the Third World believe that they are not allowed to attain these standards because of "economic and technological imperialism" imposed by western nations. According to this belief, the First World does not allow economic growth and transfer of technology to reach the developing countries because it may lead to the elimination of jobs and affect the standard of living in the developed countries.

The Third-World countries have found the global economic order imposed on them by the First World intolerable and have called for a new economic order. Policy makers in the developed countries fear that unrest over unmet economic problems may lead the poor to "desperate politics" versus the rich. The people in the Third World counter

Table 45.2

Development Indicators for Low-, Middle- and High-Income Countries

	Low-Income Countries	Middle-Income Countries	High-Income Countries	World
Population in millions (1995)	3,180	1,591	902	5,673
Energy use (oil eq.) kg (1994)	369	1,475	5,066	1,433
Carbon dioxide emission per capita (metric tons) (1992)	1.3	4.8	11.9	4.0
Infant mortality rate per 1,000 live births (1995)	69	39	0.7	55
Fertility rate (1995)	3.2	3.0	1.7	2.9
Adult illiteracy				
% Female	24	19	—	—
% Male	45	23		
1990–1995 Annual growth rate (GDP)	6.8	0.1	—	2.0
Exports (million $) (1995)	245,456	893,331	3,997,288	5,144,770
Imports (million $) (1995)	251,806	987,309	4,037,671	5,246,326
External debt (1995)	534,794	1,530,883	—	—
External debt as % of GNP	38.7	39.9	—	—

Source: Worldbank, World Development Indicators (http://www.worldbank.org/html/iecdd/products.htm)

Table 45.3

Literacy, Per-Capita GNP, and Electricity Use for Selected Developed and Developing Countries

Country	Literacy % of the Population	GNP Per Capita $ US	Electricity Use Per Capita (kWh)
Afghanistan	29	NA	39
Algeria	57	3,480	587
Australia	100	20,720	8,021
Austria	99	17,000	6,300
Bangladesh	35	1,000	75
Belgium	99	17,700	6,790
Bolivia	78	2,100	250
Brazil	81	5,000	1,531
Burma	81	950	65
Canada	97	22,200	17,900
Chad	30	500	15
Chile	93	7,000	1,630
China	78	2,200	630
Comoros	48	700	50
Colombia	87	5,500	1,050
Denmark	99	18,500	6,610
Djibouti	48	1,200	580
Egypt	48	2,400	830
El Salvador	73	2,500	390
Finland	100	16,100	11,050
France	99	18,200	7,430
Gabon	61	4,800	920
Germany	99	16,500	7,160
Ghana	60	1,500	290
Haiti	53	800	75
India	52	1,300	340
Indonesia	77	2,900	200
Israel	92	13,350	4,600
Japan	99	20,400	6,700
Kenya	69	1,200	100
Laos	64	900	220
Malaysia	78	7,500	1,610
Mexico	87	8,200	1,300
Nepal	26	1,000	50
Netherlands	99	17,200	4,200
New Zealand	99	15,700	9,250
Niger	28	650	30
Norway	99	20,800	25,850
Oman	NA	10,000	3,200
Pakistan	35	1,900	350
Peru	85	3,000	760
Saudi Arabia	62	11,000	3,690
Singapore	88	15,000	6,420

Table 45.3

Literacy, Per-Capita GNP, and Electricity Use for Selected Developed and Developing Countries, *continued*

Country	Literacy % of the Population	GNP Per Capita $ US	Electricity Use Per Capita (kWh)
Spain	95	12,700	4,000
Sudan	27	750	40
Turkey	81	5,100	750
Uganda	48	1,200	30
United Kingdom	99	16,900	5,480
United States	97	24,700	12,690
Vietnam	88	1,000	130
Yemen	38	800	120
Zimbabwe	67	1,400	740

Source: CIA World Factbook 1995–96.

that the developed countries are still practicing invisible colonialism via international economic institutions.

In 1961, in order to create a more equitable international economic system, the developing countries formed the "nonaligned movement" (NAM). The efforts by NAM led to the passage of resolutions 3201 and 3202 on May 1, 1974, at the special session of the United Nations General Assembly. These resolutions defined principles and a program to improve international economic relations between the developing and the developed countries. These resolutions were also outlined in the "Charter of Economic Rights and Duties of States," passed by the General Assembly in December 1974. These documents specified six major areas of international economic reform that were needed to resolve conflict between the First World and the Third World. The demands were as follows (Mansbach, 1994):

1. Transnational corporations are to be regulated.

2. Technology should be transferred from rich to poor.

3. The trading order should be reformed to assist the development of poor countries.

4. Poor countries' debt should be canceled or renegotiated.

5. Economic aid from rich to poor countries should be increased.

6. Voting procedures in international economic institutions should be revised to provide poor countries with greater influence.

The First World's response to these demands was mostly negative. The Third-World countries seek economic and technological help from the First-World countries in order to become economically self-sufficient. On the other hand, First-World countries can learn a lot from the Third-World countries. The people in developing countries take pride in family, history, cultural, and religious values, and they consider these values to be more important than economic fulfillment and greed for materialism. For the near future, the developing world will continue to struggle for economic assistance. With the end of the Cold War, and with the former Soviet-bloc countries competing for economic assistance from the West, the future for the developing world remains bleak.

Positive change in the Third World will occur only when both the intrinsic and extrinsic factors inhibiting the development of Third-World nations are addressed. The effects of extrinsic factors (i.e., political and economic instability caused by international financial institutions and the economic exploitation of resources by multinational companies) can be rectified only when a great change takes place in the modes of thought of the policy makers of the developed world; those leaders must begin to believe that all people are created equal; no one is superior to another; and treating people with equality and justice is the key to solving man-made economic, political, and environmental dilemmas.

To address the intrinsic factors (i.e., illiteracy, poverty, social injustice, corruption, lack of resources, population

Table 45.4
State of the Environment and Technology in Selected Third-World Countries

Country	Population/ Education (Literacy, Total % of Population) July 1994 Estimate	National Product (GDP– Purchasing Power Equivalent)/ External Debt ($ Billion)	State of Technology, Electricity Production, Telecommunications (Number of Telephones)	Environment (Current Issues)
World	5,643,289,771	29 trillion, 1 trillion for less developed countries	11.45 trillion kWh	
Afghanistan	16,903,400 29%	NA	1 billion kWh, limited number of telephones	Soil degradation, overgrazing, deforestation, and desertification.
Algeria	27,895,068 57%	89, 26	16.384 billion kWh, 820,000 telephones	Soil erosion from overgrazing; desertification; dumping of untreated sewage, petroleum refining wastes, and other industrial effluents is leading to pollution of rivers and coastal waters; the Mediterranean Sea, in particular, is becoming polluted from oil wastes, soil erosion, and fertilizer runoff; limited supply of potable water.
Angola	9,803,576 42%	5.7, 8	800 million kWh, 40,300 telephones (4.1 telephones per 1,000 persons)	Population pressures contributing to overuse of pastures and subsequent soil erosion; desertification; deforestation of tropical rain forest attributable to the international demand for tropical timber and domestic use as a fuel; deforestation, contributing to loss of biodiversity; soil erosion, contributing to water pollution and siltation of rivers and dams; scarcity of potable water.
Argentina	33,912,994 95%	185, 73	51.305 billion kWh, 2,650,000 telephones (78 per 1000 persons)	Erosion from inadequate flood control and improper land use practices; irrigated soil degradation; desertification; air pollution in Buenos Aires and other major cities; water pollution in urban areas; rivers becoming polluted due to increased pesticide and fertilizer use.

Table 45.4
State of the Environment and Technology in Selected Third-World Countries, *continued*

Country	Population/ Education (Literacy, Total % of Population) July 1994 Estimate	National Product (GDP– Purchasing Power Equivalent)/ External Debt ($ Billion)	State of Technology, Electricity Production, Telecommu- nications (Number of Telephones)	Environment (Current Issues)
Bangladesh	125,149,469 35%	122, 13.5	9 billion kWh, 241,000 telephones (1 per 522 persons)	Many people are landless and forced to live on and cultivate flood-prone land; limited access to potable water; waterborne diseases prevalent; water pollution, especially of fishing areas, results from the use of commercial pesticides; intermittent water shortages because of failing water tables in the northern and central parts of the country; soil degradation; deforestation; severe overpopulation.
Bolivia	7,719,445 78%	15.8, 3.8	1.834 billion kWh, 144,300 telephones (18.7 telephones per 1,000 persons)	Deforestation, contributing to loss of biodiversity; overgrazing; soil erosion; desertification; industrial pollution of water supplies used for drinking and irrigation.
Brazil	158,739,257 81%	785, 119	242.184 billion kWh, 9.86 million telephones	Deforestation in Amazon Basin; air and water pollution in Rio de Janeiro, Sao Paulo, and several other large cities; land degradation and water pollution caused by improper mining activities.
Burkina	10,134,661 18%	7, 0.865	320,000 million kWh, NA	Recent droughts and desertification, severely affecting agricultural activities, population distribution, and the economy; overgrazing; soil degradation.
Burma	44,277,014 81%	41, 4	2.8 billion kWh, 53,000 telephones	Deforestation.
Cambodia	10,264,628 35%	6, 1.08	70 million kWh, NA	Deforestation, resulting in habitat loss and declining biodiversity. (In particular, the destruction of mangrove swamps threatens natural fisheries.)

continued

Table 45.4
State of the Environment and Technology in Selected Third-World Countries, *continued*

Country	Population/ Education (Literacy, Total % of Population) July 1994 Estimate	National Product (GDP– Purchasing Power Equivalent)/ External Debt ($ Billion)	State of Technology, Electricity Production, Telecommu- nications (Number of Telephones)	Environment (Current Issues)
Cameroon	13,132,191 55%	19.1, 6	2.19 billion kWh, NA	Waterborne diseases are prevalent; deforestation; overgrazing; desertification; poaching.
Chad	5,466,771 30%	2.7, 0.492	40,000 kWh, NA	Desertification.
Chile	13,950,557 93%	96, 19.7	22.01 billion kWh, 768,000 telephones	Air pollution from industrial and vehicle emissions; water pollution from untreated sewage; deforestation, contributing to loss of biodiversity; soil erosion; desertification.
China	1,190,431,106 78%	2.61 trillion, 80	740 billion kWh, 11 million telephones	Air pollution from the overwhelming use of coal as fuel, producing acid rain, which is damaging forests; water pollution from industrial effuents; many people do not have access to safe drinking water; less than 10% of sewage receives treatment; deforestation; estimated loss of one third of agricultural land since 1957 to soil erosion and economic development; desertification.
Colombia	35,577,556 87%	192, 17	36 billion kWh, 1.89 million telephones	Deforestation; soil damage from overuse of pesticides.
Congo	2,446,902 57%	7, 4.1	315 million kWh, 18,100 telephones	Air pollution from vehicle emissions; water pollution from the dumping of raw sewage; deforestation.
Costa Rica	3,342,154 93%	19.3, 3.2	3.612 billion kWh, 292,000 telephones	Deforestation, largely the result of land clearing for cattle ranching; soil erosion.
Cuba	11,064,344 94%	13.7, 6.8	16.248 billion kWh, 229,000 telephones	Overhunting threatens wildlife populations; deforestation.

Table 45.4
State of the Environment and Technology in Selected Third-World Countries, *continued*

Country	Population/ Education (Literacy, Total % of Population) July 1994 Estimate	National Product (GDP– Purchasing Power Equivalent)/ External Debt ($ Billion)	State of Technology, Electricity Production, Telecommu- nications (Number of Telephones)	Environment (Current Issues)
Djibouti	412,599 48%	0.5, 0.355	200 million kWh, NA	Desertification.
Dominican Republic	7,826,075 83%	23, 4.7	5 billion kWh, 190,000 telephones	Water shortages; soil eroding into the sea damages coral reefs; deforestation.
Ecuador	10,677,067 88%	41.8, 12.7	7.676 billion kWh, NA	Deforestation; soil erosion; desertification; water pollution.
Egypt	60,765,028 48%	139, 32	47 billion kWh, 600,000 telephones (11 telephones per 1,000 persons)	Agricultural land being lost to urbanization and windblown sands; increasing soil salinization below Aswan High Dam; desertification; oil pollution threatening coral reefs, beaches, and marine habitats; other water pollution from agricultural pesticides, untreated sewage, and industrial effluents; water scarcity away from Nile, which is the only perennial water source; rapid growth in population, overstraining natural resources.
El Salvador	5,752,511 73%	14.2, 2.6	2.19 billion kWh, 116,000 telephones (21 telephones per 1,000 persons)	Deforestation, soil erosion; water pollution; contamination of soil from disposal of toxic waste.
Eritrea	3,782,543 NA	1.7, NA	NA, NA	Famine; deforestation; soil erosion; overgrazing; loss of infrastructure from civil warfare.
Ethiopia	54,927,108 24%	22.7, 3.48	650 million kWh, NA	Deforestation; overgrazing; soil erosion; desertification; famine.

continued

Table 45.4

State of the Environment and Technology in Selected Third-World Countries, *continued*

Country	Population/ Education (Literacy, Total % of Population) July 1994 Estimate	National Product (GDP– Purchasing Power Equivalent)/ External Debt ($ Billion)	State of Technology, Electricity Production, Telecommu- nications (Number of Telephones)	Environment (Current Issues)
Gabon	1,113,906 61%	5.4, 4.4	995 million kWh, 15,000 telephones	Deforestation; poaching.
Gambia	959,300 27%	0.740, 0.336	65 million kWh, 3,500 telephones	Deforestation; desertification.
Ghana	17,225,185 60%	25, 4.6	4.49 billion kWh, 42,300 telephones	Deforestation; overgrazing; soil erosion; poaching and habitat destruction threatens wildlife populations; water pollution; limited supply of safe drinking water.
Guatemala	10,721,387 55%	31.3, 2.2	2.5 billion kWh, 97,670 telephones	Deforestation; soil erosion; water pollution.
Guinea- Bissau	1,098,231 36%	0.860, 0.462	30 million kWh, 3,000 telephones	Deforestation; soil erosion; overgrazing.
Guinea	6,391,536 24%	3.1, 2.5	300 million kWh, 15,000 telephones	Deforestation; inadequate supplies of safe drinking water; desertification; soil contamination and erosion.
Guyana	729,425 95%	1.4, 1.9	276 million kWh, 27,000 telephones	Water pollution from sewage, agricultural, and industrial chemicals; deforestation.
Haiti	6,491,450 53%	5.2, 0.838	480 million kWh, 36,000 telephones	Deforestation; soil erosion.

Table 45.4
State of the Environment and Technology in Selected Third-World Countries, *continued*

Country	Population/ Education (Literacy, Total % of Population) July 1994 Estimate	National Product (GDP– Purchasing Power Equivalent)/ External Debt ($ Billion)	State of Technology, Electricity Production, Telecommu- nications (Number of Telephones)	Environment (Current Issues)
Honduras	5,314,794 73%	10, 2.8	2 billion kWh, 7 telephones per 1,000 persons	Urban population expanding; deforestation results from logging and the clearing of land for agricultural purposes; further land degradation and soil erosion hastened by uncontrolled development and improper land use practices, such as farming of marginal lands; mining activities polluting Lago de Yojoa (the country's largest source of freshwater) with heavy metals as well as several rivers and steams.
India	919,903,056 52.11%	1017, 90.1	310 billion kWh, about 1 phone per 200 persons	Deforestation; soil erosion; overgrazing; desertification; air pollution from industrial and vehicle emissions; water pollution from raw sewage and runoff of agricultural pesticides; huge and rapidly growing population is overstraining natural resources.
Indonesia	200,409,741 77%	571, 100	38 billion kWh, 763,000 telephones	Deforestation; water pollution from industrial wastes and sewage; air pollution in urban areas.
Iran	65,615,474 54%	303, 30	43.6 billion kWh, 2,143,000 telephones (35 telephones per 1,000 persons)	Air pollution, especially in urban areas, from vehicle emissions, refinery operations, and industry; deforestation; overgrazing; oil pollution in Persian Gulf; shortage of drinking water.

continued

Table 45.4

State of the Environment and Technology in Selected Third-World Countries, *continued*

Country	Population/ Education (Literacy, Total % of Population) July 1994 Estimate	National Product (GDP– Purchasing Power Equivalent)/ External Debt ($ Billion)	State of Technology, Electricity Production, Telecommu- nications (Number of Telephones)	Environment (Current Issues)
Iraq	19,889,666 60%	38, 45	12.9 billion kWh, 632,000 telephones	Government water control projects drain inhabited marsh areas, drying up or diverting the streams and rivers that support a sizable population of Shia Muslims, who have inhabited these areas for thousands of years; the destruction of natural habitat also poses serious threats to wildlife populations; damage to water treatment and sewage facilities during the Gulf War; inadequate supplies of potable water; development of Tigris–Euphrates Rivers system contingent upon agreements with upstream riparians (Syria, Turkey); air and water pollution; soil degradation (salinization) and erosion; desertification.
Jamaica	2,555,064 98%	8, 4.5	2.736 trillion kWh, 127,000 telephones	Deforestation; water pollution.
Jordan	3,961,194 80%	11.5, 6.8	3.814 billion kWh, 81,500 telephones	Lack of adequate natural water resources; deforestation; overgrazing; soil erosion; desertification.
Kenya	28,240,658 69%	33.3, 7	2.54 billion kWh, 260,000 telephones	Water pollution from urban and industrial wastes; degradation of water quality from increased use of pesticides and fertilizers; deforestation; soil erosion; desertification; poaching.
Korea, North	23,066,573 99%	22, 8	26 billion kWh, NA	Localized air pollution attributable to inadequate industrial controls.
Laos	4,701,654 64%	4.1, 1.1	990 million kWh, 7,390 telephones	Deforestation; soil erosion.

Table 45.4
State of the Environment and Technology in Selected Third-World Countries, *continued*

Country	Population/ Education (Literacy, Total % of Population) July 1994 Estimate	National Product (GDP– Purchasing Power Equivalent)/ External Debt ($ Billion)	State of Technology, Electricity Production, Telecommu- nications (Number of Telephones)	Environment (Current Issues)
Lebanon	3,620,395 80%	6.1, 0.7	3.413 billion kWh, 325,000 telephones (95 telephones per 1000 persons)	Deforestation; soil erosion.
Liberia	2,972,766 40%	2.3, 2.1	750 million kWh, NA	West Africa's largest tribal tropical rain forest, subject to deforestation; soil erosion; loss of biodiversity.
Madagascar	13,427,758 80%	10.4, 4.4	450,000 million kWh, NA	Soil erosion results from deforestation and overgrazing; desertification; surface water contaminated with untreated sewage and other organic wastes; several species of flora and fauna unique to the island are endangered.
Malawi	9,732,409 22%	6, 1.8	620 million kWh, 42,250 telephones	Deforestation; land degradation; water pollution from agricultural runoff, sewage, and industrial wastes; siltation of spawning grounds endangers fish population.
Maldives	252,077 92%	0.140, 0.148	11 million kWh, 2,804 telephones	Depletion of freshwater aquifers threatens water supplies.
Mali	9,112,950 17%	5.8, 2.6	750 million kWh, 11,000 telephones	Deforestation; soil erosion; desertification; inadequate supplies of safe drinking water; poaching.
Mauritania	2,192,777 34%	2.2, 1.9	135 million kWh, NA	Overgrazing, deforestation, and soil erosion aggravated by drought are contributing to 34% desertification; water scarcity away from the Senegal River, which is the only perennial river.
Mauritius	1,116,923 80%	8.6, 0.991	630 million kWh, 48,000 telephones	Water pollution.

continued

Table 45.4

State of the Environment and Technology in Selected Third-World Countries, *continued*

Country	Population/ Education (Literacy, Total % of Population) July 1994 Estimate	National Product (GDP– Purchasing Power Equivalent)/ External Debt ($ Billion)	State of Technology, Electricity Production, Telecommu- nications (Number of Telephones)	Environment (Current Issues)
Mexico	92,202,199 87%	740, 125	120.725 kWh, 6,410,000 telephones	Natural water resources scarce and polluted in north, inaccessible and poor quality in center and extreme southeast; untreated sewage and industrial effluents polluting rivers in urban areas; deforestation; widespread erosion; desertification; serious air pollution in the national capital and urban centers along U.S.–Mexico border.
Mongolia	2,429,762 NA	2.8, 16.8	3,740 kWh, 63,000 telephones	Limited water resources; policies of the former communist regime promoting rapid urbanization and industrial growth have raised concerns about their negative impact on the environment; the burning of soft coal and the concentration of factories in Ulaanbaatar have severely polluted the air; deforestation, overgrazing, and the conversion of virgin land to agricultural production have increased soil erosion from wind and rain; desertification.
Morocco	28,558,635 50%	70.3, 21.3	8.864 billion kWh, 280,000 telephones (10.5 telephones per 1,000 persons)	Land degradation/desertification (soil erosion resulting from farming of marginal areas, overgrazing, and destruction of vegetation); water supplies contaminated by untreated sewage; siltation of reservoirs; oil pollution of coastal waters.
Mozambique	17,346,280 33%	9.8, 5	1.745 billion kWh, NA	Civil strife in the hinterlands has resulted in increased migration to urban and coastal areas, with adverse environmental consequences; desertification; pollution of surface and coastal waters.

Table 45.4
State of the Environment and Technology in Selected Third-World Countries, *continued*

Country	Population/ Education (Literacy, Total % of Population) July 1994 Estimate	National Product (GDP– Purchasing Power Equivalent)/ External Debt ($ Billion)	State of Technology, Electricity Production, Telecommu- nications (Number of Telephones)	Environment (Current Issues)
Namibia	1,595,567 38%	3.85, 0.220	1.29 billion kWh, 62,800 telephones	Very limited natural water resources; desertification.
Nepal	21,041,527 26%	20.5, 2	1 billion kWh, 50,000 telephones	The almost total dependence on wood for fuel and cutting down trees to expand agricultural land without replanting has resulted in widespread deforestation; soil erosion; water pollution (the use of contaminated water presents human health risks).
Nicaragua	4,096,689 57%	6.4, 10.5	1.118 billion kWh, 60,000 telephones	Deforestation; soil erosion; water pollution.
Niger	8,971,605 28%	5.4, 1.2	230 million kWh, 14,260 telephones	Overgrazing; soil erosion; deforestation; desertification; wildlife populations (such as elephant, hippopotamus, and lion) threatened because of poaching and habitat destruction.
Nigeria	98,091,097 51%	95.1, 29.5	8.3 billion kWh, NA	Soil degradation; rapid deforestation; desertification; recent droughts in north severely affecting marginal agricultural activities.
Pakistan	128,855,965 35%	239, 24	43 billion kWh, 7 telephones per 1,000 persons	Water pollution from untreated sewage, industrial wastes, and agricultural runoff; water scarcity; a majority of the population does not have access to safe drinking water; deforestation; soil erosion; desertification.
Panama	2,630,000 88%	11.6, 6.1	4.36 trillion kWh, 220,000 telephones	Water pollution from agricultural runoff threatens fishery resources; deforestation of tropical rain forest; land degradation.
Paraguay	5,213,772 90%	15.2, 1.2	16.2 billion kWh, 78,300 telephones	Deforestation; water pollution; inadequate means for waste disposal present health hazards for many urban residents.

continued

Table 45.4

State of the Environment and Technology in Selected Third-World Countries, *continued*

Country	Population/ Education (Literacy, Total % of Population) July 1994 Estimate	National Product (GDP– Purchasing Power Equivalent)/ External Debt ($ Billion)	State of Technology, Electricity Production, Telecommu- nications (Number of Telephones)	Environment (Current Issues)
Peru	23,650,671 85%	70, 22	17.434 billion kWh, 544,000 telephones	Deforestation; overgrazing; soil erosion; desertification; air pollution in Lima.
Philippines	69,808,930 90%	171, 34.1	28 billion kWh, 872,900 telephones	Deforestation; soil erosion; water pollution; air pollution in Manila.
Rwanda	8,373,963 50%	6.8, 0.845	130 million kWh, NA	Deforestation; overgrazing; soil exhaustion; soil erosion.
Senegal	8,730,508 38%	11.8, 2.9	760 million kWh, NA	Wildlife populations threatened by poaching; deforestation; overgrazing; soil erosion; desertification.
Sierra Leone	4,630,037 21%	4.5, 0.633	185 million kWh, 23,650 telephones	Rapid population growth pressuring the environment; overharvesting of timber, expansion of cattle grazing, and slash-and-burn agriculture have resulted in deforestation and soil exhaustion; civil war depleting natural resources.
Somalia	6,666,873 24%	3.4, 1.9	NA, NA	Use of contaminated water contributes to health problems; deforestation; overgrazing; soil erosion; desertification.
Sri Lanka	18,129,850 88%	53.5, 5.2	3.6 billion KWh, 114,000 telephones	Deforestation; soil erosion; wildlife population threatened by poaching; coastal degradation from mining activities and increased pollution; freshwater resources being polluted by industrial wastes and sewage runoff.
Sudan	29,419,798 27%	21.5, 17	905 million kWh, NA	Contaminated water supplies present human health risks; wildlife populations threatened by excessive hunting; soil erosion; desertification.
Suriname	422,840 95%	1.17, 0.18	2.018 billion kWh, 27,500 telephones	NA

Table 45.4

State of the Environment and Technology in Selected Third-World Countries, *continued*

Country	Population/ Education (Literacy, Total % of Population) July 1994 Estimate	National Product (GDP– Purchasing Power Equivalent)/ External Debt ($ Billion)	State of Technology, Electricity Production, Telecommu- nications (Number of Telephones)	Environment (Current Issues)
Syria	14,886,672 64%	81.7, 19.4	11.9 billion kWh, 512,600 telephones (37 telephones per 1,000 persons)	Deforestation; overgrazing; soil erosion; desertification; water pollution from dumping of untreated sewage and wastes from petroleum refining; lack of safe drinking water.
Tanzania	27,985,660 46%	16.7, 6.44	600 million kWh, 103,800 telephones	Soil degradation; deforestation; desertification; destruction of coral reefs threatens marine habitats; recent droughts affected marginal agriculture.
Thailand	59,510,471 93%	323, 33.4	43.75 billion kWh, 739,500 telephones (1987)	Air pollution increasing from vehicle emission; water pollution from organic and factory wastes; desertification; wildlife populations threatened by illegal hunting.
Tunisia	8,726,562 65%	34.4, 7.7	5,096 kWh, 233,000 telephones (28 telephones per 1,000 persons)	Toxic and hazardous waste disposal is ineffective and presents human health risks; water pollution from untreated sewage; water scarcity; deforestation; over-grazing; soil erosion; desertification.
Turkey	62,153,898 81%	312.4, 59.4	44 billion kWh, 3,400,000 telephones	Water pollution from dumping of chemicals and detergents; air pollution; deforestation.
Uganda	19,121,934 48%	24.1, 1.9	610 million kWh, NA	Draining of wetlands for agricultural use; deforestation; overgrazing; soil erosion.
Uruguay	3,198,910 96%	19, 4.2	5.96 billion kWh, 337,000 telephones	NA

continued

Table 45.4

State of the Environment and Technology in Selected Third-World Countries *(Continued)*

Country	Population/ Education (Literacy, Total % of Population) July 1994 Estimate	National Product (GDP– Purchasing Power Equivalent)/ External Debt ($ Billion)	State of Technology, Electricity Production, Telecommu- nications (Number of Telephones)	Environment (Current Issues)
Venezuela	20,562,405 88%	161, 28.5	58.541 billion kWh, 1,440,000 telephones	Sewage pollution of Lao de Valencia; oil and urban pollution of Lago de Maracaibo; deforestation; soil degradation; urban and industrial pollution, especially along the Caribbean coast.
Vietnam	73,103,898 88%	72, 3.4	9 billion kWh, 25 telephones per 10,000 persons	Deforestation; soil degradation; water pollution and overfishing threatening marine life populations.
Yemen	11,105,202 38%	9, 7	1.244 billion kWh, 65,000 telephones (est.)	Scarcity of natural freshwater resources (shortages of potable water); overgrazing; soil erosion; desertification.
Zaire	42,684,091 72%	21, 9.2	6 billion kWh, NA	Poaching threatens wildlife populations; water pollution; deforestation.
Zambia	9,188,190 73%	7.3, 7.6	12 billion kWh, NA	Poaching seriously threatens rhinoceros and elephant populations; deforestation; soil erosion; desertification.
Zimbabwe	10,975,078 67%	15.9, 3.5	8.18 billion kWh, 247,000 telephones	Deforestation; soil erosion; land degradation; air and water pollution.

Source: CIA World Factbook 1995–96.

pressures, and power-hungry political and military elite), the key is education and appropriate use of technology. Education is a great equalizer for changing socioeconomic conditions. It could provide solutions to many of the problems of developing countries and set them free from economic bondage, autocratic rule, poverty, and disease. The developing nations should give a higher priority to educat-

ing their people. The innovative use of telecommunications technologies can provide endless opportunities in education (e.g., distance learning), medicine (e.g., telemedicine), and business (e.g., electronic trade) in developing countries. These applications could enable developing countries to increase teledensity and literacy, improve per-capita income, and narrow the technoeconomic

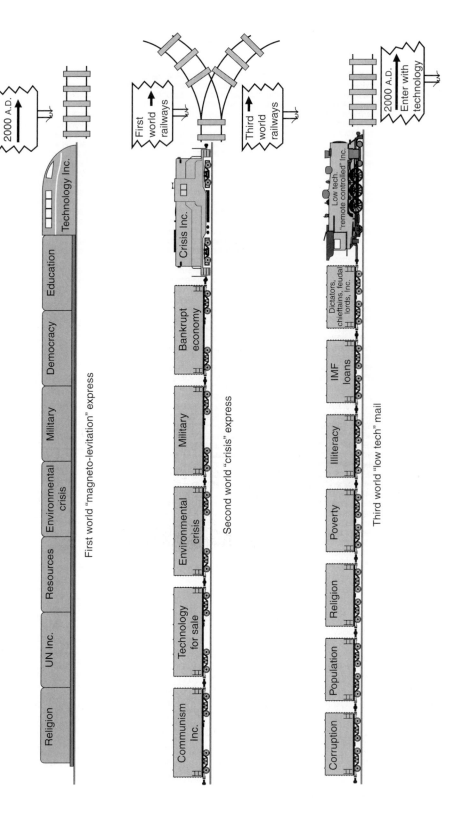

Figure 45.1

Comparison of the First World, the Second World, and the Third World: Worldwide Railways Inc.

Source: Tasneem Khan.

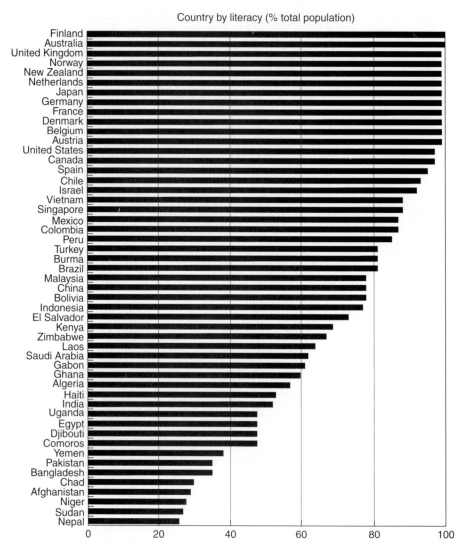

Figure 45.2
Literacy as a Percentage of Population for Selected Developed and Developing Countries
Source: CIA World Factbook 1995–1996.

gap with developed countries. And in the process of developing economic infrastructures, the Third-World countries, through proper planning and appropriate policies, could also minimize the huge cost of industrialization that the developed world has paid in terms of damage to the environment. If the developing countries fail to enter the 21st century with educated populations and technoelites, it

will be only a matter of time before they fall victim to a new era of technoslavery.

The terms "First World," "Third World," "less developed countries" (LDC), and "least developed countries" (LLDC) may have significant meanings in an economic context. However, if one looks at earth from space, it is, at this point, the only living planet in the universe, without

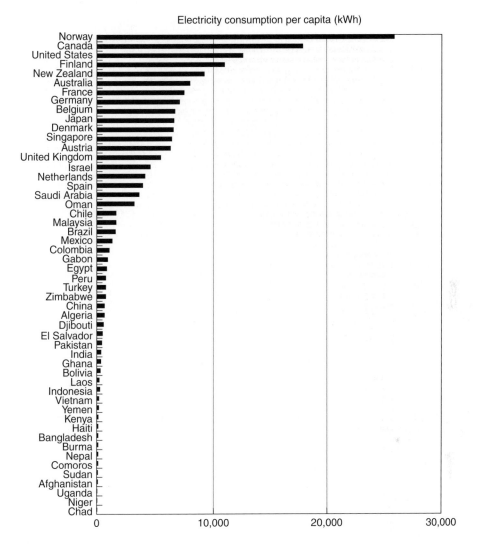

Figure 45.3
Electricity Consumption Per Capita (kWh) for Selected Developed and Developing Countries.
Source: CIA World Factbook 1995–1996.

any boundaries or barriers between the First and the Third Worlds. The people on earth are members of one big family with a common past and a common destiny. Problems like ozone depletion, global warming, soil erosion, deforestation, poverty, homelessness, illiteracy, and obstruction of world peace are universal problems and cannot be solved as long as the artificial distinction between the First World and the Third World exists. All people are created equal. No one is superior to others on the basis of color, race, or wealth. Treating people with equality and justice is key to solving man-made global economic, political, and environmental crises.

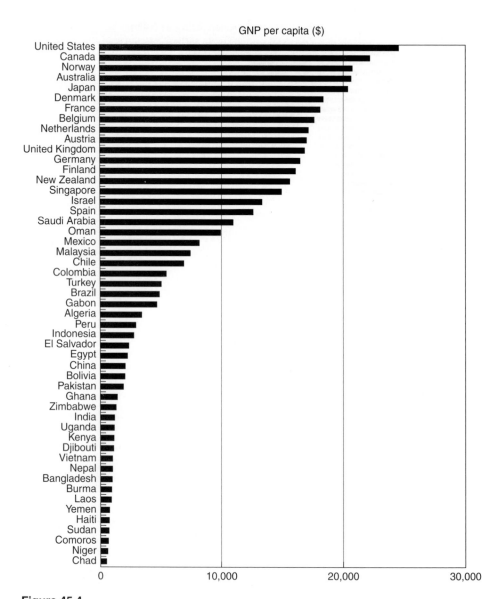

Figure 45.4

Per-Capita GNP ($) for Selected Developed and Developing Countries.

Source: CIA World Factbook 1995–1996.

REFERENCES

CIA World Factbook. (1995). Washington, DC, Brassey's.

Gardner, G. (1995, January/February). Third World Debt is Still Growing. *Worldwatch.* pp. 37–38.

Grimm, W. (1990, May 5). The Third World. *America: National Catholic Weekly.* pp. 449–451.

Mansback, R. (1994). *The Global Puzzle: Issues and Actors in World Politics.* Boston, Houghton Mifflin Company.

Mennerick, L., and Mehrangiz, N. (1991, Spring). Third World Environmental Health, Social, Technological and Economic Policy Issues. *Journal of Environmental Health.* pp. 24–29.

Myers, N. (1990) *The GAIA Atlas of Future Worlds.* New York, Doubleday.

Worldbank IED. (1997, September). *World Development Indicators.* Available: http://www.worldbank.org/html/iecdd/products.htm.

Try being poor for a day or two and find in poverty double riches.

RUMI

"The advanced Western countries completed their escape from poverty to relative wealth during the nineteenth and twentieth centuries. There was no sudden change in their economic output; but only a continuation of year-to-year growth at a rate that somewhat exceeded the rate of population growth."

NATHAN ROSENBERG AND L.E. BIRDZELL, JR.

"The gap in the world between the haves and the have-nots is becoming a gulf. Will the so-called Southerners tolerate the absurd imbalances when, to some extent (through inequitable trade terms, for example), they are poor simply because the Northerners are rich when many impoverished people among the have-not nations receive less protein per week than a domestic cat in the North?"

"INTOLERABLE IMBALANCES," THE GAIA ATLAS OF FUTURE WORLDS, NORMAN MYERS

"Sub-Saharan Africa's debt is now higher than its entire output of goods and services. Collectively, the region's debt amounts to $180 billion, three times the 1980 total, and is 10% higher than its entire output of goods and services. Debt service payment comes to $10 billion annually, about four times what the region spends on health and education combined. The burden is choking off economic development over much of the continent."

WORLD WATCH INSTITUTE'S ANNUAL REPORT, VITAL SIGN, 1995

"The ending of colonialism does not automatically inaugurate an era of peace and prosperity for liberated peoples. It could equally be the prelude to fraternal wars and new inhumanities. . . . It is my considered opinion that the third world war has already begun—in the Third World. The new war is likely to be a cumulation of little wars rather than one big war. Though the form is new, it is basically a war between the great powers and one fought for the realization of their ambitions and the promotion of their national interests. However, this is not obvious because in this new world war, the great powers are invisible."

SINATHAMBY RAJARATNAM, SINGAPORE'S SENIOR MINISTER AND VETERAN DELEGATE
TO THE NONALIGNED MOVEMENT (NAM)

QUESTIONS

1. Define the following terms:

   a. First World
   b. Second World
   c. Third World
   d. LDC
   e. LLDC

2. Many people in the Third World believe that they are the victims of technological imperialism imposed by the First World. The developed countries do not allow economic growth and transfer of technology to developing countries because it may lead to the elimination of jobs and affect the standard of living in the developed countries. Do you agree with this opinion? Explain your answer.

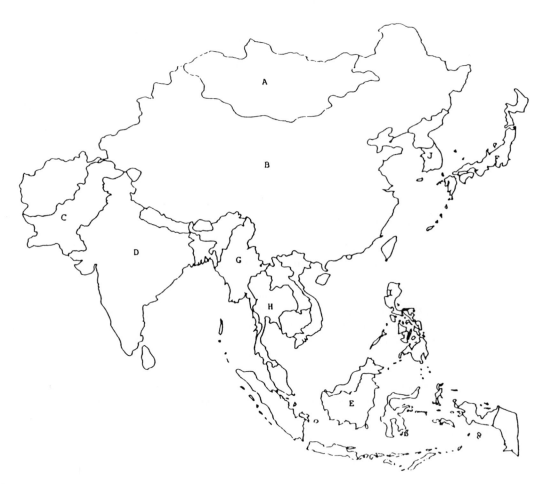

Figure 45.5
Asia

3. Identify the countries shown on the map of Asia on previous page and define the state of technology in each country.

Country	Name	State of Technology (Developed/Developing)
A		
B		
C		
D		
E		
F		
G		
H		
I		
J		

4. Identify the countries shown on the map below and define the state of technology in each country.

Country	Name	State of Technology (Developed/Developing)
A		
B		
C		
D		
E		
F		
G		
H		
I		
J		

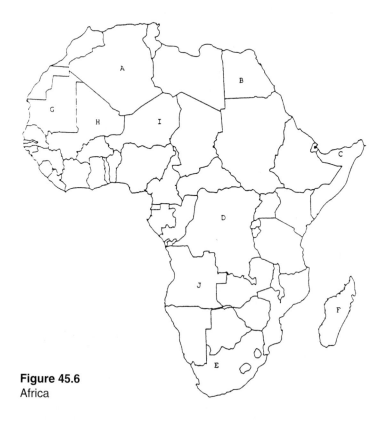

Figure 45.6
Africa

5. Identify the countries shown on the map below and define the state of technology in each country.

Country	Name	State of Technology (Developed/Developing)
A		
B		
C		
D		
E		
F		
G		
H		
I		
J		

Figure 45.7
South America

Richer and Poorer

Number of billionaires (people with the net worth of a thousand millionaires) in the world in 1989	157
Number of billionaires just five years later, in 1994	358
Number of the world's richest people whose collective wealth adds up to $762 billion	358
Number of the world's poorest people whose combined income adds up to $762 billion	2,400,000,000
Portion of global income going to the richest fifth of the population	83 percent
Portion of global income going to the poorest fifth	1 percent
Number of billionaires in Mexico in 1988	2
Number now	24
Combined income of the poorest 17 million Mexicans last year	$6,600,000,000
Wealth of the richest *single* Mexican	$6,600,000,000
Ratio of income of the richest fifth to the poorest fifth of the U.S. population in 1970	4 to 1
The same ratio in 1993	13 to 1
Combined total R&D expenditures of General Motors, Ford, Hitachi, Siemens, Matsushita, IBM, and Daimler-Benz Corporations in 1993	$31.8 billion
Military R&D expenditures of the United States in the same year	$43.5 billion
GDP of Israel in 1992	$69.8 billion
Sales of Exxon in 1992	$103.5 billion
GDP of Egypt in 1992	$33.6 billion
Sales of Philip Morris in 1992	$50.2 billion

Source: United Nations Research Institute for Social Development, *States of Disarray: The Social Effects of Globalization* (London: UNRISD, 1995). Reprinted with Permission from Worldwatch Institute (World Watch July/August 1996)

Who Dominates the World?

Number of people employed by the United Nations	53,589
Number of people employed at Disney World and Disneyland	50,000
Total budget of the United Nations in 1995–96 (two-year budget)	$18.2 billion
Revenue of a single U.S. arms manufacturer (Lockheed Martin) in 1995	$19.4 billion
U.N. peacekeeping expenditures in 1995	$3.6 billion
World military spending in 1995	$767.0 billion
Number of U.N. peacekeepers for every 150,000 people in the world	1
Number of soldiers in national armies for every 150,000 people in the world	650
U.S. contribution to the U.N. budget, per capita	$7
Norwegian contribution to the U.N. budget, per capita	$65
Number of U.S. troops serving in U.N. peacekeeping operations in 1994	965
Number of U.S. troops serving in international missions under U.S. command in 1994	86,451
Cost of the 1992 Earth Summit	$10 million
Cost of the 1994 Paris Air Show and Weapons Exhibition (U.S. portion)	$12 million

Sources: U.S. Arms Control and Disarmament Agency, *World Military Expenditures and Arms Transfers 1995;* Michael Renner, "Peacekeeping Expenditures Level Off," in Lester R. Brown et al., *Vital Signs 1996;* Michael Renner, *Remaking U.N. Peacekeeping: U.S. Policy and Real Reform,* Briefing Paper 17 (Washington, D.C.: National Commission for Economic Conversion and Disarmament, 1995); U.N. Department of Public Information. Compiled by Michael Renner. Reprinted with Permission from the Worldwatch Institute, Washington, D.C. (World Watch November/December 1996)

A Bridge to Your 21st Century Understanding

Complete the following flowchart.

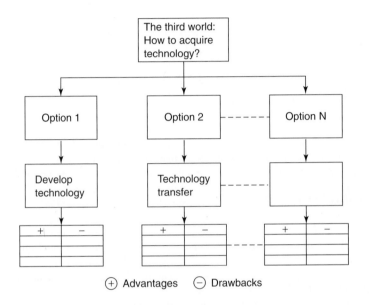

46 Income Gap Widens

Hal Kane

The gap in income among the people of the world has been widening. In 1960, according to United Nations statisticians, the richest 20 percent of the world's people received 30 times more income than the poorest 20 percent. By 1991, they were getting *61* times more. While the poorest one-fifth in 1960 received a meager 2.3 percent of world income, by 1991 that revenue share had fallen to 1.4 percent. The income share of the richest fifth, meanwhile, rose from 70 percent to 85 percent.

These disparities prevail both among countries and within them, and the large gap between individuals worldwide reflects the combination of both of those splits. Almost four-fifths of all people live in the developing world, where incomes are only a fraction of those in industrial countries. In turn, within countries in both categories, gaps in income between citizens can be even wider.

The widest income gap reported within a country is in Botswana, where during the 1980s the richest 20 percent of society received over 47 times more income than the poorest 20 percent (see Table 46.1). Brazil was second, with a ratio of 32 to 1. In Guatemala and Panama, the ratio stood at 30 to 1.

The rapidly growing economies of East Asia have had income patterns similar to those of Western Europe and North America, with the richest one-fifth often earning 5 to 10 times more than the poorest fifth. In South Asia, India, Bangladesh, and Pakistan have had relatively even distributions of income, with the richest 20 percent getting only four to five times more than the poorest quintile. Some countries that have had military conflicts apparently based in part on inequities among citizens, nevertheless have relatively even income distributions.

The split between countries and people can be seen in the marketplace. The value of luxury goods sales worldwide—high-fashion clothing and top-of-the-line autos, for example—exceeds the gross national products of two-thirds of the world's countries. The world's average income, roughly $4,000 a year, is well below the U.S. poverty line.

The poorest fifth of the world accounted for 0.9 percent of world trade, 1.1 percent of global domestic investment, 0.9 percent of global domestic savings, and just 0.2 percent of global commercial credit at the beginning of the 1990s. Each of those shares declined between 1960 and 1990.

These disparities are reflected in the consumption of many resources. At the start of this decade, industrial countries, home to roughly a fifth of the world's population, accounted for about 86 percent of the consumption of aluminum and chemicals, 81 percent of the paper, 80 percent of the iron and steel, and three-quarters of the timber and energy. Since then, economic growth in developing countries has probably reduced these percentages. China's economy, for example, is more than 50 percent larger now than it was in 1990, and developing countries have passed industrial ones in fertilizer consumption.

Uneven income distribution is shaping some of the most important trends in the world today. It raises crime rates, for example. And it drives migration. People have

Reprinted with permission from Worldwatch Institute *Worldwatch* March/April 1996 www.worldwatch.org.

Table 46.1
Income Distribution, Selected Countries, Late 1980s

Country	Ratio of Richest One-Fifth to Poorest
INDUSTRIAL COUNTRIES	
United States	9
Switzerland	9
Israel	7
Italy	6
Germany	6
Sweden	5
Japan	4
Poland	4
Hungary	3
DEVELOPING COUNTRIES	
Botswana	47
Brazil	32
Guatemala	30
Panama	30
Kenya	23
Chile	17
Mexico	14
Malaysia	12
Peru	10
Thailand	8
Philippines	7
China	6
Ghana	6
Ethiopia	5
Indonesia	5
Pakistan	5
India	5
Bangladesh	4
Rwanda	4

Source: UN Development Programs, Human Development Report 1994 (New York: Oxford University Press, 1994).

long responded to economic disparities by following a path from poor regions to richer ones, as tens of millions of workers chase higher wages and better opportunities. Some 1.6 million Asians and Middle Easterners were working in Kuwait and Saudi Arabia before they fled war in 1991, and at least 2.5 million Mexicans live in the United States.

The same is true within countries: rising disparities of income are adding to the growth of cities through rural-to-urban migration. Latin America, with some of the highest disparities of income among its citizens, is also the most urbanized region of the developing world—not entirely by coincidence. Since 1950, city dwellers there have risen from 42 percent of the population to 73 percent.

For many years, China had one of the most equal distributions of income in the world. But now that is changing, as incomes in its southern provinces and special economic zones soar while those in rural areas rise much more slowly. Also not coincidentally, the Chinese National Academy of Social Sciences forecasts that by 2010, half the population will live in cities, compared with 28 percent today and only 10 percent in the early 1980s.

In the early 1990s, developing world economies, especially in East Asia, have grown faster than the economies of the industrial countries. This has the potential to shrink disparities of income, if poorer countries continue to catch up. Yet even if the gaps among countries narrow, the gaps between *people* may not, because economic growth is distributed so unevenly within nations. Despite the recent restoration of economic growth in Latin America, for example, U.N. economists say that no progress is expected in reducing poverty, which is even likely to increase slightly.

Meanwhile, in some regions almost no one has been getting richer. The per capita income of most sub-Saharan African nations actually fell during the 1980s. After the latest negotiations of the General Agreement on Tariffs and Trade (GATT) were completed, *The Wall Street Journal* reported that "even GATT's most energetic backers say that in one part of the world, the trade accord may do more harm than good: sub-Saharan Africa," the poorest geographic region. There, an estimated one-third of all college graduates have left the continent. That loss of talented people, due in large part to poverty and a lack of opportunities in Africa, will make it even more difficult for the continent to advance.

The economic growth that has the potential to close income gaps among peoples in the developing world is instead becoming a splitting off, with some parts of societies joining the industrial world while others remain behind. Singapore, Hong Kong, and Taiwan have begun to look like wealthy industrial countries, for example. Now parts of China are following, as are the wealthier segments of Latin American society and of Southeast Asian countries. This is good news for members of the middle-income countries and for the world. But it may do little to help the poorest fifth of humanity.

QUESTIONS

1. Define the following terms.
    a. GNP
    b. GDP
    c. GATT

2. Describe the factors that are responsible for the wide gap between the per-capita GNP of the developed and of the developing countries. How could this gap be reduced?

AFGHANISTAN: A LEGACY OF THE COLD WAR

During the Cold War, in various civil conflicts around the globe, the refugee population exploded from about 4.6 million people in the early 1970s to nearly 15 million in 1989 and more than 18.9 million in 1993. In Afghanistan, more than 6 million people (one-third of its total population) alone were displaced by the Soviet invasion in 1979.

Afghanistan, a poor nation with a per-capita gross domestic product of $130 per person, life expectancy of 41 years (about half of that in the United States), and literacy rate of 10%, has suffered the following effects of the ravages of war: 50% reduction in agricultural production, a 50% reduction in livestock, damage to 70% of the paved road, 60% of rural health centers destroyed, 35% of villages destroyed, 40% of the population displaced, and 6% of the population (1 million people) killed (MacFarquhar, 1989). Over the past decade, one third of the population of Afghanistan fled the country, with Pakistan sheltering more than 3 million refugees and Iran about 3 million. About 1.4 million Afghan refugees remain in Pakistan and about 2 million in Iran. Another 1 million probably moved into and around urban areas within Afghanistan. (Brassey's, 1996)

In 1979, when the Soviet Union invaded Afghanistan, millions fled by truck, cart, donkey, and foot to Pakistan and Iran. During the early days of the occupation, Soviet helicopters and planes spread the provinces bordering Pakistan with tens of thousands of plastic "butterfly mines," colored green for vegetated areas and beige for deserts. Even more diabolically, they dropped boobytrapped toys, cigarette packs, pens, and other objects. Designed to maim rather than to kill, these antipersonnel explosives were scattered over mountain passes, caravan routes, and fields as a means of terrorizing the local population. The Hind helicopter gunship (MIL Mi24) carried out thousands of attacks on small villages, killing tens of thousands of innocent civilians (Girardet, 1985).

President Richard Nixon expressed his view about the situation in Afghanistan in the following manner: "Those who have called Afghanistan the Soviet Union's Vietnam tragically misstate the case. South Vietnam was doomed when Congress cut our aid by 80% from 1973 to 1975 and revoked the president's powers to enforce the Paris Peace Accords of 1973, while Moscow poured in massive assistance to North Vietnam. One key lesson of Vietnam is that the United States must do as much for its friends as Kremlin leaders do for theirs. We are not doing that in Afghanistan today" (Nixon, 1990).

Brigadier M. Yousuf, who played a prominent role in the war against the Soviet Union, comments on the Afghan tragedy: "I feel the only winners in the war in Afghanistan are the Americans. They have their revenge of Vietnam, they have seen Soviets beaten on the battlefield by a guerilla force that they helped to finance, and they have prevented an Islamic government replacing a Communist one in Kabul. For the Soviet Union, even their military retreat has been turned into a huge political success, with Gorbachev becoming a hero in the West. . . . The losers are most certainly the people of Afghanistan. It is their homes that are heaps of rubble, their land and fields that have been burnt and sown with millions of mines; it is their husbands, fathers, and sons who have died in a war that was almost, and should have been, won" (Yousuf and Adkin, 1992). As the rest of the world prepares to enter the 21st century, the Afghan people—the victims of technology—are struggling to cope with the socio-economic wounds inflicted by the technological might of the superpowers.

REFERENCES

CIA World Factbook (1996). Brassey's. Washington, D.C.

Girardet, E. (1985). *Afghanistan—The Soviet War.* New York, St. Martin's Press.

MacFarquhar, E. (1989, February 13). World Report. *U.S. News & World Report.* pp. 33–36.

Nixon, R. (1990, February 6). Afghanistan: Our Job Isn't Done. *Chicago Tribune.*

Yousuf, M. and Adkin, M. (1992). *The Bear Trap—Afghanistan's Untold Story.* London, Leo Copper.

BANGLADESH: NATURAL AND MAN-MADE DISASTERS

Global Warming: Impact of Sea-Level Rise

A rise in sea level is probably the most widely recognized consequence of global warming, because in a warmer climate, the oceans will expand when heated, and polar ice caps in Greenland and Antarctica may melt. Scientists calculate that expansion effects alone could raise sea levels 20–40 centimeters if the average temperature rises 1.5–4.5 degrees Celsius. A temperature rise in the middle of this range might increase the sea level by 80 centimeters. Such a rise could inundate low-lying areas, destroy marshes and swamps, erode shoreline, worsen coastal flooding, and increase the salinity of rivers, bays, and aquifers. Nearly one third of the human population lives within 60 kilometers of a coast, and thus many reside on land that would be lost. In Bangladesh, the situation would be particularly severe. Bangladesh would lose 12–28% of its area, which currently houses 9–27% of its population. Floods could penetrate further inland, leaving the nation vulnerable to the type of storm that killed 300,000 people in the early 1970s.

(Source: World Resources. (1988–89). *An Assessment of the Resources Base that Supports the Global Economy: A Report By the World Resources Institute for Environment and Development.* New York, Basic Books.)

Farakka Barrage: A Threat to the Economy and the Environment

The Farakka barrage was constructed across the Ganges river at Farakka, 11 miles upstream from the border of Bangladesh. The 70-foot-high and 7365.5-foot-long barrage, with 109 bays, was built to divert 40,000 cubic feet per second (cusec) of water from the Ganges into the Bhagirathi-Hooghly River during the low-flow period. The Indian government's objective for building the Farakka barrage was the preservation and maintenance of the port of Calcutta. The Farakka barrage has triggered a series of adverse impacts in Bangladesh:

Low water flows in the Ganges and its tributaries

Profound effects on hydrology and river morphology

Loss in agricultural production

Increase in surface and groundwater salinity

Adverse impact on navigation and fisheries

Desertification

Cyclones

Cyclones coming from the Bay of Bengal claim thousands of lives and cause major economic damage every year. In November of 1970, cyclones and tidal waves killed 200,000 people, and 100,000 were reported as missing (http://wwwinfoplease.com/ipa/). The cyclone of May 1996, the strongest cyclone in a century, killed 139,000 people and damaged or destroyed more than a million homes (http://193.67.176.1/~climate/database/records/).

China's Challenge to the United States and to the Earth

LESTER R. BROWN AND CHRISTOPHER FLAVIN

During the 1990s, China has emerged as an economic superpower, boasting the world's second largest economy. It is now challenging not only U.S. economic leadership, but the earth's environmental limits.

Using purchasing power parity to measure output, China's 1995 GNP of just over $3 trillion exceeded Japan's $2.6 trillion and trailed only the U.S. output of $6.7 trillion. If the Chinese economy continues to double every eight years, the pace it has maintained since 1980, it will overtake the United States by 2010, becoming the world's largest economy.

Over the last four years the Chinese economy has grown by 10 to 14 percent per year. As its population of 1.2 billion people moves into modern houses, buys cars, refrigerators and televisions, and shifts to a meat-based diet, the entire world will feel the effects. Already, China's rapidly rising CO_2 emissions account for one-tenth of the global total.

In recent decades, many observers noted that the United States, with less than 5 percent of the world's population, was consuming a third or more of its resources. But this is no longer true. In several areas, China has overtaken the United States. For example, China now consumes more grain and red meat, uses more fertilizer, and produces more steel than the United States.

Since China has 4.6 times as many people as the United States, its per capita demands on the earth's resources are still far less. To cite an extreme example, the average American consumes 25 times as much oil as the average Chinese citizen does.

Reprinted with permission from Worldwatch Institute. *World Watch* September/October 1996. www.worldwatch.org

Even with its still modest per capita consumption, China is already paying a high environmental price for its booming economy. Its heavy reliance on coal, for example, has led to air pollution nearly as bad as that once found in eastern Europe. As a result, respiratory disease has become epidemic in China, and crop yields are suffering.

As China, with its much larger population, attempts to replicate the consumer economy pioneered in the United States, it becomes clear that the U.S. model is not environmentally sustainable. Ironically, it may be China that finally forces the United States to come to terms with the environmental unsustainability of its own economic system. If China were to consume as much grain and oil per person as the United States does, prices of both commodities would go off the top of the charts. Carbon dioxide emissions would soar, leading to unprecedented climate instability. Together, these trends would undermine the future of the entire world.

The bottom line is that China, with its vast population, simply will not be able to follow for long any of the development paths blazed to date. It will be forced to chart a new course. The country that invented paper and gunpowder now has the opportunity to leapfrog the West and show how to build an environmentally sustainable economy. If it does, China could become a shining example for the rest of the world to admire and emulate. If it fails, we still all pay the price.

GRAIN HARVEST: CHINA

The United States, long the world's leading grain producer, was overtaken by China in 1983. Over subsequent years,

Figure 47.1
Grain Production, United States and China, 1950–95

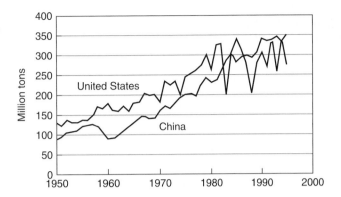

the lead changed hands several times, but since 1986, China's grain harvest has usually exceeded that of the United States. It is also much more stable, because China has 2.5 times as much irrigated land as the United States. U.S. production of corn, which dominates U.S. agriculture, is largely rainfed and, affected by both heat and drought, it fluctuates widely from year to year. In consumption, the gap is far wider. China now consumes 365 million tons per year, and has become the world's second largest grain importer, whereas the United States consumes 200 million tons and remains the world's leading exporter.

FERTILIZER USE: CHINA

U.S. fertilizer use climbed rapidly from mid century onward. By 1980, it exceeded 20 million tons. Then it leveled off, averaging less in the mid-1990s than in the early 1980s. Meanwhile, in China, the 1978 economic reforms in agriculture led to a meteoric climb in fertilizer use. In 1986, China overtook the United States to become the world leader. In 1995, Chinese farmers used 28 million tons of fertilizer, while U.S. farmers used just under 20 million tons. In the United States, and now increasingly in China, fertilizer use is constrained by the physiological capacity of crop varieties to effectively absorb more nutrients.

CONSUMPTION OF RED MEAT: CHINA

The consumption of red meat, particularly beef in the form of steak and hamburgers, has become a defining component of the U.S. lifestyle, but China's total consumption of red meat has eclipsed that of the United States. For all red meat combined—beef, pork, and mutton—China now consumes 42 million tons per year, compared with only 20 million tons in the United States. China's pork consumption of 30 kilograms per person matches the U.S. intake of 31 kilograms, but its beef consumption lags far behind—4 kilograms to 45 kilograms. If China were to close the beef gap, its people would eat an additional 49 million tons each year. Produced in feedlots, this would take some 343 million tons of grain—roughly as much as the entire U.S. grain harvest.

Figure 47.2
Fertilizer use in the United States and China, 1950–95

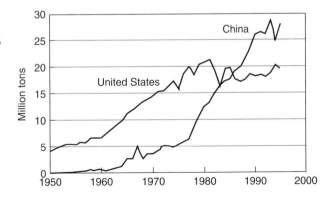

Figure 47.3
Red Meat Consumption, United
States and China, 1960–96

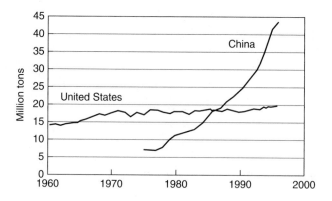

AUTOMOBILE PRODUCTION: UNITED STATES

In the production of automobiles, the United States dwarfed China by some 6.6 million to 239,000 in 1995. U.S. output is not likely to increase much in the future, since most of the automobiles made now are used for replacement rather than for expanding the fleet. China, by contrast, plans to boost production to 3 million per year by the end of the decade, building a fleet of 22 million automobiles by the year 2010. If China's ownership of automobiles were to reach 1 for every 2 people, as in the United States, its fleet of 600 million cars would exceed the 1995 *world* fleet of 480 million cars.

BICYCLE PRODUCTION: CHINA

If a crowded world is compelled to move toward a less polluting, less land consuming mode of personal transport, then the bicycle may well be the transport vehicle of the future. For this shift, China is well positioned. Its bicycle production has averaged over 40 million a year in recent years, compared with less than 8 million a year in the United States. This is perhaps the only major indicator for which the ratio of production between the countries reflects the ratio of population size. In global terms, China accounts for nearly two-fifths of world production of 110 million bicycles annually in recent years.

STEEL OUTPUT: CHINA

In the industrial world of an earlier era, steel production was perhaps the best single indicator of industrial progress. China has recently caught up with the United States in this industry, with both countries turning out 93 million tons in 1995. The big difference is that most of China's steel production is from iron ore. The United States, a more mature industrial society, now gets *roughly half* of its total steel from the reprocessing of scrap metal, a more energy-efficient, less polluting means of produc-

Figure 47.4
Automobile Production, United
States and China, 1950–95

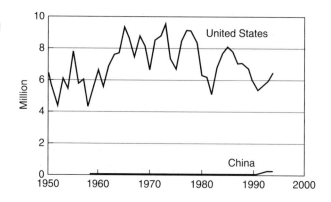

Figure 47.5
Bicycle Production, United
States and China, 1970–94

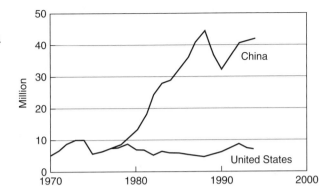

tion. If China proceeds to develop an automobile-centered transportation system as currently planned, its steel needs will soar far beyond those of the United States.

OIL USE: UNITED STATES

Despite its becoming economy, China still consumes only 3.3 million barrels of oil each day, one-fifth the 17 million consumed daily in the United States. U.S. oil use per person is a striking 25 times that in China. With its limited oil reserves, China relies on coal for 75 percent of its energy, whereas the United States relies on coal for just 22 percent. But as China becomes more dependent on automobiles and trucks, its oil use is climbing. Already, it has gone from exporting 500,000 barrels of oil per day in 1990 to *importing* 300,000 barrels per day in 1995. If China were one day to use as much oil per person as the United States does, it would need 80 million barrels daily—more than the whole world now produces or is ever projected to produce.

CARBON EMISSIONS: UNITED STATES

Carbon emissions from fossil-fuel burning totaled 1.394 million tons in the United States in 1995, 73 percent higher than the 807 million tons emitted by China. Fossil fuel burning releases carbon dioxide into the atmosphere—the main gas leading to greenhouse warming. Since 1990, U.S. carbon emissions have grown at roughly 1 percent per year while China's grew at 5 percent annually as use of coal and oil surged. Even so, the United States still emits eight times as much carbon per person as China. If China develops the sort of energy-intensive industries and lifestyles found in the United States, it would further destabilize the world's atmosphere.

COMPUTER POWER: UNITED STATES

If steel production is a key indicator of progress in an industrial society, computer use is a key indicator in the information economy of the late twentieth century. In this

Figure 47.6
Steel Production, United States
and China, 1977–95

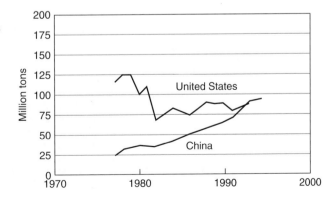

Figure 47.7
Oil consumption, United States
and China, 1950–94

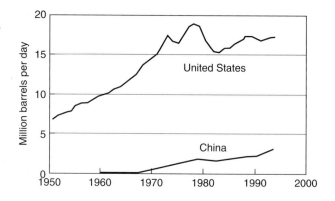

Figure 47.8
Carbon Emission from Fossil
Fuel Burning, United States and
China, 1950–95

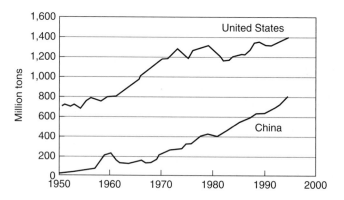

Figure 47.9
Number of Computers, United
States and China, 1993

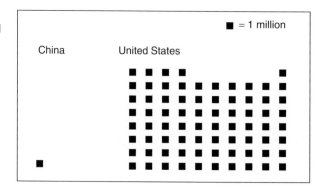

area, there is no contest: the United States has one computer for every three people; China has one for every 1,000 people. With 74 million computers out of a worldwide total of 173 million, the United States leads the world by a wide margin in the computerization of its economy. China, which has just 1.2 million computers, lags far behind not only the United States but much of the rest of the world as well.

QUESTIONS

1. Compare China with the United States in terms of the following aspects: population, amounts of carbon emissions, automobile production, and oil consumption.

2. Do you think that China, despite its large population, will emerge as an industrialized nation? Why or why not? Explain your answer.

INDIA: TAJ MAHAL THREATENED BY AIR POLLUTION

The Taj Mahal, one of the seven wonders of the world, located in Agra, Delhi, was built by the Mughal emperor Shahjahan in memory of his beloved wife Mumtaz Mahal.

Uncontrolled emissions from factories and plants around Taj Mahal are said to be damaging its marble facade. Agra's air has become so badly polluted that the Taj Mahal—a 350-year-old monument—often disappears behind the murk and smog. Environmentalists claim that clouds of pollution contain sulphur dioxide and nitrogen dioxide that are eating away the monument, turning the brilliant white marble into an ugly shade of gray.

The environmentalists are demanding the area around the monument to be designated a total pollution-free zone. This would mean relocating all of the area's 2,000 potentially polluting industries, such as cast iron foundries, glassworks, and a state-owned oil refinery, to areas located farther from Taj Mahal.

The Taj Mahal is the biggest tourist attraction in the country and the source of much-needed foreign exchange. The Indian government has recently announced plans to clean up the environmental pollution that is slowly destroying the monument. Plans call for industries that burn coal or oil to switch to natural gas or liquefied petroleum gas. People living in the city will be required to cut down on their use of kerosene and firewood. Traffic pollution will be curbed with incentives for truck drivers to use diesel with low sulphur content.

Recently, the Indian Supreme Court ordered the closing of 14 industries near the Taj Mahal that had failed to install antipollution devices. It also ordered 190 industries along the Ganges River to take steps to control their effluents.

Source: Impact International, April 1995.

A Bridge to Your 21st Century Understanding

Complete the following flowchart.

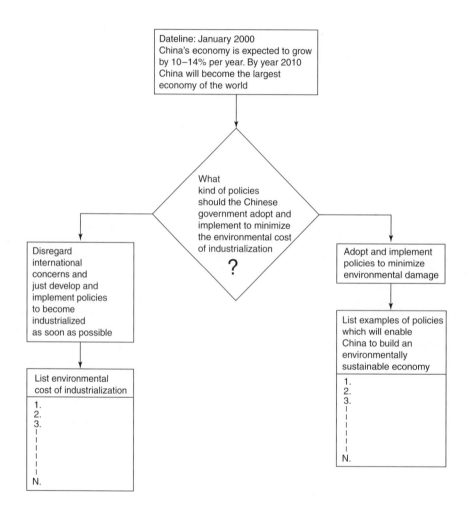

N30 WTO Showdown*

PAUL HAWKEN

When I was able to open my eyes, I saw lying next to me a young man, 19, maybe 20 at the oldest. He was in shock, twitching and shivering uncontrollably from being tear-gassed and pepper-sprayed at close range. His burned eyes were tightly closed, and he was panting irregularly. Then he passed out. He went from excruciating pain to unconsciousness on a sidewalk wet from the water that a medic had poured over him to flush his eyes.

More than 700 organizations and between 40,000 and 60,000 people took part in the protests against the WTO's Third Ministerial on November 30th. These groups and citizens sense a cascading loss of human and labor rights in the world. Seattle was not the beginning but simply the most striking expression of citizens struggling against a worldwide corporate-financed oligarchy—in effect, a plutocracy. Oligarchy and plutocracy often are used to describe "other" countries where a small group of wealthy people rule, but not the "First World"—the United States, Japan, Germany, or Canada.

The World Trade Organization, however, is trying to cement into place that corporate plutocracy. Already, the world's top 200 companies have twice the assets of 80 percent of the world's people. Global corporations represent a new empire whether they admit it or not. With massive amounts of capital at their disposal, any of which can be used to influence politicians and the public as and when deemed necessary, all democratic institutions are diminished and at risk. Corporate free market policies subvert culture, democracy, and community, a true tyranny. The American Revolution occurred because of crown-chartered corporate abuse, a "remote tyranny" in Thomas Jefferson's words. To see Seattle as a singular event, as did most of the media, is to look at the battles of Concord and Lexington as meaningless skirmishes.

But the mainstream media, consistently problematic in their coverage of any type of protest, had an even more difficult time understanding and covering both the issues and activists in Seattle. No charismatic leader led. No religious figure engaged in direct action. No movie stars starred. There was no alpha group. The Ruckus Society, Rainforest Action Network, Global Exchange, and hundreds more were there, coordinated primarily by cell phones, e-mails, and the Direct Action Network. They were up against the Seattle Police Department, the Secret Service, and the FBI—to say nothing of the media coverage and the WTO itself.

Thomas Friedman, *The New York Times* columnist and author of an encomium to globalization entitled *The Lexus and the Olive Tree,* angrily wrote that the demonstrators were "a Noah's ark of flat-earth advocates, protectionist trade unions and yuppies looking for their 1960s fix."

Not so. They were organized, educated, and determined. They were human rights activists, labor activists, indigenous people, people of faith, steel workers, and farmers. They were forest activists, environmentalists, social justice

*Reprinted with permission from Paul Hawken's book titled *Another World is Possible,* (2003, Viking).

Paul Hawken, co-author of *Natural Capitalism* and author of *The Ecology of Commerce,* and *Another World is Possible* (2003, Viking). He can be reached at Natural Capital Institute, 3B Gate Five Road, Sausalito, CA 94965.

workers, students, and teachers. And they wanted the World Trade Organization to listen. They were speaking on behalf of a world that has not been made better by globalization. Income disparity is growing rapidly. The difference between the top and bottom quintiles has doubled in the past 30 years. Eighty-six percent of the world's goods go to the top 20 percent, the bottom fifth get 1 percent. The apologists for globalization cannot support their contention that open borders, reduced tariffs, and forced trade benefit the poorest 3 billion people in the world.

Globalization does, however, create the concentrations of capital seen in northern financial and industrial centers—indeed, the wealth in Seattle itself. Since the people promoting globalized free trade policies live in those cities, it is natural that they should be biased.

Despite Friedman's invective about "the circus in Seattle," the demonstrators and activists who showed up there were not against trade. They do demand proof that shows when and how trade—as the WTO constructs it—benefits workers and the environment in developing nations, as well as workers at home. Since that proof has yet to be offered, the protesters came to Seattle to hold the WTO accountable.

THIS IS WHAT DEMOCRACY LOOKS LIKE

On the morning of November 30th, I walked toward the Convention Center, the site of the planned Ministerial, with Randy Hayes, the founder of Rainforest Action Network. As soon as we turned the corner on First Avenue and Pike Street, we could hear drums, chants, sirens, roars. At Fifth, police stopped us. We could go no farther without credentials. Ahead of us were thousands of protesters. Beyond them was a large cordon of gas-masked and riot-shielded police, an armored personnel carrier, and fire trucks. On one corner was Niketown. On the other, the Sheraton Hotel, through which there was a passage to the Convention Center.

The cordon of police in front of us tried to prevent more protesters from joining those who blocked the entrances to the Convention Center. Randy was a credentialed WTO delegate, which means he could join the proceedings as an observer. He showed his pass to the officer, who thought it looked like me. The officer joked with us, kidded Randy about having my credential, and then winked and let us both through. The police were still relaxed at that point. Ahead of us crowds were milling and

moving. Anarchists were there, maybe 40 in all, dressed in black pants, black bandanas, black balaclavas, and jackboots, one of two groups identifiable by costume. The other was a group of 300 children who had dressed brightly as turtles in the Sierra Club march the day before.

The costumes were part of a serious complaint against the WTO. When the United States attempted to block imports of shrimp caught in the same nets that capture and drown 150,000 sea turtles each year, the WTO called the block "arbitrary and unjustified." Thus far in every environmental dispute that has come before the WTO, its three-judge panels, which deliberate in secret, have ruled for business, against the environment. The panel members are selected from lawyers and officials who are not educated in biology, the environment, social issues, or anthropology.

Opening ceremonies for the World Trade Organization's Third Ministerial were to have been held that Tuesday morning at the Paramount Theater near the Convention Center. Police had ringed the theater with Metro buses touching bumper to bumper. The protesters surrounded the outside of that steel circle. Only a few hundred of the 5,000 delegates made it inside, as police were unable to provide safe corridors for members and ambassadors. The theater was virtually empty when US trade representative and meeting co-chair Charlene Barshevsky was to have delivered the opening keynote. Instead, she was captive in her hotel room a block from the meeting site. WTO executive director Michael Moore was said to have been apoplectic.

Inside the Paramount, Mayor Paul Schell stood despondently near the stage. Since no scheduled speakers were present, Kevin Danaher, Medea Benjamin, and Juliet Beck from Global Exchange went to the lectern and offered to begin a dialogue in the meantime. The WTO had not been able to come to a pre-meeting consensus on the draft agenda. The NGO community, however, had drafted a consensus agreement about globalization—and the three thought this would be a good time to present it, even if the hall had only a desultory number of delegates. Although the three were credentialed WTO delegates, the sound system was quickly turned off and the police arm-locked and handcuffed them. Medea's wrist was sprained. All were dragged off stage and arrested.

The arrests mirrored how the WTO has operated since its birth in 1995. Listening to people is not its strong point. WTO rules run roughshod over local laws and regulations. It relentlessly pursues the elimination of any restriction on the free flow of trade including local, national, or international laws that distinguish between products based on how they are made, by whom, or what happens during production.

The WTO is thus eliminating the ability of countries and regions to set standards, to express values, or to determine what they do or don't support. Child labor, prison labor, forced labor, substandard wages and working conditions cannot be used as a basis to discriminate against goods. Nor can a country's human rights record, environmental destruction, habitat loss, toxic waste production, or the presence of transgenic materials or synthetic hormones be used as the basis to screen or stop goods from entering a country. Under WTO rules, the Sullivan Principles and the boycott of South Africa would not have existed. If the world could vote on the WTO, would it pass? Not one country of the 135 member-states of the WTO has held a plebiscite to see if its people support the WTO mandate. The people trying to meet in the Green Rooms at the Seattle Convention Center were not elected. Even Michael Moore was not elected.

While Global Exchange was temporarily silenced, the main organizer of the downtown protests, the Direct Action Network (DAN), was executing a plan that was working brilliantly outside the Convention Center. The plan was simple: insert groups of trained nonviolent activists into key points downtown, making it impossible for delegates to move. DAN had hoped that 1,500 people would show up. Close to 10,000 did. The 2,000 people who began the march to the Convention Center at 7 a.m. from Victor Steinbrueck Park and Seattle Central Community College were composed of affinity groups and clusters whose responsibility was to block key intersections and entrances. Participants had trained for many weeks in some cases, for many hours in others. Each affinity group had its own mission and was self-organized. The streets around the Convention Center were divided into 13 sections and individual groups and clusters were responsible for holding these sections. There were also "flying groups" that moved at will from section to section, backing up groups under attack as needed. The groups were further divided into those willing to be arrested and those who were not.

All decisions prior to the demonstrations were reached by consensus. Minority views were heeded and included. The one thing all agreed to was that there would be no violence—physical or verbal—no weapons, no drugs or alcohol.

Throughout most of the day, using a variety of techniques, groups held intersections and key areas downtown. As protesters were beaten, gassed, clubbed, and pushed back, a new group would replace them. There were no charismatic leaders barking orders. There was no command chain. There was no one in charge. Police said that they were not prepared for the level of violence, but, as one protester later commented, what they were unprepared for was a network of nonviolent protesters totally committed to one task—shutting down the WTO.

THE VICTORY THAT WASN'T

Meanwhile, Moore and Barshevsky's frustration was growing by the minute. Their anger and disappointment was shared by Madeleine Albright, the Clinton advance team, and, back in Washington, by chief of staff John Podesta. This was to have been a celebration, a victory, one of the crowning achievements to showcase the Clinton administration, the moment when it would consolidate its centrist free trade policies, allowing the Democrats to show multinational corporations that they could deliver the goods.

This was to have been Barshevsky's moment, an event that would give her the inside track to become Secretary of Commerce in the Gore Administration. This was to have been Michael Moore's moment, reviving what had been a mediocre political ascendancy in New Zealand. To say nothing of Monsanto's moment. If the as-yet unapproved draft agenda were ever ratified, the Europeans could no longer block or demand labeling on genetically modified crops without being slapped with punitive lawsuits and tariffs. The draft also contained provisions that would allow all water in the world to be privatized. It would allow corporations patent protection on all forms of life, even genetic material in cultural use for thousands of years. Farmers who have spent thousands of years growing crops in a valley in India could, within a decade, be required to pay for their water. They could also find that they would have to purchase seeds containing genetic traits their ancestors developed, from companies that have engineered the seeds not to reproduce unless the farmer annually buys expensive chemicals to restore seed viability. If this happens, the CEOs of Novartis and Enron, two of the companies creating the seeds and privatizing the water, will have more money. What will Indian farmers have?

But the perfect moment for Barshevsky, Moore and Monsanto didn't arrive. The meeting couldn't start. Demonstrators were everywhere. Private security guards locked down the hotels. The downtown stores were shut. Hundreds of delegates were on the street trying to get into the Convention Center. No one could help them. For WTO delegates accustomed to an ordered corporate or governmental world, it was a calamity.

Up Pike toward Seventh and to Randy's and my right on Sixth, protesters faced armored cars, horses, and police in full riot gear. In between, demonstrators ringed the Sheraton to prevent an alternative entry to the Convention Center. At one point, police guarding the steps to the lobby pummeled and broke through a crowd of protesters to let eight delegates in. On Sixth Street, Sergeant Richard Goldstein asked demonstrators seated on the street in front of the police line "to cooperate" and move back 40 feet. No one understood why, but that hardly mattered. No one was going to move. He announced that "chemical irritants" would be used if they did not leave.

The police were anonymous. No facial expressions, no face. You could not see their eyes. They were masked Hollywood caricatures burdened with 60 to 70 pounds of weaponry. These were not the men and women of the 6th precinct. They were the Gang Squads and the SWAT teams of the Tactical Operations Divisions, closer in training to soldiers from the School of the Americas than local cops on the beat. Behind them and around were special forces from the FBI, the Secret Service, even the CIA.

The police were almost motionless. They were equipped with US military standard M40A1 double-canister gas masks, uncalibrated, semi-automatic, high velocity Auto-cockers loaded with solid plastic shot, Monadnock disposable plastic cuffs, Nomex slash-resistant gloves, Commando boots, Centurion tactical leg guards, combat harnesses, DK5-H pivot-and-lock riot face shields, black Monadnock P24 polycarbonate riot batons with Trum Bull stop side handles, No. 2 continuous discharge CS (orchochlorobenzylidenemalononitrile) chemical grenades, M651 CN (chloroacetophenone) pyrotechnic grenades, T16 Flameless OC Expulsion Grenades, DTCA rubber bullet grenades (Stingers), M-203 (40 mm) grenade launchers, First Defense MK-46 Oleoresin Capsicum (OC) aerosol tanks with hose and wands, .60 caliber rubber ball impact munitions, lightweight tactical Kevlar composite ballistic helmets, combat butt packs, .30 cal. 30-round magazine pouches, and Kevlar body armor. None of the police had visible badges or forms of identification.

The demonstrators seated in front of the black-clad ranks were equipped with hooded jackets for protection against rain and chemicals. They carried toothpaste and baking powder for protection of their skin, and wet cotton cloths impregnated with vinegar to cover their mouths and noses after a tear gas release. In their backpacks were bottled water and food for the day ahead.

Ten Koreans came around the corner carrying a 10-foot banner protesting genetically modified foods. They were impeccable in white robes, sashes, and headbands. One was a priest. They played flutes and drums and marched straight toward the police and behind the seated demonstrators. Everyone cheered at the sight and chanted, "The whole world is watching." The sun broke through the gauzy clouds. It was a beautiful day. Over cell phones, we could hear the cheers coming from the labor rally at the football stadium. The air was still and quiet.

At 10 a.m. the police fired the first seven canisters of tear gas into the crowd. The whitish clouds wafted slowly down the street. The seated protesters were overwhelmed, yet most did not budge. Police poured over them. Then came the truncheons, and the rubber bullets.

I was with a couple of hundred people who had ringed the hotel, arms locked. We watched as long as we could until the tear gas slowly enveloped us. We were several hundred feet from Sgt. Goldstein's 40-foot "cooperation" zone. Police pushed and truncheoned their way through and behind us. We covered our faces with rags and cloth, snatching glimpses of the people being clubbed in the street before shutting our eyes.

The gas was a fog through which people moved in slow, strange dances of shock and pain and resistance. Tear gas is a misnomer. Think about feeling asphyxiated and blinded. Breathing becomes labored. Vision is blurred. The mind is disoriented. The nose and throat burn. It's not a gas, it's a drug. Gas-masked police hit, pushed, and speared us with the butt ends of their batons. We all sat down, hunched over, and locked arms more tightly. By then, the tear gas was so strong our eyes couldn't open. One by one, our heads were jerked back from the rear, and pepper was sprayed directly into each eye. It was very professional. Like hair spray from a stylist. Sssst. Sssst.

Pepper spray is derived from food-grade cayenne peppers. The spray used in Seattle is the strongest available, with a 1.5 to 2.0 million Scoville heat unit rating. One to three Scoville units are when your tongue can first detect hotness. (The habanero, usually considered the hottest pepper in the world, is rated around 300,000 Scoville units.) This description was written by a police officer who sells pepper spray on his website. It is about his first experience being sprayed during a training exercise:

"It felt as if two red-hot pieces of steel were grinding into my eyes, as if someone was blowing a red-hot cutting torch into my face. I fell to the ground just like all the others and started to rub my eyes even though I knew better not too. The heat from the pepper spray was overwhelming. I could not resist trying to rub it off of my face. The pepper spray caused my eyes to shut very quickly. The only way I could open them was by prying them open with my fingers. Everything that we had been taught about pepper spray had turned out to be true. And everything that our instructor had told us that we would do, even though we knew not to do it, we still did. Pepper spray turned out to be more than I had bargained for."

As I tried to find my way down Sixth Avenue after the tear gas and pepper spray, I couldn't see. The person who found and guided me was Anita Roddick, the founder of the Body Shop, and probably the only CEO in the world who wanted to be on the streets of Seattle helping people that day.

When your eyes fail, your ears take over. I could hear acutely. What I heard was anger, dismay, shock. For many people, including the police, this was their first direct action. Demonstrators who had taken nonviolent training were astonished at the police brutality. The demonstrators were students, their professors, clergy, lawyers, and medical personnel. They held signs against Burma and violence. They dressed as butterflies.

The Seattle Police had made a decision not to arrest people on the first day of the protests (a decision that was reversed for the rest of the week). Throughout the day, the affinity groups created through Direct Action stayed together. Tear gas, rubber bullets, and pepper spray were used so frequently that by late afternoon, supplies ran low. What seemed like an afternoon lull or standoff was because police had used up all their stores. Officers combed surrounding counties for tear gas, sprays, concussion grenades, and munitions. As police restocked, the word came down from the White House to secure downtown Seattle or the WTO meeting would be called off. By late afternoon, the mayor and police chief announced a 7 p.m. curfew and "no protest" zones, and declared the city under civil emergency. The police were fatigued and frustrated. Over the next seven hours and into the night, the police turned downtown Seattle into Beirut.

That morning, it was the police commanders who were out of control, ordering the gassing and pepper spraying and shooting of people protesting nonviolently. By evening, it was the individual police who were out of control. Anger erupted, protesters were kneed and kicked in the groin, and police used their thumbs to grind the eyes of pepper-spray victims. A few demonstrators danced on burning dumpsters that were ignited by pyrotechnic tear-gas grenades (the same ones used in Waco).

Protesters were defiant. Tear gas canisters were thrown back as fast as they were launched. Drum corps marched using empty 5-gallon water bottles for instruments. Despite their steadily dwindling number, maybe 1,500 by evening, a hardy number of protesters held their ground,

seated in front of heavily armed police, hands raised in peace signs, submitting to tear gas, pepper spray, and riot batons. As they retreated to the medics, new groups replaced them.

Every channel covered the police riots live. On TV, the police looked absurd, frantic, and mean. Passing Metro buses filled with passengers were gassed. Police were pepper spraying residents and bystanders. The mayor went on TV that night to say, that as a protester from the '60s, he never could have imagined what he was going to do next: call in the National Guard.

LAWLESSNESS

This is what I remember about the violence. There was almost none until police attacked demonstrators that Tuesday in Seattle. Michael Meacher, environment minister of the United Kingdom, said afterward, "What we hadn't reckoned with was the Seattle Police Department, who single-handedly managed to turn a peaceful protest into a riot." There was no police restraint, despite what Mayor Paul Schell kept proudly assuring television viewers all day. Instead, there were rubber bullets, which Schell kept denying all day. In the end, more copy and video was given to broken windows than broken teeth.

During that day, the anarchist black blocs were in full view. Numbering about one hundred, they could have been arrested at any time but the police were so weighed down by their own equipment, they literally couldn't run. Both the police and the Direct Action Network had mutually apprised each other for months prior to the WTO about the anarchists' intentions. The Eugene Police had volunteered information and specific techniques to handle the black blocs but had been rebuffed by the Seattle Police. It was widely known they would be there and that they had property damage in mind. To the credit of the mayor, the police chief, and the Seattle press, distinctions were consistently made between the protesters and the anarchists (later joined by local vandals as the night wore on). But the anarchists were not primitivists, nor were they all from Eugene. They were well organized, and they had a plan.

The black blocs came with tools (crow-bars, hammers, acid-filled eggs) and hit lists. They knew they were going after Fidelity Investments but not Charles Schwab. Starbucks but not Tully's. The GAP but not REI. Fidelity Investments because they are large investors in Occidental Petroleum, the oil company most responsible for the violence against the U'wa tribe in Columbia. Starbuck's because of their non-support of fair-traded coffee. The

GAP because of the Fisher family's purchase of Northern California forests. They targeted multinational corporations that they see as benefiting from repression, exploitation of workers, and low wages. According to one anarchist group, the ACME collective: "Most of us have been studying the effects of the global economy, genetic engineering, resource extraction, transportation, labor practices, elimination of indigenous autonomy, animal rights, and human rights, and we've been doing activism on these issues for many years. We are neither ill-informed nor inexperienced." They don't believe we live in a democracy, do believe that property damage (windows and tagging primarily) is a legitimate form of protest, and that it is not violent unless it harms or causes pain to a person. For the black blocs, breaking windows is intended to break the spells cast by corporate hegemony, an attempt to shatter the smooth exterior facade that covers corporate crime and violence. That's what they did. And what the media did is what I just did in the last two paragraphs: focus inordinately on the tiniest sliver of the 40–60,000 marchers and demonstrators.

It's not inapt to compare the pointed lawlessness of the anarchists with the carefully considered ability of the WTO to flout laws of sovereign nations. When "The Final Act Embodying the Results of the Uruguay Round of Multilateral Trade Negotiations" was enacted April 15th, 1994, in Marrakech, it was recorded as a 550-page agreement that was then sent to Congress for passage. Ralph Nader offered to donate $10,000 to any charity of a congressman's choice if any of them signed an affidavit saying they had read it and could answer several questions about it. Only one congressman—Senator Hank Brown, a Colorado Republican—took him up on it. After reading the document, Brown changed his opinion and voted against the Agreement.

There were no public hearings, dialogues, or education. What passed is an Agreement that gives the WTO the ability to overrule or undermine international conventions, acts, treaties, and agreements. The WTO directly violates "The Universal Declaration of Human Rights" adopted by member nations of the United Nations, not to mention Agenda 21. (The proposed draft agenda presented in Seattle went further in that it would require Multilateral Agreements on the Environment such as the Montreal Protocol, the Convention on Biological Diversity, and the Kyoto Protocol to be in alignment and subordinate to WTO trade polices.) The final Marrakech Agreement contained provisions that most of the delegates, even the heads-of-country delegations, were not aware of, statutes that were drafted by sub-groups of bureaucrats and lawyers, some of whom represented transnational corporations.

The police mandate to clear downtown was achieved by 9 p.m. Tuesday night. But police, some of whom were fresh recruits from outlying towns, didn't want to stop there. They chased demonstrators into neighborhoods where the distinctions between protesters and citizens vanished. The police began attacking bystanders, residents, and commuters. They had lost control. When Président Clinton sped from Boeing airfield to the Westin Hotel at 1:30 bananas at below cost to muscle its way into the European market. Its stock was at a 13-year low, the shareholders were angry, the company was up for sale, but the prices of bananas in Europe are really cheap. Who lost? Caribbean farmers who could formerly make a living and send their kids to school can no longer do so because of low prices and demand.

Globalization leads to the concentration of wealth inside such large multinational corporations as Time-Warner, Microsoft, GE, Exxon, and Wal-Mart. These giants can obliterate social capital and local equity, and create cultural homogeneity in their wake. Countries as different as Mongolia, Bhutan, and Uganda will have no choice but to allow Blockbuster, Burger King, and Pizza Hut to operate within their borders. Under WTO, even decisions made by local communities to refuse McDonald's entry (as did Martha's Vineyard) could be overruled. The as-yet unapproved draft agenda calls for WTO member governments to open up their procurement process to multinational corporations. No longer could local governments buy preferentially from local vendors. The WTO could force governments to privatize healthcare and allow foreign companies to bid on delivering national health programs. The draft agenda could privatize and commodify education, and could ban cultural restrictions on entertainment, advertising, or commercialism as trade barriers. Globalization kills self-reliance, since smaller local businesses can rarely compete with highly capitalized firms who seek market share instead of profits. Thus, developing regions may become more subservient to distant companies, with more of their income exported rather than re-spent locally.

On the weekend prior to the WTO meeting, the International Forum on Globalization (IFG) held a two-day teach-in at Benaroya Hall in downtown Seattle on just such questions of how countries can maintain autonomy in the face of globalization. Chaired by IFG president Jerry Mander, more than 2,500 people from around the world attended. A similar number were turned away. It was the hottest ticket in town (but somehow that ticket did not get into the hands of pundits and columnists). It was an extravagant display of research, intelligence, and concern, expressed by scholars, diplomats, writers, academics, fishermen, scientists, farmers, geneticists, businesspeople, and lawyers. Prior to the teach-in, non-governmental organizations, institutes, public interest law firms, farmers' organizations, unions, and councils had been issuing papers, communiqués, press releases, books, and pamphlets for years. They were almost entirely ignored by the WTO.

A CLASH OF CHRONOLOGIES

But something else was happening in Seattle underneath the debates and protests. In Stewart Brand's new book, *The Clock of the Long Now—Time and Responsibility,* he discusses what makes a civilization resilient and adaptive. Scientists have studied the same question about ecosystems. How does a system, be it cultural or natural, manage change, absorb shocks, and survive, especially when change is rapid and accelerating? The answer has much to do with time, both our use of it and our respect for it. Biological diversity in ecosystems buffers against sudden shifts because different organisms and elements fluctuate at different time scales. Flowers, fungi, spiders, trees, laterite, and foxes all have different rates of change and response. Some respond quickly, others slowly, so that the system, when subjected to stress, can move, sway, and give, and then return and restore.

The WTO was a clash of chronologies or time frames, at least three, probably more. The dominant time frame was commercial. Businesses are quick, welcome innovation in general, and have a bias for change. They need to grow more quickly than ever before. They are punished, pummeled and bankrupted if they do not. With worldwide capital mobility, companies and investments are rewarded or penalized instantly by a network of technocrats and money managers who move $2 trillion a day seeking the highest return on capital. The Internet, greed, global communications, and high-speed transportation are all making businesses move faster than before.

The second time frame is culture. It moves more slowly. Cultural revolutions are resisted by deeper, historical beliefs. The first institution to blossom under *perestroika* was the Russian Orthodox Church. I walked into a church near Boris Pasternak's dacha in 1989 and heard priests and *babushkas* reciting the litany with perfect recall as if 72 years of repression had never happened. Culture provides the slow template of change within which family, community, and religion prosper. Culture provides

identity and in a fast-changing world of displacement and rootlessness, becomes ever more important. In between culture and business is governance, faster than culture, slower than commerce.

At the heart, the third and slowest chronology is Earth, nature, the web of life. As ephemeral as it may seem, it is the slowest clock ticking, always there, responding to long, ancient evolutionary cycles that are beyond civilization.

These three chronologies often conflict. As Stewart Brand points out, business unchecked becomes crime. Look at Russia. Look at Microsoft. Look at history. What makes life worthy and allows civilizations to endure are all the things that have "bad" payback under commercial rules: infrastructure, universities, temples, poetry, choirs, literature, language, museums, terraced fields, long marriages, line dancing, and art. Most everything we hold valuable is slow to develop, slow to learn, and slow to change. Commerce requires the governance of politics, art, culture, and nature, to slow it down, to make it heedful, to make it pay attention to people and place. It has never done this on its own. The extirpation of languages, cultures, forests, and fisheries is occurring worldwide in the name of speeding up business. Business itself is stressed out of its mind by rapid change. The rate of change is unnerving to all, even to those who are supposedly benefiting. To those who are not, it is devastating.

What marched in the streets of Seattle? Slower time strode into the WTO. Ancient identity emerged. The cloaks of the forgotten paraded on the backs of our children.

What appeared in Seattle were the details, dramas, stories, peoples, and puppet creatures that had been ignored by the bankers, diplomats, and the rich. Corporate leaders believe they have discovered a treasure of immeasurable value, a trove so great that surely we will all benefit. It is the treasure of unimpeded commerce flowing everywhere as fast as is possible. But in Seattle, quick time met slow time. The turtles, farmers, workers, and priests weren't invited and don't need to be because they are the shadow world that cannot be overlooked, that will tail and haunt the WTO, and all its successors, for as long as it exists. They will be there even if they meet in totalitarian countries where free speech is criminalized. They will be there in dreams of delegates high in the Four Seasons Hotel. They will haunt the public relations flacks who solemnly insist that putting the genes of scorpions into our food is a good thing. What gathered around the Convention Center and hotels was everything the WTO left behind.

In the Inuit tradition, there is a story of a fisherman who trolls an inlet. When a heavy pull on the fisherman's line drags his kayak to sea, he thinks he has caught the "big one," a fish so large he can eat for weeks, a fish so fat that he will prosper ever after, a fish so amazing that the whole village will wonder at his prowess. As he imagines his fame and coming ease, what he reels up is Skeleton Woman, a woman flung from a cliff and buried long ago, a fish-eaten carcass resting at the bottom of the sea that is now entangled in his line. Skeleton Woman is so snarled in his fishing line that she is dragged behind the fisherman wherever he goes. She is pulled across the water, over the beach, and into his house where he collapses in terror. In the retelling of this story by Clarissa Pinkola Estes, the fisherman has brought up a woman who represents life and death, a specter who reminds us that with every beginning there is an ending, for all that is taken, something must be given in return, that the earth is cyclical and requires respect. The fisherman, feeling pity for her, slowly disentangles her, straightens her bony carcass, and finally falls asleep. During the night, Skeleton Woman scratches and crawls her way across the floor, drinks the tears of the dreaming fisherman, and grows anew her flesh and heart and body. This myth applies to business as much as it does to a fisherman. The apologists for the WTO want more-engineered food, sleeker planes, computers everywhere, golf courses that are preternaturally green. They see no limits; they know of no downside. But Life always comes with Death, with a tab, a reckoning. They are each other's consorts, inseparable and fast. These expansive dreams of the world's future wealth were met with perfect symmetry by Bill Gates III, the co-chair of the Seattle Host Committee, the world's richest man. But Skeleton Woman also showed up in Seattle, the uninvited

The Storm Over Globalization*

Dispatches from the global trade wars, in the days before and after the WTO, World Bank, and IMF protests

More than 700 organizations and between 40,000 and 60,000 people took part in the protests against the Third Ministerial of the World Trade Organization on November 30. These groups and citizens sense a cascading loss of human, labor, and environmental rights in the world. Seattle was not the beginning but simply the most striking expression of citizens struggling against a worldwide corporate oligarchy—in effect, a plutocracy. Oligarchy and plutocracy are not polite terms. They often are used to describe "other" countries where a small group of wealthy people rule, but not the "first world"—the United States, Japan, Germany, or Canada. But already, the world's top 200 companies have twice the assets of 80 percent of the world's people. Global corporations represent a new empire whether they admit it or not. With massive amounts of capital at their disposal, any of which can be used to influence politicians and the public as and when deemed necessary, they threaten and diminish all democratic institutions.

Corporations are using the World Trade Organization, however, to cement into place their plutocracy. When the "Final Act Embodying the Results of the Uruguay Round of Multilateral Trade Negotiations" was enacted on April 15, 1994, in Marrakech, it was recorded as a 550-page document that was then sent to Congress for approval. Ralph Nader offered to donate $10,000 to any charity of a congressman's choice if any of them signed an affidavit saying they had read it and could answer several questions about it. Only one—Senator Hank Brown, a Colorado Republican—took him up on it. After reading the document, Brown changed his opinion and voted against the agreement. There were no public hearings, dialogue, or education. What was approved was an agreement that gives the WTO the ability to overrule or undermine international conventions, acts, treaties, and agreements when it arbitrates trade conflicts between nations. The WTO directly violates "The Universal Declaration of Human Rights" adopted by member nations of the United Nations, not to mention Agenda 21 of the 1992 Earth Summit.

---

*Paul Hawken in *The Amicus Journal,* written shortly after he woke up lying on his back on the pavement in Seattle, after being pepper-sprayed by the Seattle police.
Reprinted with permission of worldwatch magazine (www.worldwatch.org) Worldwatch Institute, Washington D.C.

guest, and the illusion of wealth, the imaginings of unfettered growth and expansion, became small and barren in the eyes of the world. Dancing, drumming, ululating, marching in black with a symbolic coffin for the world, she wove through the sulfurous rainy streets of the night. She couldn't be killed or destroyed, no matter how much gas or pepper spray or how many rubber bullets were used. She kept coming back and sitting in front of the police and raised her hands in the peace sign, and was kicked and trod upon, and it didn't make any difference. Skeleton Woman told corporate delegates and rich nations that they could not have the world. It is not for sale. The illusions of world domination have to die, as do all illusions. Skeleton Woman was there to say that if business is going to trade with the world, it has to recognize and honor the world, her life, and her people. Skeleton Woman was telling the WTO that it has to grow up and be brave enough to listen, strong enough to yield, courageous enough to give. Skeleton Woman has been brought up from the depths. She has regained her eyes, voice, and spirit. She is about in the world and her dreams are different. She believes that the right to self-sufficiency is a human right; she imagines a world where the means to kill people is not a business but a crime, where families do not starve, where fathers can work, where children are never sold, where women cannot be impoverished because they choose to be mothers and not whores. She cannot see in any dream a time where a man holds a patent to a living seed, or animals are factories, or people are enslaved by money, or water belongs to a stockholder. Hers are deep dreams from slow time. She is patient. She will not be quiet or flung to sea anytime soon.

QUESTIONS

1. Using the Internet resources, define the following:
   a. WTO
   b. IMF
   c. World Bank

2. What is meant by oligarchy and corporate plutocracy?

3. What are the threats of globalization from the point of view of:
   a. People of the developed countries
   b. People of the developing countries
   c. Small businesses
   d. Environmental groups

4. What is your opinion about the N30 WTO showdown? Do you support Paul Hawken's stand against globalization? Why or why not? Explain your answer.

49

Pakistan: Karachi's Informal "Recycling Network"

Karachi, with an estimated population of 10 million in 1995, is Pakistan's largest city. The city has been growing at more than 5 percent per annum in recent years and is expected to reach 12 to 15 million inhabitants by the year 2000. Karachi was the world's twenty-fifth largest city in 1985; it is projected to be the fifteenth largest by the end of the century.

Karachi's high rate of population growth, coupled with scarce resources and inadequate planning, have created a number of environmental problems such as water pollution, air pollution, waste disposal, etc.

According to Karachi Municipal Corporation (KMC), the city generates 7,000 tons of domestic waste daily. The city has a nonmechanized system of waste disposal which is able to remove only about one third of the solid waste that is generated daily (Shah, 1993). The waste is removed manually by the city's 9,000 municipal sweepers and then transported by a fleet of 150 refuse vans to a single dumping site located 30 kilometers from Karachi, where it is disposed of by burning. The remaining garbage lies uncollected in city alleys and is processed by the informal "recycling network." This informal network consists of scavengers, middle men commonly known as "Kabaria," and the dealer. The scavengers hunt for recyclable garbage, in community dumpsters and city streets, and sell it to a middle man, (Karabia), who sells the material to the big dealer, and finally the dealer sells recyclable waste to recycling industries.

Unlike the West, where the process of recycling has been initiated in response to environmental concerns, Pakistan's recycling industry is motivated purely by economic motives. The informal recycling network of scavengers, middlemen, dealers and recycling units constitutes an amazing and not widely known phenomenon which performs an essential service under very adverse conditions. This informal recycling effort not only helps KMC in waste removal but also provides a source of inexpensive raw material to local small industry, and sustenance to thousands of inhabitants (Ahmad, 1993).

According to a report by National Training and Consultancy Services (NTSC) for the United Nations Center for Human Settlements (HABITAT), about 40 percent of Karachi's solid waste is recycled in the informal sector, and about 2 percent of the city's population is engaged full-time in the recycling industry. Considering the other participants, like reprocessing workers, housewives, servants, entrepreneurs, and craftsmen, it seems that almost half of Karachi's population is participating in recycling (Shah, 1993).

It is estimated that there are about 25,000 scavengers operating in Karachi. Almost 95 percent of the scavengers are Afghan refugees, (who fled to Pakistan after the Soviet invasion of Afghanistan in 1979), the remainder being Burmese or Bengali. Although no accurate statistics are available, it is estimated that these scavengers reduce the work of government agencies by as much as two-thirds, saving the Karachi Municipal Corporation (KMC) up to 200 million rupees annually in garbage collection and disposal costs. Karachi's informal recycling network has created one of the largest recycling industries in Asia. Paper, glass, metal, plastics, even bones, are all recycled (Table 49.1) and hence provide an inexpensive source of raw materials for domestic and overseas industrial needs (Ahmad, 1993).

Table 49.1

Karachi's Informal "Recycling Network"

Waste Category	Efforts of "Recycling Network"
Bones	About a dozen bone processing industries have been set up in Karachi, each with a processing capacity of 5 to 10 tons per day. About 80 percent or 30,000 tons of bones find their way to the far eastern markets, particularly Japan. Bones are a vital raw material in the manufacture of photographic film, and Japan's $20 billion photographic film industry makes it the largest market for this Pakistani export.
Iron	Iron, especially scrap recycling, has been known to build fortunes. The leading tycoons of Indo-Pakistan, from Mians of Lahore to the Tatas of Bombay, are basically scrap dealers and scrap recyclers. According to conservative estimates, about 3,000 tons of iron scrap is generated in Karachi daily.
Glass	About 80 percent of the country's demand for glass is also met through recycling locally generated glass waste, saving Pakistan millions of rupees in imports. It is estimated that about 80,000 tons of glass scrap is produced each year in Karachi alone, from both industrial and domestic sources.
Plastic	Research conducted by the NED University's Environment Department, and the Pakistan Council for Scientific and Industrial Research (PCSIR) reveals that more than 1,000 tons of plastic waste is generated daily in Karachi. The material includes discarded polythene bags and the waste from hundreds of plastic factories manufacturing PVC pipes and household items. The scale of plastic recycling can be gauged from the fact that there are about 8,000 polythene bag manufacturing units in Karachi alone.
Paper	Several tons of paper is recycled into paper and cardboard daily. The recycling process involves the low tech process of dumping large quantities of paper scrap in a pool of water and later sifting it through a sieve.

Karachi's informal recycling network has emerged as an indigenous industry of miraculous proportions. About 60 to 70 percent of Karachi's reusable waste is recycled, generating employment for over 2 percent of the city's population and simultaneously saving energy and cutting down on pollution. But the industry has its pit falls, too. More than anything else, unregulated recycling practices pose health hazards especially for those engaged in the separation of waste and reprocessing. Laborers melt batteries to extract lead, electroplated utensils to extract aluminum, photographic films to extract silver, without using any masks. Laborers in the bone industry breathe in bone powder and hence suffer from respiratory disease. In plastic recycling units, laborers breathe toxic emissions (Shah, 1993).

The existence of the vast informal recycling network suggests that, contrary to popular notions about develop-

Table 49.2

Daily Garbage Generation per Person for Selected Cities

City	Daily lbs. of Garbage/Person
New York	4
Paris	2.4
Hamburg	1.9
Rome	1.5
Cairo	1.1
Karachi	1.4
Calcutta	1.1
Hong Kong	1.9
Jakarta	1.3
Singapore	1.9
Tokyo	3

Shah, an Afghan Refugee, Collects 50–60 Kilograms of Paper Waste Daily at Karachi's Clifton Beach and Earns About Rs. 90 ($2) by Selling it to Kabaria. Photo Courtesy of Ahmed S. Khan.

ing countries, Pakistan is a highly waste conscious country where a staggeringly large proportion of garbage is recycled (Ahmad, 1993).

REFERENCES

Ahmed, F. (1993, July). The Waste Merchants. *The Herald (Karachi).* pp. 107–112.

Carson, W. (1990). *The Global Ecology Handbook: What You Can Do about the Environmental Crisis.* Boston, Tomorrow Coalition, Boston Press.

Population Growth and Policies in Mega-Cities: Karachi, Population policy paper No. 13. (1988). New York, United Nations.

Shah, N. (1993, June). Our Wonderful World of Waste. *Newsline (Karachi).* pp. 71–77.

QUESTIONS

1. Discuss the pros and cons of Karachi's informal "Recycling Network."

Karachi's Informal "Recycling Network"	
(+) Pros	*(−) Cons*

2. What steps need to be taken by the government agencies of Pakistan to make the processes of waste disposal and recycling safer and more efficient?

Karachi's Informal "Recycling Network"

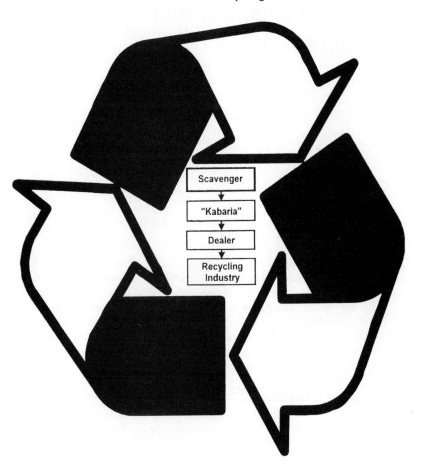

Scavenger → "Kabaria" → Dealer → Recycling Industry

50

Sri Lanka: Technology and Agriculture, Tractor versus Buffalo

The replacement of traditional animal power by modern tractors could pose serious ecological and environmental problems in Third-World countries. The use of modern technology in agriculture offers the benefits of efficient farming, but also poses the danger of eliminating jobs and a threat to the environment. A good example of the ramifying environmental consequences of technology is the case presented by Senanayake. He studied the effects of replacing buffalo power by tractor power for agriculture (Figure 50.1). At first sight, the substitution of tractor for buffalo power in the villages of Sri Lanka seems to appear a straightforward efficient trade-off between timely planting and labor saving at the expense of milk and manure. But it is not only milk and manure at stake. The substitution of buffalo with tractors could lead to a chain of events that could have an adverse impact on the environment. The buffalo and buffalo's wallows offer a number of benefits that the tractors could not offer. Figure 50.1 illustrates the these benefits.

In the dry season, the wallows are refuge for fish, who then move back to the rice fields in the rainy season. Some fish are caught and eaten by the farmers; others eat the larvae of mosquitoes that carry malaria. The thickets harbor snakes that eat rats that eat rice, and lizards that eat the crabs that make destructive holes in the ricebunds. The wal-

lows are also used by the villagers to prepare coconut fronds for thatching. If the wallows go, so do these benefits.

If pesticides are used to kill rats, crabs, or mosquito larvae, then pollution or pesticide resistance can become a potential problem. Similarly, if tiles are substituted for the thatch, forest destruction may be hastened, since firewood is required to fire the tiles.

REFERENCES

Senanayake, R. (1984). The Ecological, Energetic and Agronomic Systems of Ancient and Modern Sri Lanka in Douglas, G.K. Ed. *Agricultural Sustainability in a Changing World Order,* Boulder, Colorado, Westview Press.

Gordon, C. (1990). Agriculture and the Environment: Concepts and Issues in Huq, S., et al. Eds. *Environmental Aspects of Agricultural Development in Bangladesh,* Dhaka, University Press Limited.

QUESTIONS

1. What strategy would you suggest for employing tractors for farming in Sri Lankan villages?

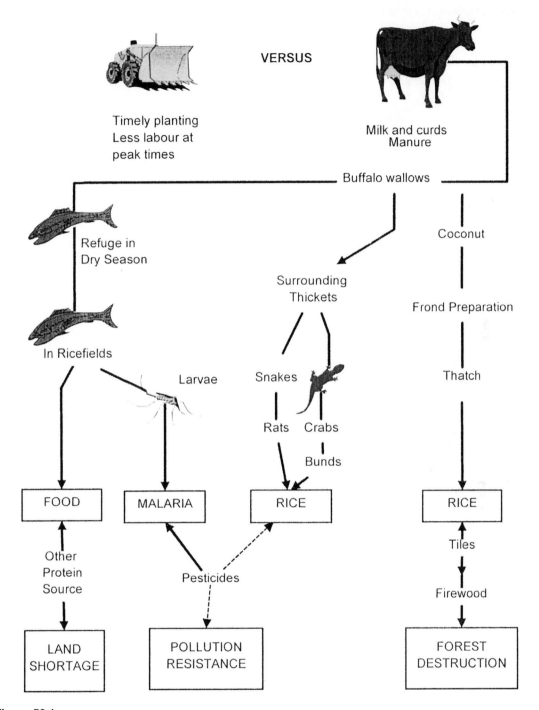

VERSUS

Timely planting
Less labour at
peak times

Milk and curds
Manure

Buffalo wallows

Refuge in
Dry Season

Coconut

In Ricefields

Surrounding
Thickets

Frond Preparation

Larvae

Snakes

Thatch

Rats   Crabs

Bunds

FOOD

MALARIA

RICE

RICE

Other
Protein
Source

Pesticides

Tiles

Firewood

LAND
SHORTAGE

POLLUTION
RESISTANCE

FOREST
DESTRUCTION

Figure 50.1
Technology and Agriculture: Tractor vs. Buffalo
Source: Agriculture and the Environment: Concepts and Issues, *Environmental Aspects of Agricultural Development in Bangladesh,* University Press Limited, Dhaka, Bangladesh.

A Bridge to Your 21st Century Understanding

Complete and discuss the following flowchart.

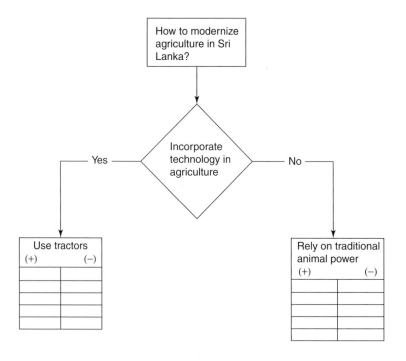

Appropriate Technologies

GADI KAPLAN

The ending of the Cold War has released more energy for tackling the concerns of underprivileged populations. "One billion people in the developing world still lack access to clean water . . . nearly 2 billion lack adequate sanitation . . . Electric power has yet to reach 2 billion people," the World Bank noted in its June *World development report 1994: infrastructure for development.*

A country's infrastructure includes services based on electrotechnology, such as electric power, telecommunications, and road, sea, and air transport. Developing countries in round figures pour US $200,000 million every year into new infrastructure, amounting to 4 percent of their national output and a fifth of their total investment, according to the World Bank report.

In purely human terms, the situation of these people is unacceptable. Technologically, it dares electrical and electronics engineers to devise remedies by applying proven technology at an affordable cost.

One challenge may well be to upgrade the infrastructure in developing countries. At present, for example, about 40 percent of their power-generating capacity is on average unavailable when needed, as a result of malfunction or scheduled maintenance; elsewhere, in the best performing power sectors in the world, the percentage is half that figure, according to the 1994 World Bank report.

To cut back on waste and improve efficiency, the bank is embarking on a thorough reform. Among its goals are to see the infrastructure managed as a business, and not as a

© 1995 IEEE, Reprinted, with permission, from (*IEEE Spectrum,* pp. 32–43, Oct '94).

bureaucracy; to see competition introduced; and to give those who use the services a stronger voice in and responsibility for their operation than has been the custom.

But what exactly is a developing country? The World Bank often refers to "developing economies" as those with low or middle per-capita income, derived by dividing the gross national product (GNP) by the population count.

Low-income countries, with US $675 or less per capita, include Burundi and Benin in sub-Saharan Africa, Egypt in North Africa, Cambodia in East Asia, Tadjikistan in Central Asia, and Haiti in the Americas. Senegal, though, and Thailand, Turkey, Iran, Algeria, and Chile are representative of middle-income economies, with $675–$8356 per capita income (in 1992). (All dollar figures are for U.S. currency.)

Indeed, a developing country quite often seeks to leapfrog straight into the 21st century, particularly if the leap makes sound economic sense. Indonesia is doing just that. Looking to a rapid expansion of its urban telephone network, it is installing a radio-based, cellular telephone system in Jakarta and West Java for a total of 280,000 subscribers.

The first topic covered in the appropriate technology series is photovoltaics, which has the potential of bringing electric energy to millions of remote rural households, greatly enhancing their quality of life.

So rosy, in fact, is its promise that the 12th European Photovoltaics Solar Energy Conference and Exhibition devoted a parallel symposium to photovoltaics in developing countries. Inspired in part by that symposium, which was held in April in Amsterdam, the Netherlands, the following report includes an overview of the technology's applications and two regional studies, one on a large project in West Africa, and the other about India.

Photovoltaics is attractive because it can help countries avoid the huge expense of expanding electric grids into rural areas, at an estimated cost of $20,000–$30,000 per kilometer, reports Erik H. Lysen, the head of a department at Novem, the Netherlands agency for energy and the environment located in Utrecht. But if affordable power is to be supplied by this means to millions of rural households, the worldwide manufacturing capacity for photovoltaic modules and systems will have to be substantially enlarged, he noted. Furthermore, funding for photovoltaics applications by national and international organizations would have to be increased by orders of magnitude, he said.

International cooperation counts heavily toward the success of photovoltaics applications. Evidence of that comes from projects in Vietnam and the republic of Kiribati in the mid-Pacific.

In West Africa, high grades are scored by a program for installing photovoltaic systems with 1200 peak kilowatt capacity for water pumping, refrigeration, and lighting. And in India, about 62,000 rural photovoltaic systems have already been installed under a government-assisted program, and the installation of 60,000 units that power VHF remote radio links is under way.

From these and other reports it is clear that designing "appropriate" technology may sound easier than it is. In the West Africa project, for instance, inverters for water pumps had to be flexible enough to accommodate various pump capacities, pipe sizes, and water heads, but this flexibility could not be bought by sacrificing efficiency. Protection of the photovoltaic system against lightning and the safety of maintenance personnel also came into the picture in a big way.

Even so, the obstacles that photovoltaics must overcome are more institutional in nature than technological, as became very clear at the Amsterdam meeting. More than 30 years in the making, photovoltaics is mature enough in itself, but will require innovative breakthroughs in financing as well as changes in energy policy, the experts believe.

PHOTOVOLTS FOR VILLAGES: ERIK H. LYSEN

Two billion people are without an electrical connection, but the cost of hooking up their homes to a conventional grid system will be too high for most developing countries.

Photovoltaic solar systems are a cheaper alternative and can reach virtually any site on earth. Granted, new and bold initiatives are necessary to realize the required 20,000-MW peak power of photovoltaic capacity. But the breakthroughs needed are less technical and more in the areas of financing and energy policy.

Energy is not the most pressing problem of the developing countries, partly because oil prices are at present so low. Instead, they have to worry about the poverty of most of the inhabitants, their food supply, the creation of sufficient and meaningful jobs, competition in the international markets for their products, and, sadly enough in some cases, the presence of war. The main energy problem in rural areas still is to find enough wood for cooking meals. In some developing countries, biomass for cooking accounts for over half of the national energy consumption.

For some decades now, photovoltaics (PV) has been on the energy scene in the developing world, particularly in established niche markets such as telecommunications, marine beacons, railway signaling, and cathodic protection of pipelines. The challenge will be to apply PV to the provision of energy to rural homes for lighting, refrigeration, and TV.

The most successful to date is the individually owned solar home system: typically a 50-W solar panel that charges a battery by day and power loads after dark. Owners are proud of having their own power system and are not susceptible to grid failure or inconsiderate neighbors, while the equipment itself can be expanded when required. Of course, the systems must be well designed, properly maintained, and paid for by the owner, because giveaways seldom last long.

On the face of it, it looks strange that people in rural areas and nonindustrialized countries are satisfied with a fraction of a kilowatt-hour a day, whereas residents of industrialized countries "need" 10 kWh or more per day. The explanation is that people are not interested in kilowatt-hours as such, but in light, a working TV set, or a refrigerator. In other words, they want the services of the energy regardless of whether it amounts to 0.1 kWh or 10 kWh per day. And as only a limited number of solar kilowatt-hours are available, people cannot afford to waste them and must use efficient lighting, low-power TV sets (a golden market for future flat-screen TV sets with liquid-crystal displays), and efficient refrigerators.

The only thing that matters to the owner is: how much do I have to pay for these services every week or month? As a first estimate, a reasonable part of the rural population can raise $5 to $8 per month. This is what they are paying now for kerosene lighting and for battery charging at grid-connected centers. (All dollar figures in this article refer to U.S. currency.)

As is well known, nearly all utilities in developing countries are losing money on rural electrification. Investment capital is not the problem, being available at comparatively low rates from large multilateral banks like the World Bank and Asian Development Bank. High costs are the issue, stemming from long transmission lines to remote customers, low consumption, the need to charge only low tariffs, and high technical and nontechnical losses (transmission and distribution losses and illegal connections)—20–40 percent in some countries. And these high costs are usually cross-subsidized by urban consumers.

Line costs are the heaviest burden, usually accounting for 80–90 percent of the budget of a rural electrification project They typically run $20,000–$30,000 per kilometer. Another factor is the widely variable number of users per kilometer of line, which may be as low as two but in Bangladesh must be at least 75. As a result, the average cost per rural connection also varies wildly: Mohan Munasighe, an energy expert with the World Bank, quotes a range from $200 to $3650 (in 1983 prices). The monthly power consumption of rural consumers is usually low: 20–40 kWh, leading to high per-kilowatt-hour costs of 10–20 cents to $1.

Power to the People

A recent report from the World Bank on its role in the power sector notes that the number of power connections in developing countries in the period 1969–89 grew by 9 percent per year, or 2.5 times the average population growth. In spite of this rapid growth, rather few members of the population are yet connected to an electricity supply. Average real power tariffs declined from 5.2 to 3.8 cents per kilowatt-hour in the period 1979–88 (in 1986 dollars).

Nevertheless, governments and utilities in developing countries have impressive plans for expanding their power sectors. The idea is to increase the total power supply capacity from the 471 GW of 1989 to 855 GW in 1999. It is estimated that no less than $1 trillion will be needed to achieve the desired 384-GW increase. Approximately 40 percent of this sum is in foreign exchange, and it is clear that the $40 billion needed annually cannot be mobilized, even by the large multilateral banks (the present level of World Bank lending for the power sector is around $7 billion per year).

The conclusion seems inescapable. Governments in the developing world will face great difficulties in expanding their power sectors as planned. Obviously, too, priority will be given to industry and the urban sector. This prospect underlines the need to consider other alternatives for rural electrification, such as pre-electrification through PV or other renewable sources.

The Story Till Now

One of the earliest references to photovoltaics in a rural setting is to a system installed in Chile in 1960. In the '60s, PV cell development was dominated by research for space applications (the first satellite was launched in 1957).

Rural energy developments in the '70s were typically the so-called "integrated energy systems" promoted by the United Nations. These projects characteristically exploited various sources of renewable energy (such as the sun and wind, biomass, and organic waste), and they distributed the electricity thus obtained through a regular grid system in the village. In practice these systems never came up to expectation; they demanded too much maintenance, were poorly designed, gave unreliable service and, most importantly, were more or less forced upon the village inhabitants.

Individual solar home systems were introduced more or less independently in the Philippines, the Dominican Republic, and Indonesia in the early 1980s.

The Philippine-German Solar Energy project (1982–88) started with a 13-kW plant for a small village. The plant was found to be not economical for wide dissemination. The second phase emphasized small PV systems for use in rural areas, namely, solar home systems and communal battery-charging stations. The first 100 of these home systems were installed at Burias Island. A would-be owner had to make a down payment of $140 and 36 monthly payments of $13.

An important lesson was learned. Solar home systems have become a status symbol because they are one of the few high-tech systems available outside urban areas and they open the door to radio, TV, and video. At present, about 105 kW of solar systems are installed in the Philippines, of which 70 kW is in residential systems and 35 kW is in systems intended for telecommunications, pumping, and other purposes.

In the Dominican Republic, a true catalyst for the development of photovoltaics was the presence of Richard Hansen, founder of Enersol Associates (USA). In April 1984 the first PV system was installed in Bella Vista. Right from the beginning, owners had to pay for the system (48 monthly installments of $10). Soon a solar credit fund was created (Adesol) with seed money from the U.S. Agency for International Development, and from the installments paid by its clients new systems could soon be bought. Local entrepreneurs were trained for servicing and sales of new systems.

By mid-1987 more than a hundred systems were up and running. That year, the Solar-Based Rural Electrification Concept was introduced, as a model for intervention by a nongovernmental organization, and gradually spread. By 1992 the number of PV systems grew to 4000. More than 10 installation businesses are active in the Dominican Republic.

The development of solar home systems in Indonesia was the fruit of cooperation between individuals and the Dutch PV company R&S (Renewable Energy Systems), the Indonesian Ministry of Research and Technology, and the Indonesian Ministry for Cooperatives. A start was made in 1987 and in 1988 the systems were demonstrated in Sukatani, a village 110 km from Jakarta. The seed money was provided by the three organizations listed above plus the Netherlands Ministry of Foreign Affairs.

Sukatani has a lower than average amount of sunshine. The idea was that if the system worked there (which it did), it would work throughout Indonesia. The local cooperative took responsibility for fee collection and simple maintenance.

The President of Indonesia became so enthusiastic about the Sukatani project that he started the Banpres project, with interest-free credit for 3000 solar home systems, which have since been successfully installed. Additional credit schemes through revolving funds have been started, one example being the $50,000 revolving fund grant from the North Holland utility PEN, for solar home systems in the village of Lebak.

By 1993 more than 10,000 home systems had been installed throughout Indonesia. The average investment has dropped below $400, and with a down payment of 35 percent the solar home system owner pays the equivalent of $8 per month.

Five Types of Problems

As with every new technology, all was not smooth sailing. Problems that arose during the systems' introduction fall into the following five categories: financial, institutional, interpersonal, infrastructural, and technical.

Financing the purchase is still the most forbidding hurdle. The average price of a solar home system is about $500. But in many countries, only 10–20 percent of rural families earn more than $100 per month. Assuming that 10 percent can be spent on a solar home system, this means they can afford a maximum of $10 per month, and preferably less. The problem therefore is how to reduce the monthly installments to, say, $5 per month.

Institutional conflicts are the next concern. Electric utilities are traditionally either hostile or at best indifferent to small autonomous systems such as solar systems for the home. It can happen that a nongovernmental organization pushes hard to install solar home systems in a village, only for the utility to show up with a grid extension a few months later. Confusion is created and part of the investment in home systems suffers. Rural electrification policies seldom include pre-electrification options such as solar home systems, or wind or microhydro supply systems.

Experience has shown that interpersonal relations were often mishandled. People were not properly involved in a timely fashion in the decision to introduce PV systems. They were not informed about the performance of the solar home system and its pluses and minuses compared to a grid connection. Training of the owners was sometimes inadequate, so that batteries were discharged too deeply if poor (or no) controllers were present. Simple but essential repairs took much too long, so that people lost confidence in the system. The key factor here is proper communication with the users.

This situation is closely linked to infrastructural weakness. Often after-sales service was either nonexistent or poorly organized. Publicity was insufficient, wrongly targeted, or even negative. Stories about failures tend to stick in people's minds for a long time, and 10 times as many successes are needed to eradicate them.

Then of course there were the technical problems associated with a new technology. Interestingly, there were hardly any problems with the PV modules. The trouble came from conventional parts of the system—batteries, controllers, and lights and switches. Controllers worked poorly or were omitted to save money, so batteries had a short life. Cheap fluorescent light tubes were used, which blackened quickly; electronic ballasts failed and switches malfunctioned.

In essence, the financing problem can be solved by lowering the investment cost of the PV system and by enabling the customers to pay smaller amounts over longer periods. The first part of the solution is largely for the manufacturers to implement and is influenced by Western development programs. The developing countries' governments can help by lowering or waiving import duties.

Several options for lowering the monthly costs for consumers have been proposed and are being practiced: revolving funds (started with grants); presidential loans (as in Indonesia); local bank loans; the Finesse approach (*fin*ancing of *e*nergy services for *s*mall-*s*cale *e*nergy consumers); and supplier's credit.

Energy experts made a good point during the Finesse workshop held in October 1991 in Kuala Lumpur with the support of the World Bank, the U.S. Department of Energy (DOE), and the Netherlands Ministry of Foreign Affairs. Given that large power companies have access to very cheap capital for new power plants and rural electrification projects, they argued, smaller-scale power options that complement the grid supply should also be allowed to tap just a fraction of those funds, and on the same conditions. Solar home systems were seen as a case in point.

This is in essence the aim of the Asia Technical Alternative Energy unit of the World Bank, which is supported by (among others) the DOE and the Netherlands Ministry of Foreign Affairs. The unit at present is active in Indonesia, India, Sri Lanka, and China.

On the institutional level, clashes with the electricity companies can be avoided if the initiatives in solar home systems are coordinated or even channeled through them. The utilities should permit, spur on, or even contract with the private sector and nongovernment organizations to start offering solar home systems in certain areas. They should make it clear to the customers that this is a pre-electrification option, and that if in the future the utilities have the means to reach the village by the grid, the PV panels can be resold to the private sector (or kept as an emergency option).

Events in the Philippines, Dominican Republic, and Indonesia prove that the early involvement of potential customers is crucial to the success of any PV project. People should be properly taught about system operation and properly informed about comparative performance and costs. A warning system for larger breakdowns should be set up, to ensure quick repairs and maintain confidence in the system. Local cooperatives should be used to collect fees and carry out basic maintenance. Local youths should be trained as PV technicians and paid for their work.

Improvements in product quality have occurred over the last few years, thanks to lessons learned from simple technical problems. Controllers and lamp ballasts have improved, some of them locally produced; battery indicators have now become available so people can "see" how much is left in their battery (as was their custom with bottles of kerosene).

The wattage needed for rural electrification of the 2 billion people as yet without electricity can be estimated. The average bad load of a rural connection is around 350–500 W, so with an average family size of five and 400 million as yet unwired households, an additional capacity of 140–200 GW is needed. At a conservative value of $2500/kW, this capacity would cost $350 billion to $500 billion. In addition, the annual fuel bills of the developing countries will be increased by $5000 to $10,000 million.

Governments in developing countries are already hard pressed to expand their industrial and urban capacity as fast as necessary. They will find it impossible to invest also in rural electrification. If, however, the latter task could be achieved through individual solar home systems, the total investment would be lower, although still considerable. Assume that a 50-Wp solar home system in the future will cost $250 on average, then 400 million solar home systems (20 GWp) will require $100,000 million, or $4000 million annually for 25 years. These systems would have to be financed through long-term loans, not grants.

To put things in perspective, it is perhaps useful to mention the size of the predicted PV market and required capacities. Note that the present market in photovoltaics is about 60 MWp, of which about 5 MWp is for rural off-grid applications. Market analyst Paul D. Maycock, president of Photovoltaic Energy Systems Inc., Casanova, Va., expects the following markets for the off-grid rural segments in the years 1995, 2000, and 2010:

- For a business-as-usual scenario: 8, 20, and 40 MWp.
- For an accelerated growth scenario: 15, 40, and 600 MWp.

Recall the estimate that the electrification of all 400 million households currently without electricity would require 20 GWp of PV capacity, or 800 MWp on average for a period of 25 years. This is even more than is predicted in Maycock's accelerated growth scenario. So probably this target will not be met unless bold new initiatives are taken.

Bold Ideas

One initiative of this nature is the Power for the World proposal put forward by Wolfgang Palz, division head of renewable energies in the European Commission's Directorate General XII, which handles science research and development. To reduce system costs and increase production volume, plants with an annual production capacity of 10 MWp must be built soon; by the year 2000, annual capacity must reach 100 MWp. This can only be done if specific conditions are met, namely, policy changes, better financing options, and system improvements. But the environmental issue must be considered right now.

There are signs of policy changes and a beginning of acceptance by electricity companies, as in Mexico and the

Philippines. But there is still a long way to go before solar PV systems are accepted by utilities in developing countries as a reliable means of pre-electrification or even electrification. Sources of capital can be instrumental in changing this attitude by requesting such policy changes (as well as financing renewable options) during the negotiations for conventional power loans.

Several financing initiatives have been offered by both national and international donors; but in terms of total funds required, the efforts have to be increased by orders of magnitude. The easy terms for large power loans should be made available for small off-grid options as well.

Manufacturers, importers, and distributors should ensure the quality of their PV products. National product standards should be established, leaving enough room for product improvement.

With the introduction on a large scale of PV battery-charging systems, the number of batteries in rural areas will soar. Environmental problems could ensue if no measures are taken beforehand. Manufacturers, together with local counterpart companies, should work out optimum and least-cost solutions. National battery recycling is already in effect in Indonesia, for example.

Acknowledgment

This article summarizes the plenary paper presented by the author at the 12th European Photovoltaic Conference, held in Amsterdam in April 1994. The full paper has been published in the conference proceedings.

To Probe Further

Costs of rural electrification are addressed in *Electricity for rural people* by G. Foley (Panos Publications Ltd., London, 1990) and in *Rural Electrification for Development: policy analysis and applications* by M. Munasinghe (Westview Press, Boulder, Colo., 1987).

R.J. Saunders is the author of "The World Bank's role in the electric power sector: policies for effective institutional, regulatory and financial reform," a 1993 World Bank Policy Paper, published in Washington, D.C. The costs of electrification are discussed by R. Turvey and D. Anderson in *Electricity Economics,* a World Bank Research Publication (Johns Hopkins University Press, Baltimore, Md., 1977).

A reference to an early photovoltaic system in Chile is included in *Photovoltaics for Development,* by R. Hill (ed.), United Nations ATAS Bulletin No. 8, 1993.

Integrated Rural Energy Planning was discussed by Y. El Mahgary and A.K. Biswas in a United Nations Energy Planning (UNEP) publication (Butterworth Scientific, England, 1985).

Solar home systems are discussed in *The Philippines' Rural Photovoltaic Electrification Scheme* by G. Santianez-Yeneza and H. Böhnke in a publication of the National Electrification Administration, Manila, 1992. Photovoltaics applications were addressed by R. Schröer and P. de Bakker in their article, "It All Began on Burias Island," *GATE* magazine, July 1989. *Solar Rural Electrification in the Developing World; Four Country Case Studies: Dominican Republic, Kenya, Sri Lanka, Zimbabwe* was dealt with by M. Hankins, Solar Electric Light Fund, Washington D.C., 1993.

The financing of energy services for small-scale energy consumers (Finesse was discussed during the World Bank workshop, Kuala Lumpur, October 1991). Paul Maycock examines *Photovoltaic technology, performance, cost and market forecast 1975–2010,* in a publication of PV Energy Systems Inc., Casanova, Va., June 1993.

Bold initiatives in rural electrification were proposed by Wolfgang Palz in his paper "Power for the world," which is to be found in the proceedings of the International Solar Energy Society (ISES) Solar World Congress, held in Budapest in 1993.

Energy sources are regularly written about in the papers in *IEEE Transactions on Electron Devices,* while energy conversion by renewable sources is regularly addressed in *IEEE Transactions on Energy Conversion.*

WATER FROM THE AFRICAN SUN: SERGE MAKUKATIN

The most ambitious worldwide program in photovoltaics ever financed by the European Union (until a year ago, the European Community) is nearing realization. Its goal is to better the living conditions of those who live far from population centers. Both short-term and permanent improvements are envisaged.

The first task is to make available drinkable water, the next to irrigate fields under cultivation. A third aim is to supply small communities with electricity for lighting and essential refrigeration. As part of this 1200-kW (peak power) project, Siemens Solar GmbH faced special requirements involving a variety of systems and key components. The project had several important social aspects as well.

The project's beneficiaries are the members of a body set up in 1973 to fight drought in West Africa's Sahel zone, the Comité Inter-Etats de Lutte contre la Sécheresse dans le Sahel (Cilss). Present membership consists of nine countries that suffered greatly from drought in 1968 and 1974: Burkina Faso, Cape Verde, Chad, Gambia, Guinea-Bissau, Mali, Mauritania, Niger, and Senegal.

It was to help these nations that the Programme Régional Solaire (PRS) was established to make use of photovoltaic energy and to that end given 34 million ECU in funding by the then European Community as nonrepayable aid.

The program was divided into three parts, with Siemens Solar GmbH the first to be awarded a contract. The contract called for systems to be installed in five Cilss states—Gambia, Guinea-Bissau, Cape Verde, Mauritania, and Senegal. All in all, Siemens Solar will supply 550 kW, peak, of solar power. Most of this power will be used by 330 pumping systems, the largest number of high-capacity pumping systems of any part of the program.

Also to be provided are a total of 339 so-called community systems—240 for lighting, 63 for refrigeration, and 36 for recharging lead-acid and nickel-cadmium batteries. The DM 30-million contract covers parts supply, installation maintenance, and after-sales support.

Under this contract, Siemens Solar had in addition to establish a service network, train local partners, and support their activities. The execution and coordination of this all-embracing contract required the establishment of a complex project organization. For instance, Siemens was obliged to find a local partner in each of the five countries.

These partners would be responsible for maintaining contact with the local building supervisors, coordinating all essential documents (dealing with orders for installations and technical information resulting from alterations) and dispatching them to Siemens, and stocking spare parts for repairs. The importance of reliable local partners cannot be overemphasized. A lack of spare parts of technical support has wrecked many a past project.

Numerous companies volunteered for the project. The challenge was to identify those that would be most suitable as partners. In the end, there were only a few serious candidates with adequate financial backgrounds.

The partnership benefits both parties. By being involved in the greatest PV project ever carried out, the local partner can earn a good sum of money for the services rendered. Siemens Solar, in turn, can extend its activities in the African market through the aid of the local partner.

For its partner in the Sahel region, Siemens Solar chose a French company that had already been a subcontractor on African projects, successfully performing installation and maintenance services. Wherever it has a local subsidiary in the five regions, Siemens' partner takes care of any tasks there directly. Otherwise, it picks a company in the area to do so.

Future Upkeep

As the main contractor, Siemens has to guarantee all systems and to deliver spare parts for five years free of charge (although consumable materials, such as light bulbs, have a one-year-warranty). After the termination of the five-year guarantee, funds will be needed for maintenance and replacement.

Once a village has inspected and provisionally accepted the PV system, those of its inhabitants who use it pay an annual sum determined by the individual system's maintenance costs. For example, a medium-sized P4 pumping system used in Mauritania will deliver 9000 m^3 of water in a year. Dividing the total US $2000 yearly costs of maintenance ($900) and replacement ($1100) by the annual 9000 m^3 water output, results in 22 cents per cubic meter of water. This is the charge users must be able to afford to keep the system operational; it does not include money for new investments in future PV plants.

Siemens chose its own M50 module, a 50-W, peak, monocrystalline solar panel with an efficiency of 12.4 percent, as the major power component. A requirement of the EC project is that all components be of European origin.

The key components for pumping systems are the pumps themselves and their ac inverters. Standard centrifugal pumps and motors are used for high water output. These pumps are equipped with 3.5-kVA Simovert-P-Solar inverters. This inverter model—a slightly modified version of the more than 10,000 units thus far produced by Siemens—is designed for highly reliable operation even under severe conditions. It has an efficiency of up to 95 percent.

The Simovert-P's microprocessor allows it to be programmed for special site conditions. It drives common three-phase, 220-Vac pumps up to 2.2-kW rated power. There are no known competitive inverters with these features.

Centrifugal pumps with modified motors are employed for the smaller pumping systems. The Grundfos Solartronic SA1500 inverters drive the modified three phase, 65-Vac pumps. (The P1 types are for surface-pumping irrigation systems.)

Both Simovert-P and the Solatronic inverter have dry-running protection and can produce frequencies from 1 to 60 Hz. The actual output frequency at which the inverter operates depends on several factors: the input power from the solar modules, type of pump, piping, and the delivery head (defined as the height of the inlet to the cistern above the well's water level). The efficiency differs with frequency and also with pump type, piping, and the slope from the delivery head. Typical efficiencies of the combined inverter, pump, and motor are in the area of 40 percent at 50 Hz and 10–20 percent at 25 Hz.

Danger Free

The installations need to be safe in two senses. They must be as immune as possible to external events, like lightning strikes, as well as harmless for people to approach and maintain.

There are two types of lightning protection: external and internal. The external variety consists of connecting all parts (not only the modules and inverters, but also the support structures and fences) to earth ground. This is accomplished by burying a ring of copper cable 1 meter deep in the ground around the system and inside its protective fence, and connecting the ring to an earthpole, or plate, in the ground. These measures ensure that, in case of an external overvoltage, no significant difference in electric potential exists between the different components.

The internal protection keeps any overvoltage from damaging the electronic devices—the inverter and solar-cell array. It consists of varistors in the junction box and in the inverter.

Since the operating voltage of the pumping system lies in a hazardous range, measures must be taken to protect the public as well as system maintenance personnel. For this reason, the systems were designed using German industrial safety standards (DIN/VDE-Standards). In addition, each pumping system is fenced off from the public.

The EC office awarding the contract imposed strict quality requirements on components and material for this project. It wanted to ensure that systems would be rugged enough to operate in harsh environments throughout a long service life.

To find accessories and components of suitable quality at acceptable prices, extensive market research was necessary. The materials and components selected had to pass severe tests performed at one or other of the four European research laboratories that are officially accepted by the Cilss organization: Technical Inspection Authority (TUV) Rheinland, Cologne, Germany; LVT/CEA, Cadarache, France; Global Renewable Energy Services (GRES), Swindon, England; or Ciemat, Madrid, Spain. The main purpose of the tests was to ensure that the technical data indicated by the suppliers reflected the specifications required.

For instance, in the case of the inverters, the test criteria used by TUV Rheinland included performance, efficiency, temperature and climate tests, electrical safety, and the examination of packaged devices after a drop from a height of 1 meter. Further, the fabrication processes of the inverters' suppliers were audited. During these tests, some possibilities for optimizing the devices came to light and were later integrated during production.

For instance, the degree of electrical protection was increased from IP54 to IP55, the mechanical mounting of several components was optimized for greater ruggedness, and the steering program within the microcontroller was improved. A 10-g vibration test was also added to production testing.

Although this testing raised the supplier's costs, in the end it was justified by the way the components operated; all devices work perfectly in the field and are highly reliable, which is not typically the case for photovoltaic inverters.

For all PV modules, a certificate (CEC specification 502) from the Ispra Institute of Ispra, Italy, is proof of qualification for the project. No problems arose in complying with the module data for any of the PV systems during the verification of the delivered power.

For the West-African region, the basis for the system design is 6 kWh/m^2 per day (which lasts from 6 AM until 6 PM there) and a daily output of water within a range of 30 percent (-10 to $+20$ percent). All design values (300 W, peak, to 3.8 kWp, rated generator power) were checked and confirmed by the customer.

For a specific site on an average day in a recommended month, a nomograph can be used to determine the characteristics of a photovoltaic pumping system (of the solar generator, the well, and the water output). Note that when water is drawn, its level in the well will drop; this so-called draw down and the rate at which the water returns to its previous level depend on the conditions of the site. No realistic statements on the delivery capacities of the systems can be made without these data. On this basis, characteristic variations for various frequencies were determined for all centrifugal pumps used.

First Results

The first of the project's PV pumping systems were installed in early 1991. As of April of this year, 55 out of the total of 330 deep-well pumping systems had been installed and had provided approximately 270,000 hours of trouble-free operation. The total efficiency (from sun to water) of the completed systems is about 3–4 percent. The efficiency of other systems is about 2–3 percent.

All systems work without any notable problems, so that the local populations can rely on receiving the precalculated quantities of water. The yield of the systems until now amounts to approximately 1,200,000 m^3 of water and is probably unique in its volume. In several villages the water output is so great (50–60 m^3 per day) that the cisterns can be filled up within two or three hours.

Cilss has installed several test PV plants that store all key data: delivery head, water output, insolation, voltage/current of solar generator, and water output. The goal is to measure performances and water output within a limited time period. So far, the stored data has shown that the systems' output exceeds the Cilss specifications by 10–20 percent (water output per day at a specific insolation).

As of April 1994, 30 of the 339 community systems had been installed, among them 11 cooling systems used for storing vaccine in Mauritanian first-aid hospitals and 6 lighting systems in Cape Verde schools. All the installed systems have satisfied their users and work without trouble.

Essential to this success were the contributions of the local companies. Following a period of well-structured instruction, these local companies assembled, installed, and maintained the pumping systems.

In this way, an excellent base of technical knowledge for system and component has been established within the Cilss organization. The local firms involved with the systems are now familiar with PV systems and their experience should form the basis for further successful applications in this and future projects.

Minor Problems

As with every project, there were some problems, but they were minor in nature. Some Gambian systems are in a region with frequent lightning strikes. To date, four lightning barriers at the input of the 3.5-kVA inverters have been destroyed by indirect lightning strikes. Nevertheless, the devices fulfilled their function because the inverters remained fully operational. The barriers were replaced with locally available spare parts.

The second event, and to date the last, was the failure of four float switches in boreholes, caused by corroding well casings. The problem was remedied by using an additional float switch filter.

Choice of Village

The selection of a village to be equipped with photovoltaic systems is governed by several criteria. There must be a real demand for water supply (population) and the basic infrastructure must be there (water hole, a small hospital). In addition, the villagers must be able to pay some money for replacement and maintenance. The water committees of the local villages take care of the administration of the money. Fees for the water can be individually paid if somebody takes water. Alternatively, it is also possible to pay an annual average per family.

Payment for the water is essential—it ensures the survival of the PV system. In Cape Verde, droughts have often lasted for as long as two years. When this happened, the villagers had to pay for water brought in tanks from the capital or places with higher rainfall. It was not difficult for these people to understand that water is not free of charge.

The Podor people of Senegal, on the other hand, have traditionally taken drinking water from the river free of charge. Even though this water was often polluted and caused many diseases, it is difficult to explain to the people that they must pay for clean drinking water. So the need to pay for water must often be clarified in lengthy discussions.

PV systems within the region have completely changed the lives of the African village people. Infectious diseases caused by polluted water have greatly diminished. Pure drinking water is available at water taps, and women and children no longer have to walk 3–5 km for water. Freed from this chore, children can go to school while women can tend a kitchen garden watered by the PV system. Because of the electrification of schools and public buildings, it is now possible to educate people after the workday ends at 6 PM.

In each case, the local population must recognize the need for a village organization that is responsible for seeing that the PV installation is kept in good condition. They must also realize that they, too, have an obligation to care for the systems; there cannot always be a solution from an external source for everything.

To solve the problem of collecting money, several sociologists, who are working on the problem of forging an "identification of people with the project," have in the meantime worked out insurance concepts. For example, in Gambia—a relatively well developed country with good business opportunities—a contract being negotiated between the Department of Water Resources and an insurance company covers the possible risks of maintenance and replacement. This had the advantages of lowering costs for maintenance and replacement and of placing financial administration outside the village.

Preliminary Findings

Three years into the project, some interim conclusions can be drawn. Because of its comprehensiveness—including the high requirements set on system quality and the reliance on local firms and local after-sales service, the training of local partners, and the stocking of spare parts within each country—the project is very likely to succeed and to set an example for others like it in the future.

Photovoltaic systems have proved to be a reliable way of offering essential services to rural communities off the electric grid. But for the successful implementation and operation of such systems, well-trained partners located in the developing communities are imperative.

That the components used in the project have shown themselves to be reliable and of high quality indicates that authorized inspection institutions play an important role in the success of large-scale projects. At the same time, projects like Cilss allow manufacturers and suppliers to optimize their systems and components and prove their reliability. This in turn is beneficial for the next customers, increasing confidence in the products.

Last, but by no means least, the Cilss project clearly demonstrates the benefits of appropriate technology to people in remote areas.

To Probe Further

More information about this project can be found in the paper, "PV Energy for a Sustained and Social Development in the Sahel Region," by F. Kabore, in *The Yearbook of Renewable Energies,* a publication of Eurosolar in collaboration with UNESCO sponsored by the Commission of the European Communities, 1994, pp. 146–149.

Acknowledgment

The author thanks Mr. Cunow and Mr. Theissen, engineers from the technical office of Siemens Solar, for their valuable support.

QUESTIONS

1. What type of energy technologies are most appropriate for improving the general well-being of Third-World countries?

2. Compare the energy consumption (kW-hour per capita) for the following countries:
   a. USA vs. Bangladesh
   b. UK vs. China
   c. France vs. Nepal
   d. Germany vs. Yemen

3. List the advantages (+) and limitations (−) of various energy technologies available to the developing countries on the table below.

4. A number of developing countries are planning to construct nuclear power plants to meet growing energy needs. Due to scarce resources in the Third World, nuclear energy seems to be the best viable option. The developed countries, however, are trying to prevent the transfer of nuclear technology to the Third World. The developing countries consider this attitude to be one of technological imperialism.
   a. Do you agree that Third-World nations should be denied access to nuclear technology?
   b. What alternative sources would you recommend for Third-World countries to acquire instead of nuclear technology?

5. What is the future of solar power in developing countries? What are the obstacles to its implementation and use?

6. Discuss the status of solar power projects in the following nations:
   a. Indonesia
   b. Chad
   c. Gambia
   d. Mali
   e. Niger
   f. Senegal
   g. Guinea-Bissau
   h. Mauritania

Energy Technology	*(+)*	*(−)*
Solar		
Wind		
Hydroelectric		
Nuclear		
Biomass		

Harnessing the Wind

JAY JAYADEV

As the new world order takes shape, energy demand in developing countries is growing at an awe-inspiring rate. At present, China and India, with nearly 2 billion people (more than 30 percent of the world's population) consume less than one-tenth the energy of the United States (in which only about 4 percent reside). By the year 2020, economists predict that the energy consumption of a typical Chinese villager will grow by 85 percent and of his Indian counterpart by 145 percent, assuming a high-growth scenario. Even then, the per-capita energy used will be less than one-fifth the amount the average U.S. citizen consumes today.

This need for more energy could lead the developing countries to devour all cheap and available energy resources to feed the furnaces of rapid industrial growth. Such a prospect is alarming. According to projections made by the United Nations' International Energy Agency, based in Paris, these countries would by the year 2010 be producing more than half of the world's emissions of carbon dioxide, a large part of which would come from India and China alone.

Fortunately, developing countries may avoid the paths the developed world took earlier. A renewable energy technology, by now mature enough for wide-scale deployment, can contribute substantial power without spoiling the environment. The world can now feasibly reap energy from the wind.

Only fairly recently has wind power been recognized as the one renewable energy source that will be economically viable in the near future. The Electric Power Research

© 1995, Reprinted, with permission, from *IEEE Spectrum,* pp. 78–83, (Nov 1995)

Institute (EPRI), in Palo Alto, Calif., which represents cooperative R&D interests of the U.S. utility industry, stated in a 1992 report: "Alone among the alternative energy technologies, wind power offers utilities with good wind resources, pollution free electricity that is nearly cost-competitive with today's conventional sources."

Since then, several wind power projects from Antarctica and Mexico to India and Indonesia have shown how practical its use can be. Not only is the power supply reliable and easily maintained, but it can also be more competitive in the long run than traditional, more heavily polluting alternatives, as case studies will shortly show.

POSSIBLE PRICES

The market for wind power is alluring. A study entitled "International Wind Power Markets," conducted by Arthur D. Little in November 1994 for the American Wind Energy Association, Washington, D.C., estimated an astounding US $2 billion–$3 billion market by the year 2000. Moreover, wind-energy costs are expected to decline in the near future. A goal of groups such as the U.S. Department of Energy and EPRI is to reach $0.04/kWh by 2000, given 21-km/h winds. (See Figure 52.1.) This achievement would mean a 20 percent reduction in cost compared with today's best technology.

Until 1994, installed capacity of wind turbines in four large developing countries—India, Brazil, China, and Mexico—amounted to only 340 MW, the equivalent of the output of a medium-sized conventional power plant. But today the ongoing and projected wind capacity additions for

421

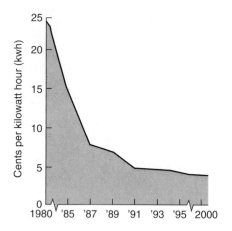

Figure 52.1
The cost of Producing Electricity from Wind Power
has Taken a Dive in Recent Years because of Better
Engineering, Materials, Manufacturing, and Power
Electronics
Courtesy of American Wind Energy Association, 1995.

the leading developing countries are quite impressive. (See Table 52.1) In addition, vigorous growth is seen for such countries as Brazil, the Philippines, Sri Lanka, and others.

Among the most ambitious plans are those of India, which has created a Ministry of Non-conventional Energy to lead activities there in solar and wind energy. The country wants to become the world's leading producer of wind energy (after the United States) within three years from now. By the end of this year, plants producing more than 250 MW of wind power are expected to be installed, and plans for a further 900 MW are in the works. The Indian ministry justifies the expense of the plants on the grounds

Table 52.1
Projected Additions to Wind Energy Capacity,
1994–2000

Country	Capacity, MW
India	700–1200
China	350–1000
Mexico	150–300
Chile	100–200
Argentina	100–150

Source: *Arthur D. Little*, 1994

that the capital costs of new wind energy projects are about the same as for new thermal power plants.

Another country with aggressive plans is China, which in the 1980s began to experiment with wind farms—that is, collections of large or medium-sized wind turbines working together as a power plant. The country announced at an energy conference in Beijing last June that it had set a goal of having available 1 GW of wind power by the year 2000. Meeting this objective would increase China's present wind power capacity of 29 MW by a factor of more than 30. The Government's interest in wind power was sparked by the prediction that pollution would surge if the country's attempt to expand its generating capacity by 10 GW annually overall relied heavily on coal-fired plants.

Additional interest in wind power worldwide has picked up with a steady stream of business announcements of contracts between Western wind energy companies and developing countries. For example, Advanced Wind Turbines, Seattle, Wash., is setting up a wind farm with 220 machines, each rated at 275 kW, in India, while Enercon of Aurich, Germany, is to deploy its wind turbines next year for the first large (130-MW) project in Mexico. For a municipality in Guatemala, New World Power Corp., Waitsfield, Vt., is building a 60-MW wind power plant, and for Inner Mongolia, FloWind Capital Corp., San Rafael, Calif., has signed an agreement with Inner Mongolia Electric Power Co. in China to form an $85 million joint venture for a 110-MW wind-power plant. And a Franco-Danish group has also won a bid to construct a 50-MW wind farm in Morocco.

WIND POWER BASICS

In all wind turbines, the wind rotor is what converts the wind's kinetic energy into rotational energy. The blades of this impeller are turned by the wind, and this rotation is coupled directly or through gears to a generator of some sort.

Wind turbines come in two basic designs: vertical-axis (or "egg-beater" style) and the more common horizontal-axis, used in both small and large wind plants. (Vertical-axis machines are less proven and less efficient at lower wind speeds, but seem to be promising for large-capacity applications.)

In the case of either horizontal- or vertical-axis turbines, good-quality electricity can be generated by two methods. In the well-established approach, induction generators are connected to the grid, resulting in the turbine's being mechanically constrained to a very narrow speed range and therefore producing a constant-frequency output suitable for pumping power to the grid. That means when high velocities or gusts of wind try to speed up the rotation of the shaft, the extra torque must be absorbed by the drivetrain and the tower. This approach causes torsional stresses on the mechanical parts and also wastes wind energy (although often the pitch of the rotor blades can be changed to reduce stress during high winds).

A newer approach, made feasible by advances in power semiconductors, is to allow the turbine speed to vary and then eliminate the fluctuations in amplitude and frequency electronically. (See "Technical Options," p. 428.) By letting the generator speed up during gusts and sustained high winds, this approach both increases the amount of electricity generated and reduces the stresses on the drivetrain.

In general, such technology as high-efficiency blades and generators, along with power electronics for variable-speed drives, has contributed to reducing the cost of wind energy. Wind turbines can produce electricity at $0.05/kWh at a good wind site, with an average wind speed of greater than 5 meters per second.

But in any situation, the fickleness of the wind must be considered. According to the American Wind Energy Association, most modern wind farms operate with a capacity factor of 25–35 percent—that is, the actual power produced over time as compared to the theoretical production of the turbines operating at maximum output 100 percent of the time. (Conventional power plants typically operate at 40–80 percent capacity factors.)

HARVESTING THE BREEZE

An advantage of wind power plants is that they are modular and can be sized to adjust to many situations and accommodate future growth in big or small increments, depending on the needs of a developing country.

The versatility of wind turbines includes both their size and use. Large turbines range from 1 to 5 MW, medium machines fall between 50 kW and 1 MW, while machines of less than 50 kW are considered small. All of these turbines can be used either as utility-connected machines or as independent, stand-alone systems. (See Figure 52.2.) Especially in the latter case, when stand-alone installations are used to provide power for isolated villages, socioeconomic considerations can be as important as technical ones—as will be

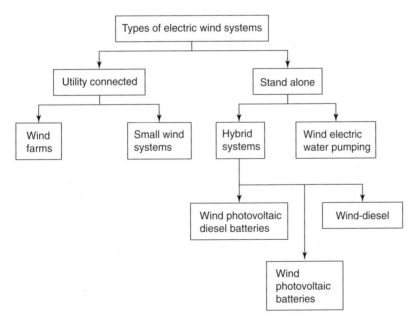

Figure 52.2
Two of the Many Ways in which Developing Countries Use Wind Turbines are as Utility-Connected Machines and as Independent, Stand-Alone Systems
Courtesy of Jay Jayadev.

shown later in illustrating how villagers in Xcalac, Mexico, updated their energy supply from wind power. Wind farms are in effect centralized power plants situated in windy areas. Because each farm individually assembles a large number of turbines, costs are often reduced.

The farms tie into a grid for their own benefit, too. Because the output of a wind turbine varies with the prevailing winds, the utility serves to absorb power as it becomes available (instead of requiring storage in batteries, for instance). A utility offering favorable buy-back rates is crucial to establishing the economic viability of wind farms.

Many of the wind farms now taking root in developing countries are connected to utilities and vary in power from a few megawatts to a few tens of megawatts. The technology they use is often identical to that of wind farms in California or Europe.

The way a wind farm is run in India, for example, illustrates some technical and economic aspects of the technology's application in developing countries. An arid plain in Gujarat is home to a joint venture supported by state and central governments but run by Vestas-Danish Wind Technology A/S, a large wind-energy company based in Lem, Denmark, and RBB Consultants, a local engineering company. The two concerns have teamed up to run a 10-MW Lamba wind farm, for which Vestas is supplying 50 horizontal-axis wind turbines churning out 220 kW each.

The Lamba project has produced not only energy but also jobs. Just as consequential as providing the turbines, Vestas has trained local engineers to install and maintain them. The project generated more than 200 new engineering and support jobs.

In 1992, two years after the farm's start-up, daily operation and maintenance were taken over by Indian companies and engineers, and the farm produced an impressive 13.4 GWh of electricity.

As in many developing countries where demand grows faster than generating capacity, networks in India are often over-loaded and subject to blackouts and burnouts. This situation adversely affects wind turbines directly connected with the grid because of reduced efficiency and capacity. Engineers are thus challenged by voltage and frequency variations unlike those in industrialized countries. Consequently, the Lamba wind farm's availability in the early years was around 90 percent. In contrast, California wind farms routinely obtain 98 percent availability levels.

Funds spent on Lamba have been modest. The cost of installing the 50 Vestas turbines was $1.7 million per megawatt. The average production cost, based on accumu-

lated data for four years, is 7.4 cents per kilowatt-hour. This amount is comparable to the life-cycle cost of conventional generators using fuels like diesel oil in India. (And in many developing countries, the wind is more to be depended on than are fuels.)

Another benefit gained from the Lamba project has been the prevention of pollution. In lauding the environmental benefits of wind power over fossil fuel power, Louis Ebler, marketing manager for Vestas, estimates that one 500-kW wind turbine at a good wind site in one year avoids the generation of 1.1 million kilograms of carbon dioxide, 82,000 kg of slag and fly ash, 9000 kg of sulfur dioxide, and 7000 kg of nitrogen oxides.

Similar findings were obtained in the United States. Paul Gipe, a well-known author on wind energy, estimates that if the 2.7 terawatt-hours of power generated annually by wind farms in California were instead generated by modern coal power plants, the atmosphere would be sullied by 2600 million kilograms of carbon dioxide, 13 million kilograms of NO_x and SO_2, 4.1 million kilograms of particulates, and 0.6 million kilogram of solid waste per year. (The estimates are based on data from the Department of Energy on pollution from modern coal power plants.)

END OF THE LINE

Developing countries are using wind turbines in tandem with the utility power supply in another way—by connecting the end of a transmission line to small wind systems to supply power for villages. Individual turbines can be built on windy spots close to the villages and the grid. So the villages' energy resource is increased without adding costly transmission lines.

Such situations are common in countries where the energy demand is growing, but generation and transmission capacity cannot keep up. (This happens in industrialized countries, too. For example, Pacific Gas and Electric Co. of California found that by reducing the loads at the end of heavily used lines, it could avoid having to construct new transmission capacity.)

Like wind farms, such utility-connected wind systems generally use an induction generator, which needs an external source of excitation to operate. The grid absorbs excess power and also serves as an excitation source and frequency reference. Reactive power correction can be made at the individual turbine site with static capacitors. Or a synchronous condenser can be connected at a strategically located distribution center. If the utility lines suffer from frequent blackouts or brownouts, the wind system owner should have a backup generator to ensure availability and meet the loads on demand.

STAND-ALONE TURBINES

To supply villages with electricity, it is often more feasible to give them an independent source of power than to invest in transmission lines to connect them to the utility grid. For the many villages that are farther than 3 km from the nearest transmission line, stand-alone wind systems are usually more economical.

Estimates by the World Bank, Washington, D.C., claim that as much as 40 percent of the world's population—some 2 billion people—live in villages that are not tied to a utility grid. When these villagers move to city areas, as they often do, the urban infrastructure is strained in many ways, not least in its ability to meet the demand for electricity.

The International Energy Agency predicts that if city populations in developing countries double in the next 15 years, energy demand will increase 45 percent, even if national income levels and population remain constant. But social and economic problems are caused by such urbanization, which might be prevented by more rural development.

Thus creating jobs at the rural level in developing countries—somewhat along the lines of village industries envisioned by leaders like Mohandas Gandhi and others—is imperative. These jobs in the modern context may be stuffing printed-circuit boards or producing textile products for export. The prerequisite is energy.

Stand-alone installations are appropriate technology for such applications. Because storage is a necessity in those isolated systems, they are technically more challenging to design and more expensive than are grid-connected generators that simply augment existing power infrastructures.

To boost overall reliability of the electricity supply, engineers can design systems that tie isolated wind machines to other types of generators. These hybrid systems often combine a wind turbine, a battery bank, and a diesel-powered backup generator.

Hybrids may electrify villages, powering lamps and small appliances, small industries, health clinics, schools, and community centers. If the loads are in tens of kilowatts, several small wind turbines, rated at 10–20 kW, and photovoltaic modules of a few kilowatts with a few hundred kilowatt-hours of batteries will suffice.

Photovoltaic panels are also employed in hybrid systems because of the complementary nature of solar and wind energy. But since wind power is lower in cost by a factor of five or so, it gets the lion's share of responsibility in hybrids. In many developing countries where biomass—for example, sugar cane husk or other crop residues—is already used to provide thermal energy, wind-biomass hybrids to furnish electricity should be attractive.

If job creation is an objective a hybrid system of 100 kW or more may be necessary. For sites with an average wind speed of just 4 meters per second, a village power system can be designed using medium-sized wind machines and diesel generators.

Coupling a wind machine to a diesel generator offers several benefits. It not only saves fuel, but is also more dependable because of its reliance on multiple power sources. Further, the hybrid arrangement allows for the expansion of loads (in contrast to the diesel operating alone) and the diesel generator lasts longer because it runs for fewer hours.

Even in a country as technologically advanced as Denmark, farmers often collectively own and operate a medium-sized wind machine of 100–250 kW. An intriguing possibility for a cluster of villages in developing countries is to share a power plant of a few hundred kilowatts.

A MEXICAN VILLAGE'S EXPERIENCE

Having a dependable power supply system can boost the demand for electricity. That experience was shared by Mexican villagers in Xcalac, a fishing village on the Yucatan peninsula that updated its power system with a hybrid wind system. Confident that they could rely on their power system, villagers bought more modern appliances and the village itself became a more attractive place to live in.

Xcalac is located about 50 km from the nearest town of Chetumal, which is where the public grid stops. The estimated cost of a power line from Chetumal to Xcalac was $3 million. But, as developers recognized, Xcalac has average wind speeds of 5.2 meters per second. So in 1991 the State of Quintana Roo decided to electrify the village with a centralized renewable hybrid-energy system, which was retrofitted to the existing diesel system.

Among the institutions participating in the construction, operation, and monitoring of the Xcalac project were Condumex, the Instituto de Investigaciones Eléctricas of Mexico, and New Mexico State University, as well as Bergey Windpower (a maker of small wind turbines) and Sandia National Laboratories.

Xcalac's prior system had proved to be problematic, to say the least. As is the case in many Latin American villages dependent on diesel electricity, power was limited to lights, television, and radio for only 4–6 hours during the evening. Village refrigerators and ice makers were powered by propane. Fuel was not always available, with the villagers sometimes appropriating the diesel oil for the community fishing vessel, leaving the electric generator high and dry.

Xcalac's hybrid solution consists of six 10-kW Bergey wind turbines, 11.2 kW of a total of 234 Siemens photovoltaic modules, 1738 A·h of 220-V GNB 6-7C23 deep-cycle flooded lead-acid batteries, and a 40-kW inverter built by Advanced Energy Systems, and the existing 125-kW Selmec diesel generator as backup.

The hybrid system produces up to 260 kWh per day, much more than was expected. Since 80 percent of the energy is produced by wind, system availability is dominated by whether the wind blows rather than the amount of sunshine or the diesel fuel supply.

During the project's monitoring period, the total energy provided to the village in October, when there is least wind, averaged 140 kWh per day, close to the predicted performance of 150 kWh per day. In high-wind months, the total energy provided averaged about 240 kWh per day. (See Figure 52.3.)

Wind turbines had a utilization factor that varied from 68 percent in the high winds of May through September to 97 percent during the doldrums of October through December, when the batteries reached full charge less

often. (Conversely, in high-wind periods, the utilization factor was lower because the batteries were often fully charged; excess wind power was therefore wasted.)

The total cost for the Xcalac system was about $450,000 (including shipping and installation)—less than one-sixth of the $3.2 million estimated by the public utility for grid extension. Xcalac proved that a total capital cost of $369,000 could result in an average of 77.3 MW-h of wind-generated energy per year, as was shown during the project's monitoring period. At reasonable interest rates, this amount would roughly translate into $0.40/kWh, a cost that is quite competitive with diesel or any other energy source for Xcalac.

By far the most serious problems raised by the Xcalac project were socio-economic concerns. Even though villagers now enjoy over 150 kWh per day of electricity, loads have grown, due in part to the return of villagers and the arrival of newcomers, both groups drawn by the availability of power.

Per-capita usage has risen, too. A load survey conducted in August 1993 by the Instituto de Investigaciones Eléctricas revealed two groups of consumers. One group had minimum facilities with low demand, such as a few lights, a radio, and a television set. The other group had acquired appliances such as refrigerators, freezers, and washing machines, using 10 times more energy than the first.

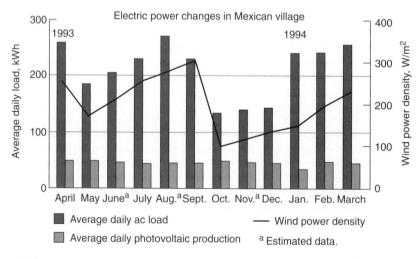

Figure 52.3
Even during the Doldrums of October to December, Wind Power was Able to Supply Enough Energy for the Entire Mexican Village of Xcalac
Courtesy of Sandia National Laboratories.

This huge growth has overloaded the Xcalac system. Consequently, a tariff system has been instituted in which the first 36 kWh used per month costs the consumer $0.34/kWh and thereafter the tariff doubles to $0.68/kWh. The money raised will pay three of the local people to operate the system and collect electricity bills. Enough funds are expected to be left over to buy fuel for the diesel engine, replace the battery bank when necessary, and maintain the overall system.

The experience at Xcalac could serve as a guide to village power planners. Still, as Ron Pate, a researcher at Sandia National Laboratories, Albuquerque, N.M., who studied the Xcalac system for a year, warned: a good program does not end after installation. It must have an ongoing institutional structure so that loads are managed, resources are generated for its operation and maintenance, and responsibilities are assigned for continued operation. If these tasks are not performed, as happens in many village projects, the program will fail.

INDEPENDENT OPTIONS

In designing stand-alone systems, engineers can choose from several options. Usually the systems operate in a variable-speed mode. To generate clean constant-amplitude, constant-frequency electricity, they therefore require a rectifier-inverter combination. For dc loads, of course, the rectifier is needed.

Generally, if the average load is greater than 25 kW or so, using a single medium-sized turbine rather than several small ones is better—as long as the logistics are practical. For example, if a village requires 100 kW, and there is good access to the site, a medium-sized machine would be a good choice. Energy from these machines costs a half to a third of that from an array of small machines.

Medium-sized machines have the additional advantages of technical support from stable companies, ease of obtaining spare parts, and so on. A number of villages might even cooperate to use the power from one medium-sized turbine located at the best wind site in the area.

If a medium-sized turbine is not practical, however, a number of small machines will do the trick. Among the 20 companies making smaller turbines are Northern Power Systems, Moretown, Vt., and Bergey Windpower Co., Norman, Okla.

Small wind turbines usually have alternators with permanent magnet excitation. They produce power according to wind speed, reaching maximum power at 12 meters per second. Modern small wind machines have achieved a high level of integration and simplicity, and have proved that they can work well in the field under extreme conditions. Bergey Windpower, for example, integrates the rotor and the hub into one component and the stator and the main frame into another, thus reducing the number of parts and boosting reliability.

WELL, WELL, WELL

A widespread application of wind power is pumping water. For hundreds of years, wind-driven pumps have helped to draw water from wells. Even in the first part of the century, windmills were a common sight in the rural United States.

Recent improvements in wind-powered pumping are a good omen for the developing world. In particular, the use of variable-voltage and -frequency electric turbines without a rectifier-inverter package makes electric pumping especially attractive and affordable.

With the new technology, developed by Bergey Windpower and the U.S. Department of Agriculture, the varying generator output is applied directly to the pump, whose load characteristics match well with such generators. Better yet, electric pumping decouples the location of the pump from its power supply so that the wind turbine may be located at the best and most practical site, such as a nearby hill.

The repercussions of these advances could be enormous. In an average year, the Indian state electricity boards (which oversee state-owned utilities) connect about 500,000 new water pumps to the grid, according to a report, "The Renewables Opportunity in India," by Shering Energy Associates, Princeton, N.J., 1994. During that year, in one state alone, the waiting list for grid extensions to existing wells exceeded half a million.

If the economics can be justified, the market potential for renewables for this application could be huge. A national photovoltaics program is already targeted to occupy 10 percent of the market, and it is a safe bet that wind electricity can supply another 10 percent wherever farms are near sites having an average wind speed of at least 4 meters per second.

IRRIGATION IN INDONESIA

Most of the farms in eastern Indonesia are so small (approximately 20,000 m^2) that 1.5-kW kerosene-fueled pumps have been used for water pumping. These pumps are inexpensive and readily available, but they are costly to operate and last less than five years.

TECHNICAL OPTIONS

Wind rotors in small turbines spin 100–300 or so revolutions per minute. Large turbines intended for grid inter-connection spin about 30–60 revolutions a minute and sometimes even more slowly.

In small systems, direct drive is preferred to geared drive for its higher reliability. Any standard, off-the-shelf generator can be used with a speed-increasing gear, but reliability is reduced. As a result, manufacturers of small turbine systems develop their own custom generators, and couple them directly to the wind rotor.

For example, in a machine developed by wind turbine manufacturer Bergey Windpower Co., Norman, Okla., a permanent-magnet ring is directly attached to the wind rotor, which envelops and rotates about the stator.

Large wind turbines connected to the grid commonly use induction machines that operate as generators running at higher than their synchronous speeds. The same machines are also used to start the wind turbine by running as induction motors, drawing power for this function from the grid.

Even though wind velocity is fickle, any power output for grid-connected machines needs to be of high quality. A grid-connected wind turbine can operate in a fixed- or variable-speed mode. In either case, however, a controlled-frequency output is required.

For fixed-speed systems that can be hooked up to a grid, maintaining a constant frequency is no problem. Such systems typically employ induction or synchronous machines connected directly to the grid. If induction machines are used, as is generally the case in the wind industry today, they are usually high-efficiency, low-slip machines. In fact, self-excitation of induction machines is possible in isolated systems, but it has not found wide application.

Variable-speed wind turbines have benefited from recent progress in power semiconductors—power MOSFETs and insulated-gate bipolar transistors, for example. These devices enable the turbines to deliver high-quality electricity. As a variable-speed system can absorb the torque transients better, it has the potential for reduced cost and increased reliability.

Currently, two generation schemes are preferred for variable-speed wind systems. One employs the popular induction motor speed control, using power electronics to operate the induction machine in the generating mode by varying voltage and frequency.

The other scheme uses a dc-excited alternator, whose varying output is rectified and then inverted to produce stable ac suitable for grid connection.

In medium-sized and large wind turbines, the controversy over running a turbine at constant speed or at variable speed persists. Nearly constant-speed systems are widely used because of their simplicity and the low cost of induction generators. But proponents of variable-speed systems claim that when the wind rotor speed is allowed to vary with the wind, the turbine can be operated at peak efficiency.

Another benefit of variable speed is that the mechanical stresses caused in structural elements by gusts and varying wind speed are diminished by letting the rotor follow the wind. The necessary power electronics, however, can be expensive—up to four or five times the cost of the generator.

—J.J.

To overcome this problem, Indonesia's Ministry of Public Works has installed a 1.5-kW wind-electric water pumping system supplied by Bergey Windpower at its demonstration farm in Oesao, 30 km east of Kupang on Timor Island. This proof-of-concept system, installed in October 1992, was supported by the Japanese Overseas Economic Cooperation Fund and the U.S. Agency for International Development.

Oesao irrigation calls for 150 m^3 of water per day pumped from a depth of 4–6 meters. On the basis of that requirement and available wind data, a 1.1-kW multi-stage, surface-mounted centrifugal pump was chosen, along with the 1.5-kW wind-electric system.

Performance predicated on scaled wind data yielded an average daily water delivery of 60–350 m^3, depending on the time of year. This variation was fully acceptable because the strongest winds blow (and the most water flows) during the dry season.

The wind resource in Timor has proven to be better than the initial data indicated, and the system has had nearly 100 percent availability since installation. Average energy production has been some 130–180 m^3 per day. So far, windless periods have not presented a big problem at this site. The soil holds enough water not to require the pumps to operate during calm periods. So no energy storage system is required.

Advantages of the wind-electric water-pumping system include its being nearly automatic, with less maintenance and lower replacement costs than the kerosene pumping systems needed. And, of course, it requires no fuel. While kerosene pumps must be replaced every two to four years, wind-electric systems require only about $25 worth of scheduled maintenance each year and have a design life of 30 years.

The equipment cost for the Desao was $7000. This amount is roughly six times the cost of a kerosene system designed to deliver the same amount of water. But these initial costs are more than offset by the reduced costs for fuel, operations, and maintenance. A value calculation comparing the two systems shows that wind-electric water pumping is about 50 percent less expensive than the kerosene systems over a life of 30 years.

FUTURE MANUFACTURERS?

In their recent book, *Power Surge,* Christopher Flavin and Nicholas Lenssen of the WorldWatch Institute, Washington, D.C., construct a scenario in which renewable energy systems play a major role in an energy revolution involving both the developed and developing countries. In this prediction, developing countries are seen as taking the lead in the utilization of renewables and even emerging as suppliers of renewable energy components.

Whether or not this comes true, at least part of the scenario may be unfolding. Already, a British company is setting up a joint venture with an Indian partner to develop and manufacture turbine blades in India.

On the horizon, wind power costs will further decline, and business is expected to accelerate because of technical advances in wind machines, cost-cutting improvements in manufacturing techniques, and environmental concerns. Moreover, wind power technology is straightforward—unlike solar cells, for example. It does not involve sophisticated processing of high-purity materials. Because of that advantage, developing countries, with their relatively cheap labor, might turn to building their own manufacturing capability in the technology.

QUESTIONS

1. List the benefits gained from the Lamba project.
2. Describe the advantages of the wind–electric system used in eastern Indonesia.
3. What benefits does the Xcalac hybrid system offer to the inhabitants of villages in Mexico?
4. Discuss the limitations of the wind-power systems.

53

Application of Telecommunications Technologies in Distance Learning

AHMED S. KHAN

ABSTRACT

Recent advances in telecommunications technologies (computer networks, satellite communications, fiber-optic systems, the Internet, and so forth) have transformed the modes of learning and teaching. The dissemination of knowledge is no longer confined to constraints of physical premises, with an instructor and a textbook, and it is no longer the only educational resource. The use of telecommunication technologies in distance learning overcomes the barriers of distance and time, and allows students in developed as well as in developing countries to learn in a synchronous or asynchronous manner. This paper presents an overview of various telecommunications technologies that are used in distance education. The discussion includes evolution of distance education, synchronous and asynchronous learning, and characteristics of broadcast television, instructional television fixed service (ITFS), microwave systems, satellite systems, direct broadcast satellite (DBS), cable systems, private fiber, and the Internet.

INTRODUCTION

Distance education or learning is a discipline that links people with information through a variety of technologies.[1] Distance learning has been practiced for over 100 years,

Reprinted with permission from The Annual Review of Communications, 1998. Copyright by the International Engineering Consortium (IEC).

primarily in the form of correspondence courses. These print-based courses solved the problem of geographic dispersal of students for specialized courses of instruction. In the late 1960s and early 1970s, emerging technologies afforded the development of a new generation of distance learning, aided by open universities, which combined television, radio, and telephone with print.

Today, advances in telecommunications technologies are again changing the face of distance learning. With the explosion of Internet tools over the last few years, distance learning has become more accessible to people whether they live in urban or rural areas. In this time of increasing technological advances, it is no surprise that college enrollment is on the rise. The increasing enrollment, coupled with yearly budget cut requests, opens the door for distance learning.

With the advent of distance-learning technologies, the dissemination of knowledge is no longer confined to the constraints of physical premises. Distance-learning technologies overcome the barrier of distance to allow face-to-face communication between students and teachers from different locations. With rapid technological change in developed countries, educational institutions are challenged with providing increased opportunities without increased budgets. In developing countries, the policy-makers face a dilemma: how to provide educational opportunities to an increased population while considering lack of resources. Distance education offers a solution to the problems of educators in developed as well as developing countries. In developed countries, there are three major motivations for an educational institution to incorporate distance learning in its program offerings:

- to compensate for the lack of specialist courses at its own institutions
- to supplement an existing curriculum to make it more enriched
- to compete with other institutions which offer distance-learning courses via the Internet

For developing countries, distance education offers a unique approach to promote literacy and enhance higher education in a cost-effective way.

MODES OF DISTANCE LEARNING

Distance learning can be employed in a synchronous as well as an asynchronous mode. In a synchronous mode, the student and teacher interact in real time, whereas in an asynchronous mode, the student learns at a convenient time and place.

Advances in computer and communication technologies have greatly contributed to asynchronous access. Asynchronous learning is time- and place-independent; learning takes place at the convenience and pace of the learner. A student can contact a college or teacher via e-mail or engage in discussion with a group through a conferencing system or bulletin board. A learner can participate interactively in a team project with other students that requires problem analysis, discussion, spreadsheet analysis, or report preparation through modern commercial groupware packages. Similarly, lectures can be transmitted through computers, videotape, or CD-ROM. Learning becomes a distributed activity, and participants in these distributed classes access resources and interact asynchronously, more or less at their own convenience.[2]

Asynchronous learning has been categorized into three different levels. They are on-campus, near campus, and far from campus. The on-campus level meets in the traditional classroom, but through computer labs students participate in asynchronous communication through listservs, bulletin boards, and the World Wide Web (WWW). The near campus level (50–60 miles from campus) requires students to meet on campus occasionally for tests, teacher consultation, or labs, but use the Internet or another electronic medium for lectures, class information, and communication. The far from campus level is the

"true" asynchronous delivery. This is where the student never sets foot on the campus, and the entire course is delivered through tools such as the Internet, videos, or CD-ROMs. All three levels meet the needs of different learners. One of the benefits of asynchronous learning is that it meets the needs of many "non-traditional" students who would normally not have access to an education due to distance or time constraint.[3]

From Britain to Thailand and Japan to South Africa, distance learning is an important part of national strategies to educate large numbers of people rapidly and efficiently. In the United States, educators are finding distance learning effective not only for outreach to new populations, but also as an important medium for new instructional models. Distance learning, used as a term associated with new technologies offering a full-fledged alternative to classroom education, got its biggest boost internationally with the founding of the British Open University (BOU) in 1969. BOU gained rapid visibility and recognition by broadcasting its video course components weekly throughout the United Kingdom on the BBC network.[4]

Table 53.1 lists the enrollment for distance learners seeking university education in various developed and developing countries. The addition of distance learners, who receive programming in elementary, secondary, training, and noncredit areas would almost quadruple these numbers.[5]

Many agree that distance learning is the fastest growing instructional pattern in the world. Technologies of delivery, particularly those related to telecommunications, have had a "greening" effect on the distance-learning enterprise. The potential to solve access, cost, time, place, and interactivity considerations that have plagued education since the beginning of time has never been greater.[6]

TELECOMMUNICATIONS TECHNOLOGIES

Telecommunications technologies provide opportunities for basic education as well as advanced education for those disadvantaged by time, distance, physical disability, and resources. Table 53.2 lists the evolution of telecommunication technologies. Table 53.3 lists the breakdown of voice, data, and video technologies used in distance education.[7]

Distance learning blends technological infrastructure and learning experiences. The experiences and applications that can be provided depend on the appropriate infrastructure. Current technologies have favored applications where

Table 53.1
Major Open Universities in Developed and Developing Countries

Institution	Country	Year Established	Enrollment
University of South Africa	South Africa	1951	50,000
Open University	United Kingdom	1969	50,000
Universidad Nacional	Spain	1972	83,000
Fernuniversitat	Germany	1974	37,000
Open University of Israel	Israel	1974	12,000
Allama Iqbal Open University	Pakistan	1974	150,000
Athabasca University	Canada	1975	10,000
Universidad Nacional Abierta	Venezuela	1977	29,000
Universidad Estatal a Distancia	Costa Rica	1977	11,000
Sukhothai Thammathirat OU	Thailand	1978	200,000
Central Radio and TV University	China	1978	1,000,000
Open University of Sri Lanka	Sri Lanka	1981	18,000
Open Universiteit	Netherlands	1981	33,000
Andrha Pradesh Open University	India	1981	41,000
Korean Air and Correspondence University	South Korea	1982	300,000
University of the Air of Japan	Japan	1983	22,000
Universitas Terbuka	Indonesia	1984	70,000
Indira Ghandi Open University	India	1986	30,000
National Open University of Taiwan	Taiwan	1986	48,000
Al-Quds Open University	Jordan	1986	not available
Universidade Aberta	Portugal	1988	3,800
Open University of Bangladesh	Bangladesh	1992	not available
Open University of Poland	Poland	Proposed	
Open University of France	France	Proposed	

a central source controls information flow. Emerging networked technologies allow for both centralization and decentralization. While centralized and decentralized approaches each have advantages and disadvantages, networked computer technologies and the distributed-learning applications they enable have not only practical benefits for information management but also attributes that are more congruent with newer educational paradigms based on cognitive approaches to learning.[8]

Broadcast Television

Broadcast TV involves the transmission of video and audio over standard VHF and UHF channels to reach a large number of sites in a limited geographic area (e.g., campus, metropolitan, county, multi-county, and so forth). Most educational programming broadcast over television does not allow real-time interaction with the television instruc-

tor but interactively can be designed into live or recorded telecourses using the telephone, computer networks, and Internet tools. This mode of transmission lacks security and confidentiality.

Instructional Television Fixed Service (ITFS)

ITFS refers to a band of microwave frequencies originally set aside by the Federal Communications Commission (FCC) in 1963 exclusively for the transmission of educational and cultural programming. ITFS is similar to broadcast television but is used in a more limited geographic area for simplex point-to-multi-point transmission. ITFS uses omni-directional microwave signals in the 2.5-GHz band to transmit standard 6-MHz video signals to remote locations. An ITFS network can serve as a stand-alone distance education-delivery system, transmitting locally orig-

Table 53.2
Evolution of Telecommunication Technologies

Year	Technological Development
1832	Telegraph
1875	Telephone
1895	Radio
1945	Audiotape
1953	Broadcast Television
1960	Videotape
1960	Audio Teleconferencing
1965	Cable Television
1975	Computer Assisted Instruction
1980	Audiographic teleconferencing
1980	Satellite Delivery
1980	Facsimile
1980	Videoconferencing
1984	Videodisc
1985	CD-ROM Compact Disc
1988	Compressed Video
1989	Multimedia
1990s	Lightwave systems, LANs, HDTV, Internet

Table 53.3
Telecommunication Technologies
in Distance Learning

	Technology
Voice	Telephone
	Radio
	Short-wave
	AM
	FM
	Audiotapes
Data	LANs
	WANs
	Internet
	Computer-assisted instructions (CAI)
	Computer-mediated education (CME)
	Computer-managed instruction (CMI)
Video	Pre-produced videotapes
	Digital video disc (DVD)
	Videoconferencing
	Broadcast television
	ITFS
	Satellite System
	DBS
	Internet

inated programming directly to local schools or cable companies for redistribution through their network. It offers limited or no interaction and a moderate level of security and confidentiality.

Microwave Systems

Terrestrial microwave systems require a line-of-sight transmission between the transmitter and the receiver site. Microwave systems allow simplex or half duplex/duplex point-to-point audio, data, and video transmission. There are two types of point-to-point microwave systems: short haul and long haul. Short-haul systems typically have a range of 5 to 15 miles, suitable for local communications between two schools or campuses. Long-haul systems typically have a range of up to 30 miles between repeaters, depending on transmitter power, terrain, dish size, and receiver sensitivity. Microwave systems offer a high level of interaction, security, and confidentiality.

Satellite Communications

Satellite communication involves the transmission of a broadcast signal from an earth station via an uplink signal

to a geosynchronous satellite, where the signal is processed by the transponder and sent, via a downlink signal, to a receiving dish antenna. Satellite communication is used to cover a large number of sites over a wide geographic area (e.g., statewide, nationwide, and continent-wide) for simplex point-to-multipoint audio and video transmission. Analog satellite systems offer low levels of security and confidentiality, and digital systems offer high levels of security and confidentiality. Satellite systems offer limited or no interaction.

Direct Broadcast Satellite (DBS)

These high-powered satellites transmit programming directly to the general public. The received dish antennas used in DBS systems are very small (<1 meter). DBS systems allow programmers to beam educational programming directly to homebound students, providing an alternative to over-the-air broadcast or cable television. DBS technology employs data-compression technologies that enhance the efficiency of video channels.

Televised education offers the following advantages:[9]

- Televised courses can actually enhance instruction by allowing instructors to present material in novel ways and thus to increase interest in it.

- Televised courses provide access to education for those who might not otherwise have it.

- Televised courses can help the instructor become a better teacher.

- Televised courses are often better organized and more highly developed than traditional courses.

- Televised courses offer an opportunity for faculty members to learn from their students as well as teach them.

Satellite communications is one of the most cost-effective modes to promote literacy and enhance higher education in developing countries. Table 53.4 lists the typical cost for leasing a video channel for satellites that provides large footprints in Asia and Africa.

Cable Systems

Cable-television systems use coaxial and fiber-optic cable to distribute video channels to local subscribers. Programming is received from local broadcast channels and national

Table 53.4
Typical Cost for Leasing a Video Channel for Satellites Providing Large Footprints in Asia and Africa

Satellite	Footprint Coverage Area	Typical Hourly Cost for Leasing a Video Channel
Panamsat	South Asia	$1850
TDRS-5	South Asia	$1680
Chinasat 1	Southeast Asia	$990
Asiasat 2	Southeast Asia	$990
Measat	Southeast Asia	$990
Palapa-B2	South Asia	$885
Intelsat-k	Western Africa	$2000
GESTAR-4	Western Africa	$2000
Inmarsat-3	Western Africa	$2000
Intelsat	South Africa	$1000
Orion 2	South Africa	$1000
Panamsat PAS3	South Africa	$1000

programming services at the cable "head end" and is sent out over the cable in a tree configuration. A cable head end can receive many types of signals, such as satellite or microwave transmission, which can then be retransmitted to schools over the normal cable system.

Cable-television systems are primarily one-way (simplex) broadcast (point-to-multipoint) type transmission systems. Many systems also have a limited number of reverse channels, providing some measure of two-way interactivity. There is a high-level inter-system security and confidentiality.

Private Fiber

Optical fiber is used for full-duplex point-to-point audio and video transmission. It allows a high degree of interaction and offers a high level of security and confidentiality. It also has the highest information-carrying capacity. The bandwidth for mono-mode step index fiber is 10–100 Gbps; multimode step index fiber is around 200 Mbps.

Public Telephone Service

Public telephone service is used over a limited to wide geographic area (local, regional, national, and international) for full-duplex point-to-point or point-to-multipoint audio transmission. It is used in conjunction with other technologies (television, satellite, etc.) to provide a feedback channel for students at remote sites to interact with the instructor and other sites. Table 53.5 compares the characteristics of distance-learning technologies.

The Internet

The Internet offers full-duplex point-to-point and point-to-multipoint audio, video, and data transmission over a wide geographic area (regional, national, and international). It allows a high degree of interaction. Generally, it has a low level of security and confidentiality but, with the use of data encryption, it has a high level of security and confidentiality.

With a computer, minimal software, and an Internet connection, people anywhere in the world can access online sites and programs. In addition, these programs can be accessed almost instantaneously through a user-friendly

Table 53.5
Comparison of Distance-Learning Technologies

Technology	Characteristics
Broadcast Television	Simplex mode of transmission One site to multi-site video and audio transmission (point-to-multipoint transmission) Limited or no interaction Public system Lack of security or confidentiality
Instructional Television Fixed Site (ITFS)	Similar to broadcast TV Semi-public system Moderate level of security and confidentiality
Broadband Cable	Similar to broadcast TV Public and private systems High degree of interaction possible High level of inter-system security and confidentiality
Microwave	Point-to-point audio and video transmission over limited area (one site to another) FDX mode of transmission High level of interaction Private system High level of security and confidentiality
Satellite	Simplex mode of transmission Point-to-multipoint audio and video transmission over wide area Limited or no interaction Private or public system Low level of security and confidentiality for older analog systems; higher level of security and confidentiality for digital systems
Private Fiber	Point-to-multipoint transmission over limited area FDX audio/video transmission High degree of interaction possible Private system High level of security and confidentiality
Public Switched Digital Service	Point-to-point and point-to-multipoint audio and video transmission over short and long distances Moderate to high degree of interaction possible Public system High level of security and confidentiality

Public Telephone Service	Point-to-point and point-to-multipoint audio transmission over short and
	   long distances. Used in conjunction with satellite and other simplex
	   technologies to provide a way for remote site to interact with instructor
	   and other sites

Public Packet Network Services	Point-to-point SPX transmission Used in conjunction with other technologies to provide a way for students to interact with the instructor

Source: http://www.oit.itd.umich.edu/reports/DistanceLearn/sect3.html

"point and click" interface, using almost any type of computer without producing and distributing printed materials, floppy disks, or CD-ROMs and with no training delivery cost. Furthermore, training can be delivered to a potentially unlimited audience, updates can be easily made on-line, and programs can link to a vast collection of other on-line resources.[10]

Although the Internet can be accessed worldwide, current access for many people is still limited. Even when access is available, bandwidth for accessing the Internet is frequently low. Low bandwidth severely limits capabilities such as on-line interactive multimedia training. Other disadvantages are that students need a moderate degree of computer literacy; sensitive/classified training requires additional security measures; and the high level of current Internet "hype" makes it difficult to determine what is practical for the near future.[11]

In addition to these technologies, a number of high-speed non-switched and switched lines could be incorporated in wide-area networking strategies in distance learning. Table 53.6 lists various switched and non-switched services provided by U.S. carriers.

Table 53.6
High-Speed Wide-Area Networks

Non-switched (Leased)	
Analog	4.8–19.2 kbps
Digital Data Service (DDS)	2.4–56 kbps
T-1	1.54 Mbps
T-3	44.736 Mbps
Frame Relay	1.54 Mbps–44.736 Mbps
Synchronous Optical Network (SONET)	51.84 Mbps–2.488 Gbps

Switched	
Dial-up/modem	1.2–28 kbps
X.25 packet switching	2.4–56 kbps
Integrated Services Digital Network (ISDN)	64 kbps–1.544 Mbps
Frame Relay	1.54 Mbps–44.736 Mbps
Switched multimegabit data service (SMDS)	1.54 Mbps, 44.736 Mbps
Asynchronous Transfer Mode (ATM)	25 Mbps–155 Mbps
B-ISDN (broadband ISDN)	155 Mbps, 600 Mbps

A recent report entitled "Distance Education in Higher Education Institutions" issued by the National Center for Education Statistics highlights the following facts:[12]

- By fall 1998, 90 percent of all institutions with 10,000 or more students and 85 percent of those with enrollments of 3,000 to 10,000 expect to offer at least some distance education courses. Among the smallest institutions (those with fewer than 3,000 students), only 44 percent plan to offer distance education courses.

- More than 750,000 students were enrolled in distance education courses in 1994–95. That year, some 3,430 students received degrees exclusively through distance education.

- More students enroll in distance education courses through public two-year institutions than through any other type of institutions.

In terms of the type of technology used to deliver courses, 57 percent of the institutions offering distance education courses in 1995 used two-way interactive video, and 52 percent used one-way prerecorded video (institutions frequently used more than one type of distance-learning technology). Twenty-four percent used two-way audio and one-way video, 14 percent used two-way on-line interactions, 11 percent used two-way audio, 10 percent used one-way audio, and 9 percent used one-way live way.

A key point to note, however, is that because of the many logistical factors involved, distance education courses typically require considerable investments of money and time. Distance education courses can be cost-effective when a large number of students take the courses over the long run. It is important to realize that such courses are costly to produce and deliver. The costs rise even more when the courses are presented with advanced one-way or two-way real-time video.[13]

The incorporation of new telecommunications technologies in distance education has abolished the barriers of location and time. However, the success of distance education in higher education depends on:[14]

- the availability of high-quality, full-motion video via the Internet

- the electronic accessibility at a reasonable cost of content currently available only in print format

- development and documentation of effective teaching/learning processes that rely on advanced technology

- the ability of distance learning to comply with an accreditation process for engineering education that is outcome-based

CONCLUSION

Distance education has come far since its humble beginnings as correspondence courses conducted by mail. To excel in the 21st century, higher education must undergo a paradigm shift. It must be transformed from an environment and culture that defines learning as a classroom process shaped by brick-and-mortar facilities and faculty-centered activities to an environment defined by "learner-centered" processes and shaped by telecommunications networks with universal access to subject content material, learner support services, and technology-literate resource

personnel.[15] The use of telecommunication technologies in distance learning will promote literacy and enhance higher education. Thus, distance learning will narrow the techno-economic gap between the developed and developing world of the 21st century.

NOTES

1. *www.fsu.edu/~lis/distance/brochure·html#program*
2. Mayadas, F. and P. Alfred (1996) *Alfred P. Sloan Foundation Home Page.* (Online). Available at *www.sloan.org/Education/ALN.new.htm.*
3. *Linking for Learning: A New Course for Education,* Office of Technology Assessment (OTA), U.S. Congress, Washington, D.C, 1989.

Satellite dish antennas are popping up on rooftops everywhere in developing countries. However, satellite broadcasts are used purely for entertainment purposes. The educational use of satellite technology is very limited. The use of satellite communications coupled with computer networks is not only a very cost-effective delivery mode to educate millions, but also has the potential to revolutionize the educational structure in developing countries in order to narrow the technoeconomic gap that exists between the First and the Third Worlds. The appropriate use of telecommunications technology is the key to solving problems of illiteracy, poverty, and disease in the developing world. Photo coutesy of Ahmed S. Khan.

4. Granger, D., "Open Universities: Closing the Distance to Learning, Change," *The Magazine of Higher Learning,* 22 (4), pp. 42–50.

5. Brown, B., and Y. Brown, "Distance Education Around the World," *Distance Education Strategies and Tools,* edited by Barry Willis, Educational Technology Publications, NJ, 1994.

6. IBID.

7. Willis, B., *Distance Education Strategies and Tools,* Educational Technology Publications, Englewood Cliffs, NJ, 1994.

8. Locatis, C. and M. Weisberg, "Distributed Learning and the Internet," *Contemporary Education,* vol. 68, no. 2, Winter 1997.

9. Richard, Larry G., "Lights, Camera, Teach," *ASEE Prism,* February 1997, p. 26.

10. Collis, B., *Tele-Learning in a Digital World: The Future of the Distance Learning,* International Thompson, London, 1996.

11. IBID.

12. *ASEE Prism,* February 1998.

13. Martin, B., P. Moskal, N. Foshee, and L. Morse, "So You Want to Develop a Distance Education Course?" *ASEE Prism,* February 1997, p. 18.

14. *ASEE Prism,* February 1997, p. 4.

15. Dubols, Jacques R., "Going the Distance, A National Distance Learning Initiative," *Adult Learning,* September/October 1996.

QUESTIONS

1. Define the following terms:
   a. Distance learning
   b. Asynchronous learning
   c. Open university
   d. DBS
   e. Synchronous learning

2. What telecommunications technologies could be used to educate millions of illiterate children and adults in the developing countries?

3. Compare the pros (+) and cons (−) of distance education.

Pros (+)	Cons (−)

MATTERS OF SCALE

The Plight of the Displaced

Number of people in the past decade displaced by infrastructure projects, such as road and dam construction	80 to 90 million
Number of people in the past decade left homeless by natural disasters including floods, earthquakes, hurricanes and landslides (based on an annual average over 25 years)	50 million
Number of people in 1981 who were landless or near-landless	938 million
Number of people expected to be landless or near-landless in 2000	1.24 billion
Number of people per square mile in the five boroughs of New York City	23,700
Number of people per square mile in Jakarta, Indonesia	130,000
Number of people per square mile in Lagos, Nigeria	143,000
Number of international refugees in the early 1960s	1 million
in the mid-1970s	3 million
in 1995	27 million
Number of refugees displaced within the borders of their own countries, in 1985	9.5 million
in 1995	20 million
Number of people currently living in coastal areas vulnerable to flooding from storm surges	46 million
Number of people living in vulnerable areas if global warming produces a 50-centimeter rise in sea level	92 million
Number of people living in vulnerable areas if global warming produces a 1-meter rise in sea level	118 million

Reprinted with permission from Worldwatch Institute, Washington, D.C. (*World Watch* Jan/Feb '96)

MATTERS OF SCALE

Human Health and the Future

Number of people now living within the reach of malaria-transmitting mosquitoes (potential transmission zone), worldwide	2.5 billion
Number of people who will live in the transmission zone by the latter half of the 21st century, after projected expansion of the transmission zone by global warming	4.8 billion
Average number of heat-related deaths in Atlanta each summer	78
Average number projected by 2050, assuming no change in the city's population size or age profile, but with projected global warming	293
Predicted decline in production of cereal grains (rice, corn, wheat, etc.) as a result of climate change (climate effects only) by 2060,	
in the developed countries	23.9 percent
in the developing countries	16.3 percent
Additional decline in grain production from physiological effects of CO_2,	
in developed countries	3.6 percent
in the developing countries	10.9 percent
Number of people at risk of hunger in 2060, without global warming	640 million
Number at risk with projected warming	680 to 940 million
Life expectancy in the most developed countries in 2000	79
Life expectancy in the least developed countries in 2000	42
Share of worldwide AIDS/HIV cases that are in developing countries	90 percent
Share of AIDS stories with non-U.S. settings in the U.S. media	4 percent

Reprinted with permission from the Worldwatch Institute, Washington, D.C. (*World Watch* September/October 1996)

WHO OWNS INDIGENOUS PEOPLES' DNA?

Aboriginal leaders have long struggled to control native lands. Now some have begun to worry that they may have to fight for control of native genes.

In August 1993, Pat Mooney, the President of Rural Advancement Foundation International (RAFI), a nonprofit concerned with third-world agriculture, discovered that the U.S. government was trying to patent a cell line derived from a 26-year-old Guaymi woman. The Guaymi people are native to western Panama. The cell line, a type of culture that can be maintained indefinitely, came from a blood sample obtained by a researcher from the U.S. National Institutes of Health in 1990. The application claimed that the cell line might prove useful for the treatment of the Human T-lymphotropic virus, or HTLV, which is associated with a form of leukemia and a degenerative nerve disease.

RAFI notified Isidro Acosta, President of the Guaymi General Congress, who demanded that the United States withdraw its claim and repatriate the cell line. Acosta also appealed to the General Agreement on Tariffs and Trade, and to an intergovernmental meeting on the Rio Biodiversity Convention. But GATT does not forbid the patenting of human material, and Acosta's case before the Biodiversity Convention fared no better. The convention does provide for sovereign rights over genetic resources, but the meeting did not rule on whether the Guaymi cell line came within its jurisdiction. As a growing number of nongovernmental organizations voiced their disapproval, however, the United States dropped its patent claim last November.

The story might have ended there had not a European researcher uncovered two similar claims in January of this year. Miges Baumann, an official at Swissaid, a Swiss NGO that supports rural initiatives in developing countries, discovered that the U.S. government had filed applications on a cell line derived from the Hagahai people of Papua New Guinea, and another from the Solomon Islanders. These lines might also prove useful for treating HTLV. Baumann's discovery came as a shock to the governments concerned but despite their protests, the U.S. has refused to withdraw the applications. In a letter dated March 3, 1994, Ron Brown, the U.S. Secretary of Commerce, explained the U.S. position to a Solomon Island official. "Under our laws, as well as those of many other countries," Brown wrote, "subject matter relating to human cells is patentable and there is no provision for considerations relating to the source of the cells that may be the subject of a patent application."

Patenting indigenous peoples' genes invites an obvious comparison with the patenting of the developing world's other biological resources, and native leaders have tended to take a dim view of the entire trend. "I never imagined people would patent plants and animals. It's fundamentally immoral, contrary to the Guaymi view of nature," said Acosta, who considers the patenting of human material a violation of "our deepest sense of morality."

But the rapid growth of biotechnology is driving a boom in human patents that may prove difficult to resist. The patenters are looking for genes that could be used to produce substances with commercial potential, usually for treating a disease. To patent a "product of nature," patent laws generally require some degree of human alteration. But in the United States, the simple act of isolating a DNA sequence removes it from nature, as far as the law is concerned.

The accessibility of the patent has fueled a growing commercial interest in the field. Companies that prospect in the human genome use a highly automated process called sequencing to decode bits of DNA from large numbers of samples. One company, Human Genome Sciences, of Rockville, Maryland, is reported to have sequenced over 200,000 chunks of DNA thus far. Patent claims may follow if the sequences obtained look novel—and in some cases, even if they don't. In one of the more spectacular instances of "driftnet patenting," as critics call the practice, Incyte Pharmaceuticals, of Palo Alto, California, filed claims on 40,000 sequences.

Observers say it's a good bet that other applications on indigenous DNA have already been filed. "I'm not aware of any others," says Hope Shand, RAFI's Research Director, "but it would surprise me if there weren't any more of them."

Reprinted with permission from Worldwatch Institute, Washington, D.C. (*World Watch* November/December 1994)

Conclusion

No great improvement in the lot of mankind are possible until a great change takes place in fundamental constitution of their modes of thought.

JOHN STUART MILL, PHILOSOPHER

The world may have become a global village as a result of technological advancements, but billions of people in the developing countries will continue to live with poverty, disease, and illiteracy. Changes in the Third World will occur only when both the intrinsic and extrinsic factors inhibiting its development are addressed. The effects of extrinsic factors (e.g. political and economic instability caused by international financial institutions, and economic exploitation of resources by the multi-national companies) can be rectified only when a great change takes place in the modes of thought of the policy makers of the developed world. These leaders must begin to believe that all people are created equal, no one is superior to another and treating people with equality and justice is the key to solving man-made economic, political and environmental dilemmas that the world faces today.

And to address the intrinsic factors (e.g. illiteracy, poverty, social injustice, corruption, lack of resources, population pressures, and power-hungry political and military elite) the key is education. Education is a great equalizer for changing the socio-economic conditions. It could provide solutions to many of the problems of developing countries and set them free from economic bondage, autocratic rule, poverty and disease. The developing nations should give a higher priority to educating their people. By incorporating innovative telecommunications technologies, the developing countries could increase their literacy rates and therefore produce a humanistic techno-elite that could then promote social justice, economic growth, and a sustainable environment, and thus narrow the technoeconomic gap with the developed countries.

Throughout this book, a spectrum of issues related to a wide array of topics (e.g. energy, ecology, population, war and technology, social responsibility, health and technology, and the technology and the Third World) has been presented. The next and final part discusses the state of technologies of the future.

INTERNET EXERCISE

1. Visit NASA's Multimedia Gallery at http://www.nasa.gov. (The site contains a large collection of photos of earth, taken from space by various space shuttle missions, depicting the impact of various human interactions, such as deforestation, desertification, pollution, oil well fires, soil erosion, mining, urbanization, irrigation, and oil drilling.) Complete the following table by listing the impact of human interaction in the developed and developing countries listed in the left-hand column.

Country	Impact of Human Interaction
Argentina	
Australia	
Bangladesh	
Brazil	
Bolivia	
Cameroon	
Canada	
China	
Ethiopia	
Egypt	
India	
Indonesia	
Japan	
Kuwait	
Libya	
Madagascar	
Malaysia	
Mexico	
Nepal	
Panama	
Philippines	
Sudan	
Tajikistan	
Tanzania	
Turkey	
USA	
Venezuela	
Vietnam	

2. The Chinese government is building an enormous dam—The Three Gorges Dam—on the Yanatze River to control flooding and to transform its hydropower into electric power. The project is expected to be completed in 2009 to have a power-generating capacity of 18,200 megawatts. This enormous capacity will enable China to move into the 21st century with a hydropower bang.

    Use any of the Internet search engines (e.g., Alta Vista, Yahoo, Infoseek, etc.) to research the following questions:
    a. How many people will be displaced due to The Three Gorges Dam project?
    b. How many towns and villages will be resettled?
    c. What are the environmental costs of this project?
    d. What are the benefits and the drawbacks of this project?

3. Use any of the Internet search engines (e.g., Alta Vista, Yahoo, Infoseek, etc.) to research the following topics:
    a. Define the following terms:
        i. GNP
        ii. GDP
    b. Compare the per-capita GNP for the following groups of developed and developing countries:

Developed Country	Per-Capita GNP	Developing Country	Per-Capita GNP
United States		Afghanistan	
France		Bangladesh	
United Kingdom		China	
Japan		Ethiopia	
Australia		Ghana	
Norway		India	
Sweden		Indonesia	
Canada		Kenya	
Switzerland		Malaysia	
Italy		Pakistan	
Singapore		Venezuela	

c. Describe the factors that are responsible for the wide gap between the per-capita GNP of developed and developing nations.
d. List five benefits of recycling.
e. Compare the following groups of developed and developing countries in terms of their recycling efforts:

Developed Country	Garbage Produced Per Year	% of Garbage Recycled	Developing Country	Garbage Produced Per Year	% of Garbage Recycled
United States			Bangladesh		
France			China		
United Kingdom			Chile		
Japan			Ethiopia		
Australia			Ghana		
Norway			India		
Sweden			Indonesia		
Canada			Malaysia		
Switzerland			Pakistan		
Italy			Philippines		
Singapore			Thailand		

f. What strategies are being used at personal, national, and international levels to promote recycling of this garbage?

Personal Level	National Level	International Level

g. Define the following terms:
   i. Photovoltaic
   ii. Inverters
h. Determine the solar (photovoltaic) power–generating capacity for the following developed and developing countries:

Developed Country	Solar Power–Generating Capacity (megawatts)	Developing Countries	Solar Power–Generating Capacity (megawatts)
United States		Bangladesh	
France		China	
Japan		Ethiopia	
Australia		Ghana	
Norway		India	
Sweden		Indonesia	
Switzerland		Kenya	
Italy		Malaysia	
Singapore		Pakistan	
United Kingdom		Venezuela	
Germany		Zimbabwe	

i. Define the following terms:
   i. kWh
   ii. MOSFET transistor
   iii. Bipolar transistor
   iv. Kinetic energy

j. Determine the wind power–generating capacity for the following developed and developing countries:

Developed Country	Wind Power–Generating Capacity (Megawatts)	Developing Countries	Wind Power–Generating Capacity (Megawatts)
United States		Bangladesh	
France		China	
Japan		Ethiopia	
Australia		Ghana	
Norway		India	
Sweden		Indonesia	
Switzerland		Kenya	
Italy		Malaysia	
Singapore		Pakistan	
United Kingdom		Venezuela	
Germany		Zimbabwe	

A Bridge to Your 21st Century Understanding

1. The history of the 20th century reveals that the First World has always imposed its will via its technological might on the Third World in dealing with the economic and political problems. (a) Do you believe that the level of confrontation between the First and Third World has increased during the last decade of the 20th century? Explain your answer. (b) How could the conflicts between the Third World and the First World be resolved in order to achieve a win–win situation in the 21st century?

The third-world predicaments

Lack of technology

Political interference by superpowers

Economic instability caused by World Bank and IMF debt

Economic exploitation of resources by multinational corporations

Illiteracy

Poverty

Social injustice

Corruption

Lack of resources

Power-hungry elite

Population pressures

The third world

2. What actions must be taken by the developed countries as well as the developing countries to improve the human suffering in the Third World?

3. What kind of relationship (cooperation vs. confrontation) do you envision will emerge between the First and the Third World in the 21st century? Explain your answer.

USEFUL WEB SITES

URL	Site Description
http://www.worldbank.org	World Bank
http://www.un.org	United Nations
http://www.worldwatch.org	World Watch Institute
http://www.envirolink.org	Environment-related links
http://www.igc.apc.org/worldviews	
http://www.sipri.se	Military expenditure statistics
http://www.citinet.com	Country information database
http://www.sunsite.unc.edu/lunarbin/worldpop	World population estimates
http://www.undp.org	United Nations Development Program
http://www.nrel.gov/business/international	Alternative energy sources
http://www.lib.umich.edu/libhome/Documents.center stats.html	Statistical resources on the Web
http://www.worldvillage.org	What if the world were a village?

PART

VIII

Technology of the Future

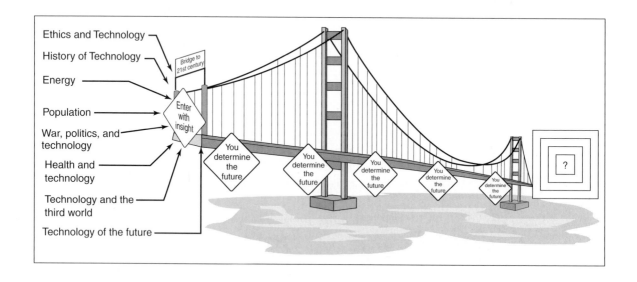

61 *Future Ecology: Is the Kyoto Climate Agreement Still Alive Due to the Bonn, 2001 Meeting?*

62 *Future: Population Growth Estimates*

OBJECTIVES

This chapter will help you to:

- Realize future impacts in society from the use and growth of today's technologies
- Review specific new technologies and understand their growth implications
- Discern ethical considerations in the use and expansion of present and future-oriented technologies
- Understand the changing trends in technology and their implications to the work-place, education, world markets, and global and local implications
- Realize the strength and potential of planned future technological strategies
- Appreciate the importance and potential contributions of individual awareness, ethical action to the planning of the future.

INTRODUCTION

As man proceeds into the future with unprecedented technological powers, it is time to pause and ponder the technological and ethical issues that bridge the past to the present, and the present to the future. The decisions made today will determine the directions of

Photo courtesy of Ahmed S. Khan

decision making into the new century. These decisions will carry implications of historical, scientific, ecological, and sociological significance as man moves rapidly through space and time. Part VIII presents a time and space approach to our examination of technology, enabling the reader to reflect on the issues of the previous eight parts and project their future development and applications to the future. The purpose of this approach is to refocus and converge the thoughts across the "bridge of now" so that they guide us to the "bridge of the 21st century."

This part begins by crossing over the bridge into the 21st century and envisioning and creating a preferred future. Specific case studies examine new technological applications that will affect the medical, military, and ecological aspects of our lives. Chapters also focus on predictions of future developments. In "Creating the Future," Norman Meyers dicusses ways to evaluate technology and its "multifarious impacts" in the new century. In "Trends Now Changing the World" authors Marvin Cetron and Owen Davies further predict specific trends and developments in economics, population, information industries, values and concerns, consumerism, energy and the environment. The authors then discuss the implications of these predictions to society, the job market, and to the future in general. This selection is followed by "Predictions: Technology of the New Century and the New Millennium," which gives a broad overview of the major growth technologies and their associated industries of the 21st century. These first three selections then focus and paint a future horizon and perspective as to the future's values, needs, predictions and technologies.

Part VIII turns to exploring specific technologies in medicine and the military, and concludes by examining the progress of global future ecology efforts and re-examining population growth estimates. "Telemedicine: The Health System of Tomorrow" by Thomas Blanton and David Balch explores the use of telemedicine in prisons, rural settings, and other areas where medical treatment has been sporadic because doctors cannot see patients on a regular basis. Even though telemedicine has been around since the late 1950s and has been a theme of science fiction films, it has only recently emerged as a viable way to reach out to medically underserved areas.

Ellison C. Urban explains in "The Information Warrior" how the world of the military will be transformed in the next century. Laser rangefinders, live videos of battle scenes, body-hugging computers, computer-generated maps, and humionics will drastically change the battlefield. Predictions for the new century and the new millennium tie together many issues that have been presented in this text. These predictions relate to the part that will be played by hyperlinks, microprocessor miniaturization, optical networks, gene alteration, artificial intelligence, and nanotechnology in the 21st century.

The Technology of the Future section comes to a close with discussion of new World Summits, Agenda 21 and other world recovery and ecological efforts and perspectives. This book concludes with commentaries from each of the authors, describing our own views of how to interpret our steps into the future as we walk across our very significant bridge, the bridge of the 21st century.

Photo courtesy of
Barb Eichler.

Creating the Future

Norman Myers

Editor's note: "Creating the Future" was written in 1990, and the goals were constructed for the year 2000. However, these goals and their relevancy remain urgent and provide vital priorities and goals to construct our future for the following decade FROM the year 2000!

The future can still be *our* future. We can possess the future, provided we do not let it possess us. All we have to do is to choose. Or rather, to choose to choose.

To do that we shall need to keep an eye better fixed on the long-term future. This is hard, since we have grown adapted to thinking about the short-term future. When the marketplace decrees an interest rate of 10 percent, investors must recoup their investment within seven years, which is akin to saying there is no future to bother about thereafter. Our political systems proclaim the future extends only till the next election. To enable us to cater for the future beyond this foreshortened perspective, we shall have to engage in much restructuring of society. Otherwise we run the risk of an unguided future descending on us, a future with upheaval abounding. Make no mistake: for the first time in human history, the next generation could find itself worse off than the previous one to an extent without parallel. A worst-case outcome—a prospect far from impossible—would entail the end of civilization as we know it. The only way to avoid that is to reconstruct our civilization from top to bottom.

Many of our options are already foreclosed through our penchant for muddling through. As a result, we face an immediate bottleneck of restricted choices. Much of Part VIII presents ways for us to squeeze through this bottleneck. We look at measures to contain the greenhouse effect, to

move to new energy paths, to reverse biomass shifts, to halt deforestation, to defuse the population bomb, to refashion our economies, to promote "soft" technologies, to build an organic agriculture, to establish green governance, to devise planetary medicine, to foster dynamic peace, and to undertake the many other prerequisites of a sustainable world.

Most of all, these changes in our outer world depend on still more revolutionary changes in our inner world. So the concluding pages look at the greatest transitions of all: from knowledge to wisdom, and from constant chatter to a readiness to listen—to listen to each other, to ourselves, and to Gaia.

We shall gain the future we deserve. Part VIII points the way, a way with portents, pitfalls, and hopes. Whatever tough times lie immediately ahead, this book, like the expansive promise of the longer-term future, is for optimists.

TOWARD 2000

As we head toward the new millennium, a number of organizations have been making ambitious plans. Well might they be ambitious: only by thinking a great deal bigger than we have in the past will we be ready to tackle the challenges ahead.

Health for All

This World Health Organization plan aims to: counter the malnutrition suffered by at least 1 billion people; supply primary health care to everyone; ensure life expectancy

From THE GAIA ATLAS OF FUTURE WORLDS by Norman Myers. Copyright © 1990 by Gaia Books Ltd, London. Used by permission of Doubleday, a division of Bantam-Doubleday Dell Publishing Group, Inc.

reaches 60 years; cut infant mortality to 50 per 1000 live births; ensure all infants and children enjoy proper weight (half of developing-world children are poorly nourished); and supply safe water within 15 minutes' walk. All this is to be accomplished by the year 2000. Just to provide primary health care for all would cost $50 billion a year, and safe water plus sanitation would cost $30 billion a year. The plan also proposes much greater attention to the greying communities in the developed world.

Education for All

This UNESCO plan aims to supply basic education for every person on Earth by the year 2000. Today, 105 million developing-world children enjoy no schooling at all, and more than 900 million adults, or one out of every four, are illiterate. Cost: around $5 billion a year.

Population 2000

According to the International Forum on Population in the 21st Century, convened in late 1989 by the UN Fund for Population Activities, the aim is to "hold the line" on the medium population projection of 6.25 billion people by the year 2000. To achieve this, we need to: reduce developing-world fertility to 3.2 children per family; increase developing-world use of contraceptives to 56%; upgrade the status of women, especially their literacy; and achieve universal enrollment of girls in primary schools. In order to accomplish just the first two items, we shall need to increase the budget for family-planning services in developing countries to $9 billion annually, or twice as much as today.

Anit-Desertification Action Plan

The United Nations' plan seeks to halt the spread of human-made deserts, now threatening one-third of the Earth's land surface. There has been next to no progress on this issue since it was first formulated in 1978, due to government indifference. Yet it would be an unusually sound investment at $6 billion a year for 20 years, against the continuing costs of inaction worth $32 billion a year in agricultural losses alone. In 1986 the world spent more than $100 billion in drought-relief operations that saved lives but did nothing to halt further desertification.

Tropical Forestry Action Plan

The plan, already in action, seeks to slow deforestation by supplying: tree farms for commercial timber; village woodlots and agroforestry plots for fuelwood; more protected areas for threatened species; more research and training, plus public education; and better management all round. Cost: $1.6 billion per year. But the plan does next to nothing to address the main source of deforestation, the slash-and-burn cultivator, who reflects an array of non-forestry problems such as population growth, pervasive poverty, maldistribution of existing farmlands, lack of agrarian reform, and hopelessly inadequate development for rural communities generally.

Climate Change Programme

Focused on global warming, and undertaken by the Intergovernmental Panel on Climate Change leading up to a proposed global-warming treaty, the Programme proposes to stabilize the concentration of greenhouse gases in the atmosphere by cutting emissions of carbon dioxide by 80%, of methane by 15–20%, of nitrous oxide by 70–80%, and of CFCs and related HCFCs by 70–85%. In the case of carbon dioxide, which accounts for roughly half the greenhouse effect, developed nations would have to reduce their fossil-fuel burning by 20% by 2005, by 50% by 2015, and by 75% by 2030. Developing nations would have to stabilize their emissions by 2010, allowing them to rise to no more than double today's levels as they continue to industrialize (developing nations would also have to halt deforestation by 2010 and engage in grand-scale tree planting). All this would mean that global emissions of greenhouse gases would fall to one-quarter of their present levels by 2050, limiting the rate of warming to 0.1° C per decade, and eventually holding it at 2° C above today's average temperatures—still a highly disruptive outcome.

Global Change Programme

Formerly known as the International Geosphere-Biosphere Programme, this research effort aims to provide the information we need to assess the future of the planet for the next 100 years, with emphasis on such areas as biogeochemical cycles, the upper layer of the oceans, Earth's soil stocks, and the sun's radiation. With an operational phase beginning in the early 1990s, the plan will last for ten years.

Beyond 2000?

As worthy as all these organizations' plans are, they all suffer from a serious shortcoming. They are targetted at the year 2000: what about longer-term goals, which must be established *now* if we are to avoid much costlier efforts later on, with much less prospect of success? Regrettably the plans reflect the timid and short-term attitudes of the governments that approve the strategies. It is not in the nature of governments to see beyond the end of their noses, let alone beyond the end of the decade.

GREENHOUSE TACTICS: OPTIONS FOR RESPONSE

The greatest environmental upheaval of all, the greenhouse effect, offers much scope for positive response. While it is not possible to tame the problem entirely—there is already too much change in the pipeline—there is still time to slow it, perhaps even to stabilize it, and buy ourselves precious decades in which to devise longer-term answers. But this will require immediate and vigorous action on a broad front, plus international collaboration and individual initiatives on an unprecedented scale.

First and foremost we need to consume less energy, through energy conservation and more efficient use of energy. In fact these two strategies represent our best energy source: from 1973 to 1978 some 95 percent of all new energy supplies in Europe came from more efficient use of available supplies—an amount 20 times more than from all other new sources of energy combined.

Since the oil crises of 1973 and 1979, most of the fuel-guzzling industrialized countries have expanded their economies by a full 30 per cent while actually cutting back on total energy consumption. In 1985 came a third oil shock, this one of plunging oil prices, whereupon we quickly reverted to our energy-wasting ways.

But the energy-saving technologies are still there, waiting to be mobilized (see below). And at the same time as we cut back on fossil-fuel burning, so we reduce acid rain, among many other forms of environmental pollution.

Burning fossil fuel for energy, however, is only about one-third of the greenhouse problem. Another source is agriculture: nitrous oxide from nitrogen-based fertilizers; methane from rice paddies and ruminant livestock. With more mouths to feed around the world, we cannot allow food production to plateau—though we could do it differently and more productively with genetically engineered breeds. A halt to tropical deforestation would prevent at least 2.4 billion tons of carbon from entering the atmosphere each year, while reforestation would soak up carbon as well as restore vital watershed functions among many other purposes.

Through multiple linkages, then, the greenhouse problem reflects the myriad ways we all live. It will only be through myriad shifts in economic sectors and lifestyles alike that we shall get on top of the problem.

Public Perception

Look out of a window and you view a world in the thrall of climatic upheaval. Although nothing can be seen, the world is undergoing an environmental shift of a type and scale to rival a geological cataclysm—and one of the most rapid ever to overtake the Earth. To confront it we need a parallel change in our inner world, our world of perception and understanding.

Changing Energy Policies

To mitigate the effects of the greenhouse we must reduce fossil-fuel burning. In the short term the greatest savings will come from buildings and products that use energy more efficiently. Many U.S. electricity utilities offer their customers energy surveys, rebates on energy-efficient appliances, and loans to finance energy-saving improvements. In the long term we need to replace fossil fuels with environmentally benign renewables, such as wind, wave, and solar. As for nuclear power, and leaving aside the unacceptable risks in its use, plus its uncompetitive price in the marketplace, to replace the world's coal-burning stations would entail building one nuclear plant every three days for the next 36 years at a cost of $150 billion annually.

A Climate Convention

A worldwide convention will need to decide how much greenhouse effect we are prepared to live with and the degree of remedial action to be taken, and by whom, in order to reduce the problem. This will mean a cap on greenhouse-gas emissions by all nations. But how hefty a cap for such disparate cases as Britain and Bangladesh, the U.S. and Brazil, Germany and China, the Soviet Union, and India? Should the industrialized nations indemnify the industrializing nations in order to safeguard the climate of all nations? Should there be special dispensations on the part of the entire community of nations to help those worst affected? The questions multiply and ramify, and we have no precedents to guide us.

Rethinking Agricultural Practices

Population growth and the intensification of food production are likely to set limits to reductions in certain greenhouse gases. Actively reducing the extent of intensive cattle-rearing operations would curtail some methane production; reducing fertilizer use, or implementing more efficient means of fertilizer application would do the same for nitrous oxide. An even more important answer lies with new varieties of crops that require less fertilizer.

DYNAMIC PEACE

Just as health is more than the absence of disease, so peace is much more than the absence of war. It is a state of dynamic stability that has to be actively maintained. It is an end to violence among human beings and, by extension, to the planet. When this relationship is understood and accepted and when basic needs are met, there exists the opportunity for positive peace. While many people talk about peace building, they rarely specify what it means. Not that they have had much chance to practice it, since peace-building activities have not received even 1 percent of 1 percent of what has been spent on war activities.

First and foremost, peace on Earth means peace with the Earth. We are engaged in World War III, a war against the planet, a war that is no contest. We must negotiate terms of surrender to the new environmental dictates, recognizing that victory over the Earth would be a no-win outcome. In turn, this postulates a new form of security, environmental security. As past president Gorbachev was one of the first to point out, the threat from the sky is no longer missiles but climate change. Stealth bombers cannot be launched against this uniquely threatening adversary—just as tanks cannot be mobilized to counter rising sea levels, nor troops dispatched to block the advance of the encroaching desert.

So peace with the Earth implies peace with each other, a peace comprising three essential components: relations of harmony among people in society; co-operation for the common good; and justice based on the concept of equity. Much environmental ruin derives from inequity. Economic and political imbalance drives the marginal Third Worlder into marginal environments—ones that are too dry, too wet, too steep for sustainable livelihoods, and where vast environmental damage is caused through desertification, deforestation and soil erosion.

To achieve dynamic peace at global level, there must be a submerging of the narrow interests of individual nations in an effort to conduct international affairs in a manner that befits a global community. But unless it is based on general consensus and it is environmentally and economically sound, there is scant hope of success.

Green Security

With the thawing of the Cold War, some of the erstwhile military outlays could be directed toward building a more secure world. According to Robert McNamara and other experts, NATO countries could soon release $175 billion a year from military budgets. Soviet leaders indicate they could release perhaps $100 billion a year, funds needed to rebuild the Soviet economy. According to Lester Brown, President of the Worldwatch Institute, to restore environmental security would cost: protecting topsoil around $9 billion in 1991; restoring forests $3 billion; halting the spread of deserts $4 billion; supplying clean water $30 billion; raising energy efficiency $10 billion; developing renewable energy $5 billion; slowing population growth $18 billion; and retiring developing world debt $30 billion. This total of $109 billion would need to rise to $170 billion a year by 2000—still no more than eight weeks' military spending at late 1980s levels.

The Role of Nonviolence

Nuclear power has forced us to contemplate both extinction and new forms of conflict resolution. The latter could soon become much more common. Nonviolence however goes far beyond ensuring peace by reducing military budgets or halting environmental destruction. As well as the weapons, we must also eliminate the "mind" that brought them into being and contemplated their use.

"Vulnerable" to Peace

The arms race has tied the economies of many nations to the military. Now there is a fear that an outbreak of peace will take jobs, shareholders' dividends too. To counter this, communities are developing "economic conversion projects." The objective is to retrain workers and adapt factories to civilian production. To this end, citizens, workers, and managers plan their future with local politicians, technocrats, and economists. In the US, 70 cities have passed

"Jobs with Peace" initiatives and are beginning the conversion process.

PLANETARY CITIZENSHIP

We started out on our human enterprise with loyalty to a hunting band of a few dozen people. From there we successively expanded our allegiance to the village, town, city, region, and eventually the nation. At each stage our sense of community grew, until today we feel a part of societies of millions. Yet the greatest loyalty leap awaits us. Can we now raise our vision to embrace the whole of humanity? So great is this challenge that it will rank as the second true step away from the cave's threshold.

First, we need to identify with a global community of individuals whose names we do not know, whose faces we shall never see, whose traditions we may not share, and whose hopes we shall not know, but who are all, whether they are aware of it or not, *de facto* members of a single society. Second, we must foster a super-allegiance to Gaia, and frame our actions accordingly. Can we learn in time to identify with these two ultimate communities?

This need not present any conflict of interest. When we salute our countries' flags, we are not thereby denying bonds with families or neighborhoods. We are simply acknowledging a greater context of kindred spirits, a loftier level of allegiance. So our new planetary loyalty will not diminish established links; rather it will enhance them by adding perspective to local attachments.

There could, however, be some exceptions. What when a country asks us to do something against the global good? To hold back for the sake of the national economy on anti-global-warming measures, for example, or to consume for the sake of the balance of payments, when that means supporting unduly polluting activities? As we head farther into an interdependent future, there may well arise numerous occasions when we shall feel torn by planetary obligations that should outweigh "state dues." On such occasions it will not be easy to rise above past practices or to keep our attention fixed firmly on the universal need. We cannot live in isolation, whether from one another or from the planetary home. Either we shall become involved through joint effort as global citizens, or we shall become involved through jointly suffering global catastrophe.

Transcending Politics

Traditional politics concerns itself with the managing of social systems. This view will need to be considerably broadened to take into account the concept of planetary citizenship. The best politics will enable people and communities to create their own solutions to their own problems within a larger context. But a prerequisite will be the moulding of a new environmental/political consensus from the present anti-environmental world. Among initiatives that could soon become commonplace is an Earth Corps, an organization providing a framework for people, young and old, to make a personal contribution to the planet. The potential is vast, not only in terms of work to be done but the reservoir of people and energy waiting to be mobilized on behalf of the Earth.

Rising Above the Nation-State

We are suspended at a hinge of history, between two ages—that of competitive nationalism and that of cooperative internationalism. Nationhood is becoming a pernicious anachronism, a primitive phenomenon like feudalism or slavery. Future generations will surely consider that nationhood was a transient phase in society's development, a holding measure until the emergence of planetary citizenship.

Our Evolutionary Conditioning

When we attempt to identify with the global community, we may find that our evolutionary conditioning stands in the way. Our individual nurturing has derived primarily from the 99% of human history spent as hunter-gatherers, dependent for survival on sinking individual interests with those of a few dozen others. So we have inherited a set of sensibilities that, however capable they were for our formative years, may have left us deficient for our future worlds. Can we develop the extra faculties we need to co-operate as a band of more than 5 billion people?

A LAW UNTO OURSELVES

In a new-age world we shall learn to be our own legal experts, in that we shall have to devise rules for living in a crowded global community without treading on anybody else's prerogatives. We shall need to recognize that planetary citizenship entails responsibilities as well as rights; and we shall have to learn that it is in our own best interests to be our own private law enforcer. Not that this concept is anything new in itself. The rule of law has always depended on a strong supporting consensus of the citizenry, without which the best-intentioned laws fail.

Consider, for example, the debacle of the Prohibition years in America. But this time around there will be such an abundance and complexity of laws, regulations, and rules, whether formal or unwritten, that there cannot be enough police to keep everyone on the straight and narrow. We must devise our own path ahead, and follow it because it is in our own best interests.

This mode of behavior will be in stark contrast to the free-wheeling years of a simpler and less vulnerable world, one where there was no threat of terminal breakdown in society through outright environmental collapse. With multiplying numbers of people, multiplying demands, and multiplying linkages through our increasing interdependency, both environmental and economic, there will be multiplying scope for disruption and dislocation. The answer will lie with the dictum of "Mutual coercion, mutually agreed upon". But this need not be a fraction as "Big Brotherish" as it may sound. It will replace the unlicensed liberty of yesterday with new forms of freedom for tomorrow—an expansive and disciplined freedom.

Above all, the new "world of laws" will be all the more acceptable in that it will not be a top-down affair, by contrast with the situation in the past. It will be a grass-roots process, a home-grown homage to largely local imperatives. Communal laws will be more like social codes, finely tuned to local needs—a world away from the rigidity of conventional laws.

Local Control in Sweden

Flexible, local systems of sharing and control are becoming a feature of codes of conduct of certain smaller nations. In Sweden, for example, most of the unionized workforce is solidly behind its government's new program of economic decentralization. In each one of Sweden's two dozen counties, a proportion of the workforce's earnings is automatically paid into a public fund administered by an elected board. The accumulated funds are used to purchase shares in local industries, which are then publicly controlled. Local economic control is thus back in the hands of the community, where decisions can be made in light of local needs.

Roundabouts or Traffic Lights?

Law-abiding societies of the future must operate by consensus, rather than by the threat of punishment. Communally agreed codes will be the order of the day, not dictates passed down from some distant national assembly.

When we all agree to drive on the same side of the road we make no sacrifice of personal freedom—rather our freedom is enhanced by communal consensus. As we progress into the future we shall find that we will be driving ever-more sophisticated models (of the figurative type) at higher speeds, and sharing the road with ever-more drivers. This will necessitate more "traffic control" to facilitate everybody's journey. These controls can take the form of roundabouts (communal codes with which people have licenses to assess, evaluate, and make decisions) or traffic lights (specific laws that require simple acquiescence).

THE GENIE OUT OF THE BOTTLE

Our technological capacities are such that it is now possible to reshape our world from top to bottom—not only our planetary living space, but our social relations, our individual inner worlds, all that we are and do. The record to date does not presage a future as constructive and bountiful as we might wish. Unbridled technology already threatens our very life-support systems. Yet technology could be one of the greatest boons for the human condition, provided we ensure that it serves our overall interests. We need to take a long look at the role of technology in our future world and to ask how we can take systematic control of its multifarious impacts. We have released the genie from its bottle, and even if we perversely wished, we cannot return it. The present challenge is to control the genie before it controls us in unwitting ways that we cannot remotely discern.

What then is to be the future role of technology, as of its scientific underpinnings? The new physics shows us that what scientists observe in nature is intimately conditioned by their minds. Hence scientific and technological applications are also conditioned by the mind and thus by human values. Scientists are not only intellectually responsible for their research, they are also morally responsible. In turn, should not scientists and technologists now be required to take methodical cognizance of the impacts of their endeavours, whether environmental, social, or even political?

True, this would mark a profound change in our attitudes to science, as to its role in society. Many scientists would be aghast at the notion that their research should somehow be trammelled. But this is not to assert that all science should be subject to detailed constraints. There must always be abundant place for the pursuit of knowledge and understanding, whatever it reveals. Yet the overall context of science should surely be examined to see whether we can determine some limits to its unfortunate technological by-products. It is this new dimension, placing a check on undesirable fall-out, that is the key to controlling the genie.

Human Hubris

Our overweening attitude toward the natural world is a recurring theme in cultures right from the Ancient World. For stealing fire from the gods, Prometheus was chained and tortured. It seems always to have been accepted that there are some areas of enquiry that are simply off limits in view of the potential costs they entrain. The question now is to determine which precisely these areas are. While we understand so little of the world about us, especially the expanding world of the future, we must move from hubris to humility and practice a cautious rather than a Promethean approach.

Self-Imposed Audits

Already some ecology professors are proposing that students' dissertations should include a chapter on the social implications of their findings. In some cases, these implications will be virtually nil, in many others they will be of marginal consequence, in certain others they will be significant. Whatever the outcome, the exercise will induce an explicit awareness that science and technology can no longer be practiced in a vacuum.

Genetic Engineering

To date, genetic engineering has caused no regrettable spillovers on to the environment. The potential, however, is certainly there as new organisms are introduced into natural communities at an increasing rate. After all, an earlier effort at improved breeding of livestock in the form of a goat variety adapted to harsh environments—hailed as an undoubted success at the time—led to semi-arid lands becoming arid lands. The new breeders, for example, gene splicers using recombinant DNA technology, need to exercise far more scrupulous care as they release multitudes of entirely new organisms. While it is unlikely that the newcomers could ever become dominant on a broad scale, they could well cause local ecological disasters that could not readily be controlled. Unfortunately there is a tendency for expectations and benefits to be overestimated, while costs

and problems receive short shrift—as is often the case during the early stages of the development of any new technology. The prospect of hosts of newly minted organisms warrants exceptional caution rather than a "rush-in" approach. Yet we have scarcely started to assess the legal, let alone ethical, issues at stake.

RESPONSIBLE TECHNOLOGY

Modern society is hooked on technology, one that gives us life-saving hospital units and liberating communications, also nuclear weapons and soulless production lines. We are led to believe that there is a technofix for every social ill and global problem—and if technology goes wrong, then technology will put it right again. Yet we continue to feed off ecodestructive agriculture, to use products from energy-wasteful industries, and to live lives in technology-ravaged environments.

Our notions of wealth and welfare, competition and efficiency, are grounded in technology. The progress of technology is supposed to be unstoppable: it represents the crowning achievement of human enterprise, and other notions of progress come second. But our resource-intensive, overcentralized technology is making itself obsolete. Petrochemical agribusiness, nuclear power, fuel-guzzling cars are environmental disasters. Yet we still have to devise guidelines for our present technology, let alone the fast-expanding technology of the future. What technology is "right"?

The answer lies with technology that supports humankind in our need to live in accord *with* the planet rather than in dominion *over* it: technology with a planetary face. Many technologies for sustainability already exist in the form of so-called "soft technologies," for instance those that utilize renewable resources and recycled materials. A good number of these technologies are already familiar to us: energy-saving devices; ultra-efficient motors; solar energy conversion; wind, wave, and tidal power utilization; organic farming; semiconductors; and superchips. The challenge is to create a flexible, benign, and humane technology—a process that will allow us to exercise our full creativity. Indeed a crucial factor in this new technology will be its readiness to draw much more on a resource we already possess in abundance—human ingenuity.

In this regard, the soft technologies are often ideal, since they tend to be small scale and decentralized. And, being generally labor intensive, they help to establish a local economy, one that is flexible and sensitive to local conditions.

A New Orientation

To solve the multiple crises we face, we don't need more energy-intensive technologies. We need to shift our emphasis from nonrenewable to renewable resources, from hard to soft technologies. But this alone will not be sufficient without a thorough-going cultural change—a move away from the mechanistic "we can fix it" mentality to a more careful and caring approach. This in turn will require the most basic retooling of all: wholescale shifts in our attitudes, lifestyles, and values.

The Solar Future

The planet's principal energy source is the sun, with its potential for limitless, nonpolluting energy. Life has evolved to make optimum use of this form of energy. Plants photosynthesize food with the aid of the sun's radiation, and in turn provide the conditions, directly or indirectly, for practically all other forms of life. Solar energy can either be "passive", as when the fabric of a building heats up and then releases its energy, or "active", as in the example of solar collectors. As well, the sun's energy raises air masses that drive wind turbines, and it also energizes the water cycle, which we harness as hydropower. Bear in mind that solar energy is not only available everywhere, it is a source of diverse energy types, hence adapted to decentralized technologies in local communities.

People's Technology

Grass-roots technology is not always accorded the recognition it deserves. In the 1960s, traditional fishermen of the Arabian Sea protested that the introduction of mechanized trawling would destroy local fishing stocks. Their detailed ecological insights were dismissed by fisheries experts. Today, and through their own technological expertise, the fishing communities are pioneering the use of artificial reefs and species-specific baits, exploiting them in a manner that is both sustainable and finely tuned to local circumstances. More, their efforts are starting to be supported by the government, strengthening their organization.

FROM KNOWLEDGE TO WISDOM

As T. S. Eliot once asked: "Where is the knowledge we have lost in information? Where is the wisdom we have lost in knowledge?" Obviously knowledge is an essential component of wisdom, but there is more to it than that. While we possess knowledge in abundance, the world about us proclaims that we are falling woefully short of the step from knowledge to wisdom.

Our long tradition of reductionist analysis has brought us to a state where we often find that we are learning more and more about less and less. Result: our attention is diverted from the whole that is more than the sum of its parts. Confronted with the disjunct pieces of a watch, would we guess that when reassembled they would make up a mechanism for telling the time? Yet we meddle with the intricate components of our planetary ecosystem, no less, and suppose we can do so with impunity, even concluding that certain pieces are forever dispensable. We are unwittingly engaged in global-scale experiments, not only environmental but technological and social experiments, without a thought for their global consequences. Yet these grandiose experiments demand the most scrupulous care, and would tax the faculties of the wisest.

Moreover there is much that we, whether scientist or not, can discern in terms of values without understanding the details. We do not need to understand the physical or chemical make-up of a weapon to realize it is designed to destroy. We do not need endless reports on the influence of the media to see their power to misinform.

Thus far we have tended to head blithely toward a "universal horizon" and have limited ourselves to doing a better job in getting from here to there—"there" being something that reflects the common understanding of most people. We make sure that we travel along smoothly in a car that is in good order and with enough fuel. Beyond our present crossroads, however, that will not suffice. Often enough we shall need to consider the proper road to follow. The landscape ahead is no longer preordained by "where we have always been going before." Increasingly we shall find that we need to choose a different route, change the car, even ask whether we want to travel at all. All the fast-growing knowledge in the world will not help us if we do not have the wisdom to look out all over the world, and to decide where we feel most at home.

Dinosaur of the Mind

"But we learn no lesson, give no moral to our children. We go on not having understood what is meant by our technological act of knowing. Some day, in the far, far future, the evolutionary processes will perhaps have gotten rid of this failing attempt at gaining knowledge; either we will be the ancestral progenitors (of a wiser species), or else we will be discarded, an unsuccessful offshoot, a dinosaur of the mind, one of nature's failures." Professor Phillip Siekevitz, Rockefeller University.

Analysis and Vision

The knowledge business—not only science but telecommunications, media, advertising—means we can generate more information than ever before, and mobilize it more efficiently. This is far from enough, even though it is the credo of government, business, and the general ethos of "advancement". For we see only what we are disposed to see. When we analyse information to produce new ideas, we do no more than find new slots for old ones. To generate truly new ideas—to strive toward "wise" insights—we must bring into play our creativity, our powers of inspiration, above all our sense of vision. All these are a world beyond the nose-to-the-microscope approach of conventional understanding.

Knowledge as Private Property

The trend toward privatized knowledge is regrettable. While the patenting of ideas is a valid safeguard, the wholesale locking away of knowledge as private property, walled by legal defenses, can only redound to our collective detriment. Instance the North-South gap: whether wittingly or not, the technological underpinnings of the knowledge explosion in the developed nations are acting to exclude developing nations from one of the most productive phenomena of our age. How ironic if the knowledge resources in the North serve to impoverish further the South. Surely we will recognize that knowledge should be generally available in the public domain, serving the needs of everyone. Only as the walls around knowledge crumble away will the shared experience of all contribute to dawnings of wisdom.

THE FUTURE OF THE FUTURE

For however much is uncertain about the future, one thing is definite: new worlds will constantly unfold, with their new problems *and* their new possibilities. In some senses, in fact, there will be as many possible futures as there are people. We are all deeply involved, whether we appreciate it or not; we shall all play our individual parts. Indeed there has never been a time like the present when the individual can *count*. So a prime aim has been to show the reader how to envision the future, and to decide how to contribute—if only at the local level, which will often be the best level.

The shadows over our future remind us that the optimist proclaims this is the best of all possible worlds, while the pessimist responds that is regrettably true. We must stay hopeful, otherwise we might as well go to the beach until the sky falls in. But let us not be seduced by an airy hope that somehow all will work out in the wash: the laundry water likely contains too many pollutants. There are prophets who assert we should not worry about the prospect of feeding 10 billion people when it is theoretically possible to feed four times as many. As Paul and Anne Ehrlich point out, it is theoretically possible too for your favorite football team to win every game for the next ten years. In any case, the Ehrlichs continue, what is the sense in converting the Earth into a gigantic human feedlot? How about more quality of life for fewer people? Moreover a future of "the same as usual, only more so" would be a future that for many people would simply not arrive.

If further persuasion is necessary, recall that of the 31 major civilizations in history, only one remains a dominant force, so-called Western civilization. As the historian Arnold Toynbee demonstrates, the rest disappeared because they tried to become dominant on every side—over their neighbors, their environments, whatever else they cast their eyes on. Western civilization, materialist to beat any other, shows plenty of signs of dominating the entire world into an ultimate crunch.

No doubt about it, we stand at a hiatus in the human enterprise. The present is so different from the past, and the present so different from the future, that it is as if we are at a hiatus in the course of human affairs. It is a unique time: a time of breakdown or breakthrough.

To break through into a future of undreamed potential, we must enable a new sort of society to be born. Indeed the stresses of the present are like the stresses of being born, a time of utmost threat yet with new life ahead. Or, to shift the analogy, our society is like a human being growing up. From the start it shows boundless appetite for resources of every type. This appetite continues throughout the first two decades of physical growth, expanding all the while. Then it suddenly levels off. This does not mean the person's growth is at an end. On the contrary, the richest stage of growth begins—mental, intellectual, emotional, and spiritual growth, growth that extends many times longer than the early phase.

Our global society is still adolescent—lusty, vigorous, and assertive. Can we move on to maturity—assured, stable and sensitive, displaying all the attributes of adulthood? Are we ready to grow up? To shift from egosystem to ecosystem, to social compact and whatever else is needed for us to become citizens of the globe?

Are we ready, in fact, to engage in the most salient experience of adulthood, the mutuality of love? Nothing less will do. Finally we can recognize that there is no longer any "we" and "they." For the first time, and for all time, there is only "us"—all of us humans, together with all our fellow species.

This will be the greatest of our global experiments. As we measure up to the challenge, we shall need to become giants of the human condition. As we approach our climacteric—never attempted before, surely never to be repeated with such instantaneous speed—let us count ourselves fortunate to be living at this hour. No generation of the past has been presented with such a chance to rise above the tide of human history. No generation of the future will have our chance, because if we do not do the job they will have little left but to pick up the pieces. What a privileged generation we are.

WHAT'S POSSIBLE; WHAT'S PROBABLE

What lies ahead? This section looks at an array of possibilities that may prompt us to speculate on what else is in store. The following list is highly selective and no more than illustrative. The items derive from analyses of experts who use techniques such as scenario planning to think methodically about the future. Futurists based at universities have set up entire departments for futures studies. There are commissions on the future, established by individual governments, United Nations agencies, the City of Tokyo, and the like. The World Future Society, based in Washington DC, organizes regular conferences on all aspects of the future. But the most remarkable feature of this futurist community is not that there are so many people thinking about what the future holds, but that there are so few. Among some prospective developments are the following.

Global Environment

As the greenhouse effect takes hold, we could witness persistent droughts over much of North America, sub-Saharan Africa, and China. Together with repeated failures of the Indian monsoon and other climatic quirks, plus the most expansive phase of the population explosion, there could ensue a greater outbreak of starvation in a single year early next century than throughout the 1980s, even culminating in the deaths of one billion people in just one decade. This worse-case scenario, by no means implausible, would amount to a human catastrophe of unique proportions.

In the longer-term future, a one-metre rise in sea level by the middle of next century could, when combined with storm surges reaching far inland, plus saltwater intrusions up rivers, threaten a total area of 5 million sq km—or one-third of today's croplands and home to one billion people already. Also on the cards is the prospect of mass migrations in the wake of the greenhouse effect. On the Chinese side of the Sino-Soviet border there are already acute pressures from 1300 persons per square kilometre, by contrast with only one person to every 2.5 square kilometres on the Siberian side. What when China starts to suffer the full rigors of global warming?

Geopolitics

Consider the following scenario. In the year 2000 the world has 350 billionaires, at least 4 million millionaires, and 250 million homeless. The average income of the top 1 billion people has reached 50 times more than that of the bottom 1 billion. More than 50 nations no longer qualify as developing nations; they are disintegrating nations. Americans spend $10 billion per year on slimming diets, while 1 billion people are so undernourished that they are semi-starving. Water from a single spring in France is still shipped to the affluent around the world, while a full 2.5 billion people lack access to clean water for basic needs. It has become plain that poverty is a luxury we can no longer afford.

Furthermore, there has been a series of crises on top of widespread starvation. Chernobyl-type accidents have occurred in four nations. the North Sea has been declared beyond foreseeable recovery. The most bountiful marine ecosystem on Earth, the Southern Ocean, has collapsed after UV-B radiation knocks out the phytoplankton. A nuclear terrorist has destroyed Cairo. Drug barons have declared jurisdiction over much of South America. The Pope has been added to Willy Brandt, Stevie Wonder, and Steffi Graf as hostages held by extremists. The latest Live World concert has been watched by two and a half billion people, and has led to mass protests throughout the world.

This all pushes governments into finally acknowledging that there is only one track ahead: global collaboration with a vengeance, and for the first time ever. As European President Joan Ruddock puts it at the World Conference for the World: "The biggest problem is no longer others, it is ourselves. We are suspended at a hinge of history, between two ages—that of competitive nationalism and that of cooperative internationalism. At long glorious last, we recognize that traditional nationhood is an anachronism, a pernicious phenomenon like feudalism or slavery."

What emerges is a system of government based on concentric circles: local councils, regional assemblies, national governments, groupings such as the European Community, and global bodies such as the United Nations (supplied with teeth). This vertical structure is paralleled by a horizontal structure of NGOs with real power, made up of professions, trade unions, Friends of the Earth, major charities, academics, service clubs, Oxfam, and the like. These NGOs receive collective representation through the long proposed Second Assembly of the Untied Nations. Under these twin structures of government, citizens can cheer equally for Edinburgh, Scotland, Britain, Europe, the Commonwealth, the world, tropical forests, Antarctica, and Action Aid.

Science and Technology

As a result of scientific advances, techno-jumps could include a breakthrough with photovoltaics that transforms the energy prospect from top to bottom, especially for tropical countries; backpack nuclear weapons, with all that means for terrorism; sex selection on the cheap, leading to a massive majority of male babies; and genetically engineered trees that sprout like mushrooms ("plant the seedling and jump aside"), allowing reforestation to do a better job of soaking up excess carbon dioxide from the atmosphere. But note that genetic engineering could lead to some unfortunate consequences of economic and social sort. It will soon be possible to "grow" cocoa in the laboratory, which could devastate those developing-world countries that now earn $2.6 billion a year from the field-grown crop.

A related technology is nanotechnology, enabling us to redesign cellular structures. This will lead to, for example, exceptionally strong and lightweight alloys, leading in turn to organically manufactured aircraft that fly much more speedily and cheaply than today's dinosaur-style devices. Among more "way out" applications of nanotechnology could be steaks from hay, without the help of cows. These, like many other techno-jumps, will derive in part from the fast-growing capacity of supercomputers. Already the latest Cray model, standing no taller than its human operator, can solve problems at a sustained rate better than one "gigaflop," or one billion calculations every second (the term gigaflop comes from "giga" meaning one billion, and FLOP for "floating point operation," a common form of computerized arithmetic). Soon to become available is a computer capable of 22 gigaflops per second, while we should soon see a machine speeding along at 128 gigaflops per second.

As for the car of the future, that is already with us. A Volvo prototype, with lightweight synthetic materials, weighs only half as much as a conventional model. Its lean-and-clean engine, backed up by a continuously variable transmission and a flywheel energy-storage unit, achieves almost 150 km per gallon in average traffic conditions. A further prospect for the petrol-driven car is that there will simply be far fewer of them in urban areas, their place having been taken by vehicles powered by electricity or hydrogen fuel. In any case, there will be far less need for them in the face of competition from efficient and cheap mass-transit systems (in Tokyo today, only 15% of commuters drive cars to the office). Moreover there will be an increasing trend for offices to be connected by electronic lines rather than crowded highways. Note too that if China were to devote as much land to asphalt as the United States does per head, it would lose over 40% of its croplands.

Health

What price an end to drug addiction? There is prospect of a final solution in the form of "opiate antagonists" that block the effects of, for example, cocaine for a month or so, whereafter the euphoric impact dissipates and the addict loses interest.

More broadly, we can anticipate a growing disaffection with established ideas about health. As more people recognize that health is intrinsically a holistic affair, so they will be inclined to accept personal responsibility for cures of "disease." There will thus be an increase in wholesome diets, exercise, and sports, stress-reducing activities, self-rehabilitation, and recreation in the sense of recreation. Sooner than we might suppose, self-help health will become mainstream.

In developing countries the technique of oral rehydration therapy (ORT) could soon become as familiar as cola—otherwise we shall witness 25 million children die of dehydration during the 1990s. Ironically, as we bring about an end to the human hemorrhage represented by child mortality in developing countries—one of the great success stories in human history—we may well witness another scourge overtaking hundreds of millions, the ravages of tobacco, as cigarette corporations of the developed world, losing their clientele at home, peddle their wares to the last corners of the developing world.

Lifestyles

Already many households feature a personal computer. A child has hardly learned to read and write before he or she starts to work with a device that, a couple of decades earlier, would have had to be as big as a house to contain the computing power of the tabletop model. It is this new skill, backed by global-scope telecommunications networks, that is opening up worlds for youngsters way beyond the dreams of their parents.

Within just a few years a majority of women will be engaged in paid work, whether in a formal workplace or in homes linked to offices by computer networks. At the same time, many men will accept a greater role in the family and home. In response to the cocktail party question, "What do you do?" the answer will increasingly be, "I'm a househusband."

QUESTIONS

1. What are the problems and challenges with our current thinking for the future?

2. What are some of the immediate needs that we must address for the future?

3. In order for population growth to be held at 6.25 billion people by the year 2000, what must be the average number of children per family worldwide? What is the current average?

4. What is the problem with current plans for the year 2000? These plans were written in 1990; do you feel that they are realistic? Explain from our current historic view.

5. Compare the Climate Change Program with the outlined plans of the Kyoto Climate Treaty of December 1997. What are the differences between the two?

6. When discussing the greenhouse effect or global warming, what is the main issue that has to be addressed? Name three approaches suggested to deal with the global warming issue.

7. Identify three approaches for developing world peace strategies.

8. Explain two views of global politics that expand planetary political unity.

9. What are some of the considerations and concepts behind planetwide laws?

10. What is being suggested of engineers, scientists, and technologists when they design new systems and technologies?

11. What is meant by the term "responsible technology"? Give two examples of it.

12. How can we attain wisdom for the future?

13. What is the biggest challenge to sustaining our world and ourselves in the future?

14. Comment on your impressions of the predictions of the future with regard to the global environment, geopolitics, science and technology, health, and lifestyles. Do you agree with these future predictions? Give some of your own predictions as a contrast.

A Bridge to Your 21st Century Understanding

Complete and discuss the following flowchart.

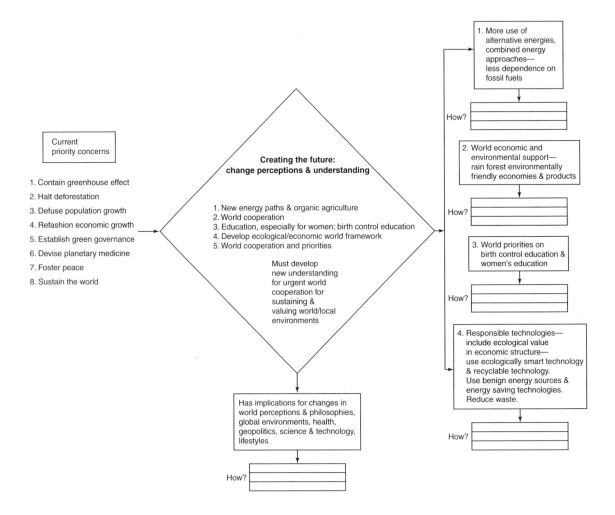

Current priority concerns

1. Contain greenhouse effect
2. Halt deforestation
3. Defuse population growth
4. Refashion economic growth
5. Establish green governance
6. Devise planetary medicine
7. Foster peace
8. Sustain the world

Creating the future: change perceptions & understanding

1. New energy paths & organic agriculture
2. World cooperation
3. Education, especially for women; birth control education
4. Develop ecological/economic world framework
5. World cooperation and priorities

Must develop new understanding for urgent world cooperation for sustaining & valuing world/local environments

Has implications for changes in world perceptions & philosophies, global environments, health, geopolitics, science & technology, lifestyles

How?

1. More use of alternative energies, combined energy approaches— less dependence on fossil fuels

How?

2. World economic and environmental support— rain forest environmentally friendly economies & products

How?

3. World priorities on birth control education & women's education

How?

4. Responsible technologies— include ecological value in economic structure— use ecologically smart technology & recyclable technology. Use benign energy sources & energy saving technologies. Reduce waste.

How?

Trends Now Changing the World: Economics and Society, Values and Concerns, Energy and Environment

MARVIN J. CETRON

OWEN DAVIES

EDITOR'S NOTE

An "exceptional" economy, increasingly integrated cultures, and stepped-up action on environmental concerns are among the key trends that will shape the world of the next two decades and beyond. Veteran forecaster Marvin J. Cetron of Forecasting International Ltd. and science writer Owen Davies describe the implications of these trends for our long-term future.

For some four decades, Forecasting International Ltd. has conducted an ongoing study of the forces changing our world. About 10 years ago, Cetron and Davies condensed their observations into reports for *The Futurist* covering key aspects of the economy, technology, business, and society.

Their expectations proved to be quite accurate: For instance, they believed that the economy of the developed world would be much more vibrant than most commentators imagined possible, and so it has been. In all, no fewer than 95% of their projections have proved correct.

Those early forecasts have often been updated and extended. In this article, the first of two excerpts from their latest report, the authors reconsider the trends from their previous work and focus on major trends that are now changing the world. For each trend, they also offer a succinct conclusion about its implications for the future and how it may affect individuals and organizations, including policy makers.

The Futurist, January-February 2001. Used with permission of the World Future Society.

GENERAL LONG-TERM ECONOMIC AND SOCIETAL TRENDS

The economy of the Developed World will Remain "Exceptional," as Fed Chief Alan Greenspan described the United States, for at Least the Next Five Years. Widespread Affluence, Low interest Rates, Low Inflation, and Low Unemployment will be the Norm

- Inflation-adjusted output in the United States grew by no less than 5.8% in the fourth quarter of 1999, and with more than 17 million new jobs created since 1991. With wages averaging $17.90 per hour, unemployment shrank to just 4%.

- The recession in Europe is nearing its end, and Asian economies are now being stabilized. This will improve global trade generally and should boost American exports in the next five to 10 years.

- National debts were brought under control throughout most of Europe, in preparation for the recent monetary unification.

- Real per capita income in the United States is rising at its strongest rate in decades. It should stabilize at a growth rate of about 1.45% per year through most of the next decade.

- At the same time, wages and benefits have remained under control throughout the industrialized world.

- Relaxation of borders within the European Union has brought new mobility to the labor force and is making for a more efficient business environment on the Continent.

- Japanese banks will finally write off their bad debts. Coupled with recent tax cuts and other reforms, this will set the stage for an economic recovery by 2002. Thereafter, Japan should provide a much healthier trading partner for the West. The three largest Japanese banks are combining into a $1.23 trillion institution. By reducing the number of branches and personnel and installing automatic teller machines, they expect to save $50 billion per year.

- Many nations of the former Soviet Union are bringing order to their economies. As they do so, they are proving to be viable markets for goods from western Europe. Whether Russia itself can stabilize its economy remains to be seen. This is one of the critical issues for the future of global prosperity, and perhaps for political stability throughout the world.

- Consumer inflation in the United States has just barely begun to be felt. The price of consumer goods other than food and energy was actually declining in the late 1990s.

- Long-term interest rates throughout most of the industrialized world are likely to remain relatively low for the foreseeable future.

- Improved manufacturing technology will continue to boost productivity and reduce the unit cost of goods. At the same time, workers who remain on the job longer will offset slow growth in the labor force, while the globalization of business will keep pressure on salaries in the developed countries. Thus, both prices and wages should remain under control.

- Automobile sales will slow as the useful life of cars stretches from its current average of about nine years to a bit more than two decades. American cars should continue to regain market share, because the mean time before failure is virtually the same for all cars now being sold. Whether cars are American, German, or Japanese, the same robots are building them. The 20-something echo boomers, or "generation dotcom," will continue to energize the market for sport utility vehicles.

Implications   Economic unification will boost all manner of trade within Europe. In the long run, the newly capitalist lands of the former Soviet Union should be one of the fastest-growing new markets. In the longer term, India will be the single fastest-growing market in the world.

Labor markets will remain tight, particularly in skilled fields. This calls for new creativity in recruiting, benefits, and perks, especially profit sharing. This hypercompetitive business environment demands new emphasis on rewarding speed, creativity, and innovation within the work force.

Part of society's affluence rests on the use or overuse of credit cards. Extension of excessive credit could result in government-imposed limitations, especially on credit rates.

The growing concentration of wealth among the elderly, who as a group already are comparatively well off, requires an equal deprivation among the young and the poorer old. This implies a loss of purchasing power among much of the population; in time, it could partially offset the forces promoting economic growth.

The World's Population Will Double in the Next 40 Years

- The greatest growth will occur in those countries least able to support their existing populations. Pakistan, for example, will have a growth rate of 2.68% per year through 2030; its population will grow from 141 million in 2000 to nearly 199 million in 2020. Ethiopia's growth rate of 3.17% per year will push the population from just under 61 million in 2000 to some 90 million 20 years later. India's population will grow by more than 220 million over the same period.

- In contrast, birthrates below the replacement level mean that populations will decline significantly in much of the developed world, not counting the uncertain effects of immigration.

- A severe, continuing, and unexplained decline in men's sperm counts in most of the developed world could eventually impair fertility enough to reduce birthrates and populations even further than current estimates anticipate.

- To meet human nutritional needs over the next 40 years, global agriculture will have to supply as much food as has been produced during all of human history.

Implications   Unless fertility climbs dramatically, either would-be retirees will have to remain on the job, or the industrialized nations will have to encourage even more immigration from the developing world.

Barring enactment of strict immigration controls, rapid migration will continue from the Southern Hemisphere to the North, and especially from former colonies to Europe. A growing percentage of job applicants in the United States and Europe will be recent immigrants from developing countries.

Russia is unlikely to attract many new workers from the rest of the world. Without radical reform of its economic and social policies, so as to provide a more appealing environment for migrants, it is doomed to growing poverty and social unrest, which its leaders will blame on the West.

Culture clashes between natives and immigrants are likely to destabilize societies throughout the developed world. Germany, Britain, and other lands traditionally welcoming to refugees and other migrants already are experiencing strong backlashes against asylum seekers.

As the customer base grows ever more diverse, services will have to be tailored to the unique needs of new markets.

Growing populations are forcing some developing countries to abandon restrictive trade practices in order to compete more effectively in the global marketplace. This will foster the development of a market-oriented global culture.

The Population of the Developed World is Living Longer

- In the developed lands, healthier diets, more exercise, the decline of smoking in the United States, and the trend toward preventive medicine are extending life-spans. Life expectancies in Japan are entering the 90s, and those in parts of Europe are not far behind. Medical advances that slow the aging process now seem within reach; they could well help today's middle-aged baby boomers to live far longer than can be predicted even today.

- The elderly population is growing fastest throughout the developed world. In Europe, the United States, and Japan, the aged also form the wealthiest segment of society.

- These twenty-first-century old folks are much healthier and more active than the elderly of previous generations. At the same time nostalgia also is a strong influence on them. Many older people still want to indulge in the same activities and entertainment they enjoyed in their youth, and they now have more disposable income to spend on them.

Implications   Research to date suggests that any practical extension of the human life-span will also prolong health and reduce the incidence of late-life diseases such as cancer. Whether Alzheimer's disease will also be delayed

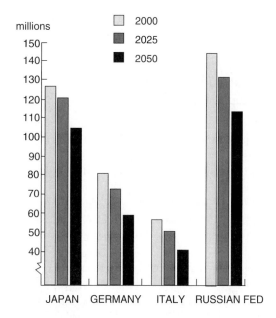

Though the world's population will double, many countries will experience population declines.

Figure 55.1
Population Will Decline in Many Developed Countries
Source: Forecasting International, Ltd.

or will arrive "on schedule" is a critical question for the future.

Global demand for services aimed at the elderly can only grow quickly in the coming decades. Medical care in particular will prosper.

With above-average wealth and relatively few demands on their time, the elderly will make up an ever-larger part of the tourist and hospitality market. This industry will prosper by catering to their needs for special facilities and services. Hotels will offer easy-to-read shop signs and brighter public areas suited to the needs of older visitors. Club Med will become "Club Medic," with doctors on call and a nursing staff for sickly vacationers.

For those who are not wealthy, the cost of retirement and medical benefits will rise sharply, even as the number of working-age people to pay for them declines.

If medicine does dramatically extend our life-span, and if preventive medicine receives government funding, the cost of health care will plummet; retirement and social security plans will have to be revised or scrapped.

The Elderly Population is Growing Dramatically Throughout the Developed World

- Those over age 65 made up 12.4% of the American population in 2000. By 2010, they will be 13%; by 2020, more than 16%.

- In Germany, the retirement-age population will climb from under 16% of the population in 2000 to nearly 19% in 2010 and 20% a decade later.

- Japan's over-65 population made up 17% of the total in 2000, rising to 22% in 2010 and nearly 27% in 2020.

- This is also true of certain developing countries. India's over-60 population is rising from 56 million in 1991 to 137 million in 2021 and 340 million in 2051.

- The number of centenarians in the world will grow from 135,000 in 2000 to 2.2 million by 2050.

Implications   Not counting immigration, between 2000 and 2050, the ratio of working-age people to retirees needing their support will drop from 5.21 to 2.57 in the United States, from 4.11 to 1.75 in Germany, from 3.72 to 1.52 in Italy, from 5.51 to 2.41 in Russia, and from 3.99 to 1.71 in Japan. Over all, the "support ratio" in the European Union will decline from 4.06 to 1.89.

Workers in the traditional retirement years represent the fastest growing employment pool, which has yet to be fully tapped.

Without dramatic advances in geriatric medicine, the cost of health care could skyrocket throughout the developed lands.

The Growth of the Information Industries is Creating a Knowledge-Dependent Global Society

- Telecommunications is removing geographic barriers.

- Information is the primary commodity in more and more industries today.

- By 2005, 83% of American management personnel will be knowledge workers. Europe and Japan are not far behind.

- By 2005, half of all knowledge workers (22% of the labor force) will opt for "flextime, flexplace" arrangements, which allow them to work at home, communicating with the office via computer networks.

- In the United States, the so-called "digital divide" seems to be disappearing. In early 2000, a poll found that, where half of white households owned computers, so did fully 43% of African-American households, and their numbers were growing rapidly. Hispanic households continued to lag behind, but their rate of computer ownership was expanding as well.

Figure 55.2
Projected Population of Seniors
(Age 65+)
Source: Forecasting International, Ltd.

2000
2020

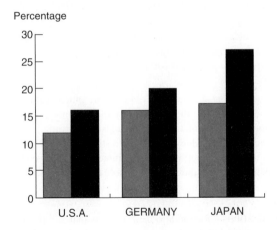

Percentage

Elderly populations will grow dramatically in proportion to total in developed countries.

- The "integrated information appliance" will combine a computer, a fax, a picture phone, and a duplicator in one unit for less than $2,500 (in 1995 dollars) by 2003. The picture will appear on a flat screen of $20'' \times 30''$. By 2005 or so, such units will include real-time voice translation, so that conversations originating in one of seven or eight common languages can be heard in any of the others.

- Company-owned and industry-wide television networks are bringing programming to thousands of locations. Business TV is becoming big business.

- Computer competence will approach 100% in U.S. urban areas by the year 2005, with Europe and Japan not far behind.

- Eighty percent of U.S. homes will have computers in 2005, compared with roughly 50% now.

- No fewer than 80% of Web sites are in English, which has become the common language of the global business and technology communities.

- In the United States, five of the 10 fastest-growing careers between now and 2005 will be computer related. Demand for programmers and systems analysts will grow by 70%. The same trend is accelerating in Europe, Japan, and India.

- By 2005, nearly all college texts and many high school and junior high books will be tied to Internet sites that provide source material, study exercises, and relevant news articles to aid in learning. Others will come with CD-ROMs that offer similar resources.

- Internet links will provide access to the card catalogs of all the major libraries in the world by 2005. It will be possible to call up on a PC screen millions of volumes from distant libraries. Web sites enhance books by providing pictures, sound, film clips, and flexible indexing and search utilities.

- Encyclopedic works, large reference volumes, and heavily illustrated manuals already are cheaper to produce and sell on the Internet or as CD-ROMs than in print form.

Implications   Anyone with access to the Internet will be able to achieve the education needed to build a productive life in an increasingly high-tech world. Computer learning may even reduce the recidivism rate of the growing American prison population.

Knowledge workers are generally better paid than less-skilled workers. Their proliferation is raising overall prosperity. Even entry-level workers and those in formerly unskilled positions require a growing level of education. For a good career in almost any field, computer competence is mandatory. This is one major trend raising the level of education required for a productive role in today's work force. For many workers, the opportunity for training is becoming one of the most desirable benefits any job can offer.

Growing Acceptance of Cultural Diversity, Aided by the Unifying Effect of Mass Media, is Promoting the Growth of a Truly Integrated Global Society. However, this is Subject to Change

- Our beliefs and values are shaped by what we see and hear. Throughout the United States, people have long seen the same movies and TV programs. These media are achieving global reach. In 1999, American films took in about $29.8 billion of the $33.4 billion earned by the world's movie industries.

- Information technologies promote long-distance communication as people hook up with the same commercial databases and computer networks, and above all with the Internet.

- New modes of transportation, automated traffic-management systems and other highway technologies, more and better accommodations (thanks to the growth of the hospitality industry), more leisure time, and greater affluence will encourage more-frequent travel. Common-carrier passenger miles have grown by nearly 4% per year since the late 1980s and show no sign of slowing down. This will produce a greater interplay of ideas, information, and concerns.

- Within the United States and Europe, regional differences, attitudes, incomes, and lifestyles are blurring as business carries people from one area to another.

- Intermarriage also continues to mix cultures geographically, ethnically, socially, and economically.

- Minorities are beginning to exert more influence over national agendas as the growing number of blacks, Hispanics, and Asians in the United States is matched by the expanding population of refugees and former "guest workers" throughout Europe.

Implications   Over the next half-century, growing cultural exchanges at the personal level will help to reduce some of the conflict that plagued the twentieth century.

This is likely to produce a reactionary backlash in societies where xenophobia is common. Some of the most fervent "culturist" movements will spring from religious fundamentalism. Would-be dictators and strong-men will use these movements to promote their own interests, ensuring that ethnic, sectarian, and regional violence will remain common. Terrorism especially will be a continuing problem.

Companies will hire ever more minority workers and will be expected to adapt to their values and needs.

In the United States, small businesses are increasingly owned by Asians and African-Americans. European countries are seeing a similar pattern among refugees and former "guest workers." Cultural conflicts may become more common, and dealing with them will require awareness and sensitivity. For example, American business traditions hold that negotiations are over once an agreement has been reached; in the traditions of some entrepreneurs newly arrived from Asia, negotiations continue until payment has been received.

The Global Economy is Growing More Integrated

- Rather than paying salaries and benefits for activities that do not contribute directly to their bottom line, companies are farming out secondary functions to suppliers, service firms, and consultants, which increasingly are located in other countries.

- In the European Union, the relaxation of border and capital controls and the use of a common currency are making it still easier for companies to farm out support functions throughout the Continent.

- Western corporations are having to adapt to Asian priorities. Where the West emphasizes "resource capital"—money and equipment—Eastern societies are more concerned with "human capital"—education, cooperation, and other practices that make the best use of people.

- New industrial standards—for building materials, fasteners, even factory machines—allow buyers to order from almost any supplier, rather than only from those with whom they have established relationships. The proliferation of standards is one of today's most important industrial trends.

- To aid in "just-in-time" purchasing, many suppliers are giving customers direct online access to their computerized ordering and inventory systems. Increasingly, the order goes directly from the customer to the shop floor, and even into the supplier's automated production equipment. Many manufacturers will no longer deal with suppliers who cannot provide this access, and the number grows daily. Thanks to the Internet, this form of integration is possible worldwide.

- The Internet and cable-TV home shopping channels are bringing retailers and manufacturers closer to distant customers, who have been out of reach.

- New procurement regulations and standards also promise to open the government market to suppliers who previously found the bidding process too difficult, costly, or just confusing.

Implications   Demand for personnel in far-off lands will increase the need for foreign-language training and documentation, employee incentives suited to other cultures, aid to executives going overseas, and the many other aspects of doing business in other countries.

Eastern Europe is likely to require a major investment in personnel development over the next 10 years.

Consolidation of standards makes it more practical for manufacturers in one country to shop for parts in another. In Europe, especially, this is quickly changing established business patterns and creating new demand.

The growth of commerce on the Internet makes it possible to shop globally for all manner of raw materials and supplies, thus reducing the cost of doing business. In niche markets, the Internet also makes it possible for small companies to compete with giants worldwide with relatively little investment.

In the wake of the "Asian flu," Western companies may have to accept that proprietary information will be shared, not just with their immediate partners in Asian joint ventures, but with other members of the partners' trading conglomerates.

TRENDS IN VALUES AND CONCERNS

Societal Values are Changing Rapidly

- Society will increasingly take its cue from generations X and dot-com, rather than the baby boomers who have dominated its thinking for most of four decades.

- In the future, both self-reliance and cooperation will be valued—self-reliance because we will no longer be able to fall back on Social Security, pensions, and other benefits; cooperation because group action often is the best way to optimize the use of scarce resources, such as retirement savings.

- Family issues will continue to dominate American society at least through 2008: long-term health care, day care, early childhood education, antidrug campaigns, and the environment. Companies are now required to grant "family leave" for parents of newborns or newly adopted children and for care of elderly or ill family members.

- Narrow, extremist views of either the left or the right will be unpopular. Moderate Republicans and conservative Democrats will lead their respective parties.

- Some liberal views will return to the mainstream after 2000, thanks to the 30-year Hegelian swing in which liberal and conservative philosophies vie for dominance in American society, eventually reaching stable compromises on most issues.

- Drugs eventually will be decriminalized. Funds saved from the criminal-justice system will be used for antidrug education and for the treatment of drug users—which will prove to be a more humane and effective approach to the problem.

Implications   The highly polarized political environment that has increasingly plagued the United States in the 1980s and 1990s will slowly moderate as result-oriented generations X and dot-com begin to dominate the national dialogue.

Young People Place Growing Importance on Economic Success, which They Have Come to Expect

- Generations X and dot-com have known only good economic times, and, while most expect hardship on the national level, they both want and expect prosperity for themselves.

- Growing numbers of people now become entrepreneurs.

- In the United States especially, most young people have high aspirations, but may lack the means to achieve them. Only one in three high-school graduates goes on to receive a college degree. Many of the rest wish to go, but cannot afford the high cost of further schooling.

- Without higher education, expectations may never be met. In 1996, male high-school graduates not enrolling in college earned an average of 28% less, in constant dollars, than a comparable group in 1973.

- In addition, more young people report no earnings—up from 7% of all 20- to 24-year-old men in 1973 to a relatively constant 12% since 1984.

Implications   The emphasis on economic success will remain powerful. Stress will keep step with it.

This will prove to be a global trend, as members of generations X and dot-com tend to share values throughout the world.

If younger-generation workers in developing lands find their ambitions thwarted by local conditions, they will create growing pressure for economic reform and deregulation. Entrepreneurialism is likely to spread to parts of the world where corporate careers have been the rule. This will represent yet another major change for Japan, where the recent loss of lifetime job security still has not been fully absorbed.

If reforms do not come fast enough in the developing world, disappointed expectations will drive underemployed young men into fringe political and religious movements. This could cause a new wave of terrorism and instability in the years after 2005 or so.

Tourism, Vacationing, and Travel (Especially International) will Continue to Grow by About 5% per Year for the Next Decade, as it did Throughout the 1990s

- People have more disposable income today, especially in two-earner families.

- The number of Americans traveling to foreign countries (excluding Canada and Mexico) increased at 5% per year from 1981 through 1996. Growth will continue at that rate for the foreseeable future.

- Through at least 2002, depressed Asian currencies will make it cheaper to visit the Far East.

- By 2010, air travel for both business and pleasure will reach triple the 1985 rate.

- Tourism will benefit as Internet video replaces printed brochures in promoting vacation destinations. Web sites cover not only popular attractions, but also current, detailed information on accommodations, climate, culture, currency, language, immunization, and passport requirements.

- Multiple, shorter vacations spread throughout the year will continue to replace the traditional two-week vacation.

- More retirees will travel off-season, tending to equalize travel throughout the year and eliminate the cyclical peaks and valleys typical of the industry.

Implications   The hospitality industry will grow at a rate of at least 5% per year for the foreseeable future, and perhaps a bit more. Tourism offers growing opportunities for out-of-the-way destinations that have not yet cashed in on the travel boom.

The Physical-Culture and Personal-Health Movements will Remain Strong, But Far from Universal

- Emphasis on preventive medicine is growing. By 2001, some 90% of insurance carriers in the United States will expand coverage or reduce premiums for policy-holders with healthy lifestyles.

- Personal wellness, prevention, and self-help will be the watchwords for a more health-conscious population. Interest in participant sports, exercise equipment, home gyms, and employee fitness programs will create mini-boom industries.

- Sixty-six percent of those answering a recent Louis Harris poll claimed to have changed their eating habits in the past five years. Americans today eat lighter fare than in 1970, consuming nearly twice as much chicken, over 25% more fish and four times as much low-fat and skim milk per capita. However, this trend has not yet had a similar impact on Europe.

- Consumer purchases show a per capita decline in annual liquor sales. Consumption of distilled liquors has declined, on average, for some two decades, while that of beer and wine accounts for more of the market. Younger drinkers have revived the once passé taste for mixed drinks, but have proved to be uncommonly responsible drinkers. Most limit themselves to one or two drinks with a meal, and "designated drivers" are standard practice.

- Smoking is also in general decline. Only 29% of American men smoke, down from a peak of 50%; 23% of women smoke, down from 32%. With state and federal cigarette-tax increases, further declines of 10% are expected. Europe, on the other hand, has yet to kick the habit, while smoking is spreading rapidly in Asia.

- There are many more magazines on health care and fitness than in the past. Again, this trend is limited to North America.

- People will be more inclined to take steps to control stress as they realize that 80% to 90% of all diseases are stress related.

Implications   Better health in later life will make us still more conscious of our appearance and physical condition. Thus, health clubs will continue to boom. Diet, fitness, stress control, and wellness programs will prosper. American tobacco companies could eventually look back on the litigation-filled 1990s as the good old days, at least in their U.S. market.

The cost of health care for American baby boomers and their children could be much lower in later life than is now believed. However, Asia faces an epidemic of cancer, heart disease, and other chronic and fatal illnesses related to health habits.

The Nutrition and Wellness Movements Will Spread, Further Improving the Health of the Elderly

- Since the beginning of the twentieth century, every generation has lived three years longer than the last, even without the antiaging treatments that research seems likely to bring.

- The average child born in 1986 will live to be 74.9 years old—71.5 years for males, 78.5 years for females.

Implications   Again, this promises a greater supply of post-retirement workers to compensate (but only partially) for the shortage of entry-level hires from the new generations.

Consumerism is Still Growing Rapidly

- A networked society is by definition a consumerist society. Shoppers increasingly have access to information about pricing, services, delivery time, and customer satisfaction from the reports of their peers "published" on the Internet. Marketers, of course, can also check the competition's offerings. This may gradually halt the decline of prices and shift competition increasingly to improvements in service and salesmanship.

- Consumer agencies and organizations will continue to proliferate.

- Better information—unit pricing, better content labels, warning labels, nutrition data, and the like—will spread through packaging, TV, and special studies and reports.

- Discount stores such as Home Depot and Wal-Mart, factory outlets, and food clubs will continue to grow in the United States, a trend that has just begun to spread to Europe and Japan.

Implications In the next 20 years, Europe and Japan can expect to undergo the same revolution in marketing that has replaced America's neighborhood stores with cost-cutting warehouse operations and "category killers."

Ultimately, fixed prices will be history, with most goods and services sold through online auctions to the highest bidders.

As prices fall to commodity levels and online "stores" can list virtually every product and brand in their industry without significant overhead, service is the only field left in which marketers on and off the Net can compete effectively.

Branded items with good reputations are even more important for developing repeat business.

The Women's Equality Movement is Becoming Less Strident, But More Effective

- Generations X and dot-com are virtually gender-blind in the work-place, compared with older generations.
- "Old girl" networks will become increasingly effective as women fill more positions in middle and upper management. The growing entrepreneurialism of women will allow the formation of really entrenched old girl networks.
- An infrastructure is evolving that allows women to make more decisions and to exercise political power, especially where both spouses work. One indication of the growing dependence on the wife's income: Life insurance companies are selling more policies to women than to men.
- More women are entering the professions, politics, and judicial positions.

Implications Whatever careers remain relatively closed to women will open wide in the years ahead.

Demand for child care and other family-oriented services will continue to grow, particularly in the United States, where national services have yet to develop.

In the long run, the need to work with female executives from the developed countries will begin to erode the restrictions placed on women's careers in Asia and other developing regions.

Now that the glass ceiling is finally broken, the women's movement will not be as important for new generations.

Family Structures are Becoming More Diverse

- In periods of economic difficulty, children and grandchildren move back in with parents and grandparents, to save on living expenses. One-third of Generation X returns home at some point in their early lives.
- Growing numbers of grandparents are raising their grandchildren, because drugs and AIDS have left the middle generation either unable or unavailable to care for their children.
- Among the poor, grandparents are also providing live-in day care for the children of single mothers trying to gain an education or build a career.
- In the United States, Vermont has just enacted the country's first law granting partners in same-sex relationships most of the legal rights formerly reserved to married couples. Similar proposals in Britain have wide support.
- Yet the nuclear family is also rebounding, as baby-boom parents adopt "family values" and grandparents retain more independence and mobility.
- Older people remain healthier longer, thanks to better medical care and more healthful lifestyles.
- The older person's family can be near, but not next door.
- Prefabricated (manufactured) housing will be cheaper than conventional construction, enabling older persons to afford housing in the suburbs or wherever they want to live.

Implications Tax and welfare policies need adjustment to cope with families in which heads of households are retired or unable to work.

In the United States, the debates over homosexuality and the "decline of the family" will remain hot-button political issues for at least two more election cycles.

ENERGY TRENDS

Despite All the Calls to Develop Alternative Sources of Energy, Oil Consumption is Still Rising Rapidly

- The world used only 57 million barrels of oil per day in 1973, when the first major price shock hit. By 1999, it was using more than 73 million barrels daily. Consumption is expected to reach 110 million barrels daily by 2020.

- However, as a fraction of the energy the world uses, oil consumption has begun to decline and is expected to drop from 40% in 1999 to an estimated 37% in 2020.

- OPEC will supply most of the world's oil. According to the U.S. Department of Energy, OPEC oil production will grow by some 24 million barrels of oil per day by 2020, to about 55 million barrels of oil per day. This is nearly two-thirds of the world's total increase in production.

- Oil production outside the OPEC nations has not yet peaked. Existing wells and refineries are operating below capacity, and at least a few non-OPEC lands have enough proven reserves to justify building more. By 2010, China, Russia, and Kazakhstan will be major suppliers—if political uncertainties in Russia and Kazakhstan do not block investment by Western oil companies.

Implications   Low oil prices in the mid- to late-1990s slowed development of fields outside the Middle East. It costs $10,000 to increase oil production by one barrel per day in most of the world, but only $5,500 for the OPEC countries. The recent surge in the price of oil—which hit a 10-year high of $37.80 per barrel in September 2000, could offer an incentive to develop new fields.

Contrary to Popular Belief, the World is Not About to Run out of Oil

- As a result of intensive exploration, the world's proven oil reserves climbed from just over 700 billion barrels in 1985 to more than 1 trillion barrels in 1990. Despite consumption, they have hovered over 1 trillion barrels ever since.

- Discoveries in the Caspian Sea area, China, and perhaps the Indian Ocean are likely to add significantly to this total.

Implications   If the price of oil rises significantly beyond current levels, new methods of recovering oil from old wells will become cost-effective. Technologies already developed could add nearly 50% to the world's recoverable oil supply.

Oil Prices are Likely to Fall Again and Remain Generally Around $20 per Barrel for at Least the Next 10 Years

- Except in times of war or economic crisis, the benchmark West Texas Crude has remained between $17 and $21 per barrel almost continuously since 1986.

- OPEC's recent decision to hold a range of $24 to $30 per barrel requires a unity of purpose that member countries have never been able to sustain for very long.

- The cost of oil production in the Persian Gulf countries is less than $1.75 per barrel.

- The Persian Gulf War showed how vulnerable oil prices really are. The attack on Kuwait more than doubled oil prices for a few weeks; they fell back to pre-conflict levels long before the war ended. Not even the destruction of Kuwait's oil fields could keep prices high. New oil supplies coming on line in the former Soviet Union, China, and other parts of the world will make it even more difficult to sustain prices at artificially high levels.

- The 20 most-industrialized countries all have at least three-month supplies of oil in tankers and underground storage. Most have another three months' worth in "strategic reserves." In times of high oil prices, customer nations can afford to stop buying until the costs come down.

Implications   In response to high (by American standards) gas prices in the summer of 2000, the U.S. government will probably boost domestic oil production and refining to build a substantial reserve of gasoline and heating oil. This stockpile would be ready for immediate use in case of future price hikes, as in the winter 2000 release from strategic reserves. This will make it easier to negotiate with OPEC.

One upward pressure on the price of American gasoline: EPA regulations mandating the production of low-sulfur fuel. They are expected to add about 4¢ per gallon to the cost of filling up. Similar regulations will add 6¢ per gallon to the price of diesel fuel.

Growing Competition from Other Energy Sources will Also Help to Limit the Price of Oil

- Natural gas burns cleanly, and there is enough of it available to supply the world's total energy demand for the next 200 years. Consumption of natural gas is growing by 3.3% annually, compared with 1.8% for oil.

- Nuclear plants will supply 16% of the energy in Russia and Eastern Europe by 2010.

- Solar, geothermal, wind, and wave energy will ease power problems where these sources are most readily available, though they will supply only a very small fraction of the world's energy in the foreseeable future. Worldwide wind-power generating capacity grew by 39% in 1999 alone.

- A new technique called muoncatalyzed fusion reportedly could produce commercially useful quantities of energy by 2020.

Implications   Though oil will remain the world's most important energy source for years to come, two or three decades forward it should be less of a choke point in the global economy. Declining reliance on oil could soon help to reduce air and water pollution, at least in the developed world. By 2060, a costly but pollution-free hydrogen economy may at last become practical.

ENVIRONMENTAL TRENDS

People Around the World are Becoming Increasingly Sensitive to Environmental Issues such as Air Pollution, as the Consequences of Neglect, Indifference, and Ignorance Become Ever More Apparent

- Soot and other particulates will come in for still greater scrutiny. Evidence has been collecting since the 1980s that they are more dangerous to human health than sulfur dioxide and other gaseous pollutants formerly believed to present the worst risks. In the United States alone, medical researchers estimate that some 64,000 people die each year from cardiopulmonary disease as a direct result of breathing particulates.

- "Acid rain" like that afflicting the United States and Canada will appear whenever designers of new power plants and factories neglect emission controls. Look for it to cover China, India, and most other industrializing countries.

Implications   Governments will take more-active measures to protect the environment. For instance, after years of ineffective gestures, Costa Rica has incorporated about 25% of its land into protected areas, such as national parks. Late in 1999, Brazil raised the maximum fine for illegal logging in the Amazon rain forest from only $2,750 to more than $27 million. It also changed legal procedures so that the fines can actually be imposed. An estimated 80% of logging in the Amazon basin is illegal.

In India, government policies consistently rate industrial development more important than the environment.

Yet in an effort to reduce air pollution, India's Supreme Court not long ago limited sales of new cars in New Delhi to 18,000 per year, less than one-fourth the average sold in recent years.

In the United States, the battle against automotive air pollution continues with an EPA mandate cutting tailpipe emissions 70% by 2004. New equipment required to meet that limit will add an estimated $100 to $150 to the sticker price of a new car, and probably more for SUVs, trucks, and vans.

Shortages of Water, and Especially Potable Water, will be a Continuing Problem for Much of the World

- According to the United Nations, most of the major cities in the developing world will face water shortages. So will one-third of the population of Africa.

- The northern half of China, home to perhaps half a billion people, already is short of water. The water table under Beijing has fallen nearly 200 feet since 1965 and 8 feet in 1999 alone.

- Water usage is causing other problems as well. For example, irrigation water evaporates, leaving minerals in the soil. By 2020, 30% of the world's arable land will be salty; by 2050, 50%. Salinization is already cutting crop yields in India, Pakistan, Egypt, Mexico, Australia, and parts of the United States.

- Pollution further reduces the supply of safe drinking water. The European Parliament estimates that 75% of the Continent's drinking water contains dangerous concentrations of nitrate pollution. Despite intensive clean-up programs, it will take between 25 and 50 years for nitrates to reach levels considered safe.

- In India alone, an estimated 300 million people lack access to safe drinking water, thanks to widespread pollution of rivers and groundwater.

- Contaminated water is implicated in 80% of the world's health problems. An estimated 40,000 people around the world die each day of diseases directly caused by contaminated water—that's more than 14 million per year.

Implications   By 2040, at least 3.5 billion people will run short of water, almost 10 times as many as in 1995. By 2050, fully two-thirds of the world's population could be living in regions with chronic, widespread shortages of water.

Water wars, predicted for more than a decade, are becoming an imminent threat. One major obstacle to the resolution of the Kashmir conflict is water: Much of Pakistan's supply comes from areas of Kashmir now controlled by India. Such problems as periodic famine and desertification also can be expected to grow more frequent and severe in the coming decades.

Recycling has Delayed the "Garbage Glut" that Threatened to Fill the World's Landfills to Overflowing, but the Threat has not Passed Simply Because it hasn't yet Arrived

- Americans now produce 4.3 pounds of trash per person per day, twice as much as they did a generation ago. Add in durable goods, such as tires, appliances, and furniture, and the U.S. waste stream has tripled since 1970.

- According to the EPA, 70% of U.S. landfills will be full by 2025; half the counties in California, home to 70% of the state's population, expect to run out of space by 2005.

- This is not simply an American problem. In London and the surrounding region, landfills will run out of space by 2012. For household trash, landfill space will be exhausted by 2007.

- In some other regions, simply collecting the trash is a major problem. Brazil produces an estimated 240,000 tons of garbage daily, but only 70% reaches landfills. The rest accumulates in city streets, where it helps to spread disease.

- Recycling and waste-to-energy plants are a viable alternative to simply dumping garbage. The United States has more than 2,200 active landfills. Europe gets by with 175.

Implications   Expect yet another wave of new regulations, recycling, waste-to-energy projects, and waste management programs in an effort to stem the tide of trash.

Industrial Development is Considered in Many Parts of the World to be Far more Important than Environmental Concerns. Broad Regions of the Planet therefore will be Subject to Pollution, Deforestation, and other Environmental Ills in the Coming Decades

- In 1999, *Samachar,* an Internet newspaper from India, asked its readers what significant problems face their country. Despite rampant deforestation, widespread air

and water pollution, loss of biodiversity, and many other such problems, environmental degradation came in next to last among 10 issues, cited by only 1% of the respondents.

- "A deep and abiding distrust of environmental imperatives has been cultivated in large segments of South Africa's population," thanks to years of apartheid-era restrictions that were often justified as environmental measures, according to a study of environmental business opportunities by Industry Canada.

- Some 70% of the energy used in China comes from coal-burning power plants, few of which are equipped with pollution controls. Scientists estimate that by the year 2025 China will emit more carbon dioxide and sulfur dioxide than the United States, Japan, and Canada combined.

Implications   Diseases related to air and water pollution will spread dramatically in the decades ahead. Already, chronic obstructive pulmonary disease is five times more common in China than in the United States. This is just a foretaste of future problems, and perhaps not the most troublesome. If global warming is still a debatable issue, it seems likely that China and India soon will produce enough greenhouse gases to prove that human activities are heating the atmosphere to destructive levels. Helping the developing lands to raise their standards of living without causing wholesale pollution will require much more aid and diplomacy than the developed world has ever been willing to devote to this cause.

Though Species Extinction May not be so Rapid as Once Believed, Loss of Biodiversity will be a Growing Worry for Decades to Come

- An estimated 50,000 species disappear each year, up to 1,000 times the natural rate of extinction, according to the United Nations Environmental Program.

- Eleven percent of birds, 25% of mammals, and 20%–30% of all plants are estimated to be nearing extinction.

- Throughout the world, amphibian populations are in decline, for reasons that, after a decade of intensive research, remain poorly understood.

- Coral reefs throughout the world are dying rapidly, again for reasons that are not entirely clear.

- The chief cause of species loss, according to University of Colorado scientists, is the destruction of natural habitats by logging, agriculture, and urbanization.

- Amazon rain forests are disappearing at a rate of roughly 25,000 square kilometers per year, twice as fast as formerly believed.

- Worldwide, some 100,000 square kilometers of rain forest are burnt each year to create farmland. Another 50,000 square kilometers are destroyed by logging. Less than 0.1% of the world's rain forest is under sustainable management.

Implications   Species loss has a powerful negative impact on human well-being. Half of all drugs used in medicine are derived from natural sources, including 55 of the top 100 drugs prescribed in the United States. So far, less than 0.5% of flowering plants have been assayed for potential pharmaceuticals.

In Indonesia, home to one-eighth of the world's coral reefs, more than 70% of reefs are dead or dying. Net losses to the Indonesian economy are estimated at between $500,000 and $800,000 per square mile of dead or damaged reef annually.

Researchers from the United Kingdom's National Environment Research Council Centre for Population Biology report that diverse ecosystems absorb more carbon dioxide than those with fewer species. Loss of biodiversity is a potential cause of global warming.

QUESTIONS

1. Why should the predictions by Cetron and Davies be considered credible?

2. What are the predictions for the future economy and the implications?

3. What are the international population growth demographics for the next 40 years?

4. What are the implications of that population growth?

5. What is the growth profile for the elderly internationally? What are the implications?

6. What is the growth of information technology internationally? What are the job implications?

7. What are the predictions for inter-society cooperation and cultural diversity? What are the implications of such cultural exchange?

8. What are the changing values of the described future society? What are the implications of these values?

9. Why will oil consumption still increase? Where are the new resources coming from?

10. What are the central international environmental concerns? What are the implications of this international concern?

11. Do you think these concerns are adequate for changing the pollution and other problems associated from consumerism and resource consumption?

A Bridge to Your 21st Century Understanding

Complete and discuss the following flowchart.

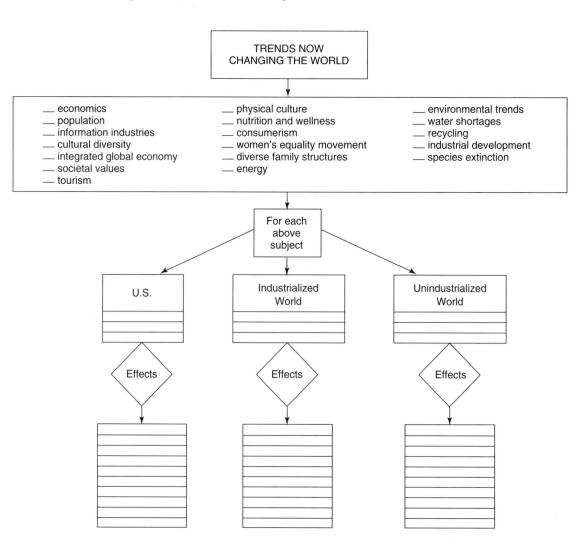

TRENDS NOW
CHANGING THE WORLD

___ economics
___ population
___ information industries
___ cultural diversity
___ integrated global economy
___ societal values
___ tourism

___ physical culture
___ nutrition and wellness
___ consumerism
___ women's equality movement
___ diverse family structures
___ energy

___ environmental trends
___ water shortages
___ recycling
___ industrial development
___ species extinction

For each
above
subject

U.S.

Industrialized
World

Unindustrialized
World

Effects

Effects

Effects

Predictions: Technology of the New Century and the New Millennium

Progress will proceed even more exponentially in everything technological. . . .

THE DIGITAL REVOLUTION

Digital World

By 2010, voice recognition will be available in computers and all sorts of digital devices. We should be able to carry on direct conversations with computers, Internet browsers, automobiles, thermostats, TVs, VCRs, microwaves, and other pieces of equipment.

Personalized World through Computers and Hyperlinks

In the future, bandwidth, the capacity of fiberoptics and other "pipelines" to carry digital communications, will become freer, which means that all types of content—including data, all forms of shows, books, and home videos—will be on-line and available to anyone in the world. The practical applications of this ability would allow marketing and subscriptions to be tailored to the specific needs and tastes of individuals. This ability to customize information then would allow the products to be specialized for each user and will change the world from a mass-market world to a more personalized one (Isaacson, 1998).

Further Miniaturization of Microprocessor Design

The following items represent future advances in the design of microprocessors:

- The use of quantum dots and other single-electron devices, for which electrons can be used as binary registers to represent the 0 or 1 of a data bit.
- The development of data storage systems using biological molecules, resulting in molecular computing. Optical computers for which streams of photons could replace electrons would be possible if a biological light-sensitive molecule were used. With biological molecules, self-synthesis would be possible through the life systems of the microorganisms. For some estimates, some photonically activated biomolecules could be linked to a three-dimensional memory system that would have a capacity 300 times greater than that of today's CD-ROMs.
- Nanomechanical logic gates where beams one atom wide would be physically moved to carry out logical operations (Patterson, 1995).

All-Optical Networks

Already, fiberoptic networks transmit data 10 to 100 times faster than standard electronic wiring, but in the future, light waves will be used for most of the transmission of voice, video, and other data along the networks' pathways. The speed and capacity of fiberoptics in the future will be used such that—in theory—a single fiber has the capacity to transmit 25 terabits (trillions of bits) each second, which is comparable to transmitting all U.S. Mother's Day telephone calls simultaneously in a second. The only time the signal would become electronic would be when it moves into the circuits of a computer or into a lower speed electronic network (Chan, 1995). At present, the state of technology allows data rates of the order of tens of billions of bits per second. Once the potential capacity is developed, the transmission rates will allow hundreds of trillions of bits per second. The future potential of optical networks will allow the transmission of 25 billion telephone calls simultaneously. In other words, consider that with the population of the earth at 6 billion people, every person on earth could talk at the same time using one fiber whose dimensions are ten times smaller than a human hair.

THE BIOTECHNOLOGICAL REVOLUTION

In the future, we will have the technological capacity to engineer and change human DNA and build on the emerging capabilities that we have already demonstrated with plants and animals. The challenges will not only be scientific, but also moral.

The Human Genome Project

With the mapping of all of our genetic material, which should be completed by the year 2005, scientists will begin a new age for biology in which we will understand biological processes and the causes of disease and behavior from the level of genes. Central control genes for the formation of organs, behavior characteristics, and other genetic human traits can be studied and understood. Genetic flaws connected to cancers, leukemia, dyslexia, Alzheimer's disease, and even alcoholism and obesity will be traceable, and their discovery contribute to the development of more effective treatments for these disorders. The moral issues of knowing genetic codes and their outcome, predicting susceptibility to disease, the relation of genetic testing to employment and insurance issues, manipulating genes, and knowing the genetic makeup of potential mates are just some of the complex ethical topics that accompany genetic knowledge and its effects (Olesen, 1995).

Plant and Animal Gene Alteration

Some scientists feel that the alteration of genes to modify and create new plants and animals has the potential to alter the 21st century even more dramatically than computers altered the 20th century. The alteration of plant and animal life-forms has the potential to feed the hungry, alleviate human suffering, provide new medicines and transplant organs, and open possible new sources of energy. But the moral decisions and consequences of such genetic procedures are complex and include the issues of genetic determination, discrimination, and unforeseen long-range problems due to the release of new genetically engineered life forms. As an example of such a public-policy and moral issue, cell biologist Stuart A. Newman and Jeremy Rifkin, president of the Washington-based Foundation on Economic Trends, have applied for a patent on genetically engineered life-forms that fuse human and animal embryonic cells. The purpose behind their patent application is to prevent or delay the use of such technology to better control and regulate such use and to give the public time to develop its moral policy on these issues (Travis, 1998).

TELECOMMUNICATIONS

In the next decade, cellular services and wireless networks will increasingly provide personalized service all over the globe, regardless of the remoteness of the location or the previous lack of telephone services. Cellular networks and other personal communications services will use digital air-interface standards and will shrink the size of the cell station, providing more compactness, quality, and efficiency. Every year and a half, the digital chips required for wireless operations shrink by 50% so that applications can also continue to shrink. Smaller and smaller cellular devices, such as high-quality wireless wristphones, tiny telephone devices in computers, and radio modems are part of this outcome. In just a few years, wireless faxes and video mail will be commonplace globally. Telephone service will be available where it has never been before; today, half of the people in the world have never made a telephone call, but in the future, anyone in any location will be able to do so. Some developing countries will soon be jumpstarting their telephone technology by eliminating the traditional network foundation and beginning with a wireless infrastructure (Zysman, 1995).

THE INFORMATION REVOLUTION
Artificial Intelligence

In the future, computers will be able to make thinking decisions and create new thoughts from stored information. Software will also become intelligent and will act independently in order to ease burdens on computers. Within just a few years, computers will build information and conclusions and automatically coordinate applications and output such as spreadsheets, databases, document preparation systems, and on-line search engines. Content will be automatically checked by word processors to determine completeness, accuracy, and the existence of adequate explanations. Searches for specific content will be possible not just from key-word searches, but also from actual content searches (Lenat, 1995).

The Internet

Information will continue to be increasingly available on the Internet and increasingly used by a predicted 30 million households by the year 2000. Cyberspace will continue to transform with new "non-PC devices" such as Internet TVs and Internet telephones. Commerce, all types of information, games, and personal buying-pattern databases, will all continue to become even more available and part of our everyday lives as cyberspace continues to develop into its global potential (The Web: Infotopia or Marketplace?, 1997).

Additional Leaps

As time goes by, software will become increasingly autonomous with the use of software "agents" that help complete users' goals. These agents will be like many digital proxies, simultaneously guiding users through technical complexities, teaching them, searching for information, performing transactions, or even representing people in their absence. They could even perform as personal secretaries—helping to remind, coordinate, and extend capabilities. Rather than manipulating a keyboard or a mouse, the user would use his or her voice or gestures to communicate with the software, and the agents would appear on the screen as people with simplified facial expressions to communicate their current state (Maes, 1995).

Virtual reality (VR) will also be used to a greater extent in a variety of applications, since it permits people to respond in a re-created environment. Presently, VR provides perceptions of sight, hearing, touch, and response to movement. This range will be extended to perceptions of force, resistance, texture, and smell as well. Applications of VR will include the performance of complex and delicate tasks in hazardous or alien environments, extensive and detailed training operations, the design of prototypes and models of all sorts of systems, and the creation of new artistic and social humanistic expressions and dimensions (Laurel, 1995).

THE WORLD OF WORK

In the future, the equivalent of today's upper crust of white-collar workers will need a graduate degree, and they will be classified as some type of "symbolic analyst," possessing some of the most intensive knowledge in the future economy. At the bottom of the ladder will be low-skill and low-paid personal service workers whose jobs will not have been too affected by the computer or information explosion. Most of the changes will occur in the middle class, made up of "technicians"—the mechanics of the computer age; their positions will require education beyond the high school diploma. This middle class will require more education and training than today's middle class. The highly educated, high-earning professionals to be known as "knowledge workers" will be exposed to downsizing with the arrival of more consultant-type, "paid by the job" employment arrangements. These knowledge workers will telecommute to their various offices and contract sites. Many people feel that workers might move to a 25-hour workweek, with higher pay and benefits allowing more time for cultural activities and nonprofit work (McGinn and Raymond, 1997–98).

NANOTECHNOLOGY

Many futurists feel that nanotechnology will provide man with the ability to participate in the creation of materials on the atomic level and that some of this capability will be realized in the next 10 to 20 years. Nanotechnology is technology that occurs on the microscopic scale of nanometers. (One billion nanometers are in a meter.) Atoms at this microscopic size level join together to form molecules that, in turn, form materials. In the nanotechnological universe, atoms can be positioned in various configurations to design different specified materials according to the material's specific atomic formula. The technology of MEMS—microelectromechanical systems—already exists. These systems are tiny micromachines that react to stimuli like light, sound, and motion. When connected to microchips, MEMS can be programmed to take actions such as those of sensing devices for human health needs and can perform the unlimited applications of control devices. "Microprocessors defined the 1980s, and cheap lasers allowed the telecommunications revolution of the 90s. MEMS and sensors generally will shape the first decade of the next century," states Paul Saffo, an analyst with the Institute for the Future in Menlo Park, California. The process of MEMS becoming less expensive and smaller will result in such applications as smart materials that react to stimuli, houses programmed for energy needs, unlimited technological sensing devices, and tiny robots that detect disease in the bloodstream—and, of course, this is only the beginning (Rogers and Kaplan, 1997–98).

THE MEDICAL FUTURE

New Surgical Techniques

In the 21st century, surgeons will develop more precise and less invasive techniques. Radio waves and lasers will replace scalpels, while light-activated drugs will pinpoint cancer treatments to be more specific and less harmful. Surgery involving small incisions and keyhole surgery are becoming developed procedures, using thin probes, cameras, miniaturized lights, and tiny and remotely controlled surgical tools. Even organ extraction and transplants can be accomplished with less invasive surgery requiring smaller incisions. Doctors are also discovering that directing radio waves by the use of electrodes and radio-wave generators can diminish injuries without the pain and difficulties of surgery. The possibilities of the use of this medicine include the tightening of ligaments, the destruction of tumors and other unwanted tissues, the cessation of abnormal uterine bleeding, the treatment of benign prostate enlargement, and the treatment of heart arrhythmias. Prescription medicines for the 21st century will also be more narrowly targeted and will have few side effects (Cowley and Underwood, 1997–98).

Gene Therapy

Gene therapy will also be part of the new technological medicine bag of the 21st century, and it will cause another medical revolution, since the use of selected corrective genes in a patient's cells can potentially cure or ease many of today's diseases. This new approach can be very effective, as so many diseases derive from the malfunctioning of one or more genes. For example, more than 4,000 conditions, such as severe combined immunodeficiency and cystic fibrosis, are caused from the developmental damage of a single gene. With gene therapy, new genes can be delivered through modified viruses to provide for new genetic coding to prevent or treat diseases. Other methods of delivery involve directly introducing the corrective gene into the tissue where it is needed. While not possible now, in the future it is hoped that physicians will simply be able to inject the new gene carriers into the bloodstream, where the new genetic carriers will locate their targets cells and simply transfer the new genetic information. Within 20 years, it is predicted that gene ther-

apy will be part of the regular approaches to treat and cure diseases. Today, many diseases are already being treated in clinical trials of gene therapy (Anderson, 1995).

OTHER NEW SCIENTIFIC TECHNOLOGIES

Microchip Sensor

A microchip sensor is an optical version of a microchip with tiny lenses and filters that use laser light. The chip is an inexpensive sensor that can be made to detect or sense just about anything. Food processing plants are interested in using this sensor to detect potentially deadly bacteria. It also could be used to detect chemicals in medical patients, replacing their need to undergo frequent blood tests. Other applications of microchip sensors would be to detect pollutants in smokestacks, streams, and automobile exhaust systems and to detect pesticides and fertilizers in farming areas (*Scientists Say Tiny Sensor Has Giant Potential,* 1996).

Molecular Design of Materials

Material design and manufacturing in the future will involve manipulating molecules to change their characteristics. A material could be made stronger or lighter or could be altered to make a new material by changing its molecular structure. The strengths of this approach to design and construction are its access to unlimited resources and its versatility, since the number of molecules in the world is virtually unlimited (Olesen, 1995).

The Futurist's Top Ten Technologies for the Next 10 Years

1. Understanding the genetic basis for human disease and characteristics on the gene level and being able to treat them on that gene level

2. Supermaterials designed on the molecular level

3. High-density energy sources that are very small and portable

4. High-definition digital television

5. A laptop computer that is interactive, wireless, and about the size of a pocket calculator, and smart cards that perform many daily activities such as those of keys, records, money, etc.

6. Smart manufacturing processes that will create individual products for individual consumers

7. Antiaging cosmetics and products

8. Diagnostic health sensors to detect diseases and physiological anomalies

9. Automobile hybrid fuel using reformulated gasoline along with some other type of fuel to be more efficient, economic, and to produce fewer emissions

10. Computer software and other products that combine education and entertainment (Oleson, 1995)

QUESTIONS

1. What do you think are the most important technological discoveries mentioned in this chapter for the 21st century? Rank the five most important ones, explain why they are important, and give the reasons for your particular ranking.

2. Of the subheadings in this article, only three include the word "revolution": the digital, the biotechnical, and information revolutions. Why do think those classifications were made in this way? Do you agree with these classifications and subheadings? Explain why or why not. What would you change about them?

3. Can you name some important developments and technologies for the 21st century that were not mentioned in this chapter?

4. Examine the technologies mentioned in this chapter, and describe in a brief narrative what would be the cumulative effect of all these technologies. Give a brief description of what you think life will be like in the future due to the effects of these technologies. What are the pros and cons of this future projection?

5. Think about and describe the specific type of education that will be needed in this future life.

6. Which new technology interests you the most? What applications do you see for it? Describe possible innovative designs or applications for that technology.

7. Research additional specific applications for the new technologies mentioned in this chapter in various periodicals and articles.

8. Find and list new projected technologies for the 21st century that are not mentioned in this chapter.

REFERENCES

Anderson, W.F. (1995, September). Gene Therapy. *Scientific American: Key Technologies for the 21st Century.* pp. 124–128.

Chan, V.W.S. (1995, September). All-Optical Networks. *Scientific American: Key Technologies for the 21st Century.* pp. 72–76.

Cowley, G., and Underwood, A. (1997–98, Winter). Surgeon, Drop That Scalpel. *Newsweek Extra 2000: The Power of Invention.* pp. 77–78.

Isaacson, W. (1998, April 13). Our Century and the Next One. *Time.* pp. 70–75.

Laurel, B. (1995, September). Virtual Reality. *Scientific American: Key Technologies for the 21st Century.* p. 90.

Lenat, D.B. (1995, September). Artificial Intelligence. *Scientific American: Key Technologies for the 21st Century.* pp. 80–82.

Maes, P. (1995, September). Intelligent Software. *Scientific American: Key Technologies for the 21st Century.* pp. 84–86.

McGinn, D., and Raymond, J. (1997–98, Winter). Workers of the World, Get Online. *Newsweek Extra 2000: The Power of Invention.* pp. 32–33.

Olesen, D.E. (1995, September–October). The Top Technologies for the Next Ten Years. *The Futurist.* pp. 9–13.

Patterson, D.A. (1995, September). Microprocessors in 2020. *Scientific American: Key Technologies for the 21st Century.* pp. 62–67.

Rogers, A., and Kaplan, D. (1997–98, Winter). Get Ready for Nanotechnology. *Newsweek Extra 2000: The Power of Invention.* pp. 52–53.

Scientists Say Tiny Sensor Has Giant Potential. (1996, November 20). *CNN Interactive.* Available: http://www.CNN.com

The Web: Infotopia or Marketplace? (1997, January 27). *Newsweek: Beyond 2000.* pp. 84–88.

Travis, J. (1998, May 9). Patenting the Minotaur? *Science News.* p. 299.

Zysman, G.I. (1995, September). Wireless Networks. *Scientific American: Key Technologies for the 21st Century.* pp. 68–71.

Telemedicine: The Health System of Tomorrow

Thomas Blanton

David C. Balch

The place: North Carolina's Central Prison in Raleigh.

The problem: Inmate No. 35271 (not his real number) has an unusual skin rash that the prison doctor can't diagnose. The prison administration does not want to be sued for inadequate medical care for its prisoners, but at the same time, it does not want to run the security risk of transporting a dangerous felon to an outside specialist. Bringing the specialist to the prison would also present problems.

The solution: A telemedicine link with East Carolina University School of Medicine. In a special telemedicine booth a hundred miles away, a dermatologist sits in front of a video console and interviews the patient while directing the prison nurse where to point the tiny dermatology camera. He diagnoses the condition and prescribes a course of treatment.

The result: The prisoner's condition is treated, the prison doesn't get sued, and the state saves potentially thousands of dollars.

A far-flung view of the future? Not at all. Telemedicine is already in practice in settings such as this and may become nearly universal in the not-too-distant future.

CREATING PORTABLE HEALTH CARE

The phrase "health care reform" has been used in recent political years to describe changes in the way health care is

Originally appeared in September–October 1995 issue of *The Futurist.* Used with permission from the World Future Society, 7910 Woodmont Avenue, Suite 450, Bethesda, Maryland 20814. 301/686-8274. (http://www.wfs.org)

financed. Even though health-care-financing reforms failed to make it through the last Congress, health care is changing in the United States. The real reform is developing in the way health care is delivered.

As technology has developed over the years, medical diagnoses have become more accurate and treatments more effective. But the x-ray, magnetic resonance imaging machines, and other tools are not portable. Treatment and therapeutic facilities have been centralized in hospitals and rehabilitation centers, so that patients have to travel distances and wait in doctors' offices to be seen. As a result, doctors don't get to see patients on a regular basis and end up treating symptoms and diseases rather than people.

Proposed solutions have been varied, from holistic health to formerly fringe ideas such as herbal medicine and massage therapy, which are gaining in popularity. But technology may soon allow the house call to return to mainstream medicine, with data being transported rather than people.

The idea of telemedicine has been around for almost as long as there has been science fiction and has been in actual practice since the late 1950s. But the recent upsurge in medical costs, combined with advances in technology, are now making telemedicine a widespread reality.

TELEMEDICINE IN PRACTICE

There are now 25 telemedicine projects in the United States, according to the American Telemedicine Association. At present, these programs tend to serve rural areas where population is sparse and income low. Areas such as western

Texas, eastern North Carolina, rural Georgia, and West Virginia are the most active in using telemedicine. A recent study by the consulting firm of Arthur D. Little, Inc., showed that healthcare costs could be reduced by as much as $36 billion a year if health information technologies such as telemedicine were widely used.

In 1989, the U.S. Public Health Service gave partial funding to Texas Tech's MedNet program, which linked health-care practitioners in hospitals and clinics from 37 rural communities in western Texas. In its first year, this project reported a net savings of 14% to 22% in the cost of health-care delivery. The agencies involved reported a decrease in salaries and ambulance costs along with an increase in earned revenue.

One important issue in developing telemedicine has been states' costs in providing health care in prisons. California alone spends $380 million per year on prison health care. With the popularity of laws such as "Three Strikes and You're Out," prison populations will increase. So will the cost of prison health care.

In 1990, a prison inmate sued the state of Florida over inadequate health care and won a $1-million settlement. That prompted Florida to begin prison telemedicine on a pilot basis.

The next year, North Carolina's largest prison—Central Prison in Raleigh—contracted with the East Carolina University School of Medicine to provide telemedicine services. Since that time, physicians from the medical school have provided more than 350 consultations with the prison. In 1994, the school built the first "telemedicine suite" in the nation—four consultation booths specially designed for telemedicine.

In addition to the prison, two hospitals in rural eastern North Carolina have consultation links with the East Carolina medical school, and two more are scheduled to come on-line in 1995. The Medical College of Georgia, too, is building a network of 60 sites around the state for telemedicine consultations.

Just as cable television began as a way of serving rural areas outside of broadcasters' range, telemedicine is beginning as a way of reaching out to medically underserved areas. As the possibilities of cable television offerings became better known, metropolitan areas became its heaviest users. Similarly, as physicians and patients see the possibilities in telemedicine, it should spread more widely and into urbanized areas.

Patients recovering from heart attacks can already put on a headset, connect the electrocardiogram (ECG) wires to their chests, and ride their stationary bikes. The patients stay at home and communicate by telephone with the hospital. The same phone wires that carry their conversation also carry the ECG information to the medical technician on the other end. Up to five patients at a time can be served this way, and they are in contact with each other as well as the hospital.

A pilot program for this type of cardiac rehabilitation took place at St. Vincent Charity Hospitals in Cleveland, Ohio, during the late 1970s and worked so well that it became a model for other areas, including Nashville, Tennessee, and Greenville, North Carolina. Doctors at East Carolina University are setting aside a cable channel so that the cardiac rehab patients can have visual contact with the hospital staff. As technology advances, there may eventually be two-way video interaction between the patient and the hospital.

HEALTH CARE AT HOME AND WORK

As the technology advances, telemedicine will spread out over the information superhighway to link rural hospitals with smaller medical centers; then local doctors' offices will come on-line; and, finally, telemedicine will reach into private homes.

QUESTIONS

1. Why is telemedicine gaining popularity and becoming more of a reality?

2. How much in annual health-care costs could be saved if technologies such as telemedicine are used?

3. Why is telemedicine an especially attractive option for use in prisons?

4. Have telemedicine sites proved to be successful? Explain.

5. What are other possible applications of telemedicine?

6. If you lived in a rural area hundreds of miles from a hospital or doctor, would you favor the use of telemedicine? Explain your position.

7. Explain the pros, cons, and implications of using telemedicine as a future approach to medicine.

A Bridge to Your 21st Century Understanding

Complete and discuss the following flowchart.

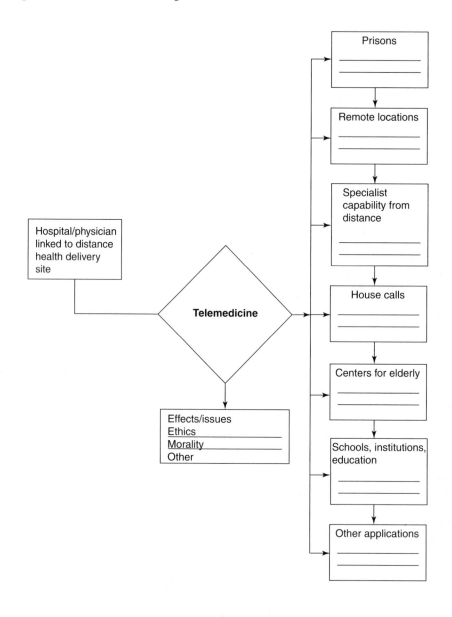

The Information Warrior

ELLISON C. URBAN

In war, in the end, it all comes down to the soldier. Electronics already equips him to fight day or night, whatever the weather, move with extraordinary speed and agility and shoot with unprecedented precision and deadliness. At the same time, it arms staff officers with the means to gather and evaluate a host of strategic and tactical information. Regrettably, the two tasks have yet to be well integrated. The soldier making life-and-death decisions in the thick of battle is seldom aware of the bigger picture. Often neither his fellows nor his commanders have good ways of dispersing what is aptly known as the fog of war.

In future wars, control of lethal weaponry and "information weaponry" will as never before be shared at all ranks of combat personnel. Combat and intelligence gathering will take place simultaneously and synergistically with the advent of the tactical information assistant—TIA, as it is called by the Advanced Research Projects Agency (ARPA) of the U.S. Department of Defense. This gentle name is applied to devices that use not bullets for ammunition, but bits.

Several types are currently in field trial. To pick out a few, there are laser rangefinders that instantly send coordinates and detailed descriptions of targets, as well as live video of the battle scene, to fellow soldiers and to remote intelligence officers and commanders; body-hugging computers and head-mounted displays with which a soldier can walk up to equipment and diagnose its status; computer-generated maps that change scale and overlay differing

types of information in response to voice requests. In design stages is a vest and display that will integrate *all* electronic systems, from voice-controlled computers to radios.

HUMIONICS ON THE MARCH

In the early days of aviation, electronic equipment was simply jammed into any available space on an aircraft without much consideration for aerodynamics, weight distribution, power requirements, interoperability, or user interface. The problems caused by this haphazard approach led to the development of a discipline called avionics. "Integrating" electronics with humans requires the creation of a similar discipline, what I have termed humionics.

The related field of ergonomics, and traditional human-factors studies in general, is primarily concerned with a person's functioning in relation to manifestly physical objects and situations. Ergonomics clearly is important for the already weight-burdened soldier, who before each mission often makes tradeoffs among food, water, bullets, and equipment. Humionics, in contrast, considers the human as a system and addresses how that system operates in its electronic environment.

Humionics is especially important for the increasing number of soldiers who carry computers, radios, Global Positioning System (GPS) displays, cameras, microphones, and sensors, along with all of their other mission-essential equipment. Soldiers commonly use straps or Velcro to affix each piece of equipment on their bodies and connect them with wires. A humionics approach eliminates duplicate components among all these devices, such

Repeated with permission from Urban E. "The Information Warrior." *IEEE Spectrum,* November 1995. ©1995 IEEE.

as keypads, displays, batteries and associated electronics, and uses only the unique front-end circuitry of, say, radios and GPS units.

The resulting system, managed by a vest-pocket computer, would use one battery, one display, one input device, and so on, with a concomitant reduction in overall size, weight, and power consumption. Equipment would be designed with uniform interfaces (either serial or wireless) to the body-worn local-area network (bodyLAN) within the vest which would eliminate the current babel of incompatible specifications. And, as discussed in more detail below, humionics covers safety issues such as placement of equipment on the body, heat dissipation, radio frequency energy, and prevention of electrical shock.

FIRST STEPS

The TIA program of the Advanced Research Projects Agency (ARPA), Arlington, Va., was started just this year. In keeping with the agency's overall mission, much of the technology has never been explored before, and is being developed with extensive cooperation between corporate and university research laboratories.

But TIAs are not blue-sky projects. Despite the short time that has been devoted to their development, all but one (the bodyLAN vest) are now undergoing field tests in the U.S. armed forces. They are critical components in warfare in the information age, part of what is being called the "horizontal battlefield." In this arena new strategy and tactics are required as decision-making becomes more and more decentralized and information flows in all directions among battle echelons.

The Technology Advanced Mini Eyesafe Rangefinder (Tamer) exemplifies a special-purpose TIA. Its prime contractor is Motorola Inc.'s Government and Space Technology Group, based in Scottsdale, Ariz. Tamer adds information technology to Melios, a widely used U.S. Army laser rangefinder. Melios, which in some versions has an internal compass and meter for vertical tilt, is hand-held or mounted on a tripod. To use the rangefinder, the operator puts its 7 × 50 optics on a target. That target is either a true military target or any landmark that, if it is marked on the soldier's paper map, will help him find out where he is.

Once he has identified his own location, the soldier presses a button to fire an eye-safe laser. The range to the target, which can be up to 10 kilometers away, is displayed on a light-emitting diode in the optics module. The operator then performs what is known in the U.S. Army as a Salute report, the word being a mnemonic for the size of the target or targets (how many of them), activity, location, unit (what type they are, say, armor), time, and equipment (say, T-72 tank). The soldier reads the Salute report information over a radio voice channel to a command post, which relays the information to a gun emplacement.

The entire operation, from self-location to ascertaining the Salute information, typically takes 5 to 10 minutes. The target's location is derived from the soldier's own position and its range bearing, and elevation in relation to the target. Further time is then spent by the recipient of the message, who calculates the coordinates of the target.

Moreover, the accuracy of the target coordinates depends on the skill of the soldier using the rangefinder and the hospitality of the terrain. Examples were demonstrated time after time in the Gulf War: it is extremely difficult to determine a location in the desert because there are few mapped terrain features to lock on to.

TAMER IN ACTION

Tamer was tested with the U.S. Army 2nd Armored Division in Fort Hood, Texas, and with the U.S. Army's Program Manager for Night Vision at Fort Belvoir, Va. (the developer of Melios). Fort Hood gave ARPA only three months to move Tamer from the drawing board to hardware. The tight deadline, fortunately enough, made for some effectively pragmatic decisions, if only for this version. The original Melios was simply sliced more or less down its middle, between its optics module and battery compartment, and the new components added.

The additional components include a 3.8-by-3.8-cm multi-chip module to communicate with the Global Positioning System, as well as a microcontroller, display driver, and input-output circuitry (the GPS circuitry was developed by the Advanced System Technology Office at ARPA). All circuitry is mounted on PCM-CIA cards (now known as PC cards) and connected via a standard bus. A 3.8-by-3.8-cm GPS antenna and display were added on the outside of Melios, as were two control buttons on the bottom.

The Global Positioning System was developed by the U.S. Department of Defense to allow simple, but accurate, navigation anywhere in the world at any time. The system depends on a constellation of 24 satellites that orbit the earth at a very high altitude. The GPS satellites transmit periodic coded messages that can be picked up by GPS receivers located on earth. A receiver can determine its own location only if it picks up at least three satellite signals.

Soldiers and officers in a future battle will use tactical information assistants (TIAs), now in trials or development. This soldier's position is monitored by a global positioning system (GPS) satellite; his position and tactical data are automatically updated on his map, which he queries orally. He aims a laser rangefinder at a target, while filling out an intelligence report on it. The reflected laser beam and GPS data provide target coordinates, which, along with the images he sees through the rangefinder, are sent to the rear, where command and control officers transmit the coordinates to an attack aircraft. The soldier's position and video data are shared with others down the line.

A Future Battle Scenario Using TIAs

The satellite signals are transmitted in two distinct forms: one for civilian use and the other for military use. The military code provides higher positional accuracies and protection against jamming. Civilian accuracies can be increased by a method called differential GPS, in which two receivers are used. One receiver is placed at a known base location while the other is free to roam. A location derived from the signals received by the satellites at the base station is compared with the known location. The difference is then transmitted to the roaming receiver so that it can modify the location that it derives from the satellites. Currently Tamer uses only civilian code accuracies, but both differential GPS and military code devices are being investigated.

When an operator positions the crosshairs on a target and fires the laser, the coordinates of his location are acquired from the satellite by the GPS unit. The Tamer computer takes the latitude and longitude of the rangefinder and its compass bearing and tilt angle and from them calculates the latitude and longitude of the target. These coordinates, along with associated time measurements supplied by the GPS satellite, are automatically loaded into a Salute report form displayed on a liquid-crystal display. Other items in the report are added by the operator, using the two thumb buttons, through menus and submenus for each item.

When the Salute report is completed, it is sent by means of an RS-232 connector to an Army radio, and then transmitted to its destination. The soldier may also fill in reports requesting an attack on the target (call for fire) or on the battlefield's Nuclear, Biological, or Chemical (NBC-1) status.

Current development work on Tamer will eliminate the external display and insert a miniature, high-resolution display in the optical path—the view through the eyepiece—so that no light will be visible at night. The form is laid over, but does not obscure, the view through the eyepiece. It will also add a solid-state compass developed by Honeywell Inc., plus a pedometer and auxiliary computational circuits. Battery life will be increased by using the GPS only to acquire a quick fix. The computerized compass and pedometer will then perform dead-reckoning calculations until another GPS fix is obtained.

Even more significant improvements to the system are planned with installation of an electrically erasable programmable ROM or miniature hard drive (currently a gigabyte can be stored on a 2.5-in. drive). Either will permit maps to be displayed along with a dot indicating current position. Adding a charge-coupled video device in the optical path will allow pictures to be recorded and then annotated with Salute data. A wireless transmitter at standard radio frequencies will be added, as will a receiver that will allow messages and updated tactical information—from soldiers on the field or from back-base intelligence officers—to be shown on the miniature display.

Other ideas, such as obtaining accurate bearing information from GPS data, improved lasers, and night vision, are under investigation by Motorola. It is not envisioned that each Tamer contain all the functions mentioned. Rather, because of the Tamer's modular architecture, an operator will be able to specify a particular configuration that meets his needs with minimum weight and cost.

A MAP THAT LISTENS

Pathfinder is a TIA program at BBN Systems and Technologies Corp., Cambridge, Mass., that should enable a small team to navigate and coordinate its movements with electronic maps. The Pathfinder is based on a Grid computer with a touch-sensitive liquid-crystal display. A handheld GPS unit and small radio are connected to the computer. Typically, 1:50,000-scale maps are displayed on the screen. Using the touch-sensitive screen, the mission commander enters the way points selected; lines linking the points complete the planned travel route. Symbols and lines can be overlaid on the map to show the current tactical battle plan.

From the GPS data, a dot appears on the map showing the user's position. As the user (and map) move, tracks are drawn on the map to indicate travel history and make visible any deviation from the plan. Information on location is transmitted over the radio to other team members to help them coordinate their movements and improve everyone's knowledge of the military situation. A limited amount of message traffic can also be transmitted and received.

Pathfinder has been demonstrated in a variety of field exercises, primarily with the Marines in the Adriatic Sea. Its most important application for the time being is in helicopter navigation. Using Pathfinder, pilots have learned to navigate over long distances without looking outside, a feat that typically cuts 20 minutes off a one-hour trip over unknown terrain. In addition, pilots sometimes become disoriented or miss visual way points when flying low and close to the earth. Pathfinder allows them to follow the way points on the map and correlate the map better with the terrain.

The light from the Pathfinder display, however, interfered with the night vision goggles of pilots. One remedy was to coat the display with a blue gel. But this only created another problem: it is hard to use a touch-sensitive screen, even with pens, when it is smeared with gel. This problem is one of the principal reasons ARPA is working towards voice input for the system.

Pathfinder also has proven useful in training, because it logs and retains all information about its position. The positions of teams in exercises are transmitted to a central site where the entire mission is observed in real time on a single screen. The information may also be replayed for debriefings after the exercise, and may be utilized for future classroom training sessions.

TALKING YOUR WAY OUT

The VoiceMap TIA takes Pathfinder to a higher level. It relies on a smaller, lighter computer from Texas Microsystems Inc., Houston, called the Grunt, which can be strapped to a pilot's leg. The pilot's hands are not tied down, because the computer accepts voice commands and queries and does so regardless of the user. (This so-called speaker-independent voice recognition is computationally tougher than training the computer to recognize just one speaker.)

One intractable problem is that large maps cannot easily be displayed on small computer screens. Instead, the user quickly scans large areas by telling VoiceMap to zoom in and out of the map display, go left or right, and so on. Another problem is the loss of detail due to the poorer resolution and inferior color range of small computer screens. However, details—of any type—can be embedded in databases carried with the unit and selectively applied as graphical overlays, which in fact is a significant improvement over single maps with a welter of information.

The soldier asks VoiceMap for what he or she wants depending on the situation. Questions such as "Where is the water tower?" or "Which way to the river?" are answered instantly by a graphical updating of the map. Other queries might involve information not contained in the user's database, but which is supplied by radio from other databases, such as "Where is Company B?"; in a related scenario, the user would send commands to update remote databases, for instance, "Send my location to the 1st Squadron."

SMART MAPS AND HYBRIDS

Under development is the addition of artificial intelligence to Voice Map. With these techniques in place, the user will be able to ask more general questions, such as, "What is the best route to way-point Delta?" Developing the artificial intelligence techniques is a formidable task. Depending on fuel level, enemy fire, mission goals, or a host of other considerations, imagine what in this example the concept "best" could signify.

Another project in the works is connecting VoiceMap to Tamer, the GPS-aided rangefinder/report-generation system. In a typical scenario where this hybrid technology would be used, Tamer grabs the coordinates of a distant object and loads them into VoiceMap. The user then annotates the data by voice, saying, for example, "enemy tank," and the object's icon appears in the correct position on the map. The information is transmitted to the command headquarters of other soldiers.

This simple scenario shows how intelligence officers can multiply their assets in the field using TIAs. Even while executing his or her combat roles, each soldier is also an intelligence agent. Moreover, the military expects that soldiers will generate entirely novel and unplanned uses for TIAs.

All military equipment must be kept in repair—a job that can also be aided enormously by TIAs. The VuMan TIA is a wearable form-fitting computer and head-up display at present used in equipment maintenance. (A head-up display can be seen through but also superimposes on the field of view computer-generated data and graphics.) The user interacts with the computer (which is hardened against shock, temperature, water, and dirt) by way of a rotary dial and a single selection ring. Menu options are presented on the head-mounted display, and the user scrolls through them by turning the dial, and choosing options with three buttons that surround the center dial. The dial and buttons are purposely made large so that they can be handled through layers of clothing.

The menu selections are simple but extensive. Different software and datasets for different pieces or classes of equipment are uploaded to the VuMan by means of a serial cable. For example, one software package now in trials contains some 600 menu and sub-menu choices. The software is based upon the VuMan Hypertext Language (VHTL), developed at Carnegie Mellon University, Pittsburgh. The VHTL simplifies the task of creating document systems that integrate forms, references, images, and complex nested menus.

When not clambering over, around and under equipment, the maintainer needs to inform logistics officers of the status of the equipment, order parts, and perform other file-keeping operations. For these tasks a plug-in serial port connects VuMan and another computer, usually a personal computer (which itself could be networked to remote databases). VuMan then simply disgorges itself of the data entered by the user as he went about the inspection and repair.

Experience using VuMan in the maintenance of heavy vehicles—predominantly tanks and tracked vehicles—has been encouraging. First off, maintenance no longer needs the usual minimum of two people, one to actually perform the repairs or inspection, and the other to guide him through it by reading aloud from the manuals. Other results have been a 40 percent reduction in the time taken to perform inspections, and a reduction of about one third in the time it takes to fill out the mandatory forms about the procedures undergone and the status of the equipment.

And, of course, the system features one of the saving graces of computer storage: the VuMan is, on average, less than one thousandth of the weight of bound documentation and manuals needed to hold the equivalent amount of data. As for VuMan's "sponsors," the prime contractor is Carnegie Mellon University, while subcontractors are Stanford University, Lockheed Martin, Telxon Corp., Apple Computer, and IBM.

THE BODYLAN VEST

Architecture and component requirements (the CPUs, memory, bus, and so forth) are now being defined for what in some ways is the most ambitious TIA program: an electronic vest that will be a TIA routing station, so to speak. The fundamental design includes the vest, a head-mounted display, and a microphone. Initially the components will include a computer, data storage devices, GPS device, wireless modem, and a plug-in diagnostic interface module.

The first application being developed is the maintenance and repair support system (Marss). The Marss vest will contain components in communication with each other by wireless radio within the vest as well as with the outside world, which ordinarily add up to the equipment being maintained plus logistics databases. Communication with the computers will be by voice, and information—manuals, status reports, reports from other sites, for example—will appear on the head-up display.

Marss is being developed by McDonnell Douglas Aerospace, headquartered in Huntington Beach, Calif. All decisions regarding the content of the software for maintenance are coordinated by the U.S. Army's Program Manager for Test, Maintenance and Diagnostic Equipment. Human-factors studies of the vest are being coordinated by the U.S. Army Soldier System Command (formerly the Natick Research, Development and Engineering Center), in Natick, Mass.

BATTERIES AND BULLETS

The design of the vest is a prototypical case of the need for good humionics. First, old-fashioned human-factors issues—traditional ergonomics—must be addressed. The vest must be lightweight, so it can be worn underneath a soldier's bulletproof Kevlar vest. If it is made waterproof, the electronics will be better protected; but it also must be porous enough to allow heat to escape and sweat to evaporate.

When the ergonomic and human factors and health issues come up against electronic design, humionics comes into play. Some issues have been studied in a general way for years—for example, the effects of close proximity with radio frequencies. New in the TIA program is the soldier's intimate proximity, one might say, with all sorts of electromagnetic radiation. Health issues related to electromagnetic radiation are under the purview of the Human Research and Engineering Directorate Command of the U.S. Army Research Laboratory, Aberdeen Proving Ground, Md.

Other humionics questions concern the vest itself. For instance, one design under review has flex cables running up either side of the chest over the breasts, with plug-in connections from components to power modules. An alternative has batteries next to the components they power.

Whichever design is adopted, a fundamental factor of war must be addressed: soldiers get shot. Being shot is bad enough under any circumstances, but consider the added dangers of just one of the vest's components, the batteries. Under review for the TIAs are lithium-ion batteries of 2.2 ampere-hours, at around 3.6 volts. Now, someone touching a 1.5-V household battery with a finger would feel nothing; touching it with the tongue would mean a relatively benign buzz. But under certain conditions, batteries put into direct contact with the conductive bloodstream, as could happen in battle, could have more serious effects.

In addition, the designers of the vest's power system naturally want the greatest energy density. But the higher the energy density, the greater the batteries' explosiveness. Finally, surprising as it may seem considering the human cost of war, the environmental impact of equipping entire armies with batteries is quite a concern. Scorched-earth tactics aside, the battlefield should not be further damaged after the end of hostilities. In addition, in field training, huge numbers of batteries could litter the arena.

These applications herald a revolutionary change in the soldier's armament. And new armaments, in turn, bring new ways of waging war. Obviously, the ranks of the military will be interconnected as never before. Surely, too, the soldier's personal weapons will be slaved to computer-assist systems. Nothing better signifies the new battlefield realities than the two vests to be worn by combat troops: the mechanical, Kevlar, variety, to protect them from incoming firepower, over an electronic one, with its own kind of firepower.

QUESTIONS

1. What will be the differences in the level of and types of information used in future wars?

2. Explain what a TIA is, and give four examples of it.

3. What is "humionics," and why is it important for the soldier?

4. What can the Tamer (Technology Advanced Mini Eyesafe Rangefinder) do for the soldier and other supportive operations?

5. What is "Pathfinder"? What are its various applications?

6. What is the "VoiceMap" TIA?

7. What can artificial intelligence add to VoiceMap applications?

8. What is the purpose of "VuMan" TIA?

9. Why is the bodyLAN vest the most ambitious TIA program?

10. What are the effects and implications of the use of TIAs by soldiers in warfare?

SCENARIO: TECHNO WARS

Dateline: 19:19:01 GMT December 21, 20XX

The Northern Economic Trading Block (NETB) Authority has accused the Southern Economic Trading Block (SETB) of illegal trading practices in winning major trading export contracts worth more than $35 billion from the Eastern Trading Zone (ETZ).

A global videoconference between the leaders of the Northern and Southern trading blocks has resulted in a stalemate and increasing frustration. The leaders of the Northern economic block have accused the Southern block of technoimperialism.

The Southern block leaders think that the Northern block is trying to pressure them to revoke the contracts they signed with the African Economic Block for the development of carefree oil fields, which are estimated to contain 30% of the oil reserves. The carefree oil fields were recently developed by the Southern economic zone after it won the development bid in the international market. The Northern block accused the Southern block of offering billions of dollars to the royal families of the African Trading Zone in exchange for the bid. During the past 10 years, the Northern economic block has lost manufacturing and high-tech jobs to the Southern block. A number of multiblock global corporations have moved operations from north to south. The gross domestic products of the Southern economic block are expected to be equal to those of the Northern economic block. The experts in the North have predicted that if the Southern development is not curbed, it will seriously impact the Northern economy, which stands to lose $100 billion exports globally. The leaders of the North have issued an ultimatum to the South to vacate the carefree oil fields by January 1; otherwise, the North will declare a technowar. The South has refused to bend to what it calls technoimperialism.

The Untied Nations has failed to find any solution to this crisis. The last meeting on the GLOBALNET between the five trading blocks resulted in futile attempts to solve the crisis, and all efforts have been in vain. The North has superior high-tech electromagnetic weapons as compared to the South, where for the last 50 years there has been little development of technoweaponry; the South has instead emphasized the development of business technology to gain an edge on global consumer businesses.

QUESTIONS

1. Is the Northern Economic Trading Block justified in issuing an ultimatum to the Southern Economic Trading Block? Discuss future economic and political scenarios for this situation.

2. What kind of high-tech electromagnetic weapons might the NETB use against the SETB? Discuss real and prospective weaponry of this type.

3. Do you think that in the future, business conflicts will be resolved by technowar?

4. What type of actions and resolutions would you recommend to prevent future technowars?

A Bridge to Your 21st Century Understanding

Complete and discuss the following flowchart.

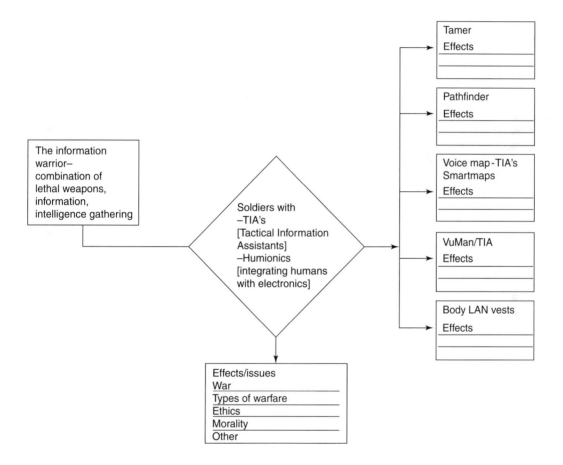

New Horizons of Information Technology

AHMED S. KHAN

The exponential growth in fiberoptics technology during the past two decades has significantly changed global telecommunications. Today, fiber is the medium of choice for short- and long-haul broadband applications. Numerous new applications of fiber are proving to be the impetus for the technological developments in the field. With rapid advances in optoelectronic devices, low-loss zero-dispersion fibers, and optical amplifiers—coupled with innovations in optical computing and switching—fiberoptics systems are transforming the 20th-century electronic era of telecommunications towards an optical era of the 21st century.[1]

Oceanic fiberoptics systems have revolutionized global telecommunications. Oceanic lightwave systems have been in service since 1988 across the Atlantic Ocean and since 1989 across the Pacific Ocean. These high-capacity digital communications systems have brought about a revolution in available system capacity and service quality compared to prior analog coaxial systems. On the drawing board are the systems with capacities of 5000+ Megabits per second (Mbps). The oceanic fiberoptics systems are transforming the world into a global village.

Approximately 60 million kilometers of fiber have been installed worldwide. The double-digit growth rate continues unabated, as nations strengthen and extend their infrastructure, and as societies enter the information age.[2] The key developments that have revolutionized light-wave communications are the advances in rare-earth optical amplifiers that are used for data transmission. Optical amplifiers have matured into one of the most significant advances in fiberoptic technology since the fiber itself. With the advent of erbium-doped fiber amplifiers (EDFA),

transmission capacities have increased from hundreds to ten thousands of gigabit-kilometers per second.[3] Optical solitons that exploit both the nonlinearity and dispersion of the fiber medium to maintain a constant pulse shape offer a potential for high-capacity ultralong distance systems. Researchers at AT&T have reported achieving error-free soliton data transmission at 10 Gbps over a distance of 1,000,000+ km using recirculating EDFA loop.[4] There exists a wide technoeconomic gap between the developed and the developing countries, especially in the area of telecommunication networks. In developed countries, telecommunications have played a key role in technoeconomic growth. In most developed countries, the teledensity is close to 50%, and the literacy rate is over 80%. In contrast, a vast majority of the developing countries have very low teledensities and low literacy rates (less than 40%). More than three billion people living in African, Asian, and South American developing countries have little access to telecommunications services.

With the globalization of business, it has become a universally acknowledged truth that telecommunications will play a critical role in the global economy. Developed nations want a sophisticated telecom infrastructure to provide a competitive edge for their business interests. Developing nations need a basic telecommunications infrastructure to provide a foundation for economic growth. According to International Telecommunications Union, an investment of $318 billion is required to add 212 million lines in developing countries. The development of a number of global, regional, and national fiberoptics systems will play an important role in meeting the demands for developed as well as developing countries and will help in narrowing the technoeconomic gap between the two.

The development of global, regional, and national light-wave systems will not only satisfy the growing telecommunication needs of developing countries, but will also provide the required infrastructure for creating endless opportunities in education (e.g., distance learning), medicine (e.g., telemedi-cine), and business (e.g., e-commerce). The lightwave systems will provide an impetus for economic and technological development and enable developing countries to increase their teledensity and literacy rate, improve public health, and increase per-capita income, thus narrowing the technoeconomic gap with the developed countries and allowing them to enter the 21st century with state-of-the-art optical networks.

NEURONS AND PHOTONS: THE LIMITS OF HUMAN KNOWLEDGE AND THE VASTNESS OF THE UNIVERSE

When the potential of fiberoptics communication technology is discussed, new perceptions of accumulated human knowledge can be made that can be visualized with our new understanding of the laws of physics. Fiberoptics technology has a potential transmission capacity of hundreds of trillions of bits of information per second. We can use the holdings of the Washington, D.C., Library of Congress as an example to provide an understanding of the enormous potential information-carrying capacity of the optical fiber.

In 1997, the Library of Congress held 17,402,100 books. If we assume that each book has 300 pages, with 450 words per page, this totals about 126,000 words per book, or 15.3 trillion words. If we further assume that each word averages about 7 letters, the information can be digitized using ASCII code, where each letter represents 7 bits. Therefore, all of the holdings of the Library of Congress would amount to 107.44 trillion bits of information. Using fiberoptics, an optical fiber that has the capability of transmitting 100 trillion bits per second, the entire Library of Congress book collection could be transmitted anywhere in the world in 1.07 seconds.

If we further assume that the total accumulation of human knowledge over the past 5,000 years was represented by 1,000 Libraries of Congress, all of the accumulation of the recorded knowledge of man could be transmitted through an optical fiber in 17.9 minutes. Therefore, in less than 20 minutes, all known human knowledge could be transmitted anywhere in the world. This analogy helps to build our awareness and appreciation of the finite nature of man's knowledge in contrast to the vast dimensions of the universe and the infinite possibilities of the time-space continuum. It also builds our awareness of how limited our knowledge really is, how finite we are on the infinite time-space axis of the ever-expanding cosmos, and how great our future dimensions of knowledge are.

SCENARIO: OPTICAL ERA OF INFORMATION SOCIETY

Date line: 10:01 AM, March 10, 20XX

The MTT Labs has announced that their scientists have successfully developed a micro-optical logic gate. A logic gate is the basic building block of a computer. The state of the current electronics technology allows CPUs to operate at speeds of 500+ MHz. The development of the optical gate will lead to the development of optical computers that will operate at 100+ THz speed ("T" = Tera = 1000,000,000,000 Hz). During the last decade of the 20th century, high-speed optical fibers have already been developed that allow data transmission rates of Tbps (tera bits per second). The present technology of optical storage permits the storage of 100 Terabytes of information per optical compact disc (OCD). The development of a micro-optical logic gate at MTT Labs will enable the development of an end-to-end optical communications system. An end-to-end optical system will use optical computing, optical data storage, and optical data transfer. The development of the miniature optical gate has transformed the electronic era of telecommunications into an optical era of information.

Optical Era of Information Society

Answer the following questions:

a. Discuss the pros and cons of the optical computer.

b. The speed of optical computers and optical data transfer will be much faster than human response times and will thereby force users to process information in their brain much faster and to interact more quickly, thus increasing the stress level of the average computer user. Do you think that optical end-to-end information networks will help or harm the health of a computer user in terms of stress levels?

c. Discuss the implications of the development of end-to-end optical information exchange systems on society.

REFERENCES

1. Khan, A. (1994). Fiber Optics Communications: An Introduction to Technology and Applications. *Annual Review of Communications Vol. 47.* p 503.
2. Li, T. (1995, January). Optical Amplifiers Transform Lightwave Communications. *Photonics Spectra.* p. 115.
3. Desurvire, E. (1992, January). Lightwave Communications: The Fifth Generation. *Scientific American.* pp. 114–121.
4. Desurvire, E. (1994, January). The Golden Age of Optical Fiber Amplifiers. *Physics Today.* pp. 20–27.

QUESTIONS

1. Why will telecommunications and fiberoptics be the information technology of the future?

2. How soon do you think we will have end-to-end optical communications systems? Do you agree with its progressive advances? How do you think it will impact society?

3. What are some of the problems of the advancement of telecommunications and fiberoptics technologies?

60

Future Ecology: The 2002 Second World Summit Focuses on a Renewed Spirit and Commitment

In 1992, the Earth Summit of the United Nations Conference on Environment and Development (UNCED), in Rio de Janeiro, Brazil with over 100 heads of state in attendance, making it one of the largest gatherings of world leaders in history, raised high hopes for the commitment of the world's communities to saving the earth. The summit was the first global effort devoted to making world progress on issues of environment and poverty. In the years that have passed since 1992, however, not too much has happened in the way of progress. A five-day review session, held in June, 1997, called "Earth Summit II," visited by about 70 heads of state, reviewed the progress over the years since the 1992 summit, but not too much was actually accomplished. Progress is urgently needed, but not much has been made.

The global environment has continued to decline with

- increasing levels of toxic pollution
- increasing levels of greenhouse gas emissions
- increasing amounts of solid waste
- increasing levels of rainforest destruction
- decreasing levels of biodiversity
- an increasing need for financing for developing nations so that they may adopt strategies and technologies that produce less pollution.

Progress from the 1992 summit included the creation of a climate change convention (held in Kyoto, Japan, in December 1997), a biodiversity convention, a collection of forest principles and other declarations, and Agenda 21, an overarching agreement for a sustainable, environmentally

sound development pattern for the 21st century. One important step taken by the summit was the creation of a recommendation, signed by 166 nations, that called for the reduction of emissions of greenhouse gases to their 1990 levels by the year 2000, but the recommendations did not contain tight guidelines and strong-enough measures to actually curb emissions. Furthermore, President Clinton disappointed environmentalists at Earth Summit II by not committing to specific reductions and firm standards for U.S. greenhouse gas emissions. President George W. Bush has shown further lack of support by not agreeing to the emissions criteria for the U.S., finding them unrealistic for continued U.S. economic progress.

The 1992 Earth Summit made history by focusing global attention (new at that time) on the concept that the planet's environmental problems were linked to economic conditions and social justice. It demonstrated that social, environmental, and economic needs must be dealt with in balance with each other by planning for long term sustainable outcomes. It further demonstrated that when people are poor and national economies weak, the environment suffers and, accordingly, if the environment is abused and resources over-consumed, people suffer and economies decline. "The conference also demonstrated that the smallest local actions have potential worldwide ramifications" ("What is Johannesburg," p. 1).

The Rio de Janeiro Summit illustrated the way social, economic and environmental factors are interdependent and influence each other. It also identified the critical elements of change, outlining that success in one area also requires action in other areas for long-range success and change to

502

be effective. "The Summit's primary aim was to produce an extended agenda and a new plan for international action on environmental and developmental issues that would help guide international cooperation and policy development into the next century" ("What is Johannesburg," p. 1).

The Earth Summit emphasized the concept of sustainable development as a workable objective for everyone around the world at all levels—local, national, regional or international. It also focused on the integration and balance of economic, social and environmental issues which would require new ways of viewing how the world consumes, lives, works, gets along with one another, and makes decisions. This concept was revolutionary and started a lively debate among governments, and between governments and their citizens, on how to achieve sustainability ("What is Johannesburg," p. 2).

A major achievement of the World Summit was Agenda 21—a broad-ranging program of future actions to reach global sustainable development in the 21st century. Its recommendations ranged from new ways to educate, and care for natural resources, to the participation in the design of a sustainable economy. "The overall ambition of Agenda 21 was breathtaking, for its goal was nothing less than to make a safe and just world in which all life has dignity and is celebrated." ("What is Johannesburg," p. 2). (Please see outline of Agenda 21).

"The Second World Summit on Sustainable Development in Johannesburg, South Africa in 2002 will aim to answer the following questions among others:

- What has been accomplished since 1992?

- What have the participating countries done so far to implement Agenda 21?

- Have they adopted the National Sustainable Development Strategies as they agreed they would by 2002?

- What obstacles have they encountered?

- What lessons have they learned about what works and what does not?

- And what new factors have emerged to change the picture?

- What mid-course corrections need to be made to reach the goals?

- Where should further efforts be concentrated?" ("What is Johannesburg 2002?," p. 3).

The World Summit of 2002 will not revise Agenda 21, but it will try to establish consensus on the general assessment of current conditions and on prioritization for further action in new areas. Decisions will seek to strengthen commitment to achieving the goals of Agenda 21 and also foster discussion of findings of environmental sectors (forests, oceans, climate, energy, fresh water, etc), economic conditions, new technologies and globalization. It will also consider the revolutionary impacts of technology, biology and communications since 1992, while realizing that one out of every six individuals on Earth has yet to make a telephone call. The World Summit of 2002 begins with the assessment and evaluation of 1992, and broad participation of individuals and governments is critical. Such assessment must begin with the facts as individuals are presently living them; they then need to be redirected to a plan and ultimate goal of sustainability. "Governments cannot do this alone if further action is to be owned by all and thus be effective in achieving the ultimate goal of sustainability" ("What is Johannesburg 2002?," p. 3).

Following is an outline of AGENDA 21 which, as stated above, was the revolutionary outcome of the Earth Summit of 1992 in Rio de Janeiro. The purpose of AGENDA 21 is to enact a broad-ranging program of actions demanding new ways of investing in our future to reach global sustainable development in the 21st century. (The full text of AGENDA 21 can be accessed on the World Wide Web through the United Nations Sustainable Development Committee http://www.un.org/esa/sustdev/agenda21text.htm).

DISCUSSION

The 1992 Earth Summit was an event for which many people had high expectations. What are some ideas you have to keep Earth Summit priorities urgent, the agendas working, the issues focused, the countries cooperating, and progress advancing on these issues? Do you think that summits of this kind are the best approach to use for such world problems? Do you think 2002 will be more successful than 1992? Why or why not? What influence do you think the United States has on the rest of the world? What level of leadership do you think the United States should assume regarding the amelioration of such environmental problems if these problems are going to be addressed by all nations? What influence can citizens have on U.S. government positions? Should these kinds of summits be sponsored by the United Nations or by a different entity or entities? How can such summits be made more effective? What other ways can you think of to address the world's pollution and other population, resource and economic problems?

Agenda 21

Contents

Chapter	Paragraphs
1. *Preamble*	1.1–1.6

Section 1: Social and Economic Dimensions
Section 2: Conservation and Management of Resources for Development
Section 3: Strengthening the Role of Major Groups
Section 4: Means of Implementation
The Rio Declaration
Forest Principles
U.N. Publications Order Form

SECTION I. SOCIAL AND ECONOMIC DIMENSIONS

2. *International cooperation to accelerate sustainable development in developing countries and related domestic policies*	2.1–2.43
3. *Combating poverty*	3.1–3.12
4. *Changing consumption patterns*	4.1–4.27
5. *Demographic dynamics and sustainability*	5.1–5.66
6. *Protecting and promoting human health conditions*	6.1–6.46
7. *Promoting sustainable human settlement development*	7.1–7.80
8. *Integrating environment and development in decision-making*	8.1–8.54

SECTION II. CONSERVATION AND MANAGEMENT OF RESOURCES FOR DEVELOPMENT

9. *Protection of the atmosphere*	9.1–9.35
10. *Integrated approach to the planning and management of land resources*	10.1–10.18
11. *Combating deforestation*	11.1–11.40
12. *Managing fragile ecosystems: combating desertification and drought*	12.1–12.63
13. *Managing fragile ecosystems: sustainable mountain development*	13.1–13.24
14. *Promoting sustainable agriculture and rural development*	14.1–14.104
15. *Conservation of biological diversity*	15.1–15.11
16. *Environmentally sound management of biotechnology*	16.1–16.46
17. *Protection of the oceans, all kinds of seas, including enclosed and semi-enclosed seas, and coastal areas and the protection, rational use and development of their living resources*	17.1–17.136
18. *Protection of the quality and supply of freshwater resources: application of integrated approaches to the development, management and use of water resources*	18.1–18.90
19. *Environmentally sound management of toxic chemicals, including prevention of illegal international traffic in toxic and dangerous products*	19.1–19.76
20. *Environmentally sound management of hazardous wastes, in hazardous wastes*	20.1–20.46
21. *Environmentally sound management of solid wastes and sewage-related issues*	21.1–21.49
22. *Safe and environmentally sound management of radioactive wastes*	22.1–22.9

SECTION III. STRENGTHENING THE ROLE OF MAJOR GROUPS

SECTION IV. MEANS OF IMPLEMENTATION

* For Section I (Social and Economic Dimensions), See A/CONF.151/26 (Vol. I); for Section III (Strengthening the Role of Major Groups) and Section IV (Means of Implementation), See A.CONF/151/26 (Vol. III).
* For Section II (Conservation and Management of Resources for Development), See A/CONF.151/26 (Vol. II); for Section III (Strengthening the Role of Major Groups) and Section IV (Means of Implementation), See A/CONF.151/26 (Vol. III).
* For Section I (Social and Economic Dimensions), see A/CONF.151/26 (Vol. I); for Section II (Conservation and Management of Resources for Development), See A/CONF.151/26 (Vol. II).

Copyright © United Nations Division for Sustainable Development 29/06/2000.

REFERENCES

No Decision on Global Warming. (1997, Fall). *The Amicus Journal,* p. 3.

Swanson, S. (1997, June 22). Was Anyone Listening? Failures of '92 Earth Summit Spur a New One. *Chicago Tribune,* p. 6.

"United Nations Sustainable Development, Agenda 21." United Nations Sustainable Development. Retrieved September 5, 2001 from the World Wide Web: http://www.un.org/esa/sustdev/agenda21text.htm.

"What is Johannesburg 2002?". United Nations Division for Sustainable Development. Retrieved September 5, 2001 from the World Wide Web: http:/www.johannesburgsummit.org/web_pages/rio+10_background.htm.

INTERNET EXERCISE
Access the site in the references for Agenda 21. Download some or all of its contents. Discuss in groups how to deal with its various topics and associated issues for world sustainability.

Future Hopes For A World Climate Agreement
Photo courtesy of Ahmed S. Khan.

Future Ecology: Is the Kyoto Climate Agreement Still Alive Due to the Bonn, 2001 Meeting?

In July, 2001 in Bonn, Germany, 178 countries agreed to implement the Kyoto Climate Agreement of 1997. This is a compromise agreement of which the U.S. is **not** a part, but it is a global historic political agreement which will be turned into a legal document that will allow the Kyoto Protocol to come into force in 2002, once individual states ratify it. This agreement is considered a rescue of the Kyoto protocol through its compromising efforts and intense negotiations. The agreement leaves the U.S. (the world's biggest polluter) as the only world power not to accept the Kyoto accord. Almost every original country involved in the Kyoto protocol stayed in the accord except the U.S. President George W. Bush rejected the Kyoto treaty in March, saying it would harm the U.S. economy and that the protocol was "fatally flawed."

The conference president, Dutch Environment Minister Jan Pronk, was successful in weaving together multinational negotiations through conflicting national interests and objections. Kyoto required industrialized countries to reduce, over the next 11 years, their emissions of six gases, believed to be contributing to global warming, by an average of 5.2% below their 1990 levels. The Bonn agreement, however, will reduce that 5.2% figure to about 2%. Even though many environmental conservationists say the deal has serious gaps, they feel the Kyoto Protocol has been rescued and countries can now start the ratification process and start to take important action on climate change. The treaty must be ratified by 55 nations responsible for 55 percent of global green gas emissions to take effect. It is expected that the treaty will be ratified in 2002 with over 30 nations already having ratified the pact to date.

At first Japan, Australia and Canada sided with the U.S. on key points and were reluctant to endorse the pact without U.S. participation. But conference president Jan Pronk successfully negotiated their cooperation after 48 hours of intense bargaining. Japan's support, along with that of other large nations, was crucial since without U.S. involvement the protocol could only be effective if ratified by the big polluters—the EU, eastern Europe, Russia and Japan. The result of President Bush's lack of agreement has left the U.S. politically isolated from the world on the issues of global warming. Aside from not supporting the world environmental progress of the accord, U.S. uninvolvement additionally could put many U.S. corporations at a disadvantage.

The Bonn agreement includes provisions for emissions credits for Japan and other countries with large carbon-absorbing forests and agricultural regions, but provides the U.S. with only minor advantages since they are not party to the agreement. Additionally, some U.S. multinational corporations would be harmed because they would have to do business under differing sets of emissions requirements at home and in other countries, with possible carbon taxes imposed there. The Bonn agreement is very definitely historic because of international agreement, cooperation and action on urgent environmental issues. But in its present form it is compromised, and without U.S. agreement and lower standards it becomes more of a symbol than a battle cry.

BACKGROUND

As an outcome of the 1992 Earth Summit, a convention on world climate change was held in December 1997 in Kyoto, Japan, with representatives from about 160 nations in attendance. These nations adopted an historic agreement that placed legally binding limits on their emissions of heat-trapping greenhouse gases in order to protect the environment. The resulting tough agreements mark this conference as one of the most important meetings in the history of the environmental movement and a critical and remarkable first step to stabilizing the earth's climate.

At the Kyoto conference, progress was slow and stalemated until the 8th hour of the 10th day of the conference, when Vice President Gore addressed the delegates and implied U.S. willingness to commit to greater emissions reduction if other nations also strengthened their positions on key issues. At this juncture, the opponents of an agreement, including some oil-producing countries and some countries known for their tendencies to pollute heavily, increased their efforts to thwart negotiations. Several environmental organizations, such as the Environmental Defense Fund, the World Wildlife Fund, Greenpeace International, the Natural Resources Defense Council, and other nongovernmental organizations (NGOs), then joined forces and issued a statement of priorities necessitating a strong stance against harming the environment, a reduction of emissions to below 1990 levels, the elimination of loopholes, and the adoption of strong compliance measures. Through the efforts of these NGOs, the intense level of negotiation was held, communications remained focused, oppositions were effectively deferred, and a binding agreement was successfully attained. The resulting treaty "was prompted by NGOs urging action . . . [and] underscores the strengthened role of NGOs in world affairs. . . . Virtually unified international NGO support for [emissions] trading and ambitious targets was key to saving the treaty. We were able to bring along a number of developing countries," stated Fred Krupp, Director of the Environmental Defense Fund (Krupp, 1998).

The final agreement that emerged from the Kyoto conference included nine critical points that the NGOs fought for, seven of which included the following measures:

1. The placement of binding limits on emissions.

2. The reduction of emissions to at least 5% below their 1990 levels.

3. The occurrence of reductions by a specified time period, between 2008 and 2012.

4. The placement of limits on all six significant greenhouse gases—not just carbon dioxide, methane, and nitrous oxide, but also hydrofluorocarbons, perfluorocarbons, and sulfur hexafluoride.

5. The awarding of credits on emissions levels for enhancing carbon "sinks," such as forests that remove carbon dioxide from the atmosphere.

6. The implementation of incentive-based mechanisms, such as emissions trading, to encourage early, cost-effective emissions reductions.

7. The institution of a provision, called the "Clean Development Mechanism," to promote greenhouse gas reductions in developing countries. Reduction projects could be financed by industrial nations as a cost-effective way of meeting some of the emission-reduction obligations.

The other two points included a compliance agreement that places penalties on industrial nations that fail to meet emissions limits and various explicit commitments for developing nations. Intensive work is still needed on these two points as well as on making the Kyoto agreement become a political and a global reality. (Kyoto Climate Agreement is a Critical First Step, 1998.)

The Kyoto Protocol stipulates that, collectively, the industrial nations must drop emissions levels of the six previously mentioned pollutants to 5.2% below their 1990 levels by the years 2008–2012. Individually, the nations agreed to differential limits: the European Union at 8% below 1990 levels; the United States at 7%; Japan at 6%; Australia at 8%; Iceland at 10%; and the Russian Federation, Ukraine, and New Zealand at 1990 levels. The treaty also divides the industrial countries into two composite "bubbles" that provide trading and selling pollution rights (not to exceed excess) to other countries. The protocol also allows industrial nations to obtain credit (in terms of allowance for emissions) for newly forested land since 1990, but these nations also must balance their reductions to account for deforested land. The treaty imposes binding emission limits on 39 countries, but it does not affect 121 less developed countries, many of which are already emitting large amounts of greenhouse gases (Raloff, 1997).

In order to meet U.S. commitments, predictions are that the United States may be cutting U.S. energy use by a third in the next 10 years. This reduction in energy use will involve not only commitments from industry, but also from the consumer. The United States already failed to comply with the stipulations made at the 1992 Earth Summit in

Rio de Janeiro, Brazil, where it promised to cut emissions to 1990 levels by the year 2000. Instead, the United States has created even more emissions and used greater amounts of fuel. (The United States alone emits one fourth of the world's total CO_2 emissions and is the world's greatest polluter of global warming gases.) U.S. fuel use increased by 3.5% in 1997, the largest annual increase on record. By the year 2000, the United States will be burning 15% more fossil fuels—and will therefore be creating 15% more emissions—than it did in 1990, thereby making it necessary to reduce emissions by 22% over 10 years rather than its original goal of 7% (McKibben, 1998).

FACTS

- Since the 1992 convention in Rio de Janeiro, the worldwide level of greenhouse gas emissions has risen by 2% percent each year.

- According to the World Watch Institute, emissions of greenhouse gases in the United States have grown by as much as 13% over the last decade. In 1997 alone, U.S. emissions rose 3.4% to a level that is 7.4% above 1990 levels.

- The United States is the world's largest producer of greenhouse gases by almost double. In 1992, the top 10 CO_2 emitters were the United States, which emitted about 5 million metric tons of CO_2; China, with under 3 million; Russia, with about 2 million; Japan, with about 1 million; and Germany, India, Ukraine, the United Kingdom, Canada, and Italy, with under 1 million.

- Between 1990 and 1994, carbon dioxide emissions from the developing world have risen rapidly, with Brazil increasing emissions by 16%, India by 24%, South Korea by 44%, and China by 13%.

- The problem is that the universal appetite for energy and coal is growing with energy demand and has been forecasted to increase by almost 40% by the year 2015 (Lynch, 1998).

HOW GLOBAL WARMING WORKS

Global warming is part of the natural protection and function of the atmosphere, but when it gets too hot, that warming becomes dangerous, causing severe weather, weather pattern changes, and climatic change. When the sun's rays reach the earth's atmosphere (a 12-mile-thick protective layer of natural greenhouse gases made up mostly of water vapor, but also of carbon dioxide, methane, nitrous oxide,

and chlorofluorcarbons, that keeps the earth's average temperature at about 60 degrees), about one third of the sun's solar radiation bounces back into space. The other two thirds of the sun's energy is absorbed into the oceans, forests, and soil, where heat energy is generated and released back into space in the form of heat, not solar radiation. Some of this heat energy is then captured by greenhouse gases. Scientists are concluding that the buildup of greenhouse gases by industrialization and slash-and-burn farming practices is responsible for the heating of the earth's natural heating system. (There has been approximately a 25% increase of CO_2 in the earth's atmosphere in the last century.) It has been concluded that the earth's average temperature has increased by 1 degree in the past century, and it is predicted that it will increase by another 3.6 degrees within the next hundred years. This kind of increase would cause serious problems for the global warming system, where such an increase represents the most rapid warming in 10,000 years. This temperature change would not only change climatic weather patterns, but would also produce a rise in sea level between 1.5 to 3 feet due to melting glaciers, flooding coasts and islands around the world (Lynch, 1998).

QUESTIONS

1. Why do you think the U.S. did not sign the Bonn agreement? Research the particulars for understanding President Bush's position.

2. How could the agreement be changed so that the U.S. would sign the agreement? Research the arguments.

3. Even though standards in the Bonn agreement have been lowered, do you think that the Kyoto-Bonn agreement is an effective significant step for future world cooperation for action with environmental urgent issues? Is it a model for future global planning?

4. How did the Kyoto conference differ from the 1992 Earth Summit in terms of its approach and results?

5. Discuss the roles and effectiveness of NGOs with regard to global policy on the environment. Do you think that coordination and involvement of grassroots organizations are necessary for humanitarian and ecological causes?

6. What does this Kyoto protocol illustrate as far as government negotiations and priorities?

7. The Kyoto Protocol is regarded as a critical and remarkable first step toward overall improvement of the environment. Why is this agreement so remarkable?

Do you feel that this agreement represents hope for the 21st century in dealing with critical global issues?

8. Why do you think the United States did not take a leading role in reducing emissions, as it is the world's number-one polluter?

9. If the Kyoto protocol is just the first step, what are the other steps that are necessary to make it an effective and binding agreement? Was the Bonn compromise necessary?

10. What do you think of the concept of selling and trading pollution rights and allowing emissions credits?

11. Discuss the issues of global warming with regard to the following aspects: increasing amounts of gases in the atmosphere; sources of emissions; and various future strategies to abate the emissions of these gases, with the possible consequences of a global temperature rise and other effects. In your opinion, is the Kyoto-Bonn protocol on the right track to begin to limit emissions? Are the limits placed on emissions by the protocol enough? What are the steps necessary for the protocol's goals to be met?

REFERENCES

Kirby, A. (2001, July 23). "Compromise Saves Climate Treaty". *BBC News On Line*. Retrieved September 7, 2001 from the World Wide Web: http://news.bbc.co.uk/hi/english/sci/tech/newsid_1452000/1452315,stm.

Krupp, F. (1998, April). A New Force at Kyoto. *Environmental Defense Fund Letter*. Vol. XXIX, No. 2, p. 3.

Kyoto Climate Agreement is a Critical First Step. (1998, April). *Environmental Defense Fund Letter*. Vol. XXIX, No. 2, pp. 1, 5.

"Kyoto Climate Deal Hopes Grow." (2001, July 19). *CNN.Com./World*. Retrieved September 7, 2001 from the World Wide Web: wysiwyg://22/http:www.cnn.com/2001/WORLD/europe/07/18/kyoto.climate/

Lynch, C. (1998, Winter). Stormy Weather. *The Amicus Journal*. pp. 15–19.

Making History in Kyoto. (1998, Spring). *The Amicus Journal*. pp. 3–4.

McKibben, B. (1998, March–April). Warming Up to Kyoto. *Audubon*. pp. 54–58.

"New Proposals Could Secure Kyoto Climate Agreement." (2001, June). *WWF. News*. Retrieved September 7, 2001 from the World Wide Web: http://www.panda.org.za/news/messages/97.html.

Pianin, E and Milbank, D. (2001, July 24). "Climate Agreement Leaves U.S. Out in the Cold." *Washington Post On Line*. Retrieved September 7, 2001 from the World Wide Web: wysiwyg://10/http://www.washingtonpost.com/wp-dyn/articles/A39341-2001Jul23.html.

Raloff, J. (1997, December 20/27). Nations Draft Kyoto Climate Treaty. *Science News*. p. 388.

"Work Starts on Kyoto Deal Details." (2001, July 23). *CNN.Com./World*. Retrieved September 7, 2001 from the World Wide Web: wysiwyg://31/http://www.cnn.com/2001/WORLD/europe/07/23/kyoto.talks/

Future: Population Growth Estimates

United Nations general projections of population growth for 150 years from now, in the year 2150, range as high as 28 billion people and as low as 4 billion people, with the medium being somewhere around 11.5 billion people. New estimates from the United Nations, however, are stating that the global population will peak at about 11 billion in the year 2200. This new estimate is 6% lower, or 600 million people lower, than 1992's projection of 11.6 billion. The reduction in numbers is caused chiefly from more sharply declining fertility rates than those estimated in 1992. It is turning out that fertility rates are falling steadily and, in some cases, dramatically. For example, in Bangladesh, the fertility rate has declined from 6.2 to 3.4 children per woman in 10 years. This decline is not a temporary decline due to war or depression, but instead seems to be a sustained decline, perhaps caused by increasing modernization. Some of the reasons demographers cite for the correlation between modernization and declining fertility rates are that urban populations do not need extra farm workers as rural populations do, women's status and education is improving as many countries modernize, and family planning programs and methods are more available in modernizing countries (Swanson, 1998).

Population growth is still a problem, however, since even with a decline in fertility rates, the world's population is still increasing by about 80–90 million people a year. This number represents about a 3% growth per year, which, when compounded over decades, is an explosive population growth that can strain and devastate local life support systems (Brown, *Full House,* 1994). Additionally, the number of people will continue to increase quickly, since most of the world's people are young, meaning that there will be more women entering childbearing years in the near future. Predictions are that the world's population will increase to 9.4 billion people in 2050, an increase in about half of a century of 65% of the current population (Swanson, 1998).

About 50 countries, most of them industrialized nations, have fertility rates below the 2.1 children-per-woman "replacement level." The challenge is then for the unindustrialized countries, with the remaining 86% of the world's population, to reach levels below the replacement level (Brown, 1996).

The 20th century has experienced an extraordinary growth in world population from 1.6 billion to today's over 6.1 billion. Since 1960, when world population was 3 billion, those who are older than 40 today have witnessed a doubling of the world population—a unique world history event (Tran, 2001, p. 4).

Wolfgang Lutz, of the International Institute for Applied Systems Analysis in Austria, released a study in 2001 which predicts that the world's population will peak at around 9 billion in 2075 and then begin to naturally and culturally decline. This new estimate is below the UN projections and decline adjustments. It is similar to UN projections to 2050, but they then diverge. The UN estimate continues the world growth numbers to 10 billion by 2185, where Lutz's numbers peak at about 9 billion in 2075. The studies differ, but both the UN studies and Mr. Lutz's studies agree on a natural cultural population decline in some countries due to modernization, education and medical improvements, but not in others in Africa and other parts of the highly populated unindustrialized world (Tran, p. 4).

Aside from slowing trends in world population, there will be a dramatic shift in the 22nd century from large numbers of youthful population under 15 to large numbers of elderly population over 60. By 2100, 40 percent of the population of North America will be over 60—an increase of 16 percent. Worldwide in the 21st century there are approximately 10 percent over the age 60, but in the 22nd century the prediction is for about 34 percent over the age of 60. This "graying" of the world will be seen very clearly in Japan, where every other person will be 60 or older by 2100. There will also be increased economic and resource tensions between those countries with shrinking populations and those with growing populations (Tran, p. 4). Whatever the figure adjustments, the 21st century will experience continued unprecedented population growth and resource use on a worldwide scale. Trends of slowing population growth will be helpful but are not adequate in themselves for future sustainability. More initiatives for population planning will be needed for future sustainable growth patterns and resource use and sharing from a world view of the 21st century.

QUESTIONS

1. Declining fertility rates seem to be a positive note in the overall picture of a population explosion that has been predicted. Discuss the positive implications of this decline.

2. Considering the readings in Part V, what do you think are some of the reasons for the declining fertility rates? Explain what you think are some of the most significant reasons for such declines and for future declines in the next 200 years.

3. Do you think that these declining fertility rates provide hope for an abatement of the population growth problem? Explain.

4. Research some of the implications of the high population-growth patterns of unindustrialized nations with regard to food supply, economic hardship, environmental degradation, and availability of seafood. Suggest some ways to help abate these countries' high population-growth patterns.

REFERENCES

Brown, L., and Kane, H. (1994). *Full house.* New York, W.W. Norton.

Brown, L. (1996). The Acceleration of History. *State of the World, 1996,* W.W. Norton. p. 12.

Swanson, S. (1998, February 17). UN Analysts Trim Global Population Forecasts. *Chicago Tribune.* pp. 1, 9.

Tran, Q. (2001, August 2). Population Forecast Sees Peak, Decline. *Chicago Tribune,* section 1, p. 4.

INTERNET EXERCISE

1. Use any of the Internet search engines (e.g., Alta Vista, Yahoo, Infoseek, etc.) to research the following topics:
   a. Nanotechnology
   b. Responsible technology
   c. Biotechnology
   d. Neuroscience
   e. Telemedicine
   f. Genetic medicine
   g. Human chromosomes and DNA
   h. Xenotransplants
   i. Virus transfer
   j. Humionics
   k. Bionics
   l. Ergonomics
   m. Tactical Information Assistants (TIAs)
   n. Any words you have listed from your research and reading of future issues

2. Use any of the Internet search engines (e.g., Alta Vista, Yahoo, Infoseek, etc.) to research the following issues:
   a. The concepts of responsible technology—what does this responsibility entail?
   b. What are the fields of neural engineering and neuroscience?
   c. Research the progress and developments of the Human Genome Project. What are some of the implications of this project?
   d. What is the capacity and capability of the latest most highly developed computer?
   e. What are the latest developments in human bionics? Are there any animal bionic developments? What are some of the ethical issues involved in bionics?
   f. Research world educational levels. Then identify world educational levels for women. Draw any possible conclusions as to correlations with population levels.
   g. Discover the criteria needed for sustainable world development. What are some of the necessary philosophies involved?
   h. What are some alternative energy systems? How effective can they be?
   i. What is meant by the concepts of population stabilization? When can that happen on a global level?
   j. Explain what is meant by the biotechnological revolution. What are some of the technologies involved, and why are they so revolutionary?
   k. Explain possible ways that ecology and ecological resources can be included in global economic planning.
   l. Research more about military Tactical Information Assistants (TIAs).
   m. Discover the latest pros and cons and progress in the field of xenotransplants.
   n. What is the U.S. space exploration plan for the next 15 years?
   o. Explain some of the newest developments in telecommunications.
   p. Research the available material on the military power of two or more unindustrialized nations and two or more industrialized nations. Discuss the implications of the differences you find.
   q. Research the ecological priorities for the United States and the world. Explain the differences.

USEFUL WEB SITES

N@N@N@N@N@N@N@N@N@N@N@N@N@N@N@N@N@N@URLIHIHIHIHIHIHIHIHIHIHIHIHIHIHIHIHIH

Site Description

http://www.ufs.org
http://www.ufs.org/futurist.html

http://www.discover.com
http://www.extropy.org/eo/
http://www.press.umich.edu/jep
http://www.nanotech.news.com/nanotech.news/nano
http://www.nanozine.com/news.html
http://nanospot.org
http://www.nasa.gov/today/index.html
http://www.nature.com/nature/
http://www.science-mag.aaas.org
http://www.sciencenow.sciencemag.org
http://www.science.org
http://www.spectrum.ieee.org

http://radburn.rutgers.edu/andrews/projects/ssit/default.htm
http://www.cnn.com/TECH/
http://www.scitechdaily.com
http://jefferson.village.virginia.edu/pmc
http://www.wired.com

http://www.worldwatch.org
http://www.amsci.org/amsci/

Authors' Commentaries

Authors/Editors: Ahmed Khan, Dr. Barbara Eichler, Linda Hjorth, Dr. John Morello.
Photographed by Evan Girord

Technology and the Future: The Military Picture

When dawn first broke on the typical battlefield several hundred years ago, it must have revealed a surreal picture. Rays of sunlight, cutting through the early morning mist, found their way to the polished steel of the sabres and spears of the armies facing each other. Bright, almost gaudy uniforms stood in stark contrast to the surrounding terrain. At a given signal, usually drums, bugles, or bagpipes, the two forces converged on a designated point, slugging it out until one side disengaged, defeated, but not destroyed. There were always other dawns and other battlefields.

Move to that next battlefield a few generations later. The sun that illuminated that same field of battle might catch not only the glare of sabres and spears, but also the spurs of cavalry and the sight of artillery. The uniforms might still have been garish, given the surrounding terrain. This time, the hostilities began not with the burst from a bagpipe, but most likely with an artillery barrage, as one side tried to use technology to soften up the opposition prior to the assault. And when that assault finally came, it was not led by the screams of men rushing headlong into the breach, but by pounding hooves, as the cavalry, attacking from the flanks, sought to collapse the enemy's perimeter, dash to the rear, and panic the troops headlong into the advancing infantry. There would be no disengagement this time—only total victory, as one side destroyed the other.

The sun that broke over the early battlefields of World War I helped to illuminate not just the familiar past, but also the wave of the future. Men on horseback, sabres and spurs glinting in the sun, prepared to race across an open field to wreak havoc on the enemy's position. However, this time the men on the other side did not quail; instead, they retaliated with machine guns, hand grenades, and poison gas. The technical advantage had shifted, if only temporarily, while the horsemen dismounted and climbed into tanks for a second attack.

And so it has been over the years. In war, technology has proven to be a decisive edge. It has helped to make offensive efforts irresistible, defensive stands impregnable, and for those who found themselves facing superior numbers, it made them equal. However, the edge did not remain so for long. In time, the other side would have the secret, or perhaps go it one better. And so, the search for technological superiority in war has become an escalating factor. Where will it all lead? Here are a few possibilities:

1. *Fewer high-tech wars:* Given the escalating cost and complexity of the new military technologies, only those nations with deep pockets and high-tech capability will be able to use these weapons effectively. Cruise missiles, like the kind used during the Gulf War, can cost $1,000,000 each—not the kind of thing to shoot off willy-nilly. Planes like the F-22 also pack budget-busting price tags. The bottom line is, or should be, that because of the cost of these weapons, nations may want to think twice about the reasons they go to war. Are these reasons sound enough to mortgage their financial present and future?

No one should be naive enough to think that swearing off high-tech weapons will put a stop to warfare. There are plenty of older, more conventional weapons to go around. Millions of antipersonnel devices, also known as land mines, are still buried on battlefields around the world. They are cheap and have a terrific shelf life. ABC News reported in September 1997 that an Egyptian girl was killed when she stepped on a land mine buried by the German Army during World War II. Nations that want weapons don't have to go high tech; low tech will do just fine. The complexity of high-tech weapons might also be a deterrent to war. Nations wanting sophisticated weapons systems might someday realize that these systems are of no use unless they have the trained personnel to use them or an infrastructure that can support them. And even then, there's no guarantee that the weapons will get the job done. After the glow of praise for the Patriot missile's achievements during Operation Desert Storm had faded, a reevaluation of the system revealed that the missiles were not as effective as everyone had thought they were. Now, if the United States cannot get the expected results out of a weapons system, especially one it built, can we honestly expect anyone else to?

2. *Fewer casualties, less political fallout:* High-tech wars, if they are fought, can produce unexpected benefits. For starters, those personnel committed to the battlefield will be better armed and better prepared than previous fighting forces. Equipment could make 1 soldier's firepower equal to that of 10. Therefore, fewer troops will have to be put in harm's way. Because of that fact, wars may become more politically sustainable. Taxpayers may be less likely to question their government's intentions if only a few hundred troops are being jeopardized. And, given the high-tech equipment available, those troops will have a better chance of survival and success than ever before. Medical technologies will make it possible for seriously injured troops to receive immediate attention right there on the battlefield, where it really counts. Communications technology will help locate the wounded and get them to a rear area for more care. "Soft Kill" weapons systems will be used not to kill, but to blind, disorient, or otherwise incapacitate the opposition. Having to deal with thousands of sick or wounded soldiers is actually more difficult than tending to the dead ones.

3. *Simulated war as conflict resolution:* Iraqi forces are on alert. All reservists have been called up, and the Republican Guard has been mobilized. Scud missiles are being repositioned, and the Iraqi Air Force has been given new targets. The front page of the *New York Times?* No, the latest data from satellites tracking Iraq's military operations. Those satellites have been watching for some time now, cataloging troop strength and weapons capability. The data produced has been sifted through by analysts who conclude what Iraq is up to, how it intends to achieve its goals, and the proposed order of battle. Furthermore, those analysts have produced a number of scenarios that the United States might execute to blunt anything Saddam Hussein tries. Iraqi intentions and U.S. responses have been loaded on a floppy disk. The U.S. ambassador at the United Nations delivers the disc to his Iraqi counterpart. "We know what you're doing," he says, "and we think we know where and when. Here's what will happen. We have developed a simulation of the impending conflict, and you will lose. Don't believe me? Load this on your computer and make your own decision." Several days later, satellites record that the Iraqi military machine has stood down from its alert. Saddam's advisors have realized that even before the war has begun, it has already been lost. Just as high-tech weapons can be the edge in conflict, they can also be an effective deterrent. Just knowing what the other guy has and what he might be contemplating is half the battle. Of course, the reverse of this scenario might be that Iraq vows to upgrade its weapons systems until they are on par with those of the United States before trying anything. But there should always be room for optimism.

John Morello, Ph.D.

John Morello, Ph.D.

Thoughts On Medical Technology

I still recall vivid memories from my childhood of Civil War stories as told by my grandfather. His grandfather was a medical doctor in the Civil War. Not only did Gramps talk about the scarcity of medicines and pain killers, gangrene and unbearable pain, his detailed accounts of a time period long gone also revolved around treatments. In his accounts, he described a small black box (approximately $4'' \times 2''$) with a handle on top and razor blades on the bottom. When a patient had an infection, Dr. Stevens, my great-great-grandfather, would pull out the box, place it on one of the patient's arteries, and push hard on the handle, forcing the blades into the skin and causing blood to gush; the goal was to extinguish the patient's ills. The theory behind the technologically simple black box was that infection and impurities would be extracted from the body through the relentless blood flow caused by the "bleeder's" gashes on the skin's surface. The treatment for gangrene, often correlated with bullet wounds, was to remove the tainted limb, aided only by primitive forms of anesthesia.

In recalling these stories, I still remain amazed at the dramatic historical changes in medical technology. When thinking of "bleeders" or primitive amputations, I feel revulsion and confusion. Because today's medicine is more humane, complex, and progressive, it remains difficult to imagine the limitations of doctors in past times. If my grandfather had told his grandfather that in the 1990s, antibiotics (penicillin was discovered in 1928), computed tomography (CAT scans were discovered in 1972), test-tube babies (the first one was born in England in 1978), eradication of smallpox (a very serious disease in his time, and one that was wiped out in 1977), organ transplants (including the successful transplant of genetically altered pig hearts into baboons in 1995), and DNA analysis for disease prevention (1999) would be available or had occurred, I am sure that he would have told his grandson that he was either "nuts" or grandiose (Nuland, 1996, pp. 12–13).

As new medical technologies are created, families are grateful because life spans are extended, low-weight babies are saved and sight can be restored, as can life after death. However, I remain convinced that my great-great-grandfather would still shake his head in dismay at some of the current ethical issues that doctors face today. His stories were gruesome, bold, and primitive, but the stories for the new millennium are amazing, technological, and sometimes ethically disconcerting. If Grandpa Doc were alive today, what would he say about baboon-heart and pig-stomach transplants, tissue engineering, or Dr. Kervorkian's "death machine"? I am not sure, but I do know that he would express a concern that all medical technology be used with caution, remembering that its purpose is to increase the quality of life, not necessarily to make life technology-dependent.

Linda Stevens Hjorth

Linda Stevens Hjorth

REFERENCE

Nuland, S.B. (1996, Fall). An Epidemic of Discovery. *Time.* pp. 12–13.

Technology and the Future of the Future

The future belongs to those who are willing to learn from the experiences of the present and the past. Indeed, no one can predict the future. But with the lessons learned during the gigantic explosion of technology in the 20th century, we can chart pathways to a future that will enable us to control the fission-like chain reaction of technology's growth for its appropriate, humane, and responsible utilization.

As we approach the new millenium, the great technological accomplishments of the 20th century in the areas of telecommunications, computers, energy, agriculture, materials, medicine, genetic engineering, and defense have transformed the world and brought people closer, yet millions of people worldwide still go to bed hungry at night. These technological advances have enabled us to design advanced early warning systems against missile attacks, but we have failed to develop an advanced early warning system to warn against and prevent global famine or the spread of disease. Thanks to state-of-the-art technologies, we are able to design spaceships to explore life on other planets, yet we have failed to preserve life on planet Earth. Millions of children worldwide continue to die due to malnutrition, disease, and poverty.

We have made tremendous strides in science and technology, but at the cost of numerous environmental, social, and moral dilemmas. Technological advances and economic expansion could become meaningless if moral and social implications are not considered in charting the course of the future.

In developed countries, technology has provided an impetus for economic growth at the cost of many social and environmental problems. In developing countries, the lack of advanced technology has resulted in low economic growth. Many developing nations aspire to become technologically advanced, but lack the required infrastructures and skilled manpower and are reluctant to consider the social and environmental costs that the industrialized nations have paid in the course of their industrialization. The key question for developing nations is how to increase the pace of industrialization without paying the same cost that the developed nations have paid for becoming industrialized. The answer to this question lies in the appropriate use of technology. For example, telecommunications technologies can provide endless opportunities in education (e.g., distance learning), medicine (e.g., telemedicine), and business (e.g., e-commerce) in developing countries. These applications could enable developing countries to increase teledensity and literacy, improve per-capita income, and narrow the technoeconomic gap with developed countries.

As we march into a new millennium with our technological might, we ought to be aware not only of the short-term gains and advantages of technology, but also of its long-term impact and associated problems. In the preceding text, we have attempted to present a spectrum of issues related to a wide array of topics (i.e., energy, ecology, population, war and technology, social responsibility, health and technology, technology and the Third World, and technology of the future) that deal with the impact of technology in the developed and the developing world. I hope that our endeavor will serve as a guide to help the reader understand, anticipate, and address the impact of technology on the various facets of society in order to make future decisions for the humane, just, and responsible utilization of technology.

Ahmed S. Khan

Ahmed S. Khan

It Is Now the Time for Wisdom

The *future* is that concept that we who have embraced Western civilization and the concepts of progress, change, improvement, and technology can hardly wait for! The future is ours. We only have to design it, and through technology, it will service us and give us greater pleasures than ever before. The future adventures, the daring, the excitement, and the new luxuries appeal unlimitedly to those of us in the industrialized societies and are dazzling and changing the unindustrialized societies. Anticipating the future gives us a telescopic sight of the accomplishment of our greater hopes, dreams, and visions . . . new reaches for our graceful hand of knowledge.

But this future represents many changes from the futures that were imagined by past generations. This future represents social and economic differences from previously imagined futures, which were more predictable, less multidimensional, less exponential, and less critical. Yes, most futures represent change, growth, and variation from the "comfortable known," but this one is really different. This future is not only a telescopic graceful vision; it is also a multimedia-accelerated production based on technology everywhere, with increasingly faster machines and tools to aid, support, and satisfy mankind. Along with the mach speed of technologies are accompanying critical concerns that have to be addressed before we blindly ride the technological acute paths. We have to ask ourselves some basic questions, for we are at a critical juncture. Critical ecological issues emerge because of our new technological development. Questions and concerns about population emerge because of our progress in the fields of health and medicine and due to our expanded life spans. War can be globally devastating both easily and quickly. Moral and ethical concerns of actions increasingly parallel all of the thoughtful use of technology. The unindustrialized world hopefully tries to adopt some of the technologies to help its economic development, but has a problem with the expense, consequences, and appropriateness of Western technology.

Part VIII's objective has been to converge the thoughts of the other seven parts to their implications for the 21st century so that issues from those parts can find a responsible bridge to the future, a future that is sustainable and fulfills its potential exciting promise. The path, however, is not automatic. Because of the steeper acceleration of technology and accompanying critical humanitarian and ecological issues, all actions have to include more than economic appetites and a gratification of needs and wants; they must also contain larger understandings of technological responsibility and global sustainability. The concern for the quality of life in the future—for a natural sustainable life—depends on our understanding and accompanying actions. This is the challenge and ultimate flow chart for the 21st century: wise decisions and wise and caring actions.

Barbara A. Eichler Ed.D.

Beyond the Bridge . . . Explore Future Options

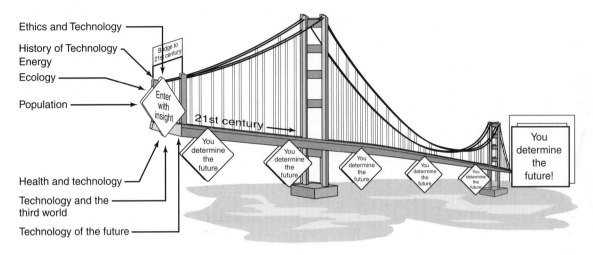

It is a great advantage that man should know his station and not erroneously imagine that the whole universe exists only for him.
—MAIMONIDES
(Dalalat al-ha'irin, Part iii, Chapter xii, c. 1190)

Index